AF616236

River-Dominated Shelf Sediments of East Asian Seas

Geological Society books refereeing procedures

The Society makes every effort to ensure that the scientific and production quality of its books matches that of its journals. Since 1997, all book proposals have been refereed by specialist reviewers as well as by the Society's Books Editorial Committee. If the referees identify weaknesses in the proposal, these must be addressed before the proposal is accepted.

Once the book is accepted, the Society Book Editors ensure that the volume editors follow strict guidelines on refereeing and quality control. We insist that individual papers can only be accepted after satisfactory review by two independent referees. The questions on the review forms are similar to those for *Journal of the Geological Society*. The referees' forms and comments must be available to the Society's Book Editors on request.

Although many of the books result from meetings, the editors are expected to commission papers that were not presented at the meeting to ensure that the book provides a balanced coverage of the subject. Being accepted for presentation at the meeting does not guarantee inclusion in the book.

More information about submitting a proposal and producing a book for the Society can be found on its website: www.geolsoc.org.uk.

It is recommended that reference to all or part of this book should be made in one of the following ways:

CLIFT, P. D., HARFF, J., WU, J. & QUI, Y. (eds) 2016. *River-Dominated Shelf Sediments of East Asian Seas*. Geological Society, London, Special Publications, **429**.

ZONG, Y., HUANG, G., LI, X. Y. & SUN, Y. Y. 2016. Late Quaternary tectonics, sea-level change and lithostratigraphy along the northern coast of the South China Sea. *In*: CLIFT, P. D., HARFF, J., WU, J. & QUI, Y. (eds) *River-Dominated Shelf Sediments of East Asian Seas*. Geological Society, London, Special Publications, **429**, 123–136. First published online September 3, 2015, updated May 26, 2016, http://doi.org/10.1144/SP429.1

GEOLOGICAL SOCIETY SPECIAL PUBLICATION NO. 429

River-Dominated Shelf Sediments of East Asian Seas

EDITED BY

P. D. CLIFT
Louisiana State University, USA

J. HARFF
University of Szczecin, Poland

J. WU
Sun Yat Sen University, China

and

Y. QUI
Guangzhou Marine Geological Survey, China

2016
Published by
The Geological Society
London

THE GEOLOGICAL SOCIETY

The Geological Society of London (GSL) was founded in 1807. It is the oldest national geological society in the world and the largest in Europe. It was incorporated under Royal Charter in 1825 and is Registered Charity 210161.

The Society is the UK national learned and professional society for geology with a worldwide Fellowship (FGS) of over 10 000. The Society has the power to confer Chartered status on suitably qualified Fellows, and about 2000 of the Fellowship carry the title (CGeol). Chartered Geologists may also obtain the equivalent European title, European Geologist (EurGeol). One fifth of the Society's fellowship resides outside the UK. To find out more about the Society, log on to www.geolsoc.org.uk.

The Geological Society Publishing House (Bath, UK) produces the Society's international journals and books, and acts as European distributor for selected publications of the American Association of Petroleum Geologists (AAPG), the Indonesian Petroleum Association (IPA), the Geological Society of America (GSA), the Society for Sedimentary Geology (SEPM) and the Geologists' Association (GA). Joint marketing agreements ensure that GSL Fellows may purchase these societies' publications at a discount. The Society's online bookshop (accessible from www.geolsoc.org.uk) offers secure book purchasing with your credit or debit card.

To find out about joining the Society and benefiting from substantial discounts on publications of GSL and other societies worldwide, consult www.geolsoc.org.uk, or contact the Fellowship Department at: The Geological Society, Burlington House, Piccadilly, London W1J 0BG: Tel. +44 (0)20 7434 9944; Fax +44 (0)20 7439 8975; E-mail: enquiries@geolsoc.org.uk.

For information about the Society's meetings, consult *Events* on www.geolsoc.org.uk. To find out more about the Society's Corporate Affiliates Scheme, write to enquiries@geolsoc.org.uk.

Published by The Geological Society from:
The Geological Society Publishing House, Unit 7, Brassmill Enterprise Centre, Brassmill Lane, Bath BA1 3JN, UK

The Lyell Collection: www.lyellcollection.org
Online bookshop: www.geolsoc.org.uk/bookshop
Orders: Tel. +44 (0)1225 445046, Fax +44 (0)1225 442836

British Library Cataloguing in Publication Data

A catalogue record for this book is available from the British Library.
ISBN 978-1-86239-740-8
ISSN 0305-8719

Distributors
For details of international agents and distributors see:
www.geolsoc.org.uk/agentsdistributors

Typeset by Nova Techset Private Limited, Bengaluru & Chennai, India
Printed and bound by CPI Group (UK) Ltd, Croydon CR0 4YY

Contents

CLIFT, P. D., HARFF, J., WU, J. & QIU, Y. Introduction to the River-Dominated Shelf Sediments of the East Asian Seas 1

Provenance proxies

CLIFT, P. D. Assessing effective provenance methods for fluvial sediment in the South China Sea 9

SHAO, L., QIAO, P., ZHAO, M., LI, Q., WU, M., PANG, X. & ZHANG, H. Depositional characteristics of the northern South China Sea in response to the evolution of the Pearl River 31

Chemical weathering

HU, D., CLIFT, P. D., WAN, S., BÖNING, P., HANNIGAN, R., HILLIER, S. & BLUSZTAJN, J. Testing chemical weathering proxies in Miocene–Recent fluvial-derived sediments in the South China Sea 45

CUI, Z., HOU, Y., XIA, Z., LIANG, K., XUE, Q. & ZHANG, L. Geochemical characteristics and palaeoenvironmental reconstruction of the sediments from the Gulf of Tonkin, South China Sea 73

Sea-level rise and sediment storage

NI, Y., HARFF, J., XIA, Z., WANIEK, J. J., ENDLER, M. & SCHULZ-BULL, D. E. Post-glacial mud depocentre in the southern Beibu Gulf: acoustic features and sedimentary environment evolution 87

CHEN, H., HARFF, J., QIU, Y., OSADCZUK, A., ZHANG, J., TOMCZAK, M., CUI, Z., CAI, G., WEN, M. & LI, L. Last Glacial Cycle and seismic stratigraphic sequences offshore western Hainan Island, NW South China Sea 99

ZONG, Y., HUANG, G., LI, X. Y. & SUN, Y. Y. Late Quaternary tectonics, sea-level change and lithostratigraphy along the northern coast of the South China Sea 123

YANG, S., BI, L., LI, C., WANG, Z. & DOU, Y. Major sinks of the Changjiang (Yangtze River)-derived sediments in the East China Sea during the late Quaternary 137

Shelf currents

WU, J., REN, J., LIU, H., QIU, C., CUI, Y. & ZHANG, Q. Trapping and escaping processes of Yangtze River-derived sediments to the East China Sea 153

Anthropogenic impacts

JIA, P., XIA, Z., YIN, Y. & XUE, Q. Lingdingyang Bay, Pearl River Estuary (China): geomorphological evolution and hydrodynamics 171

WANG, Y., LIU, X., LI, G. & ZHANG, W. Stratigraphic variations in the Diaokou lobe area of the Yellow River delta, China: implications for an evolutionary model of a delta lobe 185

Environmental impacts

ZHOU, Y., CHEN, F., WU, C., YU, S. & ZHUANG, C. Palaeoproductivity linked to monsoon variability in the northern slope of the South China Sea from the last 290 kyr: evidence of benthic foraminifera from Core SH7B 197

ZHANG, J., TOMCZAK, M., LI, C., WITKOWSKI, A., QIU, Y., CHEN, H. & GAO, H. Significance of the *Paralia sulcata* fossil record in palaeoenvironmental reconstructions of the SE Asia marginal seas over the Last Glacial Cycle 211

GAO, S., WANG, D., YANG, Y., ZHOU, L., ZHAO, Y., GAO, W., HAN, Z., YU, Q. & LI, G. Holocene sedimentary systems on a broad continental shelf with abundant river input: process–product relationships 223

Index 261

Introduction to the River-Dominated Shelf Sediments of the East Asian Seas

PETER D. CLIFT[1]*, JAN HARFF[2,3], JIAXUE WU[4] & YAN QIU[5]

[1]*Department of Geology and Geophysics, Louisiana State University, Baton Rouge, LA 70803, USA*

[2]*University of Szczecin, Institute of Marine and Coastal Sciences, Mickiewicza 18, PL-70-383 Szczecin, Poland*

[3]*Leibniz Institute for Baltic Sea Research, Seestrasse 15, D-18119 Rostock–Warnemünde, Germany*

[4]*Centre of Coastal Ocean Science and Technology, Sun Yat-Sen University, Guangzhou 510275, China*

[5]*Guangzhou Marine Geological Survey, China Geological Survey, Guangzhou 510760, China*

**Corresponding author (e-mail: pclift@lsu.edu)*

Asian shelves as environmental recorders

The marginal seas of eastern Asia are supplied by some of the largest, and most sediment-rich, rivers on Earth. Many of these rivers have their original sources on the Tibetan Plateau and are fed by the rains of the summer monsoon that are especially intense around the edge of the plateau (Fig. 1). In turn, the climate and tectonically generated topography account for the high sediment loads of the rivers that subsequently construct a number of giant deltas across the region and result in the construction of some of the widest continental shelves seen anywhere globally. Understanding the marine sedimentary records of the East Asian marginal seas has been a focus for geologists for many years. This is because the sediments can be used to constrain the origin of the sedimentary basins themselves, via subsidence analysis, and because the sediments can be used to reconstruct sedimentary conditions onshore in the terrestrial basins at the time of their deposition. Theoretically, the weathering conditions, floral assemblages, and both erosion rates and patterns within the drainage basin might be reconstructed from the sediment deposited within the deltas and under the shelves.

Although, sediment storage and recycling may result in the loss of erosional signals in many larger river floodplains (Castelltort & Van Den Driessche 2003; Jerolmack & Paola 2010), at least over timescales of less than 1 myr, it has also been recognized that there is a response to climate change in the supply of sediment to the ocean both in terms of volumes and composition over millennial timescales in several east and south Asian river systems (Goodbred & Kuehl 2000; Clift *et al.* 2008; Hu *et al.* 2013). There is a climatic response in most river systems to even seasonal discharge cycles, where floods linked to the wet season (summer monsoon in East Asia) will tend to dominate the clastic flux to the ocean over discharge during drier times and to overwhelm the effects of sediment transport by tidal currents (Gugliotta *et al.* 2016; Hoitink & Jay 2016). The resultant chronological archive of terrestrial environmental conditions could be used to address outstanding questions in Asian marine and geosciences, such as the history of the Asian monsoon, and the timing and patterns of topographical uplift in the Tibetan Plateau and surrounding areas, as well as the environmental responses to climatic and tectonic forcing factors. Shelf sediments might be used to test models for drainage evolution onshore in the global type area for tectonically induced headwater capture (Brookfield 1998; Clark *et al.* 2004), since large-scale transfer of drainage between neighbouring river basins must have an impact on the rate of sediment delivery and on the provenance of the clastic materials reaching the coastal ocean.

Transport on the shelf

Any interpretation of sediment preserved on the continental shelf requires us to understand how sediment is transported after its arrival in the marine environment (Fig. 2). The same would be true concerning deciphering the records of sediments accumulating on the continental slope and abyssal

From: CLIFT, P. D., HARFF, J., WU, J. & QUI, Y. (eds) 2016. *River-Dominated Shelf Sediments of East Asian Seas*. Geological Society, London, Special Publications, **429**, 1–8.
First published online June 21, 2016, updated June 24, 2016, http://doi.org/10.1144/SP429.15

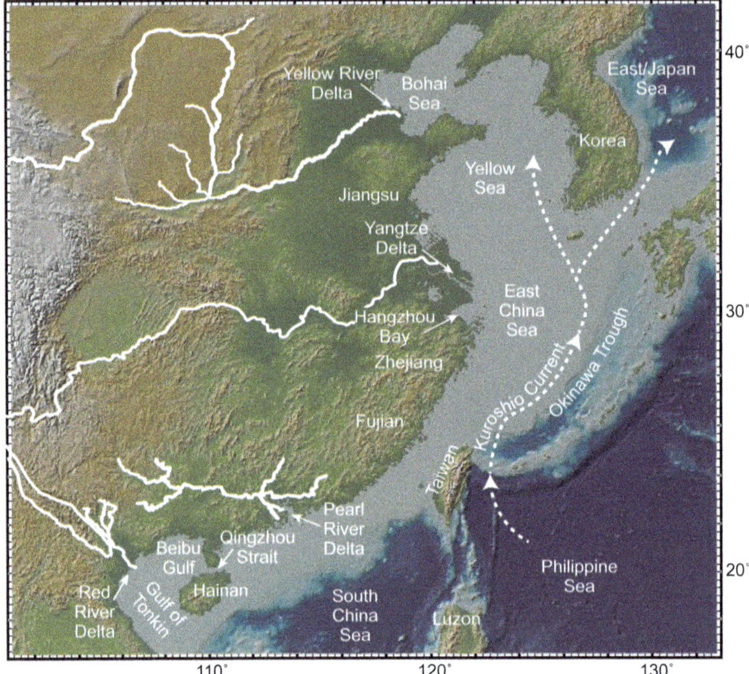

Fig. 1. Shaded bathymetric map of the area considered in this Special Publication showing the major geographical features discussed in this Introduction.

plain, because such sediments must necessarily pass through the shelf zone on their way to deep-water depocentres. High-energy conditions are common on continental shelves, which facilitates the dispersal of sediment from river mouths, as well as the reworking and mixing of new sediment with material already on the shelf. Such currents also control the form and location of shelf depocentres (Xu *et al.* 2012). Furthermore, longshore currents can transport sediment great distances parallel to the coastline, resulting in additional signal scrambling (Liu *et al.* 2009). Continental shelves are swept by powerful bottom currents, which may redistribute material across and along the shelf.

The northern shelf of the South China Sea provides a typical example of the complex nature of forces distributing sediments delivered by different riverine sources (Harff *et al.* 2013). A permanent longshore current transports the load delivered by the Pearl River to the west of the estuary. Particulate matter enters the Beibu Gulf through the bottleneck of the Qiongzhou Strait, where it mixes with the sediment load of the Nanliu and the Red rivers before it settles out and is preserved in the depocentre of the Gulf of Tonkin. The year-long east to west transport is superposed on by seasonal fluctuating transports induced by the Asian summer and winter monsoons. The switch in transport direction has induced the formation of a 'Butterfly Delta' east and west of the Qiongzhou Strait (Ni *et al.* 2014). In eastern Asia, the Kuroshio Current is especially noteworthy for its role on the East China Shelf (Andres *et al.* 2008) (Fig. 1). Reworking, recycling and mixing of fluvial sediment with older shelf deposits may cause the preserved stratigraphy of any continental margin to be at least partly homogenized over millennial timescales. Understanding the processes responsible for building clastic shelves can be applied to the oil and gas industry because high-energy, well-sorted shelf sediments are important hydrocarbon reservoirs whose nature needs to be well defined if exploitation of the preserved hydrocarbons is to be maximized. A particular consideration is the origin of shelf-edge deltas, whose origin is usually not linked to sea-level highstands, but, instead, tend to be formed during periods of forced regression (Muto & Steel 2002; Limmer *et al.* 2012), where sediment supply is better able to keep up with accommodation space. This is especially true in East Asia where the shelves are mostly wide.

Climatic feedbacks

Continental shelves may play a part in controlling global climate by providing a feedback via the

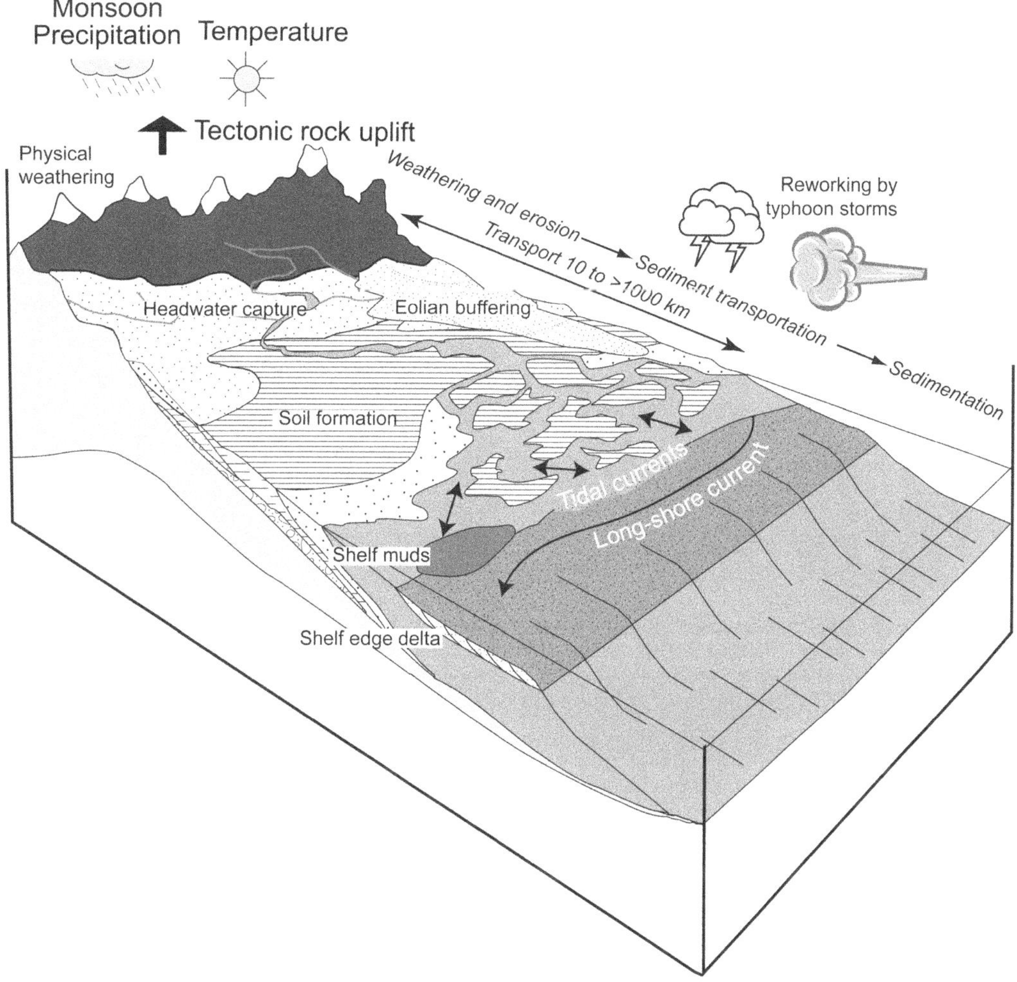

Fig. 2. Schematic block model showing the different processes that can affect fluvial sediment on its way from source to sink in East Asia. Modified from Fagel *et al.* (1994).

processes of chemical weathering. When sea level falls as glaciers expand, wide areas of sediment are exposed and potentially made available for chemical weathering by direct interaction with the atmosphere, as well as via groundwater flow under the exposed shelf. Although work in high latitudes suggests no net change in weathering flux because of the reduced rates of weathering (Foster & Vance 2006), it is unclear whether the low latitude shelves of East Asia operate in a similar fashion. This is because of their great expanse compared to most Atlantic Ocean examples and because the temperature variations during glacial times at low latitudes may not be as great at high latitudes (deMenocal *et al.* 2000; Lea *et al.* 2003), so that the reductions in chemical weathering rates are not so great.

Eustacy and stratigraphy

The wide continental shelves of East Asia are also ideal natural laboratories for the study of how sea-level rises have influenced shelf stratigraphy. Traditional sequence stratigraphy is based on data derived from basins and continental margins surrounding the North Atlantic and Gulf of Mexico (Vail *et al.* 1977), but it is open to question as to whether such models are generally applicable. Deltas feeding the wide shelves around East Asia have far greater volumes of accommodation to fill before they would be able to regain a connection with their deep-water depositional systems following a transgressive period (Liu *et al.* 2004). This consideration must be balanced by the accelerated sediment delivery during highstands that typifies those drainages fed by regions influenced by an intensified summer monsoon. The summer monsoon tends to intensify as the global climate warms and sea levels rise because this is primarily controlled by solar insolation (Clemens *et al.* 2010).

This collection has been brought together to highlight recent developments in our understanding of how fluvial sediment has been delivered to the marginal seas of eastern Asia and to show what these deposits can tell us about the various processes mentioned above. The collection represents the final product of a special session on this topic convened at the 2012 International Geological Congress (IGC) in Brisbane, Australia, and an international follow-up workshop held at Sun Yat-Sen University, Guangzhou, China, in 2013.

Provenance

Provenance reconstructions are a major driver in the study of East Asian shelf sediments. This problem is now addressed by integration of geochemical and mineral proxies, and by thermochronological methods, especially the increasing popular U–Pb dating of detrital zircons. **Clift (2015)** reviewed the general application of such methods to the tracing of sediments in the South China Sea where material may be delivered from a series of diverse tectonic blocks. This study highlights the issue that anthropogenic disruption of the landscape since the establishment of agriculture means that modern river compositions may make poor fingerprints for tracing the effects of a given river in the geological past because sediment tends to be more altered in the modern day compared to the recent past. The popular provenance-related Nd and Sr isotope systems work best at identifying sediment flux from younger volcanic islands, especially Luzon, but are problematic when attempting to separate the flux from other continental blocks around the basin. Instead, differential exhumation rates, traced by low-temperature thermochronometers, in different possible source terrains proved more effective as discriminants. Apatite fission track, in particular, was highlighted as being more effective than either U–Pb zircon or Ar/Ar muscovite dating.

Shao *et al.* (2015) applied some of these provenance methods when they analysed sedimentary rocks from Pearl River Mouth Basin, offshore southern China, as well as from various parts of the Pearl River Basin. This study considered the bulk sediment geochemistry, and also looked at the heavy mineral suites in both the depocentres and source rivers. They were able to show that the Pearl River Basin can be divided into three regions – east, central and west – and that each region generated sediment with a distinctive fingerprint. Comparison with offshore deposits indicates that in the Early Oligocene sediment sources were close to, or within, the basin, but had expanded to the coastal granites of southern China by the end of the Oligocene. After the regional 23 Ma unconformity, the river appears to have widened through time. Although the palaeo-Pearl River continued to supply sediments to most parts of the northern South China Sea during the Early Miocene, the southern part of the Baiyun Sag, a particularly deep continental margin sub-basin, was affected by basic volcanic sources close to this depression.

Chemical weathering

Geochemical proxies are also used to reconstruct the intensity of chemical weathering, but it is not always clear which proxies are the most effective and whether any of these have universal applicability. This issue was addressed by **Hu *et al.* (2015)**, who compared bulk sediment geochemical ratios and clay mineral proxies from the Pearl River delta, as well as from Ocean Drilling Program sites 1144 and 1146 (Fig. 1). Comparison with speleothem rainfall records since the Younger Dryas (*c.* 12 ka) indicates that K/Al tracks variations in precipitation most closely and out-performs the widely used Chemical Index of Alteration (CIA). Correlation of K/Al and kaolinite/illite indicates that this clay ratio is also an effective proxy of weathering intensity across all sites and timescales. In contrast, while kaolinite/smectite, and smectite/(illite + chlorite), are also indicative of weathering intensity, they show more scatter between sites that may be linked to provenance effects.

Geochemical proxies were also applied by **Cui *et al.* (2015)** to investigate Holocene environmental conditions in the Tonkin Gulf, in the NW South China Sea. Changing values of Al/Ti, K/Al and Mg/Al were used to reconstruct the intensity of chemical weathering in the source regions. Zr/Ti and SiO_2/Al_2O_3 are believed to reflect the sand content of the sediment, and thus, potentially, record the history of current velocity. La/Co and La/Sc values coupled with rare earth element characteristics were used to argue that Hainan Island was the main source of the sediment to the basin during the Holocene. Based on these results, the authors divided the Holocene into four phases of climatic development. At 10.1–6.5 ka, sea level rose as the climate warmed. From 6.5 to 4.3 ka, the depositional environment was stable, but the local climate became colder and drier. From 4.3 to 3.5 ka, the currents changed following the opening of the Qiongzhou Strait between Hainan and the Leizhou Peninsula, thus also affecting sediment provenance.

Since 3.5 ka, the depositional environment appears to have been relatively stable.

Sea-level rise and sediment storage

Closer to the Qiongzhou Strait, **Ni *et al.* (2016)** for the first time report on a shelf mud accumulation within the Beibu Gulf known as the Southern Beibu Gulf Mud Depocentre (SBGMD). Identification was based on high-resolution sub-bottom profiles. The SBGMD lies in water depths of 50–80 m, and covers an area of more than 11 000 km^2. The SBGMD is imaged as a homogenous seismic unit surrounded by an erosive area of gullies and sand/mud waves. The SBGMD unconformably overlies a parallel-bedded seismic unit. Lithological and geochemical analysis of a core within the high sedimentation rate area of the SBGMD shows that before 17.0 ka this area was exposed and had a terrestrial, fluvial environment. From 17.0 to 11.6 ka BP, conditions appear to have been brackish and lacustrine. Subsequently (11.6–8.4 ka), the area experienced marine transgression, followed by continuous shallow-marine conditions to the present day.

Further to the south, **Chen *et al.* (2015)** investigated the relationship between sea-level changes during the Last Glacial Cycle and stratigraphic sequences offshore western Hainan using a combination of coring and high-resolution seismic reflection data. An age model was developed from AMS ^{14}C, optically stimulated luminescence (OSL) and $\delta^{18}O$ dates, spanning 110 ka in a 88.3 m-long core. Correlation with a sediment core in the adjacent basin allows identification of Marine Isotope Stages (MIS) 1–5e. They identify unconformities that separate seven sea-level cycles preserved within the study area, and which correlate with regional and global sea-level change models. Seismic imaging indicates a primary control by sea-level change on the shelf stratigraphy, modified by sediment flux and tectonic uplift. High sediment discharge, as well as tectonically driven uplift of Hainan, may have resulted in the development of prograding low-stand delta wedges on the shelf margin that took place during MIS 5–2.

Recent sediments on the northern margin of the South China Sea were considered by **Zong *et al.* (2015)**. This review paper identified two marine successions offshore of the Song Hong (Red) River, Han River and Pearl River deltas. These two are dated as younger than the Last Glacial Maximum (*c.* 20 ka) and spanning the period 120–126 ka. The authors proposed that all these marine basins are bounded by active faults and because the older marine succession is recorded at a minimum of −15 m below present sea level, this requires active tectonic subsidence to account for the marine conditions after accounting for the difference in sea level at 120–126 ka and the present day. Zong *et al.* (2015) linked these vertical motions to the long-term thermal subsidence of the northern South China Sea continental shelf, but noted that active faulting has enhanced local subsidence, resulting in marine inundation during interglacial high sea levels.

The fate of late Quaternary sediment delivered by the Yangtze to the East China Sea was considered by **Yang *et al.* (2015)**. Most Yangtze sediment has been preserved on the outer shelf and in the Okinawa Trough during the last glacial maximum, which is not surprising given the lowstand of sea level. During the deglacial marine transgression, the gently dipping shelf was rapidly inundated and strong tides prevented fine-grained sedimentation on the open shelf, resulting in the development of a unique tidal sand-ridge system. As sea level stabilized close to the modern level, most Yangtze sediment accumulated in its estuary where it then built a large delta, with only a fraction reaching the inner shelf and coastal embayments. Source to sink transport of sediment in the East China Sea is heavily controlled by sea level, although monsoon-modulated sediment flux and bottom-current activity also plays a role in controlling the fate of Yangtze sediment reaching the ocean.

Sediment transport

Sediment dispersal in the Yangtze offshore regions was investigated by **Wu *et al.* (2015)**, who focused on the large-scale mud belt on the inner shelf developed south of the river mouth. The processes responsible for sediment transport in this area were investigated by field observations during the 2013 monsoon season, with special attention paid to shelf circulation currents and their interaction with the Yangtze river outflow, as well as small- and meso-scale processes, including bottom boundary-layer flows, stratification, mixing and upwelling. The role of river-borne sediment-gravity and contour currents were also considered. Turbidity maxima within the estuary trap sediment suspensions near the seafloor and move these downslope to form sediment-gravity currents that are strengthened by tidal currents. In contrast, the buoyant coastal current is not a controlling factor in the formation of the mud belt. Long-distance dispersal of material to the mud belt is made possible by near-bed sediment transport along the shelf, together with a potentially significant contour-parallel sediment channel.

Anthropogenic impacts

The more recent and future evolution of the Pearl River delta were considered by **Jia *et al.* (2016)**.

These workers compared two marine geophysical surveys from 2003 to 2012 across Lingdingyang Bay, the estuary in front of the Pearl River delta front. Within this area, they mapped three shoals and two channels that appear to be heavily impacted by anthropogenic activities. In the four river mouths that empty into Lingdingyang Bay, the sediment dispersal is controlled by tides, discharge strength and the presence of islands. The surveys show that, in the recent past, these outlets have become narrower, so that the water area of the bay is diminishing. Sediment flux into the bay is reducing and becoming finer grained because shoals and islands are blocking marine currents, so that sediments are easier to deposit as the energy in the water reduces. At the same time, water exchange between the inner bay and the open South China Sea is decreasing because of unrestricted anthropogenic activities. Without better management the bay will eventually sediment up and infill, resulting in potential hazards in the form of flooding and tides.

Further north, the Yellow River currently has the highest sediment load of any single river on Earth. The offshore growth of the Yellow Delta was constrained by **Wang *et al.* (2015)**, who used cores within the Diaokou lobe of the Yellow Delta to date the construction of this feature. This lobe was especially active during the period 1964–76, after the river changed its course to the present location. This study used grain-size data, magnetic property data and AMS ^{14}C dating of cores across the Diaokou lobe to define the Holocene stratigraphy. They identified a clear upwards succession from shallow-marine, to river and lake, to salt-marsh and, finally, to delta facies. This area started to receive deltaic deposits in 1855, and has experienced prodelta, delta-front and delta-plain environments. During the formation of the Diaokou lobe sediment transport was first by dispersed-flow deposition, then by single-channel deposition, followed by diversion deposition, and then culminated in abandonment and erosion.

Environmental impacts

Monsoon conditions since 290 ka were reconstructed by **Zhou *et al.* (2015)** using a benthic foraminiferal record from the northern continental slope of the South China Sea. Changes in oxygenation and types of organic matter reflect the palaeoproductivity linked to monsoon variability. Four assemblages, characterizing different palaeoenvironmental changes, were recognized. These show that interglacial periods MIS 7, MIS 5 and MIS 3 were associated with well-oxygenated bottom-water environments linked to a weak East Asian Winter Monsoon (EAWM). Foraminifera that depend on seasonal supplies of more altered organic matter are mainly found in interglacial periods (MIS 5, MIS 3 and MIS 1), suggesting highly seasonal palaeoproductivity associated with the intensification of the EAWM. Times of increased river run-off and increased primary productivity correlate with an enhanced EAWM, and have lead to severe bottom-water oxygen depletion. In contrast, during MIS 4, MIS 5, MIS 7 and MIS 8, there is evidence of an enhanced EAWM with low seasonality of palaeoproductivity. Changes in bottom-water environments of the northern South China Sea since 290 ka are proposed to have been driven by the fluctuating palaeoproductivity linked to the variability of the EAWM.

Palaeo-environmental reconstructions can also be achieved using other types of microfossil, such as diatoms. **Zhang *et al.* (2015)** employed *Paralia sulcata* for this task because it is often well preserved in siliciclastic sediments. These workers synthesized the distribution of *P. sulcata* in sediment cores across the Asian marginal seas. They also noted that *P. sulcata* is strongly variable in abundance through time over glacial cycles, with lower abundances during the last glacial maximum and higher values during deglaciation within the South China Sea. This diatom is very common at 8–11 ka, a phenomenon linked to the opening of the Taiwan Strait to more flow, as well as to the general coastal water circulation pattern associated with the deglaciation process. In contrast, *P. sulcata* is less abundant in the East China Sea during the Holocene, but more common during the Last Glacial Maximum and especially during 11–15 ka.

Gao *et al.* (2015) considered the wider region of the Bohai, Yellow and East China seas as a single wide continental shelf environment. Holocene sediment distribution, composition and deposition rates are related to active sediment transport processes including tides, waves and shelf currents, as well as sediment gravity flows. Theoretically, the sediments in any one place should contain a record of high temporal resolution, but each site would only have limited duration. Gao *et al.* (2015) argue that if these records are connected, then they may form a complete archive for environmental change studies. Mid-Holocene coastal deposits on the Jiangsu coast and early–middle Holocene sequences in Hangzhou Bay, as well as the Holocene mud deposits off the Zhejiang–Fujian coasts, are of importance in process–product relationship studies. Sediment may be supplied to the shelf both from seabed reworking during times of sea-level rise and from enhanced river discharges.

This collection highlights some of the complexities of fluvial sediment transport on continental shelves, demonstrating what can and cannot be deduced from such deposits in the geological record.

The scale of sediment flux in East Asia, reflecting the size of the rivers and the intensity of the summer monsoon, makes this region an ideal place to work on sediment transport issues. While these studies have advanced the science by a significant degree, they also show how much we still have to learn and so provide a stimulus for further studies.

References

Andres, M., Wimbush, M. *et al.* 2008. Observations of Kuroshio flow variations in the East China Sea. *Journal of Geophysical Research*, **113**, C05013, http://doi.org/10.1029/2007JC004200

Brookfield, M.E. 1998. The evolution of the great river systems of southern Asia during the Cenozoic India-Asia collision; rivers draining southwards. *Geomorphology*, **22**, 285–312.

Castelltort, S. & Van Den Driessche, J. 2003. How plausible are high-frequency sediment supply-driven cycles in the stratigraphic record? *Sedimentary Geology*, **157**, 3–13.

Chen, H., Harff, J. *et al.* 2015. Last Glacial Cycle and seismic stratigraphic sequences offshore western Hainan Island, NW South China Sea. *In*: Clift, P.D., Harff, J., Wu, J. & Qui, Y. (eds) *River-Dominated Shelf Sediments of East Asian Seas*. Geological Society, London, Special Publications, **429**. First published online December 1, 2015, updated January 28 and May 26, 2016, http://doi.org/10.1144/SP429.9

Clark, M.K., Schoenbohm, L.M. *et al.* 2004. Surface uplift, tectonics, and erosion of eastern Tibet from large-scale drainage patterns. *Tectonics*, **23**, TC1006, http://doi.org/10.1029/2002TC001402

Clemens, S.C., Prell, W.L. & Sun, Y. 2010. Orbital-scale timing and mechanisms driving Late Pleistocene Indo-Asian summer monsoons: reinterpreting cave speleothem ∂^{18}O. *Paleoceanography*, **25**, PA4207, http://doi.org/10.1029/2010PA001926

Clift, P.D. 2015. Assessing effective provenance methods for fluvial sediment in the South China Sea. *In*: Clift, P.D., Harff, J., Wu, J. & Qui, Y. (eds) *River-Dominated Shelf Sediments of East Asian Seas*. Geological Society, London, Special Publications, **429**. First published online September 3, 2015, updated May 26, 2016, http://doi.org/10.1144/SP429.3

Clift, P.D., Giosan, L. *et al.* 2008. Holocene erosion of the Lesser Himalaya triggered by intensified summer monsoon. *Geology*, **36**, 79–82, http://doi.org/10.1130/G24315A.1

Cui, Z., Hou, Y., Xia, Z., Liang, K., Xue, Q. & Zhang, L. 2015. Geochemical characteristics and palaeoenvironmental reconstruction of the sediments from the Gulf of Tonkin, South China Sea. *In*: Clift, P.D., Harff, J., Wu, J. & Qui, Y. (eds) *River-Dominated Shelf Sediments of East Asian Seas*. Geological Society, London, Special Publications, **429**. First published online December 9, 2015, updated March 3 and May 26, 2016, http://doi.org/10.1144/SP429.12

deMenocal, P., Ortiz, J., Guilderson, T. & Sarnthein, M. 2000. Coherent high- and low-latitude climate variability during the Holocene Warm Period. *Science*, **288**, 2198–2202, http://doi.org/10.1126/science.288.5474.2198

Fagel, N., Debrabant, P. & André, L. 1994. Clay supplies in the Central Indian Basin since the Late Miocene: climatic or tectonic control? *Marine Geology*, **122**, 151–172, http://doi.org/10.1016/0025-3227(94)90209-7

Foster, G. & Vance, D. 2006. Negligible glacial–interglacial variation in continental chemical weathering rates. *Nature*, **444**, 918–921, http://doi.org/10.1038/nature05365

Gao, S., Wang, D. *et al.* 2015. Holocene sedimentary systems on a broad continental shelf with abundant river input: process–product relationships. *In*: Clift, P.D., Harff, J., Wu, J. & Qui, Y. (eds) *River-Dominated Shelf Sediments of East Asian Seas*. Geological Society, London, Special Publications, **429**. First published online September 7, 2015, updated May 26, 2016, http://doi.org/10.1144/SP429.4

Goodbred, S.L. & Kuehl, S.A. 2000. Enormous Ganges–Brahmaputra sediment discharge during strengthened early Holocene monsoon. *Geology*, **28**, 1083–1086.

Gugliotta, M., Kurcinka, C.E., Dalrymple, R.W., Flint, S.S. & Hodgson, D.M. 2016. Decoupling seasonal fluctuations in fluvial discharge from the tidal signature in ancient deltaic deposits: an example from the Neuquén Basin, Argentina. *Journal of the Geological Society, London*, **173**, 94–107, http://doi.org/10.1144/jgs2015-030

Harff, J., Leipe, T., Waniek, J. & Zhou, D. (eds). 2013. *Depositional Environments and Multiple Forcing Factors at the South China Sea's Northern Shelf. Journal of Coastal Research*, Special Issue, **66**, 1–90.

Hoitink, A.J.F. & Jay, D.A. 2016. Tidal river dynamics: implications for deltas. *Reviews of Geophysics*, **54**, 240–272, http://doi.org/10.1002/2015RG000507

Hu, D., Clift, P.D. *et al.* 2013. Holocene evolution in weathering and erosion patterns in the Pearl River delta. *Geochemistry, Geophysics, Geosystems*, **14**, 2349–2368, http://doi.org/10.1002/ggge.20166

Hu, D., Clift, P.D., Wan, S., Böning, P., Hannigan, R., Hillier, S. & Blusztajn, J. 2015. Testing chemical weathering proxies in Miocene–Recent fluvial-derived sediments in the South China Sea. *In*: Clift, P.D., Harff, J., Wu, J. & Qui, Y. (eds) *River-Dominated Shelf Sediments of East Asian Seas*. Geological Society, London, Special Publications, **429**. First published online December 11, 2015, updated May 26, 2016, http://doi.org/10.1144/SP429.5

Jerolmack, D.J. & Paola, C. 2010. Shredding of environmental signals by sediment transport. *Geophysical Research Letters*, **37**, L19401, http://doi.org/10.1029/2010GL044638

Jia, P., Xia, Z., Yin, Y. & Xue, Q. 2016. Lingdingyang Bay, Pearl River Estuary (China): geomorphological evolution and hydrodynamics. *In*: Clift, P.D., Harff, J., Wu, J. & Qui, Y. (eds) *River-Dominated Shelf Sediments of the East Asian Seas*. Geological Society, London, Special Publications, **429**. First published online May 31, 2016, http://doi.org/10.1144/SP429.14

Lea, D.W., Pak, D.K., Peterson, L.C. & Hughen, K.A. 2003. Synchroneity of Tropical and High-Latitude Atlantic Temperatures over the Last Glacial

Termination. *Science*, **301**, 1361–1364, http://doi.org/10.1126/science.1088470

Limmer, D.R., Henstock, T.J., Giosan, L., Ponton, C., Tabrez, A.R., Macdonald, D.I.M. & Clift, P.D. 2012. Impacts of sediment supply and local tectonics on clinoform distribution: the seismic stratigraphy of the mid Pleistocene-Holocene Indus Shelf. *Marine Geophysical Research*, **33**, 251–267, http://doi.org/10.1007/s11001-012-9160-6

Liu, J.P., Milliman, J.D., Gao, S. & Cheng, P. 2004. Holocene development of the Yellow River's subaqueous delta, North Yellow Sea. *Marine Geology*, **209**, 45–67.

Liu, J.P., Xue, Z., Ross, K., Wang, H.J., Yang, Z.S., Li, A.C. & Gao, S. 2009. Fate of sediments delivered to the sea by Asian large rivers: long-distance transport and formation of remote alongshore clinothems. *The Sedimentary Record*, **7**, 4–9.

Muto, T. & Steel, R.J. 2002. In defense of shelf-edge delta development during falling and lowstand of relative sea level. *Journal of Geology*, **110**, 421–436, http://doi.org/10.1086/340631

Ni, Y., Endler, R. *et al.* 2014. The 'butterfly delta' system of Qiongzhou Strait: morphology, seismic stratigraphy and sedimentation. *Marine Geology*, **355**, 361–368.

Ni, Y., Harff, J., Xia, Z., Waniek, J.J., Endler, M. & Schulz-Bull, D.E. 2016. Post-glacial mud depocentre in the southern Beibu Gulf: acoustic features and sedimentary environment evolution. *In*: Clift, P.D., Harff, J., Wu, J. & Qui, Y. (eds) *River-Dominated Shelf Sediments of East Asian Seas*. Geological Society, London, Special Publications, **429**. First published online March 14, 2016, updated May 26, 2016, http://doi.org/10.1144/SP429.13

Shao, L., Qiao, P., Zhao, M., Li, Q., Wu, M., Pang, X. & Zhang, H. 2015. Depositional characteristics of the northern South China Sea in response to the evolution of the Pearl River. *In*: Clift, P.D., Harff, J., Wu, J. & Qui, Y. (eds) *River-Dominated Shelf Sediments of East Asian Seas*. Geological Society, London, Special Publications, **429**. First published online November 12, 2015, updated May 26, 2016, http://doi.org/10.1144/SP429.2

Vail, P.R., Mitchum, R.M. *et al.* 1977. Seismic stratigraphy and global changes of sea-level. *In*: Payton, C.E. (ed.) *Seismic Stratigraphy – Applications to Hydrocarbon Exploration*. American Association of Petroleum Geologists, Memoirs, **26**, 49–212.

Wang, Y., Liu, X., Li, G. & Zhang, W. 2015. Stratigraphic variations in the Diaokou lobe area of the Yellow River delta, China: implications for an evolutionary model of a delta lobe. *In*: Clift, P.D., Harff, J., Wu, J. & Qui, Y. (eds) *River-Dominated Shelf Sediments of East Asian Seas*. Geological Society, London, Special Publications, **429**. First published online October 2, 2015, updated May 26, 2016, http://doi.org/10.1144/SP429.8

Wu, J., Ren, J., Liu, H., Qiu, C., Cui, Y. & Zhang, Q. 2015. Trapping and escaping processes of Yangtze River-derived sediments to the East China Sea. *In*: Clift, P.D., Harff, J., Wu, J. & Qui, Y. (eds) *River-Dominated Shelf Sediments of East Asian Seas*. Geological Society, London, **429**. First published online August 27, 2015, updated May 26, 2016, http://doi.org/10.1144/SP429.7

Xu, K.H., Li, A.C. *et al.* 2012. Provenance, structure, and formation of the mud wedge along inner continental shelf of the East China Sea: a synthesis of the Yangtze dispersal system. *Marine Geology*, **291–294**, 176–191, http://doi.org/10.1016/j.margeo.2011.06.003

Yang, S., Bi, L., Li, C., Wang, Z. & Dou, Y. 2015. Major sinks of the Changjiang (Yangtze River)-derived sediments in the East China Sea during the late Quaternary. *In*: Clift, P.D., Harff, J., Wu, J. & Qui, Y. (eds) *River-Dominated Shelf Sediments of East Asian Seas*. Geological Society, London, Special Publications, **429**. First published online October 2, 2015, updated May 26, 2016, http://doi.org/10.1144/SP429.6

Zhang, J., Tomczak, M., Li, C., Witkowski, A., Qiu, Y., Chen, H. & Gao, H. 2015. Significance of the *Paralia sulcata* fossil record in palaeoenvironmental reconstructions of the SE Asia marginal seas over the Last Glacial Cycle. *In*: Clift, P.D., Harff, J., Wu, J. & Qui, Y. (eds) *River-Dominated Shelf Sediments of East Asian Seas*. Geological Society, London, **429**. First published online December 9, 2015, updated May 26, 2016, http://doi.org/10.1144/SP429.11

Zhou, Y., Chen, F., Wu, C., Yu, S. & Zhuang, C. 2015. Palaeoproductivity linked to monsoon variability in the northern slope of the South China Sea from the last 290 kyr: evidence of benthic foraminifera from Core SH7B. *In*: Clift, P.D., Harff, J., Wu, J. & Qui, Y. (eds) *River-Dominated Shelf Sediments of East Asian Seas*. Geological Society, London, Special Publications, **429**. First published online November 6, 2015, updated May 26, 2016, http://doi.org/10.1144/SP429.10

Zong, Y., Huang, G., Li, X.Y. & Sun, Y.Y. 2015. Late Quaternary tectonics, sea-level change and lithostratigraphy along the northern coast of the South China Sea. *In*: Clift, P.D., Harff, J., Wu, J. & Qui, Y. (eds) *River-Dominated Shelf Sediments of East Asian Seas*. Geological Society, London, Special Publications, **429**. First published online September 3, 2015, updated May 26, 2016, http://doi.org/10.1144/SP429.1

Assessing effective provenance methods for fluvial sediment in the South China Sea

PETER D. CLIFT

Department of Geology and Geophysics, Louisiana State University, Baton Rouge, LA 70803, USA (e-mail: pclift@lsu.edu)

Abstract: Sediment is delivered by the rivers of SE Asia to the South China Sea where it provides an archive of continental environmental conditions since the Eocene. Interpreting this archive is complicated because sediment may be derived from a number of unique sources and the rivers themselves have experienced headwater capture that also affects their composition. A number of methods exist to constrain provenance, but not all work well in this area. Anthropogenic impacts, most notably agriculture, mean that the modern rivers contain more weathered materials than they did up until about 3000 years ago. The rivers have also changed their bulk chemistry and clay mineralogy in response to climate change, so that these proxies, as well as Sr isotopes, are generally unreliable provenance indicators. Nd isotopes resolve influx from Luzon, but many other sources in SE Asia have similar values and clear resolution of end members can be difficult. Instead, thermochronology methods are best suited, especially apatite fission track, which shows more diversity in the sources than either U–Pb zircon or Ar/Ar muscovite dating. Nonetheless, even fission track is best used as part of a multiproxy approach if a robust quantitative budget is desired.

The South China Sea is one of the largest marginal basins in the Western Pacific and has been the repository of large volumes of clastic sediment delivered to the continental margins by the large rivers that drain the East Asian continent. In theory, these sedimentary records could be used to decipher the history of tectonism, surface uplift and environmental evolution in this region. Such data are essential if we are to understand the relationships between solid Earth evolution and the development of climate in the aftermath of the India–Asia collision, most notably the intensification of the East Asian monsoon whose history of activity is still not well understood (Sun & Wang 2005; Clift *et al.* 2014). However, if we are to read and interpret these sedimentary records then it is necessary to first understand where the sediment in each sub-basin was derived from because changes in sediment source may also result in changes in composition and mineralogy, which could be mistaken for changes in environmental conditions when in reality they merely reflects derivation from source bedrocks with different bulk compositions. Unless the sediment source can be properly assigned then other data sets cannot be used to their full potential in understanding how processes such as the intensification of the East Asian monsoon have impacted the continental environment over the last 65 million years. In this paper I review a number of the more commonly applied methods for constraining the source of sediment into this basin and explore the effectiveness of each in allowing us to understand how sediment is dispersed after its delivery to the ocean.

Sediment provenance is particularly important in the South China Sea because such methods have been used to track sediment transport within the basin and subsequently to infer the influence of bottom currents, which are in turn controlled by the opening and closure of gateways (Lei *et al.* 2007), especially those into the Western Pacific, such as the Bashi Straits between Taiwan and Luzon (Fig. 1). Furthermore, the three large rivers that drain into the basin, the Mekong, the Red (Song Hong) and the Pearl Rivers have all been proposed to have experienced significant amounts of headwater drainage capture, as a result of the eastwards tilting of the Asian continent during the uplift of the Tibetan Plateau (Brookfield 1998; Clark *et al.* 2004). The timing of this reorganization is controversial (Clift *et al.* 2006*a*; Robinson *et al.* 2013; Zheng *et al.* 2013), but is also important for testing models of surface uplift in Tibet and surrounding regions. If we are not able to pinpoint the source of the sediment within the delta and submarine fan systems in the marginal seas then it is impossible to fingerprint the influence of one river compared to another and thus to isolate the potential impact of drainage capture. Developing robust provenance tools is the first stage in addressing this process.

From: Clift, P. D., Harff, J., Wu, J. & Qui, Y. (eds) 2016. *River-Dominated Shelf Sediments of East Asian Seas*. Geological Society, London, Special Publications, **429**, 9–29.
First published online September 3, 2015, updated May 26, 2016, http://doi.org/10.1144/SP429.3

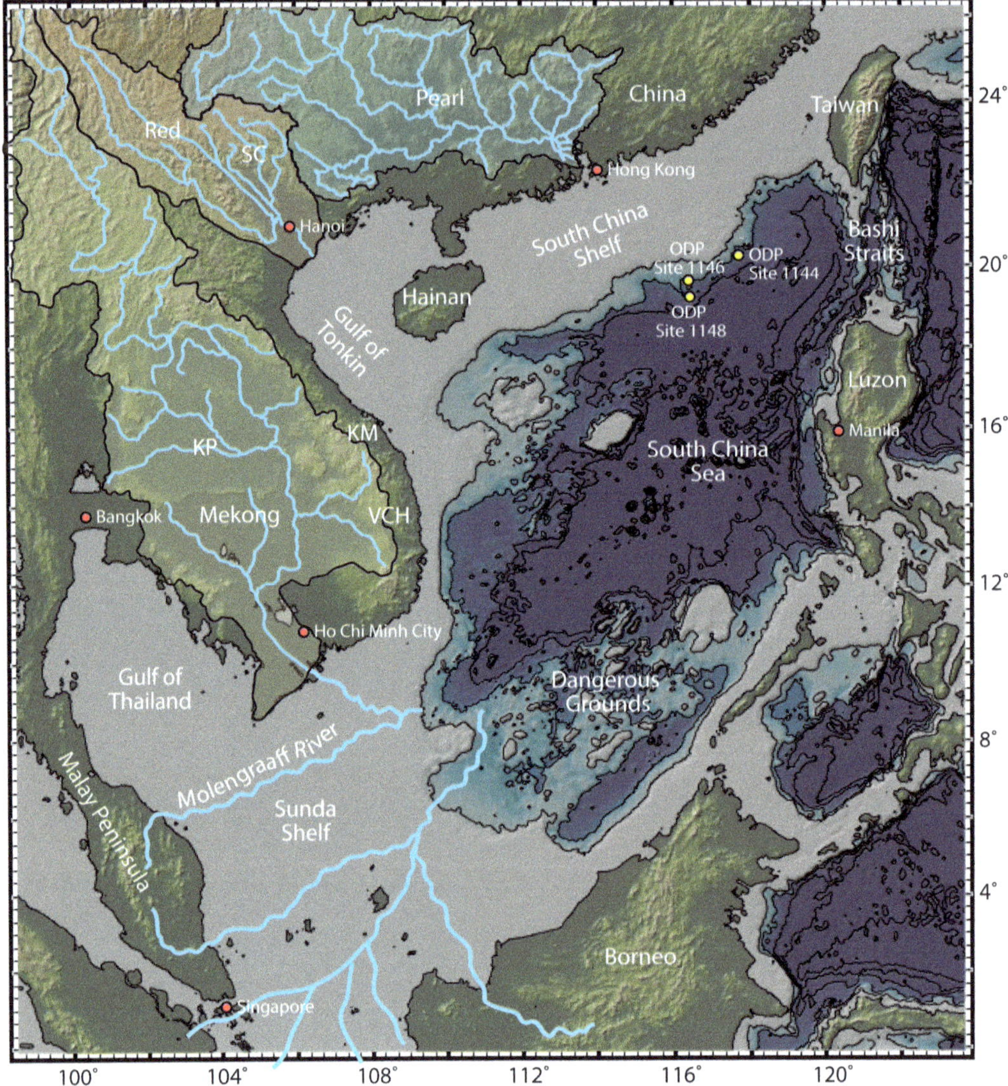

Fig. 1. Shaded bathymetric and topographical map of the South China Sea and surrounding continental areas showing the major drainage systems that are feeding sediment into the basin together with the Molengraaff River, which was active during sea level below stands at least in the recent geological past (Voris 2000). The map is from GeoMapApp. Bathymetric contours are in 1000 m increments. VCH, Vietnamese Central Highlands; KM, Kontum Massif; KP, Khorat Plateau; SC, Song Chay Massif.

Most of the methods that I discuss are not particularly novel, but represent standard methodologies that are applied to many sedimentary basins around the world. The South China Sea represents a special challenge because of the diversity of possible sources, but also provides an example of how, even in such a relatively complicated setting, effective sediment provenance reconstructions can be achieved.

Geological setting

The tectonic origins of the South China Sea have been strongly debated, but it is reasonably clear that the region began to experience significant continental extension during the Eocene (Ru & Pigott 1986; Franke 2013), culminating in the onset of seafloor spreading around 28 Ma (Briais *et al.* 1993; Barckhausen *et al.* 2014). Seafloor spreading ceased by 17 Ma, after which time the region has largely been affected by thermal subsidence, although disrupted by localized neotectonic activity, such as the volcanism and uplift in Hainan island (Shi *et al.* 2011). The most important tectonic event to have influenced sediment supply to the basin after the end of extension has been the collision between the oceanic island arc of Luzon with the passive margin of China, as manifest in the island of Taiwan (Suppe 1984; Huang *et al.* 2006). Taiwan is one of the world's great sediment sources and exceeds the Ganges–Brahmaputra delta in supplying sediment to the ocean despite its small size (Milliman & Syvitski 1992).

As well as receiving sediment from eastern and southern Taiwan, the South China Sea is supplied by sediment from smaller rivers on the Philippine Islands to the east, from Borneo to the south, but, more importantly, from three major rivers along the western and northern sides of the basin, and namely the Mekong, Red and Pearl rivers (Fig. 1). These, respectively, carry loads of 160, 170 and 69 Mt a^{-1} (Milliman & Syvitski 1992; Le *et al.* 2007). These numbers can only be considered as a rough guide, as the budgets are typically only for the suspended load, are not always taken near the river mouth and may be affected by anthropogenic disruption of the basin. Despite the large size of some of these drainage systems, the two largest rivers in SW Taiwan – the Tsengwen and the Kaoping rivers – have measured pre-modern sediment loads of 31 and 49 Mt a^{-1}, respectively, and the long-term recent discharge into the Taiwan Strait exceeds 380 Mt a^{-1} (Kao & Milliman 2008). It is noteworthy that Taiwan is also struck by many large typhoons originating in the central Pacific. These typhoons result in significant erosion and sediment discharge, with Typhoon Herb in 1996 alone being responsible for the discharge of 142 Mt into the South China Sea in a single week (Milliman & Kao 2005). Although Taiwan cannot have been a significant sediment source before the uplift of the present ranges after around 6 Ma (Huang *et al.* 2006), it is certainly worth considering the sediment-producing potential of each possible source when making volumetric assessments of the contribution from different end members to the deeper part of the South China Sea.

Basis of provenance

The ability to distinguish and estimate the amount of sediment derived from a given source is mostly based on the concept that the bedrock sources providing the sediment are distinguished from one another in different parts of the potential source area on the basis of their chemistry, geochronology or tectonic evolution and that these differences are transferred from the bedrock to the sediment in the rivers. Southeast Asia is remarkably suitable for such provenance work because of the assemblage of a number of contrasting tectonic blocks or terranes in this region (Fig. 2). These were largely brought into juxtaposition during the Triassic Indosinian Orogeny (Carter *et al.* 2001; Lepvrier *et al.* 2004; Carter & Clift 2008), with later additions, especially in the Tibetan headwaters of the large rivers during the final collision between India and Eurasia. Because of their contrasting geological histories, tectonic blocks shown in Figure 2 produce sediment of different composition or geochronological age, which can then be detected in the sediments deposited in the South China Sea. The geological evolution of each of these blocks is relatively complicated, but it is possible to say that essentially southern China, Cathaysia, represents a tectonic block that collided with the Yangtze Craton at approximately 800 Ma and that subsequently this was the host to a Mesozoic volcanic arc complex (Hutchison 1994; Fletcher *et al.* 2004). In contrast, the central part of China is dominated by the ancient crust of the Yangtze Craton, which itself collided with the North China Block during the Triassic (Hu *et al.* 2006). The Songpan Garze Terrane, which no longer provides sediment directly into the South China Sea, represents an accretionary complex formed during this collision between north and southern China (Zhou & Graham 1996; Huang *et al.* 2003; Enkelmann *et al.* 2007).

On the western side of the basin, the Indochina Peninsula is dominated by a separate tectonic block, but one that was also involved in the Indosinian Orogeny (Carter *et al.* 2001). Indochina has undergone more recent deformation as a result of the emplacement of flood basalt sequences in the Central Highlands of Vietnam during the Late Miocene (Carter *et al.* 2000). This is one of a number of rather enigmatic volcanic provinces around the basin, which provide isotopically unique sediment from newly emplaced primitive volcanic sequences. Likewise, on the eastern side of the basin, the island arc of Luzon has provided some sediment into the basin in the recent geological past, although it is worth noting that its location to the east of the main deep-water basin is a relatively recent development following the northwards drift of the arc and the collision of the arc with the southern margin of mainland Eurasia (Fig. 3). During the latest Miocene, plate tectonic reconstructions show that Luzon was positioned somewhat to the south of the basin and has only been able to supply sediment to the basin in the last few million years (Hall 2002).

The rivers draining the island of Borneo are probably the least well defined of any potential source around the sea, but they too have had a

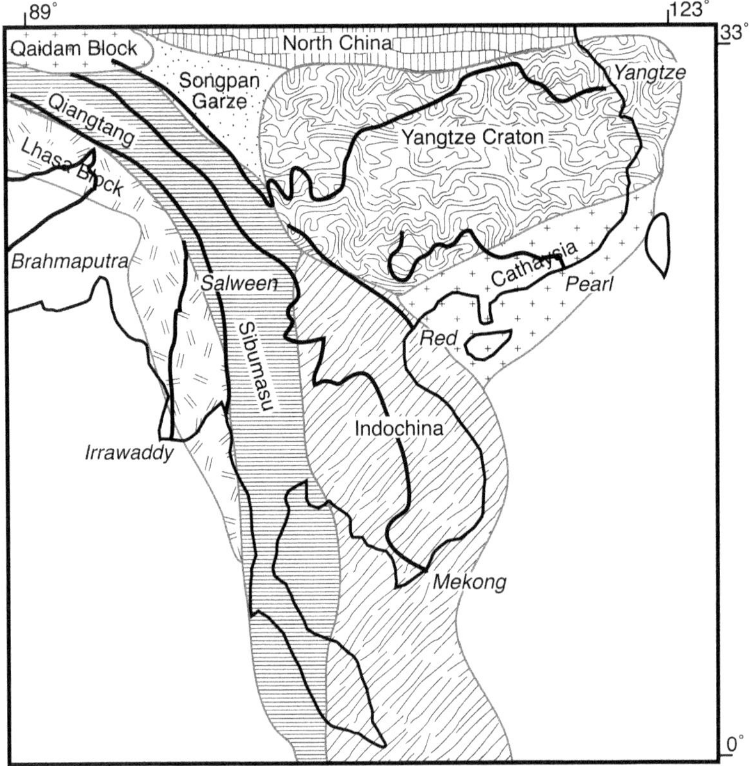

Fig. 2. Simplified tectonic terrane map of East and SE Asia showing the major blocks discussed in this paper and the courses of the rivers. After Metcalfe (1996).

limited impact on sediment flux into the ocean, partly because they were separated from the present basin by a palaeo-South China Sea until around 16 Ma when the southern margin of the basin, dominated by the Dangerous Grounds, began to collide with the northern margin of Borneo (Hutchison 2005; Clift *et al.* 2008*a*). Even since that time, direct sediment supply from Borneo into the deeper parts of the South China Sea has been restricted by the tectonic topography of the Dangerous Grounds (Hutchison & Vijayan 2010), and particularly by the long and deep North Borneo Trough that separates Borneo from the main part of the sea and which acts as a very effective sediment trap (Hutchison 2010), so that only plumes of suspended relatively fine-grained sediment can be transported deep into the basin.

Consideration of such plate tectonic reconstructions is very important when making provenance assessments because it is clearly impossible to derive sediment from a tectonic block that was not in the vicinity of the South China Sea at the time of sedimentation. Conversely, convincing provenance data can help us to better define the tectonic development of SE Asia by showing which blocks were present at any particular time.

Sediment mixing on the continental shelf

Provenance analysis of sediment in the deep basin does not necessarily reflect the contribution of a single river system to the overall basin budget because of filtering of the signal through the continental shelf. Changes in sea level and, therefore, in the position of the river mouth relative to the continental shelf edge affects how important that river will be in supplying sediment into the deep basin. For example, the Mekong River appears to have been important in supplying sediment into the deep SW part of the basin during sea-level lowstands, but has been relatively cut-off following post-glacial sea-level rise (Szczuciński *et al.* 2013). The same is true of the Red and Pearl rivers, with their wide continental shelves. In contrast, areas where the continental shelf is very narrow, such as offshore central Vietnam, may be important sources of sediment to the deep basin even during periods of

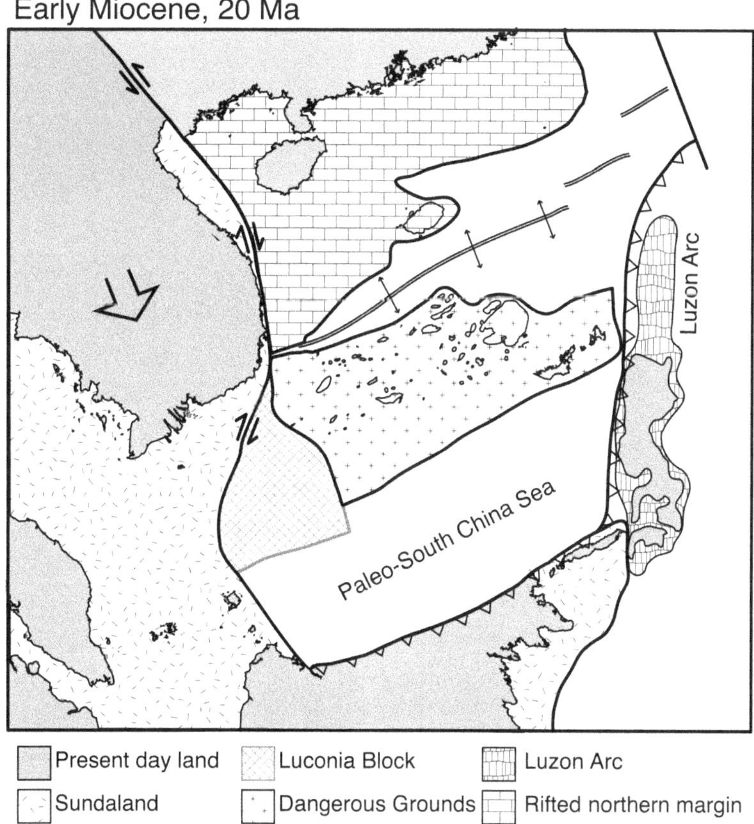

Fig. 3. Palaeogeographical map of the South China Sea at 20 Ma showing that the Luzon arc was only beginning to become a potential sediment source region at this time. Prior to this, sediment flux from such primitive arc sources would have been impossible. Modified from P. Clift *et al.* (2008*a*).

relatively high sea level (Schimanski & Stattegger 2005; Lahajnar *et al.* 2007; Szczuciński *et al.* 2009).

Reworking on the continental shelf is important in influencing the net contribution to the deep basin. Sediment originating from the Mekong River catchment to the deep basin is now delivered, not from modern sediment discharge at the river mouth, but from erosion of the lowstand Mekong delta at the shelf edge (Dung *et al.* 2013). Newly delivered sediment is, instead, sequestered close to the coast. Geochemical data from South Asia has reinforced the idea that the continental shelf is a location in which sediment from different sources may be mixed via longshore transport and that reworking of older deposits, due to storms or bottom currents, may influence the composition of sediment delivered to the deep sea, and, in particular, this does not necessarily reflect the large river mouth in close proximity (Limmer *et al.* 2012).

As a result, fingerprinting of sediments in modern rivers can help to resolve the influence of onshore basins in supplying sediment layers to the continental shelf and to the deep ocean. Sediment deposited in the offshore, even close to the river mouth, cannot be considered a reliable fingerprint of provenance.

Clay minerals

Clay minerals can provide an important source of sediment provenance data, especially in distal fine-grained sediment sequences where single grain methods may not be applicable, most sand being sequestered on the shelf in many systems. The method is based on the contrasting mineralogy of sediments in the different river systems that currently supply the basin and the documented changes in clay mineralogy of the modern seafloor across the basin (Chen 1978). The method suffers from being only ‘semi-quantitative’, in that relative proportions of different key minerals are determined by a variety of methods such as X-ray diffraction data, with different methods producing slightly different estimates that may not all be completely in accordance with one another (Holtzapffel 1985; Moore & Reynolds 1989; Hillier 2003). The accuracy of the methods is not entirely clear and is best applied to resolving relatively larger differences in assemblages (>5%). Smaller differences in clay mineral assemblage cannot be considered robust and are not considered as effective provenance tools.

A number of studies have now highlighted the fact that the Pearl, Red and Mekong rivers, as well as smaller rivers draining Borneo and the Malay Peninsula, are characterized by different clay mineral assemblages that roughly correlate with the intensity of chemical weathering in the fluvial basin (Liu *et al.* 2007*b*, 2012). Although some of the differences in clay mineral content are related to source rock compositions, much of the contrast is the result of a variable intensity of chemical weathering, which in turn is linked to the topography. The intensity of the summer monsoon, which provides much of the moisture and related warmth to the area, also modulates the rate of chemical weathering and, thus, the clay mineralogy (West *et al.* 2005), accounting for the abundance of smectite and kaolinite in the more tropical regions of the southern South China Sea, especially the Sunda Shelf (Aoki 1976; Chen 1978).

Composition is important in the case of the rivers draining Luzon because these rivers are rich in smectite, which is a product of the chemical breakdown of volcanic rocks (Liu *et al.* 2009) that are abundant in this volcanic arc, as well as in the ranges of central Vietnam (Jagodziński 2005). In contrast, although the rocks of Taiwan originated as sediments on the passive margin of China, these have experienced significant metamorphism and are now dominated by illite and chlorite, which largely represent the products of physical erosion of the high mountains in Taiwan rather than the products of chemical weathering of pre-existing bedrock. Meanwhile, sediment delivered by the Pearl River is dominated by kaolinite with some illite and chlorite influx (Liu *et al.* 2007*a*). Further south, the Mekong submarine delta, which is the primary depocentre for the modern river, is dominated by illite with lesser amounts of smectite, kaolinite and chlorite (Szczuciński *et al.* 2013). Further illite is supplied from the rivers of northern Borneo (Liu *et al.* 2007*c*).

Differences in clay mineralogy of Holocene and recent sediment in the deep basin, as well as those found in older deposits, have been used to separate the different contributions from potential sources (Boulay *et al.* 2005; Liu *et al.* 2010*b*). Figure 4 shows a triangular diagram and the type of mixing arrays that have been proposed in the past to separate the influence of these different source terrains. In this particular example, we see that sediment from Ocean Drilling Program (ODP) Site 1144 lies close to the field of Taiwan, allowing the provenance of this deposit to be constrained to this island (Hu *et al.* 2012). Similar approaches have been applied to modern sediment in the northern South China Sea (Z. Liu *et al.* 2003; J. Liu *et al.* 2014) and used to infer sediment transport directions, driven by bottom currents.

Unfortunately, this method is prone to problems because it is based on the assumption that river clay mineral assemblages have been the same in the past as they are at the present. There seems little doubt that environmental changes have caused changes in clay assemblages over long periods of geological time (Clift *et al.* 2002; Clift *et al.* 2014; Wan *et al.* 2007), largely as a result of changes in the monsoon, so that the end members cannot be considered stable over long periods of time as the climate in SE Asia has evolved. Furthermore, the clay mineral evidence for the origin of the Holocene sediment at ODP Site 1144 appears to show that the sediment is not entirely derived from Taiwan, despite the fact that other proxies indicate that it would seem to be most likely (Hu *et al.* 2012). Why then do these data not plot directly in the modern Taiwan field? The difference was attributed by Hu *et al.* (2012) to weathering of the sediment on the exposed continental shelf during the Last Glacial Maximum and then reworking during the Holocene before final sedimentation. Sediment buffering between source and sink is often associated with chemical weathering (Lupker *et al.* 2012), the net result of which is a transformation of the clay mineral assemblage relative to the modern river composition. The method is also open to error if the sequences are affected by burial diagenesis, as may be the case in deep boreholes where burial heating can be significant.

An additional complexity was recognized by Steinke *et al.* (2008), who showed that the clay mineralogy at any one site on the Sunda Shelf was largely controlled by sea level, which acted to disrupt the large drainage systems that existing during the Last Glacial Maximum. In particular, sediment rich in kaolinite derived from the south, as well as from the exposed Sunda Shelf itself, is strongly reduced during the Holocene as flux from the Mekong increased relative to the southern sources, as their river mouths retreated from proximity to the shelf edge. Moreover, these workers recognized that, as sea level rose, little of this material was reaching the deep basin, but, rather, was sequestered in terrestrial flood plains and submarine delta clinoforms. Furthermore, studies of the clays around the Mekong delta shows that these differ from east to west in the modern system, partly reflecting the preferential settling of some clays close to the river

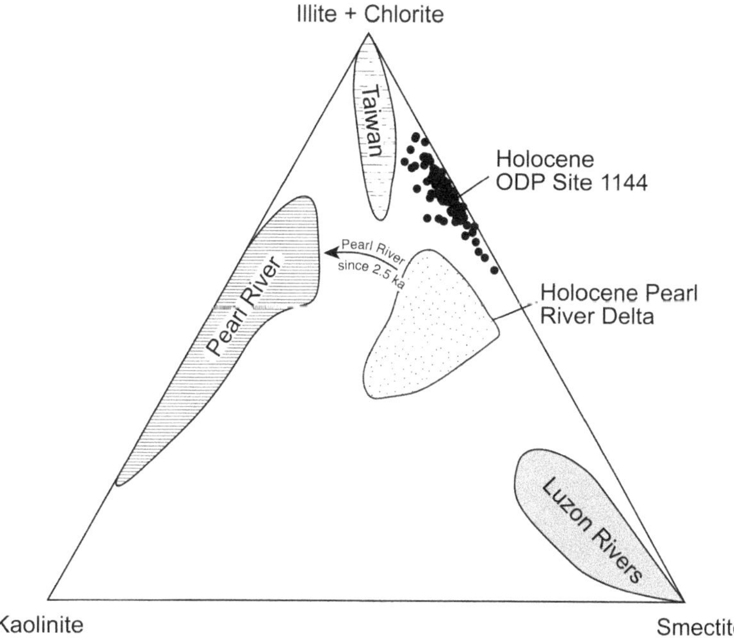

Fig. 4. Clay mineralogy of the Holocene sediments analysed from ODP Site 1144 (Hu *et al.* 2012) compared to modern river and offshore sediment compositions in the possible source regions (Liu *et al.* 2010*a*). Holocene Pearl River data are from Hu *et al.* (2013). The arrow shows how the Pearl River composition changes after 2.5 ka.

mouth (Xue *et al.* 2014). Sediments from the east, on the South China Sea side of the delta, look like the Mekong River itself, while those to the west have much less illite and chlorite, and imply reduced fine particle inputs from the Mekong River towards the Gulf of Thailand.

Most seriously of all is the recognition that the composition of the modern rivers is not in equilibrium and may bear little resemblance to the natural state of the river before a few thousand years ago. It has increasingly been demonstrated that human settlement has had a major impact on landscape evolution and, in particular, has encouraged the rapid erosion of soils as agriculture became prevalent over the last several thousand years (Syvitski *et al.* 2005; Syvitski & Kettner 2011). Data from the delta of the Pearl River suggests that human settlement in southern China had become highly disruptive to the river by around 2500 years ago (Hu *et al.* 2013) and similar patterns might be anticipated for Indochina. More recently, the development of more sophisticated industrial cultures has resulted in the damming of major river systems so that the rate of sediment delivery to the ocean has been reduced in recent times.

A number of studies have now shown that the alteration state of sediment reaching the delta since the onset of agriculture has increased (Bayon *et al.* 2012; Hu *et al.* 2013), consistent with the increased influence from anthropogenic contaminants at that time (Zong *et al.* 2010). For example, the Pearl River shows a large dominance of kaolinite in the modern system, but cores taken in the river mouth show that the situation only arose at around 2500 ka and that before that time smectite was much more important, as shown by the difference between modern Pearl River sediments and those from the Holocene delta (Fig. 4). Thus, to assume that much of the smectite in the South China Sea was derived from Luzon or from central Vietnam prior to 2500 ka would be incorrect, and before that time similar size variations can be linked to climate change during the early Holocene when the monsoon was stronger (Wang *et al.* 2001; Dykoski *et al.* 2005). Unless the impact of modern human disruption to river systems and past climate change can be accounted for, therefore, it seems that clay mineralogy by itself is a rather unreliable provenance proxy unless simply applied to the present day. Although in deep-water settings, where coarser sediment is not found, it may be one of the few methods that can even be attempted.

Bulk geochemistry

An alternative method to using weathering regime as a tool for sediment provenance is to look at the bulk geochemistry. This approach is potentially useful in sediment of many different grain sizes, including the muds that are not easily analysed using the thermochronology methods discussed below. Care must, however, be exercised in comparing geochemical data between sediments of approximately similar size as hydrodynamic sorting of minerals and preferential alteration are methods by which different grain sizes can end up with quite different chemistries, despite coming from the same original source.

The premise of this particular method is that elements that are particularly soluble in aqueous solution are depleted in rocks and sediment, which has experienced more chemical alteration, and are mobilized and removed compared to immobile elements such as aluminum or silicon (Nesbitt & Young 1982; Galy & France-Lanord 1999). This means that rivers in warmer, wetter environments tend to have more altered sediment in them than those in drier, colder places. Nonetheless, it has to be recognized that topography is also important because fast-flowing, high-energy rivers from steep mountains, such as Taiwan, tend to transport sediment quickly, leading to less chemical weathering. Liu *et al.* (2007*b*) surveyed the rivers of the South China Sea and showed that the Pearl River tended to include sediment that was more weathered than that in the Mekong, which in turn is more weathered than that in the Red River. This study was able to do this using the geochemical proxy 'chemical index of alteration' (CIA: Fig. 5), which was developed for use in soils (Nesbitt & Young 1982). Despite showing some overlap, Liu *et al.* (2007*b*) did highlight differences between the rivers when a limited grain size was considered. The CIA proxy was not developed for marine sediments and should probably not be applied in this situation, but has, nevertheless, a long history of being used to look at the alteration state of deep-water marine sediments.

Application of proxies such as the CIA require that only limited grain-size fractions should be compared with one another because of the effects of the hydrodynamic sorting of different mineral species that has a dominant control on bulk chemistry. Furthermore, marine sediments are often contaminated by biogenic calcium, which needs to be corrected for if the proxy is to have any meaning whatsoever (Singh *et al.* 2005). Likewise, sodium may be increase in marine sediments as a result of seawater in pores, which has to be flushed from

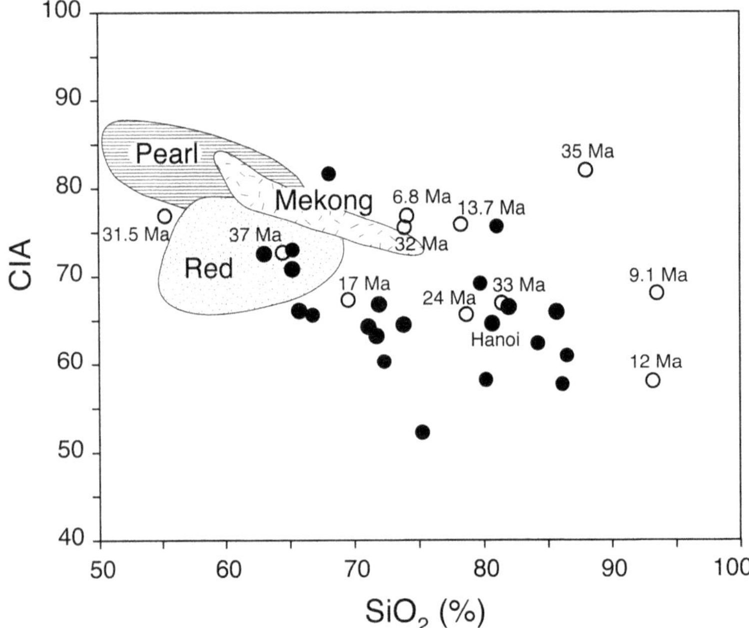

Fig. 5. Chemical index of alteration (CIA) plotted against silica content for sand and silt samples from the modern Red River (black dots) and older borehole samples from the Hanoi Basin labelled with depositional age (open circles) (Clift *et al.* 2008*b*). Fields for modern trunk Red, Mekong and Pearl rivers are from Liu *et al.* (2007*b*), and include only fine-grained sediments.

the sample if the analysis is to have any significant meaning. However, with appropriate sample preparation, these issues can be overcome.

Figure 5 shows how the data presented by Liu *et al.* (2007*b*), which was generated after decarbonation, compares with a series of more sandy sediments from the Red River alone (Clift *et al.* 2008*b*). What is apparent is that more sandy material tends to have lower CIA values, but that the scatter of data is very significant. The distribution of values within a single river is so great that it is probably unrealistic to hope to find a characteristic value for any given river system. This also suggests that using CIA or other chemical weathering proxies as a way to track provenance is probably useless unless this is restricted to the present day and/or to a very limited grain-size range. Geochemical studies of the Yellow Sea region do, however, show that with appropriate filtering for grain size and with correction for biogenic carbonates that some degree of provenance can be successfully achieved using bulk sediment major-element chemistry (Kim *et al.* 1999; Lim *et al.* 2013).

Other types of bulk sediment geochemical data have been used in the past to track provenance, most notably rare earth elements (REEs). These are more uniformly immobile to aqueous chemical weathering and might be expected to better preserve the average composition of the source bedrocks that are generating the sediment. However, much of the upper continental crust tends to have a gentle enrichment in light REEs (LREEs) and to be often similar between different basins (Rudnick & Gao 2003). Total REE content is not useful because the REEs are strongly concentrated in a number of heavy minerals so that the concentrations are largely a reflection of the absolute abundance of these minerals in the sample selected (Garçon *et al.* 2014). This in turn is related to hydrodynamic sorting during sedimentation, so that samples taken from different parts of a single bedform might show a quite different REE content. Studies of the REEs in the Red River showed a wide scatter of REE concentrations and ratios, but no systematic difference in REE character between different tributaries, suggesting that large tracts of the upper continental crust have quite similar degrees of enrichment and that, therefore, large rivers with diverse catchment geologies will often have a similar chemical character (Clift *et al.* 2008*b*).

In contrast, some studies claim that key REE ratios such as La/Yb or Gd/Yb, as well as the relative europium anomaly, can be distinctive of different source regions, although this will be less true in larger river basins where more averaging can occur (Vital & Stattegger 2000; Jung *et al.* 2012). As with other forms of chemistry, REE chemistry may be affected by variations in grain size and by fractionation of heavy minerals, and so are best used on a only a limited size fraction

In general, REEs do not seem like promising provenance proxies, with the exception of potentially finding sediment derived from the Luzon Arc, which, being a more primitive, mantle-derived piece of crust, would be associated with more LREE-depleted compositions (i.e. low La/Yb values).

Bulk isotope character

Combined Nd and Sr isotopes have an established track record in terms of resolving provenance in many basins worldwide and specifically in the South China Sea (Clift *et al.* 2002; Li *et al.* 2003). Nd, in particular, is generally recognized as being immobile to chemical weathering, and is not fractionated by erosion, weathering and transport (Goldstein & Jacobsen 1988). Sr, however, is more mobile, and is fractionated so that more weathered sediments tend to have higher $^{87}Sr/^{86}Sr$ values that represent both source composition and weathering intensity (Derry & France-Lanord 1996). Nd is a powerful provenance proxy, although, again, grain size may have an influence because Nd is largely dominated by monazite content (Garçon *et al.* 2013). The method has most successfully been applied to fine-grained sediments, where it gives an estimate of the relative age of the continental crust from which the sediment is derived (Allegre & Ben Othman 1980), although it could be applied to coarser materials provided that these data were compared with other coarse sediment measurements.

Figure 6 shows the range of measured Sr and Nd isotope ratios for a series of modern rivers draining into the basin, together with selected analyses from marine cores, largely in the northern part of the sea. It is clear that the modern Pearl River and some parts of the modern Red River are characterized by very high $^{87}Sr/^{86}Sr$ values and might be resolvable in this respect. However, this ignores the fact that these rivers are anthropogenically disrupted and are presently carrying much more altered sediment than has been typical for the Holocene.

Sediment in rivers draining Luzon is truly unique in Sr and Nd isotopes, in having both very low $^{87}Sr/^{86}Sr$ values and very high ε_{Nd} values. Because Nd is resistant to change during weathering, the influence of Luzon in providing sediment should be easily resolved using this approach. However, many possible sources cluster around $^{87}Sr/^{86}Sr$ values of 0.72 and ε_{Nd} values of -10, including the modern Mekong and Red rivers, Taiwan, and the Holocene of the Pearl River. Not surprisingly, many of the cored sediments analysed for these isotopes also plot in this region, suggesting that this

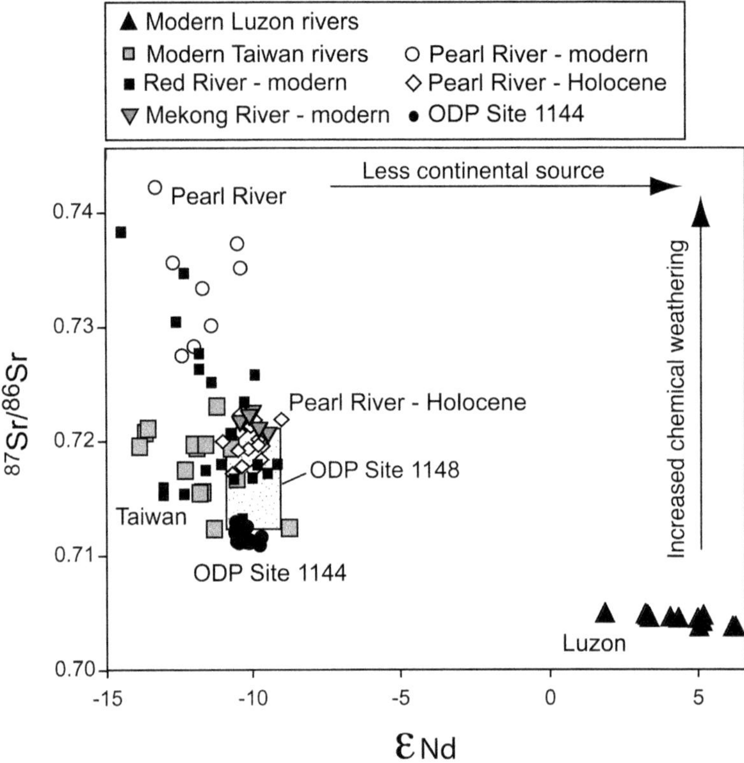

Fig. 6. Sr and Nd isotopic plot showing the variability in Holocene and modern sediments from ODP Site 1144 (Hu *et al.* 2012) and in Lower Miocene–Recent sediments at ODP Site 1148 compared to modern potential sources around the South China Sea. The diagram shows the general similarity of the sediment with modern Taiwanese rivers and bedrock samples (Chen & Lee 1990; Lan *et al.* 2002), and the differences with sediments in the modern Pearl River (Liu *et al.* 2007*b*) and with potential volcanic sources in the Philippine island of Luzon (Zhou *et al.* 2002). The Neogene sediments at ODP Site 1148 show an overlap with both Taiwanese rivers and the Holocene sediments of the Pearl River Estuary (Hu *et al.* 2013). Data from the Red and Mekong rivers are from Liu *et al.* (2007*b*) and Clift *et al.* (2008*b*).

approach may not be the best at resolving sources in SE Asia. It is noteworthy that ODP Site 1144 shows overlap with Taiwan and has the lowest $^{87}Sr/^{86}Sr$ values, which do not seem unique to that source compared to sources in southern China or Indochina. The relative lack of variation in ε_{Nd} values reflects the fact that many of the blocks in SE Asia have similar crustal residence ages, but it should be noted that when rivers have connected to sources in the much older Yangtze Craton in the past then much lower ε_{Nd} values were recognized, most notably in the Red River delta (Song Hong-Yinggehai Basin) (Clift *et al.* 2006*a*).

Apatite fission track

Comparison of fission-track ages between sediments and bedrock sources has been a powerful method for sediment provenance for some time (Hurford & Carter 1991; Carter 2007), and is one with some of the greatest potential in the South China Sea. The apatite fission-track method records cooling of rocks through approximately 60–125°C over timescales of 1–10 Ma (Green *et al.* 1989) and, provided that the sediment has not been buried and reheated again since deposition, allows single grains to be tied to sources with unique exhumation histories. This favours sedimentary systems in SE Asia because different parts of the margins have experienced different deformation and erosion histories. Fission-track methods are best applied to sediment that is fine sand or coarse in grain size, reflecting the need to measure track densities and the 16 μm length of new tracks. However, coarse silts can be used in extreme circumstances.

Luzon is less easily recognized in this approach because basalts that dominate the arc are not rich in apatite but should be Oligocene or younger if present, given the known range of magmatism

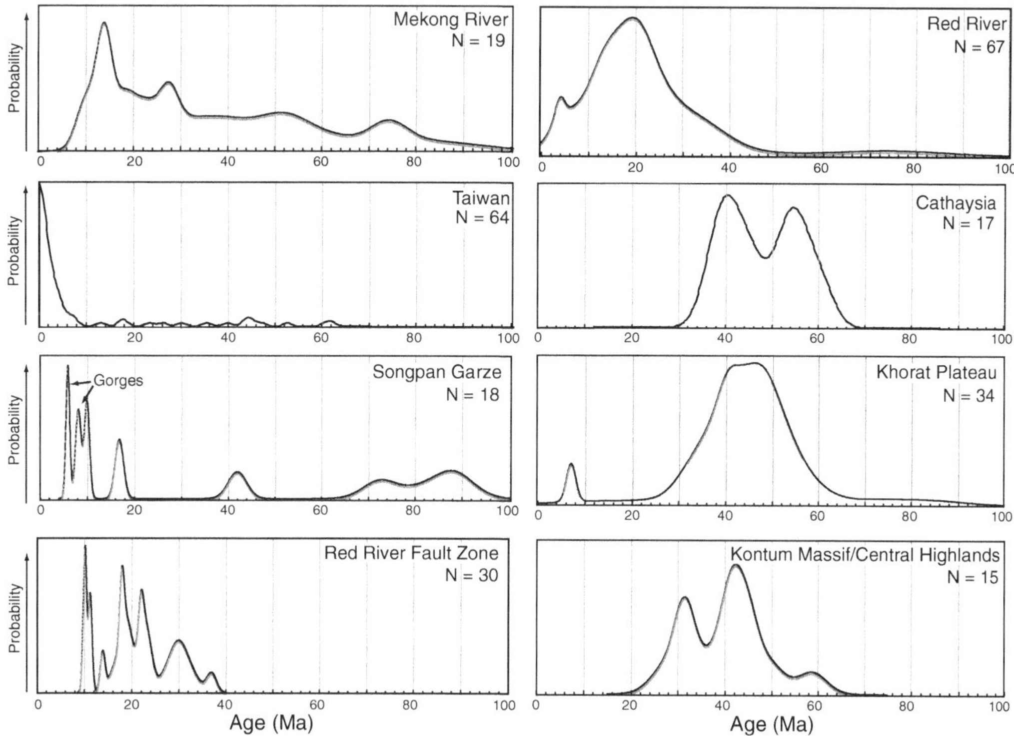

Fig. 7. Probability density diagrams for apatite fission-track ages from sands in the modern Mekong and Red rivers compared with those from possible source terrains in eastern Asia (Clift *et al.* 2006*b*). Data from the Ailao Shan/Red River Fault Zone is from Bergman *et al.* (1997) and Maluski *et al.* (2001). Data from the Khorat Plateau is from Racey *et al.* (1997) and Upton (1999). Data from the Songpan Garze Terrane is from Reid *et al.* (2005). Data from the Kontum Massif is from Carter *et al.* (2000). Data from Taiwan is from Fuller *et al.* (2006) and Willett *et al.* (2003). Data from Cathaysia is from Yan *et al.* (2009).

(Defant *et al.* 1989). Figure 7 shows a range of kernel density estimate (KDE) plots designed to demonstrate the most likely cooling ages for each source (Vermeesch 2012). Taiwan is unique in this system, with very young ages that post-date the start of collision at *c.* 5 Ma, making grains derived from this island easy to resolve within mixtures. Fission-track cooling ages from Cathaysia in southern China, and from the Khorat Plateau, the Vietnamese Central Highlands and the Kontum Massif in Indochina have a similar range of central ages, especially around 30–60 Ma. This age reflects cooling driven by the break-up of the South China Sea and the subsequent erosional degradation of the rifted margins. However, because this process is common, sediment derived from these different sources cannot be resolved with this technique, with the possible exception that the Central Highlands has slightly younger ages than other parts of the western margin because of rejuvenation of that region during emplacement of volcanic sequences at approximately 8 Ma (Carter *et al.* 2000).

An intermediate grouping of fission-track ages are those grains with cooling ages of *c.* 20 Ma. These are typical of bedrocks in the Red River Fault Zone, but are found throughout the SE flank of the Tibetan Plateau, especially in the gorges of SW China (Clark *et al.* 2005) and are not specific to the fault zone. This cooling is related to the deformation and rock uplift associated with the southeastward propagation of the Tibetan Plateau during the Miocene (Schoenbohm *et al.* 2006). This explains why grains of this age dominate the modern Red River (Fig. 7) which derives its sediment from this region (Clift *et al.* 2006*b*). In contrast, the Mekong River has a subtly different age spectrum, with a younger peak at approximately 14 Ma and a long tail of older ages. The younger aged grains in the Mekong are believed to be derived from younger sources in its upper reaches in Tibet and the older ages from the lowlands of Indochina, including the Khorat Plateau (Clift *et al.* 2006*b*). For whatever reason, the fission-track ages of these two rivers contrast with each other and the presumed spectrum for the Pearl, which must be dominated by the older ages from Cathaysia.

Apatite fission-track thermochronology thus holds great promise in being able to distinguish sediment from the three major rivers, as well as Taiwan, even if there is more ambiguity concerning grains dated at 30–60 Ma, which may be eroded from either the basin's western or northern margins.

Muscovite Ar/Ar dating

There is much less existing $^{40}Ar/^{39}Ar$ data from potential bedrock sources around the South China Sea compared to fission-track data, yet this method is also useful at pinpointing sources and can be complementary to the lower temperature method. The $^{40}Ar/^{39}Ar$ method has been applied in several Asian provenance studies (Najman *et al.* 1997; White *et al.* 2002; Szulc *et al.* 2006; Hoang *et al.* 2010) and allows the age when each single mica grain cooled through an isotherm of around 300°C to be determined (Hodges 2003). Triassic cooling ages are common in SE Asia and are related to the Indosinian Orogeny (Huang *et al.* 2003; Lepvrier *et al.* 2004). However, there are important discrepancies to this general picture, and resolvable differences between sources around the northern and western edges of the basin that make this a good provenance tool. Single-grain mica dating requires substantial crystals and even multiple grain methods mean that the method is really only applicable to sands, generally limiting its application to the proximal parts of a given sedimentary system.

Analysis of the modern Red and Mekong rivers shows that the Mekong has a generally younger set of Indosinian micas than seen in the Red River; 150–210 Ma compared to 210–250 Ma (Clift *et al.* 2006*b*) (Fig. 8). This difference must reflect contrasting cooling ages of Indosinian metamorphic rocks in the two drainage basins and when transferred to the marine realm would allow sediment from each river to be resolved within offshore depocentres. Both rivers contain muscovite micas dating to *c.* 20–30 Ma, similar to the rocks of the

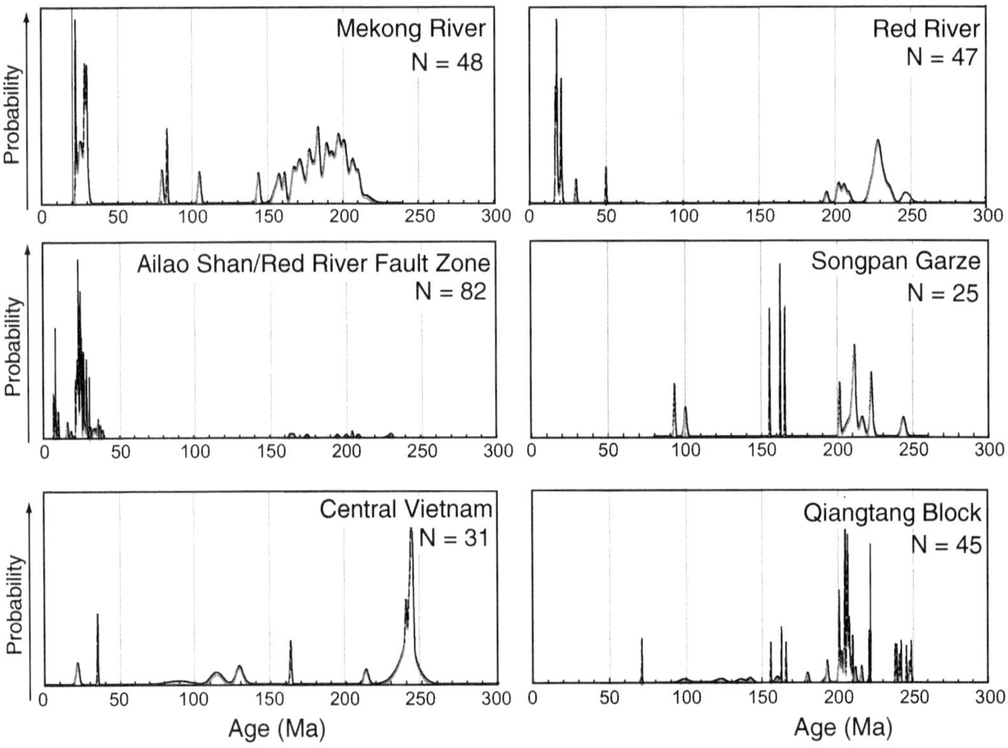

Fig. 8. Probability density diagrams for Ar/Ar ages of detrital muscovites within the Mekong and Red rivers compared to known ranges from possible source regions. Central Vietnam data are from Lepvrier *et al.* (1997) and Nagy *et al.* (2000). Ailao Shan/Red River Fault Zone data are from Leloup *et al.* (1993, 2001), Harrison *et al.* (1996), P. L. Wang *et al.* (1998), Jolivet *et al.* (1999) and Maluski *et al.* (2001), and data from Songpan Garze are from Reid *et al.* (2005). Qiangtang Block data are from Kapp *et al.* (2000).

Red River Fault Zone and equivalent structures across SE Tibet and Indochina, such as the Mae Ping and Three Pagoda faults (P. L. Wang *et al.* 1998; Morley 2002). As a result, these ages are only partially source diagnostic, and are only related to sources from SE Tibet. The timing of motion on these faults is commonly Oligo-Miocene, making it impossible to resolve between the different sources. Only the Xianshuihe–Xiaojiang Fault of SW China has continued rapid motion until recent times (E. Wang *et al.* 1998). Nonetheless, the presence of 20–30 Ma micas does at least rule out erosion from southern China. $^{40}Ar/^{39}Ar$ data from southern China–Cathaysia is sparse, but that which does exist tends to be of Indosinian age, 207–215 Ma (Wang *et al.* 2005), but critically slightly younger than dated basement rocks in central Vietnam, which tend to cluster at about 240 Ma (Lepvrier *et al.* 1997). This latter tectonic block is, however, marked by a minor population of micas dating at 110–130 Ma, which is not found further north.

Luzon is not a significant source of muscovite because of its dominant basaltic oceanic arc character and any mica derived from the Central Ranges of Taiwan must have very young ages of <5 Ma if they have been reset by the orogeny on that island. If muscovite grains from Taiwan have not been reset – for example, in the lower structural levels of the thrust wedge – then they would carry the same array of ages as seen in SE China and would not be resolvable using this method. Although the $^{40}Ar/^{39}Ar$ method is generally slower and more expensive than that of the fission track method, it can form an important tool for quantifying erosion budgets in the South China Sea because of the array of unique sources that characterize some of the major rivers, especially along the western margins of the basin.

Zircon U–Pb dating

U–Pb dating of zircon has become one of the most popular forms of provenance analysis, following the improvement in dating technology that now allows large numbers of grains to be quickly and cheaply dated using laser ablation inductively coupled plasma mass spectrometry (ICP-MS) instead of the more traditional mass spectrometry. This method can only be applied to grains that are large enough to be dated, which usually means >50 μm (coarse silt), as smaller grains are too small to be targeted by a laser ablation ICP-MS that dominates the approach. This means that the method is less useful in the most distal, deep-water deposits in the basin centre.

U–Pb ages in zircon are considered to reflect the time of zircon crystallization at around 750°C (Hodges 2003), so that they may be interpreted to record the last growth phase in a rock's history. However, zircons are known to be resistant to physical abrasion during erosion and transport, as well as to chemical weathering, making them susceptible to multiple phases of reworking (DeCelles *et al.* 2004; Campbell *et al.* 2005). Although it is impossible to say precisely where a single grain with a single age is derived from, it is possible to suggest the most likely source rock unit for many grains, making U–Pb dating a powerful provenance tool in this area. Because the different possible source terrains have unique age spectra, it is usually possible to at least exclude possible dominant sources based on the U–Pb age of the sediment.

As with the Ar/Ar ages, Triassic Indosinian ages are very common in many of the source regions and tend not to be diagnostic by themselves. Figure 9h shows that the Qiangtang Block of Tibet and its equivalent terrane, Sibumasu, in SE Asia (Fig. 2) are particularly dominated by this age range and have relatively few older grains, resulting in a potentially unique signature. Likewise, Red River Fault Zone rocks yield Cenozoic ages that are synchronous with the motion on that fault (Fig. 9g), and which are effectively unknown outside this zone and presumably equivalent fault zone rocks in the other major strike-slip zones of SE Tibet–SW China. In any case, observations from the Mekong and Red rivers indicate that the fault zone rocks are not major contributors to the net flux to the ocean (Fig. 9a, b). A high proportion of grains dated between 700 and 1000 Ma is typical of bedrock from the Yangtze Block. These grains are also common in Cathaysia-derived sediment, but these have a slightly older peak at about 900 Ma rather than about 800 Ma, and also contain significant populations clustered around 1800 and 2400 Ma, which are present but are not so abundant in the Yangtze Block. The Songpan Garze terrane shows intermediate character, having relatively few 700–1000 Ma grains, but common approximately 1800 and 2400 Ma grains, as well as a strong 400–500 Ma population (Fig. 9d), which is unknown in the Yangtze Block, rare in Cathaysia but is also known in Indochina in the Khorat Plateau, the Kontum Massif and the Central Highland areas that fringe the western edge of the South China Sea.

Consideration of the spectra of zircon ages in modern Red and Mekong river sediments shows that while they share many of the same populations there are differences between these systems that reflect the contrasting ages in the bedrock sources. The Red River has a much stronger peak at around 800 Ma compared to the Mekong River, which may be interpreted to indicate more erosion from

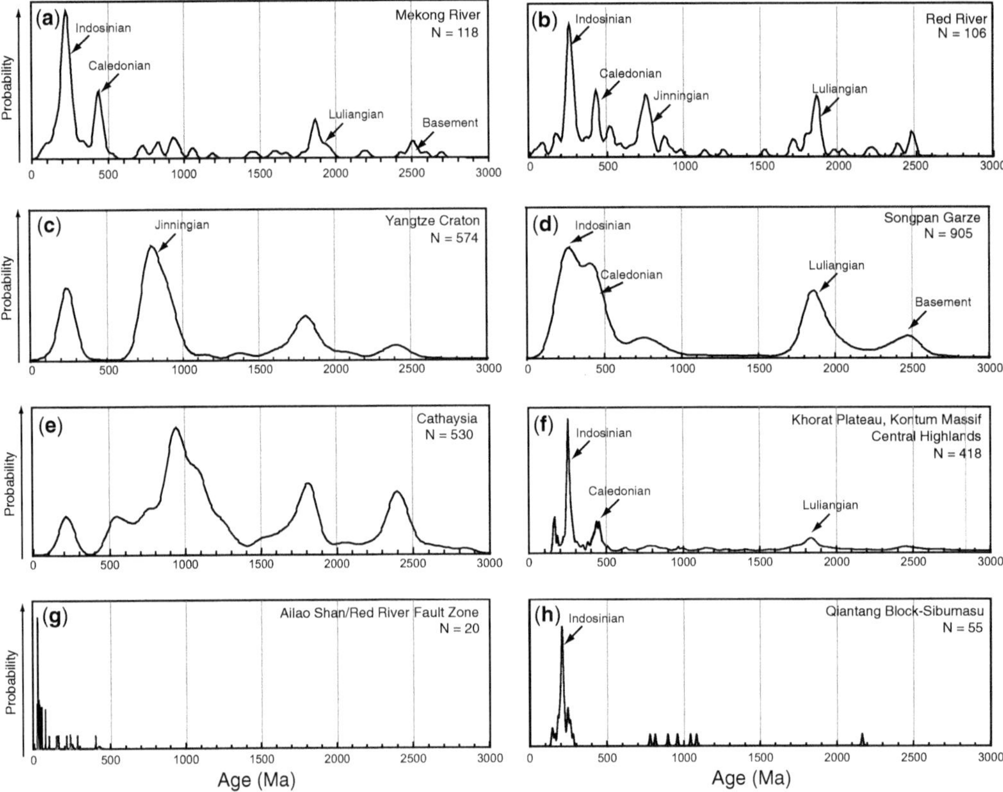

Fig. 9. Probability density diagrams showing the detrital U–Pb ages of zircon grains known from the major tectonic blocks in SE Asia. Data for Ailao Shan and Red River Fault Zone rocks are from Schärer *et al.* (1990, 1994), Zhang & Schärer (1999), Nagy *et al.* (2000) and Carter *et al.* (2001). Data for Cathaysia are from Li *et al.* (1989, 2001, 2005). Data for Indochina sources (including the Khorat Plateau, the Kontum Massif and the Central Highlands of Vietnam) are from Carter & Moss (1999), Carter *et al.* (2001), Nagy *et al.* (2001) and Carter & Bristow (2003). Data for the Songpan Garze Terrane are from Hu *et al.* (2005), Bruguier *et al.* (1997) and Weislogel *et al.* (2006, 2010). Data for the western Yangtze Craton are from Sun *et al.* (2009), Xu *et al.* (2008) and Yang *et al.* (2005). Data for the Qiangtang Block are from Roger *et al.* (2000, 2003).

Table 1. *Summary of the different provenance methods reviewed in this paper and simplified results for the different major source terrains that are supplying sediment to the modern South China Sea*

Method	Taiwan	South China	Indochina	Luzon
Bulk sediment chemistry	Variable	Variable	Variable	Volcanic
Clay minerals	Chlorite and illite	Kaolinite, smectite and illite	Kaolinite, smectite and illite	Smectite
Nd isotopes	−9 to −11	−8 to −11	−9 to −13, except the Central Highlands	+4 to +8
Apatite fission track	<3 Ma	30–60 Ma	17–23 Ma (S. Chay Massif), 27–50 Ma (Kontum)	<45 Ma
Zircon fission track	<5 Ma	90–120 Ma	90–550 Ma	N/A
Ar–Ar muscovite	<6 Ma	200–220 Ma	160–250 Ma	N/A
U–Pb zircons	<500 and *c.* 1800 Ma	800–1500 Ma	240–450 and *c.* 1800 Ma	<45 Ma

the Yangtze Craton, consistent with the known extent of the modern drainage. Furthermore, Clift *et al.* (2006*b*) demonstrated that the Indosinian peak in the Mekong River is slightly younger than in the Red River, a difference that could be exploited to resolve and quantify relative input from the two drainages.

Although rivers from Luzon have not yet been analysed for zircon U–Pb ages, those sources are not considered to be important to the zircon budget of the basin because of the relative lack of zircons in basaltic and basaltic andesite lavas in that arc. When they are present then they must be younger than the Oligocene when that arc became active and thus could only be potentially confused with sediment from the Red River Fault Zones. Metamorphic rocks in Taiwan were not heated sufficiently to reset the U–Pb system in zircon during that arc–continent collision and are anticipated to yield typical Cathaysian ages based on the sources to that part of the passive margin (Lan *et al.* 2014). Fortunately, Taiwan has a unique signature in terms of lower temperature thermochronology systems so that its contribution is not dependent on U–Pb zircon dating. It appears that U–Pb dating can be a useful component to a provenance study in the South China Sea, but often yields mixed or ambiguous signals because many of the major rivers share similar aged bedrocks.

Discussion and conclusions

In this review of the major provenance methods now applied in the South China Sea, it is clear that a single method is generally insufficient to quantify the relative sediment flux contributions from multiple terrains into a given sediment. Table 1 shows a summary of the character of the sources on three sides of the basin. There is relatively little information about the character of sources in Borneo or peninsular Malaysia, both of which may have been more important in the past when sea level was lower and when the Molengraaff River may be fed material to the Sunda Shelf and into deep-water basin, at least at its SW end (Voris 2000; Steinke *et al.* 2008; Hanebuth *et al.* 2011). When sea level was higher, sediment flux into the South China Sea from these sources was generally minimal. More work needs to be carried out to look at additional methods that should be effective in this basin, such as heavy mineral studies, but at the time of writing insufficient is known about the typical mineralogy of the modern drainages to assess whether this would actually work on ancient sequences or not. Only the continental shelf of Vietnam has been investigated for heavy minerals (Jagodziński 2005).

What is clear is that proxies that are strongly affected by chemical weathering make unreliable provenance proxies for sediments predating the last few hundred years. The modern rivers are known to be significantly disrupted in terms of bulk sediment chemistry and clay mineralogy, so that these compositions cannot be used to fingerprint ancient sediments. The same is also true of Sr isotopes. Even if the pre-modern composition can be fixed by looking at Holocene sediments under the most recent cover, it must be realized that the sediment composition is influenced by the changing climate, especially the intensity of the East Asian Monsoon.

Nd isotopes are only of limited use, except to resolve input from the young and primitive rocks of Luzon and in the case where rivers are draining sediment out of central southern China where the ancient crust of the Yangtze Craton results in very negative ε_{Nd} values. Nd compositions in Indochina and southern China are very similar and largely overlap, resulting in serious ambiguity in terms of resolving the effects of different sources. However, coherent changes may be detected and could be useful in discerning some changes depending on the size and heterogeneity of the basement rocks involved, since not all parts of Cathaysia or Indochina are identical. Erosion of the Miocene volcanic rocks from the Vietnamese Central Highlands would be anticipated to yield more positive ϵ_{Nd} values than is typical, for example, for the rest of the Vietnamese coast.

The most useful provenance proxies are those related to detrital thermochronology, but even here it is generally most effective to use more than one method to achieve a robust result. Apatite fission track is probably the best single method because of the contrasting timing of moderate amounts of exhumation around the basin, which allow most sources to be distinguished, although there is some overlap between parts of Indochina and southern China, not only in apatite but also in muscovite Ar/Ar. Fortunately, these sources can be separated on the basis of U–Pb zircon dating. Provenance in the South China Sea is complex, but it is not an impossible task and one that is essential to solve if we are to reconstruct the palaeogeography, drainage evolution and chemical weathering regimes in SE Asia during the Cenozoic. Because this is a classic area for collision tectonics, associated drainage reorganization and the development of the monsoon, the sedimentary archives in the South China Sea are invaluable to efforts to address these issues and require provenance control if the records are to be understood.

PDC thanks the Charles T. McCord Chair in Petroleum Geology at Louisiana State University for support to undertake this research.

References

ALLEGRE, C. J. & BEN OTHMAN, D. 1980. Nd–Sr isotopic relationship in granitoid rocks and continental crust development: a chemical approach to orogenesis. *Nature*, **286**, 335–342.

AOKI, S. 1976. Clay mineral distribution in sediments of the Gulf of Thailand and the South China Sea. *Journal of the Oceanographic Society of Japan*, **32**, 169–174.

BARCKHAUSEN, U., ENGELS, M., FRANKE, D., LADAGE, S. & PUBELLIER, M. 2014. Evolution of the South China Sea: revised ages for breakup and seafloor spreading. *In*: PUBELLIER, M., FRANKE, D., MCINTOSH, K., MENIER, D. & LI, C.-F. (eds) *Evolution, Structure, and Sedimentary Record of the South China Sea and Adjacent Basins. Marine and Petroleum Geology*, **58**, 599–611, http://doi.org/10.1016/j.marpetgeo.2014.02.022

BAYON, G., DENNIELOU, B., ETOUBLEAU, J., PONZEVERA, E., TOUCANNE, S. & BERMELL, S. 2012. Intensifying weathering and land use in iron age Central Africa. *Science*, **335**, 1219–1222, http://doi.org/10.1126/science.1215400

BERGMAN, S. C., LELOUP, P. H., TAPPONNIER, P., SCHÄRER, U. & O'SULLIVAN, P. B. 1997. *Apatite fission Track Thermal History of the Ailao Shan-Red River Shear Zone, China.* Paper presented at the European Union of Geoscientists, Strasbourg, 23–27 March 1997.

BOULAY, S., COLIN, C., TRENTESAUX, A., FRANK, N. & LIU, Z. 2005. Sediment sources and East Asian monsoon intensity over the last 450 ky: mineralogical and geochemical investigations on South China Sea sediments. *Palaeogeography Palaeoclimatology Palaeoecology*, **228**, 260–277.

BRIAIS, A., PATRIAT, P. & TAPPONNIER, P. 1993. Updated interpretation of magnetic anomalies and seafloor spreading stages in the South China Sea: implications for the Tertiary tectonics of Southeast Asia. *Journal of Geophysical Research*, **98**, 6299–6328, http://doi.org/10.1029/92JB02280

BROOKFIELD, M. E. 1998. The evolution of the great river systems of southern Asia during the Cenozoic India-Asia collision; rivers draining southwards. *Geomorphology*, **22**, 285–312.

BRUGUIER, O., LANCELOT, J. R. & MALAVIEILLE, J. 1997. U–Pb dating on single detrital zircon grains from the Triassic Songpan-Ganze Flysch (central China); provenance and tectonic correlations. *Earth and Planetary Science Letters*, **152**, 217–231.

CAMPBELL, I. H., REINERS, P. W., ALLEN, C. M., NICOLESCU, S. & UPADHYAY, R. 2005. He–Pb double dating of detrital zircons from the Ganges and Indus rivers; implication for quantifying sediment recycling and provenance studies. *Earth and Planetary Science Letters*, **237**, 402–432.

CARTER, A. 2007. Heavy minerals and detrital fission-track thermochronology. *Developments in Sedimentology*, **58**, 851–868, http://doi.org/10.1016/S0070-4571(07)58033-7

CARTER, A. & BRISTOW, C. S. 2003. Linking hinterland evolution and continental basin sedimentation by using detrital zircon thermochronology; a study of the Khorat Plateau basin, eastern Thailand. *Basin Research*, **15**, 271–285.

CARTER, A. & CLIFT, P. D. 2008. Was the Indosinian orogeny a Triassic mountain building or thermotectonic reactivation event? *Comptes Rendues de l'Academie Scientifique, Geoscience*, **340**, 83–93.

CARTER, A. & MOSS, S. J. 1999. Combined detrital-zircon fission-track and U–Pb dating: a new approach to understanding hinterland evolution. *Geology*, **27**, 235–238.

CARTER, A., ROQUES, D. & BRISTOW, C. S. 2000. Denudation history of onshore central Vietnam: constraints on the Cenozoic evolution of the western margin of the South China Sea. *Tectonophysics*, **322**, 265–277.

CARTER, A., ROQUES, D., BRISTOW, C. & KINNY, P. D. 2001. Understanding Mesozoic accretion in Southeast Asia: significance of Triassic thermotectonism (Indosinian orogeny) in Vietnam. *Geology*, **29**, 211–214.

CHEN, C. H. & LEE, T. 1990. A Nd-Sr isotopic study on river sediments of Taiwan. *Proceedings of the Geological Society of China*, **33**, 339–350.

CHEN, P. Y. 1978. Minerals in bottom sediments of the South China Sea. *Geological Society of America Bulletin*, **89**, 211–222.

CLARK, M. K., SCHOENBOHM, L. M. *ET AL*. 2004. Surface uplift, tectonics, and erosion of eastern Tibet from large-scale drainage patterns. *Tectonics*, **23**, TC1006, http://doi.org/10.1029/2002TC001402

CLARK, M. K., HOUSE, M. A., ROYDEN, L. H., WHIPPLE, K. X., BURCHFIEL, B. C., ZHANG, X. & TANG, W. 2005. Late Cenozoic uplift of southeastern Tibet. *Geology*, **33**, 525–528, http://doi.org/10.1130/G21265.1

CLIFT, P., LEE, J. I., CLARK, M. K. & BLUSZTAJN, J. 2002. Erosional response of south China to arc rifting and monsoonal strengthening; a record from the South China Sea. *Marine Geology*, **184**, 207–226.

CLIFT, P. D., BLUSZTAJN, J. & NGUYEN, D. A. 2006*a*. Large-scale drainage capture and surface uplift in eastern Tibet-SW China before 24 Ma inferred from sediments of the Hanoi Basin, Vietnam. *Geophysical Research Letters*, **33**, http://doi.org/10.1029/2006GL027772

CLIFT, P. D., CARTER, A., CAMPBELL, I. H., PRINGLE, M., HODGES, K. V., LAP, N. V. & ALLEN, C. M. 2006*b*. Thermochronology of mineral grains in the Song Hong and Mekong Rivers, Vietnam. *Geophysics, Geochemistry, Geosystems*, **7**, Q10005, http://doi.org/10.1029/2006GC001336

CLIFT, P., LEE, G. H., NGUYEN, A. D., BARCKHAUSEN, U., HOANG, V. L. & SUN, Z. 2008*a*. Seismic evidence for a Dangerous Grounds mini-plate: no extrusion origin for the South China Sea. *Tectonics*, **27**, TC3008, http://doi.org/10.1029/2007TC002216

CLIFT, P. D., HOANG, V. L., HINTON, R., ELLAM, R., HANNIGAN, R., TAN, M. T. & NGUYEN, D. A. 2008*b*. Evolving East Asian river systems reconstructed by trace element and Pb and Nd isotope variations in modern and ancient Red River-Song Hong sediments. *Geochemistry, Geophysics, Geosystems*, **9**, Q04039, http://doi.org/10.1029/2007GC001867

CLIFT, P. D., WAN, S. & BLUSZTAJN, J. 2014. Reconstructing chemical weathering, physical erosion and monsoon intensity since 25 Ma in the northern South China Sea: a review of competing proxies.

Earth-Science Reviews, **130**, 86–102, http://doi.org/10.1016/j.earscirev.2014.01.002

DeCelles, P. G., Gehrels, G. E., Najman, Y., Martin, A. J., Carter, A. & Garzanti, E. 2004. Detrital geochronology and geochemistry of Cretaceous-Early Miocene strata of Nepal: implications for timing and diachroneity of initial Himalayan orogenesis. *Earth and Planetary Science Letters*, **227**, 313–330.

Defant, M. J., Jacques, D., Maury, R. C., De Boer, J. & Joron, J. L. 1989. Geochemistry and tectonic setting of the Luzon arc, Philippines. *Geological Society of America Bulletin*, **101**, 663–672, http://doi.org/10.1130/0016-7606(1989)101

Derry, L. A. & France-Lanord, C. 1996. Neogene Himalayan weathering history and river $^{87}Sr/^{86}Sr$; impact on the marine Sr record. *Earth and Planetary Science Letters*, **142**, 59–74.

Dung, B. V., Stattegger, K., Unverricht, D., Phach, P. V. & Thanh, N. T. 2013. Late Pleistocene-Holocene seismic stratigraphy of the Southeast Vietnam Shelf. *Global and Planetary Change*, **110B**, 156–169.

Dykoski, C. A., Edwards, R. L. *et al.* 2005. A high-resolution, absolute-dated Holocene and deglacial Asian monsoon record from Dongge Cave, China. *Earth and Planetary Science Letters*, **233**, 71–86.

Enkelmann, E., Weislogel, A., Ratschbacher, L., Eide, E., Renno, A. & Wooden, J. 2007. How was the Triassic Songpan-Ganzi basin filled? A provenance study. *Tectonics*, **26**, TC4007.

Fletcher, C. J. N., Chan, L. S., Sewell, R. J., Campbell, S. D. G., Davis, D. W. & Zhu, J. 2004. Basement heterogeneity in the Cathaysia crustal block, southeast China. *In*: Malpas, J., Fletcher, C. J. N., Ali, J. R. & Aitchison, J. C. (eds) *Aspects of the Tectonic Evolution of China*. Geological Society, London, Special Publications, **226**, 145–155, http://doi.org/10.1144/GSL.SP.2004.226.01.08

Franke, D. 2013. Rifting, lithosphere breakup and volcanism: comparison of magma-poor and volcanic rifted margins. *Marine and Petroleum Geology*, **43**, 63–87, http://doi.org/10.1016/j.marpetgeo.2012.11.003

Fuller, C. W., Willett, S. D. & Brandon, M. T. 2006. Formation of forearc basins and their influence on subduction zone earthquakes. *Geology*, **34**, 65–68, http://doi.org/10.1130/G21828.1

Galy, A. & France-Lanord, C. 1999. Weathering processes in the Ganges-Brahmaputra basin and the riverine alkalinity budget. *Chemical Geology*, **159**, 31–60.

Garçon, M., Chauvel, C., France-Lanord, C., Huyghe, P. & Lavé, J. 2013. Continental sedimentary processes decouple Nd and Hf isotopes. *Geochimica et Cosmochimica Acta*, **121**, 177–195.

Garçon, M., Chauvel, C., France-Lanord, C., Limonta, M. & Garzanti, E. 2014. Which minerals control the Nd–Hf–Sr–Pb isotopic compositions of river sediments? *Chemical Geology*, **364**, 42–55, http://doi.org/10.1016/j.chemgeo.2013.11.018

Goldstein, S. J. & Jacobsen, S. B. 1988. Nd and Sr isotopic systematics of river water suspended material; implications for crustal evolution. *Earth and Planetary Science Letters*, **87**, 249–265.

Green, P. F., Duddy, I. R., Laslett, G. M., Hegarty, K. A., Gleadow, A. J. W. & Lovering, J. F. 1989. Thermal annealing of fission tracks in apatite; 4, quantitative modelling techniques and extension to geological timescales. *Chemical Geology; Isotope Geoscience Section*, **79**, 155–182.

Hall, R. 2002. Cenozoic geological and plate tectonic evolution of SE Asia and the SW Pacific: computer-based reconstructions and animations. *Journal of Asian Earth Sciences*, **20**, 353–434.

Hanebuth, T. J. J., Voris, H. K., Yokoyama, Y., Saito, Y. & Okuno, J. I. 2011. Formation and fate of sedimentary depocentres on Southeast Asia's Sunda Shelf over the past sea-level cycle and biogeographic implications. *Earth-Science Reviews*, **104**, 92–110, http://doi.org/10.1016/j.earscirev.2010.09.006

Harrison, T. M., Leloup, P. H., Ryerson, F. J., Tapponnier, P., Lacassin, R. & Wenji, C. 1996. Diachronous initiation of transtension along the Ailao Shan-Red River shear zone, Yunnan and Vietnam. *In*: Yin, A. & Harrison, T. M. (eds) *The Tectonic Evolution of Asia*. Cambridge University Press, Cambridge, 110–137.

Hillier, S. 2003. Quantitative analysis of clay and other minerals in sandstones by X-ray powder diffraction (XRPD). *In*: Worden, R. H. & Morad, S. (eds) *Clay mineral Cements in Sandstones*. International Association of Sedimentologists, Special Publications, **34**, 213–251.

Hoang, L. V., Clift, P. D., Mark, D., Zheng, H. & Tan, M. T. 2010. Ar–Ar Muscovite dating as a constraint on sediment provenance and erosion processes in the Red and Yangtze River systems, SE Asia. *Earth and Planetary Science Letters*, **295**, 379–389, http://doi.org/10.1016/j.epsl.2010.04.012

Hodges, K. 2003. Geochronology and thermochronology in orogenic systems. *In*: Rudnick, R. (ed.) *The Crust*. Elsevier, Amsterdam, 263–292.

Holtzapffel, T. 1985. *Les mineraux argileux, preparation, analyse diffractometriques et determination*. Societie Geologique du Nord, Lille.

Hu, D., Böning, P. *et al.* 2012. Deep sea records of the continental weathering and erosion response to East Asian monsoon intensification since 14 ka in the South China Sea. *Chemical Geology*, **326–327**, 1–18, http://doi.org/10.1016/j.chemgeo.2012.07.024

Hu, D., Clift, P. D. *et al.* 2013. Holocene evolution in weathering and erosion patterns in the Pearl River delta. *Geochemistry, Geophysics, Geosystems*, **14**, 2349–2368, http://doi.org/10.1002/ggge.20166

Hu, J., Meng, Q., Shi, Y. & Qu, H. 2005. SHRIMP U–Pb dating of zircons from granitoid bodies in the Songpan-Ganzi terrane and its implications. *Acta Petrologica Sinica*, **21**, 867–880.

Hu, S., Raza, A. *et al.* 2006. Late Mesozoic and Cenozoic thermotectonic evolution along a transect from the north China craton through the Qinling orogen into the Yangtze craton, central China. *Tectonics*, **25**, TC6009, http://doi.org/10.1029/2006TC001985

Huang, C. Y., Yuan, P. B. & Tsao, S. H. 2006. Temporal and spatial records of active arc-continent collision in Taiwan: a synthesis. *Geological Society of America Bulletin*, **118**, 274–288.

Huang, M., Maas, R., Buick, I. S. & Williams, I. S. 2003. Crustal response to continental collisions between the Tibet, Indian, South China and North

China blocks; geochronological constraints from the Songpan-Garze orogenic belt, western China. *Journal of Metamorphic Geology*, **21**, 223–240.

Hurford, A. J. & Carter, A. 1991. The role of fission track dating in discrimination of provenance. *In*: Morton, A. C., Todd, S. P. & Haughton, P. D. W. (eds) *Developments in Sedimentary Provenance Studies*. Geological Society, London, Special Publications, **57**, 67–78, http://doi.org/10.1144/GSL.SP.1991.057.01.07

Hutchison, C. S. 1994. Gondwana and Cathaysian blocks, Palaeotethys sutures and Cenozoic tectonics in South-east Asia. *Geologische Rundschau*, **83**, 388–405, http://doi.org/10.1007/BF00210553

Hutchison, C. S. 2005. *Geology of North-West Borneo: Sarawak, Brunei and Sabah*. Elsevier, Amsterdam.

Hutchison, C. S. 2010. The North-West Borneo Trough. *Marine Geology*, **271**, 32–43, http://doi.org/10.1016/j.margeo.2010.01.007

Hutchison, C. S. & Vijayan, V. R. 2010. What are the Spratly Islands? *Journal of Asian Earth Sciences*, **39**, 371–385, http://doi.org/10.1016/j.jseaes.2010.04.013

Jagodziński, R. 2005. *Petrography and Geochemistry of Surface Sediments from Sunda and Vietnamese Shelves (South China Sea)*. Adam Mickiewicz University Press, Poznań.

Jolivet, L., Maluski, H., Beyssac, O., Goffe, B., Lepvrier, C., Phan, T. T. & Nguyen, V. V. 1999. Oligocene-Miocene Bu Khang extentional gneiss dome in Vietnam: geodynamic implications. *Geology*, **27**,67.

Jung, H.-S., Lim, D., Choi, J.-Y., Yoo, H.-S., Rho, K.-C. & Lee, H.-B. 2012. Rare earth element compositions of core sediments from the shelf of the South Sea, Korea: their controls and origins. *Continental Shelf Research*, **48**, 75–86, http://doi.org/10.1016/j.csr.2012.08.008

Kao, S. J. & Milliman, J. D. 2008. Water and sediment discharge from small mountainous rivers, Taiwan: the roles of lithology, episodic events, and human activities. *Journal of Geology*, **116**, 431–448.

Kapp, P., Yin, A. *et al.* 2000. Blueschist-bearing metamorphic core complexes in the Qiangtang Block reveal deep crustal structure of northern Tibet. *Geology*, **28**, 19–22.

Kim, G., Yang, H. S. & Church, T. M. 1999. Geochemistry of alkaline earth elements (Mg, Ca, Sr, Ba) in the surface sediments of the Yellow Sea. *Chemical Geology*, **153**, 1–10.

Lahajnar, N., Wiesner, M. G. & Gaye, B. 2007. Fluxes of amino acids and hexosamines to the deep South China Sea. *Deep Sea Research I*, **54**, 2120–2144.

Lan, C. Y., Lee, C.-S., Shen, J. J.-S., Lu, C. Y., Mertzman, S. A. & Wu, T.-W. 2002. Nd -Sr isotopic composition and geochemistry of sediments from Taiwan and their implications. *Western Pacific Earth Sciences*, **2**, 205–222.

Lan, Q., Yan, Y. *et al.* 2014. Tectonics, topography, and river system transition in East Tibet: insights from the sedimentary record in Taiwan. *Geochemistry, Geophysics, Geosystems*, **15**, 3658–3674, http://doi.org/10.1002/2014GC005310

Le, T. P. Q., Garnier, J., Gilles, B., Sylvain, T. & Minh, C. V. 2007. The changing flow regime and sediment load of the Red River, Viet Nam. *Journal of Hydrology*, **334**, 199–214, http://doi.org/10.1016/j.jhydrol.2006.10.020

Lei, S., Li, X. *et al.* 2007. Deep water bottom current deposition in the northern South China Sea. *Science in China Series D: Earth Sciences*, **50**, 1862–2801, http://doi.org/10.1007/s11430-007-0015-y

Leloup, P. H., Harrison, T. M., Ryerson, F. J., Chen, W. J., Li, Q., Tapponnier, P. & Lacassin, R. 1993. Structural, petrological and thermal evolution of a tertiary ductile strike-slip shear zone, Diancang Shan, Yunnan. *Journal of Geophysical Research: Solid Earth*, **98**, 6715–6743.

Leloup, P. H., Arnaud, N. *et al.* 2001. New constraints on the structure, thermochronology, and timing of the Ailao Shan-Red River shear zone, SE Asia. *Journal of Geophysical Research*, **106**, 6657–6671.

Lepvrier, C., Maluski, H., Vuong, N. V., Roques, D., Axente, V. & Rangin, C. 1997. Indosinian NW-trending shear zones within the Truong Son belt (Vietnam) $^{40}Ar/^{39}Ar$ Triassic ages and Cretaceous to Cenozoic overprints. *Tectonophysics*, **283**, 105–127.

Lepvrier, C., Maluski, H., Vu, V. T., Leyreloup, A., Phan, T. T. & Vuong, N. V. 2004. The Early Triassic Indosinian orogeny in Vietnam (Truong Son Belt and Kontum Massif); implications for the geodynamic evolution of Indochina. *Tectonophysics*, **393**, 87–118.

Li, W. X., Li, X. H. & Li, Z. X. 2005. Neoproterozoic bimodal magmatism in the Cathaysia Block of South China and its tectonic significance. *Precambrian Research*, **136**, 51–66.

Li, X., Tatsumoto, M., Premo, W. R. & Gu, X. 1989. Age and origin of the Tanghu Granite, Southeastern China; results from U–Pb single zircon and Nd isotopes. *Geology*, **17**, 395–399.

Li, X., Wei, G. *et al.* 2003. Geochemical and Nd isotopic variations in sediments of the South China Sea; a response to Cenozoic tectonism in SE Asia. *Earth and Planetary Science Letters*, **211**, 207–220.

Li, Z. X., Li, X. H., Zhou, H. & Kinny, P. D. 2001. Grenvillian continental collision in south China: new SHRIMP U–Pb zircon results and implications for the configuration of Rodinia. *Geology*, **30**, 163–166.

Lim, D., Choi, J. Y., Shin, H. H., Rho, K. C. & Jung, H. S. 2013. Multielement geochemistry of offshore sediments in the southeastern Yellow Sea and implications for sediment origin and dispersal. *Quaternary International*, **298**, 196–206.

Limmer, D. R., Boening, P. *et al.* 2012. Geochemical record of holocene to recent sedimentation on the Western Indus continental shelf, Arabian Sea. *Geochemistry, Geophysics, Geosystems*, **13**, Q01008, http://doi.org/10.1029/2011GC003845

Liu, J., Clift, P. D., Yan, W., Chen, Z., Chen, H., Xiang, R. & Wang, D. 2014. Modern transport and deposition of settling particles in the northern South China Sea: sediment trap evidence adjacent to Xisha Trough. *Deep-Sea Research I*, **93**, 145–155, http://doi.org/10.1016/j.dsr.2014.08.005

Liu, Z., Trentesaux, A. & Clemens, S. C. 2003. Clay mineral assemblages in the northern South China Sea: implications for East Asian monsoon evolution over the past 2 million years. *Marine Geology*, **201**, 133–146.

LIU, Z., COLIN, C., HUANG, W., CHEN, Z., TRENTESAUX, A. & CHEN, J. F. 2007*a*. Clay minerals in surface sediments of the Pearl River drainage basin and their contribution to the South China Sea. *Chinese Science Bulletin*, **52**, 1101–1111.

LIU, Z., COLIN, C., HUANG, W., LE, K. P., TONG, S., CHEN, Z. & TRENTESAUX, A. 2007*b*. Climatic and tectonic controls on weathering in south China and Indochina Peninsula: clay mineralogical and geochemical investigations from the Pearl, Red, and Mekong drainage basins. *Geochemistry, Geophysics, Geosystems*, **8**, Q05005, http://doi.org/10.1029/2006GC 001490

LIU, Z., ZHAO, Y., LI, J. & COLIN, C. 2007*c*. Late Quaternary clay minerals off Middle Vietnam in the western South China Sea: implications for source analysis and East Asian monsoon evolution. *Science in China, Series D, Earth Sciences*, **50**, 1674–1684.

LIU, Z., ZHAO, Y., COLIN, C., SIRINGAN, F. P. & WU, Q. 2009. Chemical weathering in Luzon, Philippines from clay mineralogy and major-element geochemistry of river sediments. *Applied Geochemistry*, **24**, 2195–2205.

LIU, Z., COLIN, C. ET AL. 2010*a*. Clay mineral distribution in surface sediments of the northeastern South China Sea and surrounding fluvial drainage basins: source and transport. *Marine Geology*, **277**, 48–60, http://doi.org/10.1016/j.margeo.2010.08.010

LIU, Z., LI, X., COLIN, C. & GE, H. 2010*b*. A high-resolution clay mineralogical record in the northern South China Sea since the Last Glacial Maximum, and its time series provenance analysis. *Chinese Science Bulletin*, **55**, 4058–4068.

LIU, Z., WANG, H., HANTORO, W. S., SATHIAMURTHY, E., COLIN, C., ZHAO, Y. & LI, J. 2012. Climatic and tectonic controls on chemical weathering in tropical Southeast Asia (Malay Peninsula, Borneo, and Sumatra). *Chemical Geology*, **291**, 1–12.

LUPKER, M., FRANCE-LANORD, C. ET AL. 2012. Predominant floodplain over mountain weathering of Himalayan sediments (Ganga basin). *Geochimica et Cosmochimica Acta*, **84**, 410–432.

MALUSKI, H., LEPVRIER, C. ET AL. 2001. Ar–Ar and fission-track ages in the Song Chay Massif: Early Triassic and Cenozoic tectonics in northern Vietnam. *Journal of Asian Earth Sciences*, **19**, 233–248.

METCALFE, I. 1996. Pre-Cretaceous evolution of SE Asian terranes. *In*: HALL, R. & BLUNDELL, D. J. (eds) *Tectonic Evolution of SE Asia*. Geological Society, London, Special Publications, **106**, 97–122, http://doi.org/10.1144/GSL.SP.1996.106.01.09

MILLIMAN, J. D. & KAO, S.-J. 2005. Hyperpycnal discharge of fluvial sediment to the ocean: impact of super-typhoon Herb (1996) on Taiwanese Rivers. *Journal of Geology*, **113**, 503–516.

MILLIMAN, J. D. & SYVITSKI, J. P. M. 1992. Geomorphic/tectonic control of sediment discharge to the ocean; the importance of small mountainous rivers. *Journal of Geology*, **100**, 525–544.

MOORE, D. & REYNOLDS, R. 1989. *X-ray Diffraction and the Identification and Analysis of Clay Minerals*. Oxford University Press, Oxford.

MORLEY, C. K. 2002. A tectonic model for the Tertiary evolution of strike-slip faults and rift basins in SE Asia. *Tectonophysics*, **347**, 189–215.

NAGY, E. A., SCHÄRER, U. & NGUYEN, T. M. 2000. Oligo-Miocene granitic magmatism in central Vietnam and implications for continental deformation in Indochina. *Terra Nova*, **12**, 67–76.

NAGY, E. A., MALUSKI, H., LEPVRIER, C., SCHÄRER, U., PHAN, T. T., LEYRELOUP, A. & VU, V. T. 2001. Geodynamic significance of the Kontum Massif in central Vietnam; composite $^{40}Ar/^{39}Ar$ and U–Pb ages from Paleozoic to Triassic. *Journal of Geology*, **109**, 755–770.

NAJMAN, Y. M. R., PRINGLE, M. S., JOHNSON, M. R. W., ROBERTSON, A. H. F. & WIJBRANS, J. R. 1997. Laser $^{40}Ar/^{39}Ar$ dating of single detrital muscovite grains from early foreland-basin sedimentary deposits in India; implications for early Himalayan evolution. *Geology*, **25**, 535–538.

NESBITT, H. W. & YOUNG, G. M. 1982. Early Proterozoic climates and plate motions inferred from major element chemistry of lutites. *Nature*, **299**, 715–717.

RACEY, A., DUDDY, L. R. & LOVE, M. A. 1997. Apatite fission track analysis of Mesozoic red beds from northeastern Thailand and western Laos. *In*: DHEERADILOK, P., HINTHONG, C.ET AL. (eds) *Proceedings of the International Conference on Stratigraphy and Tectonic Evolution of Southeast Asia and the South Pacific, Volume 1*. Department of Mineral Resources, Bangkok, 200–209.

REID, A. J., WILSON, C. J. L., PHILLIPS, D. & LIU, S. 2005. Mesozoic cooling across the Yidun Arc, central-eastern Tibetan Plateau: a reconnaissance $^{40}Ar/^{39}Ar$ study. *Tectonophysics*, **398**, 45–66.

ROBINSON, R. A. J., BREZINA, C. A. ET AL. 2013. Large rivers and orogens: the evolution of the Yarlung Tsangpo–Irrawaddy system and the eastern Himalayan syntaxis. *Gondwana Research*, **26**, 112–121, http://doi.org/10.1016/j.gr.2013.07.002

ROGER, F., LELOUP, P. H., JOLIVET, M., LACASSIN, R., TRINH, P. T., BRUNEL, M. & SEWARD, D. 2000. Long and complex thermal history of the Song Chay metamorphic dome (Northern Vietnam) by multi-system geochronology. *Tectonophysics*, **321**, 449–466.

ROGER, F., ARNAUD, N. ET AL. 2003. Geochronological and geochemical constraints on Mesozoic suturing in east central Tibet. *Tectonics*, **1037**, 22. http://doi.org/10.1029/2002TC001466

RU, K. & PIGOTT, J. D. 1986. Episodic rifting and subsidence in the South China Sea. *American Association of Petroleum Geologists Bulletin*, **70**, 1136–1155.

RUDNICK, R. L. & GAO, S. 2003. The composition of the continental crust. *In*: RUDNICK, R. L. (ed.) *The Crust, Volume 3*. Elsevier–Pergamon, Oxford, 1–64.

SCHÄRER, U., TAPPONNIER, P., LACASSIN, R., LELOUP, P. H., DALAI, Z. & SHAOCHENG, J. 1990. Intraplate tectonics in Asia; a precise age for large-scale Miocene movement along the Ailao Shan-Red River shear zone, China. *Earth and Planetary Science Letters*, **97**, 65–77.

SCHÄRER, U., ZHANG, L. S. & TAPPONNIER, P. 1994. Duration of strike-slip movements in large shear zones: the Red River belt, China. *Earth and Planetary Science Letters*, **126**, 379–397.

SCHIMANSKI, A. & STATTEGGER, K. 2005. Deglacial and Holocene evolution of the Vietnam Shelf; stratigraphy, sediments and sea-level change. *Marine Geology*, **214**, 365–387.

Schoenbohm, L. M., Burchfiel, B. C. & Chen, L. 2006. Propagation of surface uplift, lower crustal flow, and Cenozoic tectonics of the southeast margin of the Tibetan Plateau. *Geology*, **34**, 813–816, http://doi.org/10.1130/G22679.1

Shi, X., Kohn, B. *et al.* 2011. Cenozoic denudation history of southern Hainan Island, South China Sea: constraints from low temperature thermochronology. *Tectonophysics*, **504**, 100–115.

Singh, S. K., Sarin, M. M. & France-Lanord, C. 2005. Chemical erosion in the eastern Himalaya; major ion composition of the Brahmaputra and d13C of dissolved inorganic carbon. *Geochimica et Cosmochimica Acta*, **69**, 3573–3588.

Steinke, S., Hanebuth, T. J., Vogt, C. & Stattegger, K. 2008. Sea level induced variations in clay mineral composition in the southwestern South China Sea over the past 17000 years. *Marine Geology*, **250**, 199–210.

Sun, W. H., Zhou, M. F., Gao, J. F., Yang, Y. H., Zhao, X. F. & Zhao, J. H. 2009. Detrital zircon U–Pb geochronological and Lu–Hf isotopic constraints on the Precambrian magmatic and crustal evolution of the western Yangtze Block, SW China. *Precambrian Research*, **172**, 99–126.

Sun, X. & Wang, P. 2005. How old is the Asian monsoon system? Palaeobotanical records from China. *Palaeogeography, Palaeoclimatology, Palaeoecology*, **222**, 181–222.

Suppe, J. 1984. Kinematics of arc–continent collision, flipping of subduction, and backarc spreading near Taiwan. *In*: Tsan, S. F. (ed.) *A Special Volume Dedicated to Chun-Sun Ho on the Occasion of His Retirement*. Geological Society of China, Memoirs, **6**, 21–33.

Syvitski, J. P. M. & Kettner, A. J. 2011. Sediment flux and the Anthropocene. *Philosophical Transactions of the Royal Society of London, Series A: Mathematical and Physical Sciences*, **369**, 957–975, http://doi.org/10.1098/rsta.2010.0329

Syvitski, J. P. M., Vörösmarty, C. J., Kettner, A. J. & Green, P. 2005. Impact of humans on the flux of terrestrial sediment to the global coastal ocean. *Science*, **308**, 376–380.

Szczuciński, W., Stattegger, K. & Scholten, J. 2009. Modern sediments and sediment accumulation rates on the narrow shelf off central Vietnam, South China Sea. *Geo-Marine Letters*, **29**, 47–59.

Szczuciński, W., Jagodziński, R. *et al.* 2013. Modern sedimentation and sediment dispersal pattern on the continental shelf off the Mekong River delta, South China Sea. *Global and Planetary Change*, **110**, 195–213, http://doi.org/10.1016/j.gloplacha.2013.08.019

Szulc, A. G., Najman, Y. *et al.* 2006. Tectonic evolution of the Himalaya constrained by detrital $^{40}Ar/^{39}Ar$, Sm/Nd and petrographic data from the Siwalik foreland basin succession, SW Nepal. *Basin Research*, **18**, 375–391.

Upton, D. R. 1999. *A Regional Fission Track Study of Thailand: Implications for Thermal History and Denudation*. University of London, London.

Vermeesch, P. 2012. On the visualisation of detrital age distributions. *Chemical Geology*, **312–313**, 190–194, http://doi.org/10.1016/j.chemgeo.2012.04.021

Vital, H. & Stattegger, K. 2000. Major and trace elements of stream sediments from the lowermost Amazon River. *Chemical Geology*, **168**, 151–168.

Voris, H. K. 2000. Maps of Pleistocene sea levels in Southeast Asia: shores, river systems and time durations. *Journal of Biogeography*, **27**, 1153–1167.

Wan, S., Li, A., Clift, P. D. & Stuut, J.-B. W. 2007. Development of the East Asian monsoon: mineralogical and sedimentologic records in the northern South China Sea since 20 Ma. *Palaeogeography, Palaeoclimatology, Palaeoecology*, **254**, 561–582.

Wang, E., Burchfiel, B., Royden, L., Chen, L., Chen, J., Li, W. & Chen, Z. 1998. *Late Cenozoic Xianshuihe-Xiaojiang, Red River, and Dali Fault Systems of Southwestern Sichuan and Central Yunnan, China*. Geological Society of America, Boulder, CO.

Wang, P. L., Lo, C. H., Lee, T. Y., Chung, S. L. & Yem, N. T. 1998. Thermochronological evidence for the movement of the Ailao Shan-Red River shear zone: a perspective from Vietnam. *Geology*, **26**, 887–890.

Wang, Y., Zhang, Y., Fan, W. & Peng, T. 2005. Structural signatures and $^{40}Ar/^{39}Ar$ geochronology of the Indosinian Xuefengshan tectonic belt, South China Block. *Journal of Structural Geology*, **27**, 985–998, http://doi.org/10.1016/j.jsg.2005.04.004

Wang, Y. J., Cheng, H., Edwards, R. L., An, Z. S., Wu, J. Y., Shen, C.-C. & Dorale, J. A. 2001. A high-resolution absolute-dated late Pleistocene Monsoon record from Hulu Cave, China. *Science*, **294**, 2345–2348.

Weislogel, A. L., Graham, S. A., Chang, E. Z., Wooden, J. L., Gehrels, G. E. & Yang, H. 2006. Detrital zircon provenance of the Late Triassic Songpan–Ganzi complex: sedimentary record of collision of the North and South China blocks. *Geology*, **34**, 97–100.

Weislogel, A. L., Graham, S. A., Chang, E. Z., Wooden, J. L. & Gehrels, G. 2010. Detrital zircon provenance from three turbidite depocenters of the Middle-Upper Triassic Songpan-Ganzi complex, central China: record of collisional tectonics, erosional exhumation, and sediment production. *Geological Society of America Bulletin*, **122**, 2041–2062.

West, A. J., Galy, A. & Bickle, M. J. 2005. Tectonic and climatic controls on silicate weathering. *Earth and Planetary Science Letters*, **235**, 211–228, http://doi.org/10.1016/j.epsl.2005.03.020

White, N. M., Pringle, M., Garzanti, E., Bickle, M., Najman, Y., Chapman, H. & Friend, P. 2002. Constraints on the exhumation and erosion of the high Himalayan slab, NW India, from foreland basin deposits. *Earth and Planetary Science Letters*, **195**, 29–44.

Willett, S. D., Fisher, D., Fuller, C., Chao, Y.-E. & Yu, L.-C. 2003. Erosion rates and orogenic-wedge kinematics in Taiwan inferred from fission-track thermochronometry. *Geology*, **31**, 945–948.

Xu, Y. G., Luo, Z. Y. & Huang, X. L. 2008. Zircon U–Pb and Hf isotope constraints on crustal melting associated with the Emeishan mantle plume. *Geochimica et Cosmochimica Acta*, **72**, 3084–3104.

XUE, Z., LIU, J. P. *ET AL.* 2014. Sedimentary processes on the Mekong subaqueous delta: clay mineral and geochemical analysis. *Journal of Asian Earth Sciences*, **79A**, 520–528, http://doi.org/10.1016/j.jseaes.2012.07.012

YAN, Y., CARTER, A., XIA, B., GE, L., BRICHAU, S. & XIAOQIONG, H. 2009. A fission-track and (U–Th)/He thermochronometric study of the northern margin of the South China Sea: an example of a complex passive margin. *Tectonophysics*, **474**, 584–594.

YANG, J. H., CHUNG, S. L., WILDE, S. A., WU, F. Y., CHU, M. F., LO, C. H. & FAN, H. R. 2005. Petrogenesis of post-orogenic syenites in the Sulu orogenic belt, east China; geochronological, geochemical and Nd-Sr isotopic evidence. *Chemical Geology*, **214**, 99–125.

ZHANG, L. S. & SCHÄRER, U. 1999. Age and origin of magmatism along the Cenozoic Red River shear belt, China. *Contributions to Mineralogy and Petrology*, **134**, 67–85.

ZHENG, H., CLIFT, P. D., TADA, R., JIA, J. T., HE, M. Y. & WANG, P. 2013. A Pre-Miocene birth to the Yangtze River. *Proceedings of the National Academy of Sciences of the United States of America*, **110**, 7556–7561, http://doi.org/10.1073/pnas.1216241110

ZHOU, D. & GRAHAM, S. A. 1996. Songpan–Ganzi Triassic flysch complex of the West Qinling Shan as a remnant ocean basin. *In*: YIN, A. & HARRISON, M. (eds) *The Tectonic Evolution of Asia*. Cambridge University Press, Cambridge, 281–299.

ZHOU, W., LU, X., WU, Z., DENG, L., JULL, A. J. T., DONAHUE, D. & BECK, W. 2002. Peat record reflecting Holocene climatic change in the Zoige Plateau and AMS radiocarbon dating. *Chinese Science Bulletin*, **47**, 66–70.

ZONG, Y., YU, F., HUANG, G., LLOYD, J. M. & YIM, W. W.-S. 2010. Sedimentary evidence of Late Holocene human activity in the Pearl River delta, China. *Earth Surface Processes and Landforms*, **35**, 1095–1102, http://doi.org/10.1002/esp.1970

Depositional characteristics of the northern South China Sea in response to the evolution of the Pearl River

LEI SHAO[1], PEIJUN QIAO[1]*, MENG ZHAO[1], QIANYU LI[1], MENGSHUANG WU[1,2], XIONG PANG[2] & HAO ZHANG[1]

[1]*State Key Laboratory of Marine Geology, Tongji University, Shanghai 200092, China*

[2]*China National Offshore Oil Corporation Ltd, Shenzhen Branch, Guangzhou 510240, China*

**Corresponding author (e-mail: qiaopeijun@tongji.edu.cn)*

Abstract: Geochemical data from South China Sea sedimentary rocks show the effects of both source composition and depositional environments. This enables us to link tectonic trends with erosion in the Pearl River region since *c.* 32 Ma. In particular, a shift in the geochemistry appears to signal a response to a well-recorded regional tectonic event at *c.* 23–25 Ma, probably corresponding to a jump in the seafloor spreading axis from the west to the SW within the South China Sea. This may correlate with the uplift of the West Yunnan Plateau and possibly also the eastern Tibetan Plateau. Clay mineralogy, sand–mud ratio, and major and rare earth element concentrations, also varied in response to the environment in the drainage areas of the palaeo-Pearl River. By comparing data from the modern sources and the sedimentary record from the northern South China Sea, especially the erosion–transportation–deposition patterns, three groups of index minerals (Ati, GZi, ZTR), as well as rare earth elements can be recognized. These are used to characterize the Pearl River from the east to the west, representing three different parent rock sources. The evolution of the palaeo-Pearl River can be tracked by variations of heavy minerals and key elements that are indicative of provenance.

Supplementary material: Supplementary material, including Tables S-1 and S-2 listing the sampling and analytical data, is available at http://www.geolsoc.org.uk/SUP18855

The South China Sea is located at the boundary between the Eurasian, Pacific and Indian Plates. As a marginal sea basin that is transitional between oceanic and continental structural domains, it has experienced a series of complicated magmatic and tectonic activities, sedimentation episodes, and periods of metamorphism (Liu *et al.* 2002). In particular, the South China Sea connects with the Tibetan Plateau through rivers such as the Red and Mekong, and is surrounded by plate boundaries in the east and south. These features have led to sediment supply from multiple sources and complicated basin-filling processes, making it an ideal area for advanced sedimentology studies.

The Pearl, Red and Hanjiang Rivers are three major drainage systems supplying the northern South China Sea (Fig. 1). Most of the Red River sediment is trapped in the Yinggehai (Song Hong) Basin, and only a small amount has reached the Qiongdongnan Basin, at least during recent times (Zhu *et al.* 2007; Wang *et al.* 2011). The Hanjiang River drained a limited area to the northeastern South China Sea in the Oligocene and Miocene (Zhang *et al.* 2012). In the case of recent terrigenous sediments in the northern South China Sea, the main contributors are the Pearl River and the coastal area of South China and Taiwan (Shao *et al.* 2009; Wei *et al.* 2012). To the NE of a line from Hong Kong to the Dongsha Islands, the coastal area of South China and Taiwan exerts a major impact, while to the SW of this line, the Pearl River is the dominant source (Shao *et al.* 2009). Taiwan was uplifted from *c.* 5 Ma (Huang *et al.* 2006), so there were no sediments eroded from Taiwan into the northern South China Sea during the Oligocene and Miocene.

As an extended part of the South China block, the northern South China Sea basement consists of seven distinctive rock types, similar to those cropping out in South China (Fig. 1): Palaeozoic and Mesozoic metamorphic rocks, Mesozoic sedimentary rocks, intermediate acid intrusive rocks, intermediate basic magmatic rocks, extrusive rocks and transitional crust basalt, with magma compositions changing from acid to basic moving seawards (Chinese Academy of Geological Science 1975; Wang *et al.* 2002). These basement rocks could have affected the sediment composition during the basin's early formation.

The landform of the present Pearl River drainage is complicated, because the river runs from the Yunnan–Guizhou Plateau (Fig. 1), which is an extension of the eastern Tibetan Plateau in the west through a mountainous region, hills and plains before emptying into the South China Sea (Fig. 1).

From: Clift, P. D., Harff, J., Wu, J. & Qui, Y. (eds) 2016. *River-Dominated Shelf Sediments of East Asian Seas*. Geological Society, London, Special Publications, **429**, 31–44.
First published online November 12, 2015, updated May 26, 2016, http://doi.org/10.1144/SP429.2

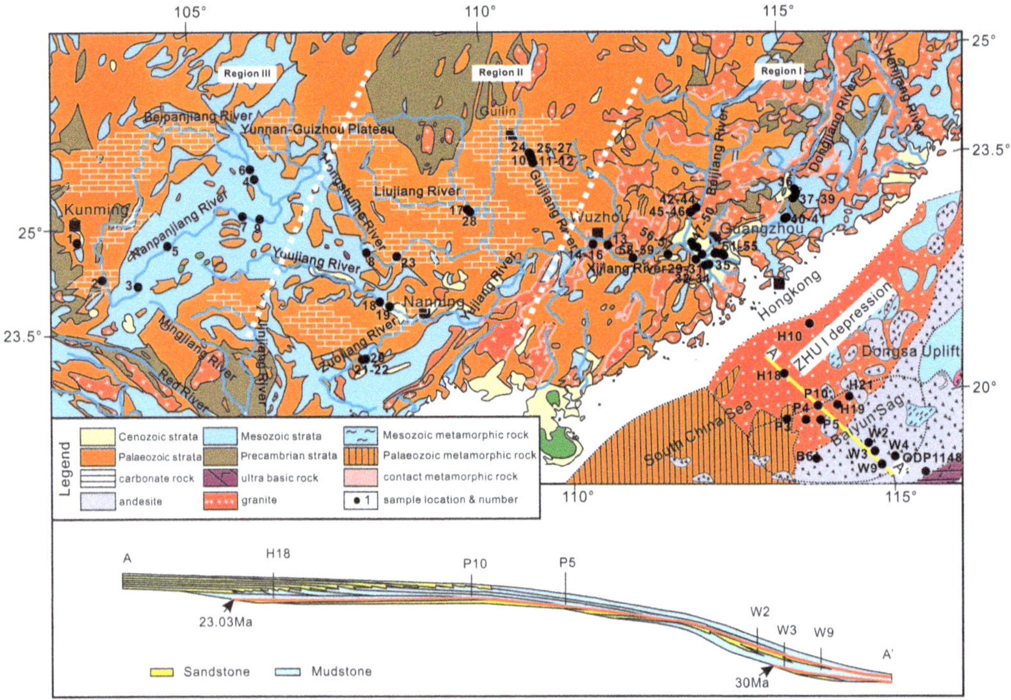

Fig. 1. (Upper panel) Geological map of the Pearl River drainage (modified from Guangdong Bureau of Geology and Mineral Resources 1988) and the basement of the northern South China Sea (modified after Wang *et al.* 2002). (Lower panel) Cross profile of the northern South China Sea (based on Shao *et al.* 2015, unpublished data). Sampling details are given in Supplementary data S-1 and S-2.

Most of the drainage area is covered by mountains and hills (94.4%), leaving only 5.6% of the drainage covered by plains with an elevation under 50 m (Cai 1981).

Covering about 44% of the total drainage area, the upper reaches of the Xijiang River (Region III in Fig. 1, including the Nanpanjiang and Beipanjiang Rivers) lie in mountains of Mesozoic limestone, with some Palaeozoic sandstone, shale and minor granitic intrusions. The middle reaches of the Xijiang River (Region II), including the Liujiang and Guijiang Rivers region, are covered by Palaeozoic strata. The Dongjiang and Beijiang Rivers, corresponding to the lower reaches of the Xijiang River, mainly cut through hills exposing widespread granite and volcanic rocks (Region I in Fig. 1; Guangdong Bureau of Geology and Mineral Resources 1988). Thus, these three regions show distinctive lithological characteristics (Fig. 1).

Some studies have argued that the Oligocene Pearl River was a short or near-source river with smaller deltas (Pang *et al.* 2006; Li *et al.* 2007; Liu *et al.* 2007, 2010; Zhu *et al.* 2010; Mi *et al.* 2011; Yu *et al.* 2012). In the late Oligocene, the northern margin of the South China Sea subsided (Ru & Pigott 1986; Clift & Lin 2001), and the region transformed from a rift basin to a thermal subsidence basin. Meanwhile the palaeo-Pearl River sediment flux greatly increased (e.g. Clift 2006), and passed over the ZHU-I depression, forming many underwater fans in the Baiyun Sag, which was once a shelf depocentre in front of the river delta (Peng *et al.* 2004, 2005; Liu *et al.* 2007; Wang *et al.* 2010; A. Li *et al.* 2011; Y. Li *et al.* 2011) . The sediment changed from being sand-rich (sand/mud = 62.3/37.7) in the Oligocene to mud-rich (sand/mud = 35.3/64.7) from the start of the Miocene, based on acoustic and resistivity logging data analyses from many industrial wells, accompanied by some sharp decreases in smectite and great variations in the $\varepsilon Nd_{(0)}$ (a measure of Nd isotopic composition relative to the present-day mantle reservoir) in sediments from Ocean Drilling Program (ODP) Site 1148 and Well PY33, indicating provenance changes near the Oligocene–Miocene boundary (Clift *et al.* 2002; Li *et al.* 2003; Tang *et al.* 2004; Pang *et al.* 2005, 2007*a*; Shao *et al.* 2007, 2009).

The timing of these prominent changes in the sediment composition coincided with expansion of the South China Sea at *c.* 23–25 Ma (e.g. Briais *et al.* 1993). Shao *et al.* (2008) suggested that the change of the sand–mud ratio may reflect an increase of the sediment transport distance, as a result of enlargement of the palaeo-Pearl River drainage which completed its transformation from a short river into a long river.

There are disputes about the provenance change from Oligocene to Miocene in sediments from the northern South China Sea. Drilling at ODP Site 1148 firstly revealed abrupt changes across the Oligocene–Miocene boundary in both the acoustic and resistivity logging curves (Wang *et al.* 2000; Shao *et al.* 2004). Seismic profiles show high amplitude reflections with two-way irregular bending-biconvex lenses, or 'double reflectors' (Wang *et al.* 2003). The sediment recovered from this section shows deformed slump structures, which are quite different from the normal deep-sea Oligocene and Miocene sequences (Clift *et al.* 2002; Li *et al.* 2005). Based on clay mineral and Nd isotope analyses, Clift *et al.* (2002) proposed that after the end of the Early Miocene, the source area of the northern South China Sea sediment gradually extended from the South China block into the continental interior, leading to the compositional changes observed. This is in addition to change in sediment composition caused by climate. However, Li *et al.* (2003) proposed that the Oligocene sediments in the northern South China Sea were dominantly derived from the SW (Indochina–Sunda Shelf and possibly northwestern Borneo) before changing to a northern provenance from South China in the Miocene.

Tang *et al.* (2004) attributed the high content of smectite in Oligocene sediments to erosion of a Mesozoic volcanic arc during the early stages of South China Sea extension, and a decreased smectite content in Miocene sediments to an enlarged source area with a more diverse source geology. In the deep-water Baiyun Sag, however, Pang *et al.* (2005, 2007*a*) and Liu *et al.* (2007) reported the dominance of shallow-water delta-shore facies of sandstone and mudstone in the Oligocene, deep-water fan mudstone in the Early and Middle Miocene, and fine-grain sediments since 10.5 Ma, which can be correlated with different stages in the extension of the South China Sea.

Rare earth element (REE) analyses by Miao *et al.* (2008) indicated that the main source area to the Baiyun Sag was from the adjacent granite massif in the Oligocene, but also included Palaeozoic sedimentary rocks from inland South China during the Miocene. Geochemical analyses by Shao *et al.* (2008) on sediments of ODP Site 1148 and two nearby wells, XJ33 and PY33, revealed a dominant silicate source for the Oligocene but a carbonate source for the Miocene, probably as a result of the Pearl River's westward headwater erosion and the related drainage enlargement, as supported also by Nd isotope data (Clift *et al.* 2002; Pang *et al.* 2007*a*; Shao *et al.* 2007; Chen 2012).

Based on single-grain-quartz oxygen isotope ratios, the grain size of isolated terrigenous materials, the terrigenous mineral accumulation rate and scanning electron microscopy analysis of isolated quartz grains from ODP Site 1148, A. Li *et al.* (2011) and Y. Li *et al.* (2011) proposed five stages for the evolution of the South China Sea basin, which were characterized by sediment source changes mainly from Palawan in the early stage to older sedimentary rocks in inland South China after 25 Ma. Similarly, Wu *et al.* (2012) found chondrite-normalized REE patterns in 11 petroleum wells from the Baiyun Sag and adjacent areas and argued that these mainly reflect deposition from the palaeo-Pearl River with a change in provenance at the Oligocene–Miocene boundary. However, intermediate volcanic components are characteristic in some samples, thus indicating erosion from the Dongsha Rise, while basic volcanic materials in samples suggest erosion from the south.

Disputes about the provenance changes of the Oligocene and Miocene sediments are partly a reflection of the poorly understood provenance of sediments accumulating in the modern northern South China Sea, although some progress has been recently reported (Liu *et al.* 2014). The work reported in this paper was designed to address these questions and started with systematic sampling of different tributaries of the Pearl River, as well as Oligo-Miocene sedimentary rocks from the shelf-slope area of the northern South China Sea, whose depositional age is based on micropalaeontology. We analysed these samples for mineralogy and geochemistry (black dots in Fig. 1). We then compared the variations of index minerals (Ati, GZi, ZTR, described further in 'Materials and methods') and REE in samples from these sections and from wells in the Pearl River Mouth Basin, and here we discuss the application of such mineral and geochemical evidence for reconstructing sedimentation and the tectonic evolution of the Pearl River since the Oligocene.

Material and methods

Samples used in this study were mainly from two sources: the modern Pearl River and industrial wells, as shown in Figure 1. A total of 58 samples were collected in March and April 2011, during the dry season. These samples cover most of the Pearl River tributaries, including the Nanpanjiang, Beipanjiang and Hongshuihe Rivers in the upper reaches, and the Zuojiang, Youjiang, Guijiang, Beijiang and Dongjiang Rivers from the middle and lower reaches. In addition, ten samples from the Red River, five samples from the upper reaches of the Mekong River and 15 samples from Central Vietnam were collected. A total of 795 samples were taken from Oligocene and Miocene strata in industrial wells and included: 393 samples in three drilling wells (Wells P3, W3, W9) analysed for trace element composition; 128 samples from 11 drilling wells (Wells P3, P4, P5, P10, H19, H21, B6, W2, W3, W4, W9) examined for their heavy mineral assemblage; 167 samples from Wells H18 and P3 analysed for Al_2O_3 and CaO content; and 35 samples from Well P3 analysed for Nd isotope composition (Fig. 1).

The age models for these drilling wells were based on micropalaeontology (Hao *et al.* 1996) and unpublished material from China National Offshore Oil Corporation (CNOOC) (Pang *et al.* 2007*b*). For geochemical analysis, samples were washed through a 63 μm sieve with deionized distilled water, and mud and fragments <63 μm were collected and dried at $<60^\circ C$, before being crushed and ignited at 620°C for 60 minutes to remove organic matter and interlayer water from clay minerals. After weighing and calculation of the ignition loss, 0.1 mol HCl was used to remove marine $CaCO_3$.

Samples were then dissolved in mixed HF and HNO_3, and one repeated sample and one blank sample were analysed for every 30 samples in order to check for consistency (Liu *et al.* 1996, Li *et al.* 2002). After dissolution, samples were analysed by inductively coupled plasma atomic emission spectroscopy (ICP-AES; THERMO ICP-IRIS Intrepid II) to measure the abundance of major elements, and ICP mass spectrometry (ICP-MS; THERMO X SERIES) for trace and REE, with 1 ppb Ru added to keep measuring consistency. Each sample was measured six times, calibrated with international rock standards (e.g. GSR-1, JSD-1), and checked with drift control samples, duplicate samples and blank samples. Relevant analytical details were described by Li *et al.* (2002, 2003) and Wei *et al.* (2006). We calculated $\varepsilon Nd(0)$ based on the method described by Rollinson (1993), using a $^{143}Nd/^{144}Nd$ value of 0.512638 for the Chondritic Uniform Reservoir (CHUR; Hamilton *et al.* 1983).

For mineral analyses, 200–300 g of the >63 μm sand and silt fractions were washed with deionized distilled water, and dried at $<60^\circ C$ before using Bromoform (2.89 cm^{-3} in density) for the separation of heavy minerals. Heavy minerals were washed two to three times with ethyl alcohol and dried at 60°C. After magnetic separation, both magnetic and non-magnetic minerals were weighed, and about 600 grains from each sample were used for heavy mineral identification using an optical microscope. Analytical details were described by CNPC (1998) and the Chinese Academy of Geological Science (1977). All analyses were done in the State Key Laboratory of Marine Geology of Tongji University.

Multivariate analyses were run using the package of Statgraphics XV.II for grouping mineral and elemental assemblages. Cluster and discriminant analyses are widely used multivariate procedures for discriminating between different sample groups and to evaluate the relationships between geologically grouped data (Davis 1986; Woronow 1997). The resulting groups can be interpreted to reflect different tectonic environments and/or sedimentary provenances. Stepwise linear discriminant analysis is one of the most useful approaches (see Jennrich 1977; Stattegger 1991). Discriminant functions maximize the differences between groups by linear classifications, which select the most discriminating variable among the groups at each step. A variable is deleted if its discriminatory power becomes too low.

The discriminant functions used in this study were:

$$1 = -4.142 + 0.074243 \times Ce - 2.91125 \times Dy + 1.59888 \times Er + 14.132 \times Eu - 5.92441 \times Gd - 8.11288 \times Ho - 0.576772 \times La + 43.6271 \times Lu - 4.07464 \times Nd + 17.16 \times Pr + 2.8574 \times Sm + 25.2153 \times Tb - 17.6412 \times Tm - 0.116488 \times Yb.$$

$$2 = -0.612751 \times Ce + 0.849413 \times Dy - 17.6013 \times Er + 1.54903 \times Eu + 3.58686 \times Gd + 7.53673 \times Ho - 1.05839 \times La + 4.66411 \times Lu + 4.43947 \times Nd + 4.02636 \times Pr - 13.9696 \times Sm + 3.39962 \times Tb + 9.33079 \times Tm - 5.40967 \times Yb.$$

Results and implications

Rare earth elements

For provenance determination, the REE patterns were compared with the North American Shale Composite (NASC) and the Post-Archean average Australian Shale (PAAS). These standards have been widely used (Gromet & Silver 1983; Taylor & McLennan 1985, 1995; McLennan 1989) because of their similarity to the average upper continental crust patterns. All sample show high light REE (LREE) enrichment, moderate or flattened heavy REE (HREE), as well as a negative Eu anomaly. The Eu-depletion may reflect shallow, intracrustal differentiation associated with the production of granitic rocks, compared to the Eu-enriched lower continental crust (McLennan *et al.* 1993).

The chondrite-normalized REE results for Oligocene–Miocene South China Sea sediments show similar patterns to the PAAS. The REE results from Wells P3 and W3 reflect the characteristics of the common granite and sedimentary rocks exposed within the Pearl River drainage and along the South China coast (Fig. 2a, b). However, in Well W9 (Fig. 2c) from the southern deep-water area, the REE patterns show a positive Eu anomaly, suggesting the influence of basic volcanic sources, such as those found in the southern and eastern part of the South China Sea (Wu *et al.* 2012). Based on seismic profile analyses, a series of deltas appear to have formed and migrated southwards from the Oligocene to the Miocene in the northern South China Sea, indicating that a palaeo-Pearl River existed since that time (Fig. 1, cross profile A–A′, based on industrial data from CNOOC).

Heavy minerals

In Well H10 from the Pearl River delta (Fig. 1), the combined abundance of heavy minerals including zircon, tourmaline, leucoxene and anatase was about 37% in the Early Oligocene, and increased to about 86% in the Late Oligocene and 95% in the Miocene (Fig. 3). The Lower Oligocene heavy mineral assemblage mainly contains zircon, garnet, apatite, epidote and a small amount of tourmaline, leucoxene and rutile. Zircon accounts for 32% and apatite for 19% of the total heavy mineral assemblage and represents erosion of igneous rocks. The presence of garnet (4%) and the high content of epidote (up to 42%) also indicate granite and metamorphic parent rocks (Wu & Zhao 1982; Liang *et al.* 1989; He *et al.* 2001). Unstable minerals such as epidote (42%) and apatite (19%) suggest near-source deposition without a long period of weathering (Morton & Hallsworth 1994).

In the Upper Oligocene mineral assemblage, zircon increases from 33% to 54%, and garnet increases from 5% to 12% (Fig. 3), while apatite decreases sharply from 19% to 2% and epidote decreases from 39% to nearly zero, reflecting longer transport distance and stronger weathering of the unstable minerals, probably driven by climate change (Xiang *et al.* 2011). Abundant zircon, garnet and tourmaline also imply strong erosion of granite and contact metamorphic rock sources, although some of them could have been reworked from older sediments.

Miocene mineral assemblages are dominated by stable minerals, with zircon accounting for up to 65% of the total. Leucoxene increases from 5% to 18%, while garnet decreases to 1% and tourmaline from 24% to 5%. Minimal amounts of garnet, coupled with a large decrease in tourmaline, may signal erosion of metamorphic rocks rather than granite as the main source, and/or a lengthened sediment transport distance, which caused stable minerals to further increase their proportion.

Morton & Hallsworth (1994, 1999) showed that the ratios or concentrations of some stable minerals can be used to track their provenance assuming similar hydrodynamic environments of sedimentation. These heavy mineral indices include GZi ($=100 \times$ garnet%/(garnet% + zircon%)), Ati ($=100 \times$ apatite%/(apatite% + tourmaline%)) and ZTR (=total zircon, tourmaline and rutile). GZi is mostly used to differentiate among parent rock types, particularly metamorphic rocks and magmatic rocks; Ati is mainly used to determine the degree of weathering during transportation; while ZTR represents the mineralogical maturity, and also correlates positively with mineral stability.

Our results show high ZTR and low Ati values in samples from Wells H21, H19, P5, P4 and P3, with a nearly constant trend from the Late Oligocene to the Miocene (Fig. 4). In contrast, Wells B6, W2 and W3 all have relative high Ati values, most likely influenced by magmatic sources. These results suggest the existence of two different sediment sources for the northern South China Sea during the Oligocene–Miocene, one correlated with possible source rocks in the palaeo-Pearl River delta in the north and the other with pre-Tertiary magmatic rocks within uplifted basement highs (including Dongsha Rise in the NE) (Fig. 4).

Major elements

Our results from Wells H18 and P3 (Fig. 5, location in Fig. 1) complement the early studies of sediments from ODP Site 1148 in showing that smectite gradually decreased from around 80% of the total assemblage in the Oligocene to *c.* 20% in the Pliocene (Clift *et al.* 2002; Tang *et al.* 2004). As a primary weathering product of magmatic rocks, smectite decrease could reflect a change of the parent rock type since *c.* 23 Ma from magmatic rocks to other rocks, or a change in climate and decrease in chemical weathering. Meanwhile, the terrigenous CaO, Al_2O_3, Th/Sc, Th/Co, as well as $\varepsilon Nd_{(0)}$ also changed at the boundary of the Oligocene and Miocene at Wells P3 and H18. The increase in the terrigenous CaO and decrease in Al_2O_3 may reflect a provenance change from silicate rocks to more carbonate in the source regions (Fig. 5).

Discussion

Characteristics of the source areas

Two source areas could have contributed sediment to the northern South China Sea during the Oligocene–Miocene: the palaeo-Pearl River and

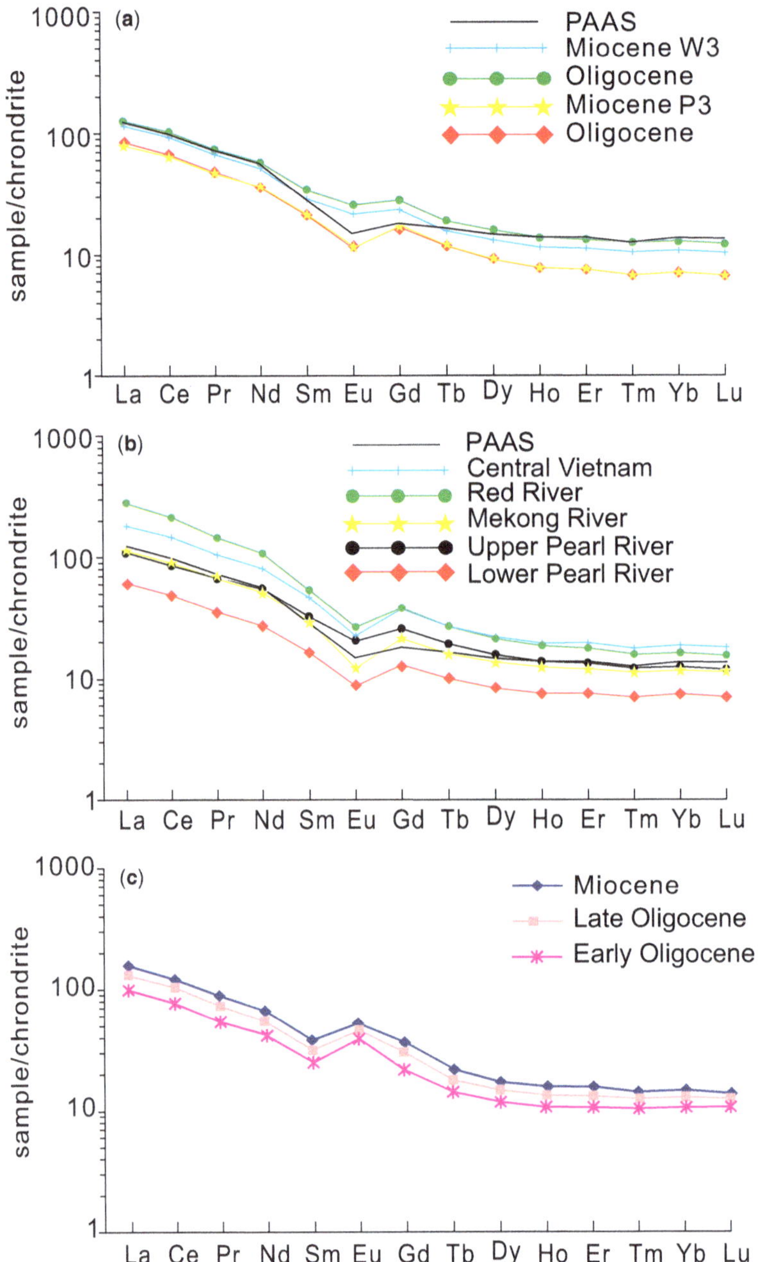

Fig. 2. Chondrite-normalized REE patterns for: (**a**) Wells W3 and P3; (**b**) major rivers in the region; (**c**) Well W9. The complete geochemical and heavy mineral data are given in Supplementary S-1 and S-2.

uplifted basement highs within the basin, including some volcanic material transported from the south. Arc magmatism along the eastern margin, as well as erosion from Taiwan, however, did not become an important source until the Pliocene–Pleistocene.

Efforts have been made to distinguish the sources of the main components in the river samples. If the Pearl River is simply separated into two parts above Wuzhou (Fig. 1), then the chondrite-normalized REE should be similar to other big

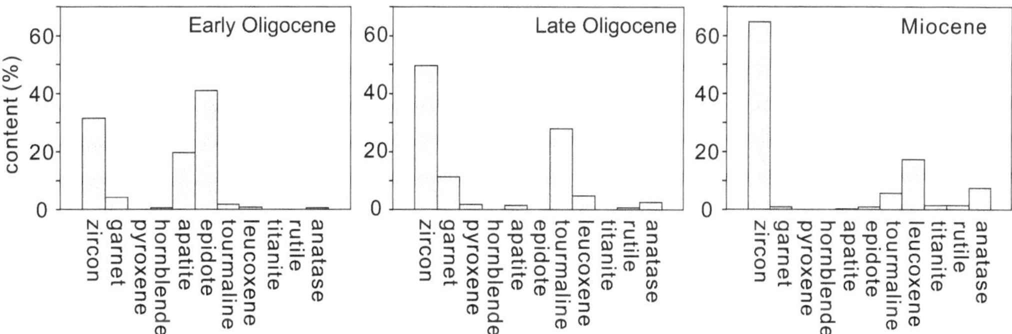

Fig. 3. Characteristics of the Oligocene–Miocene heavy mineral assemblages from Well H10.

rivers around the South China Sea, in being similar to PAAS with negative Eu anomalies, as a result of intensive hybrid transportation and mixing (Fig. 2b, the upper Pearl River and the lower Pearl River). As a result, it is critical to understand the relationship between the sediment components in different parts of the Pearl River and their parent rock types.

Cluster analysis of REE (Fig. 6) in the Pearl River sediments identifies three groups of samples from the three regions shown in Figure 1, as also supported by discriminant analysis of REE with *c.* 80% accuracy for their correlation (Fig. 7a; Table 1). We conclude that the parent rocks are mineralogically and geochemically different within each of the three main drainage regions of the Pearl River, making it possible to constrain the regional sediment evolution patterns based on the component analyses.

The heavy mineral assemblage is closely tied to the parent rock type, but also modulated by chemical weathering and diagenesis. Mixing occurs when new parent rocks are added to a basin after a river enlarges or because of modification of the assemblages during increased transport caused by physical abrasion or chemical weathering. As reported by Xiang *et al.* (2011), poorly sorted heavy minerals, mainly apatite, pyroxene, ilmenite and magnetite, are most abundant in sediments of the upper reaches of the Xijiang River, suggesting short transportation distance. In contrast, the lower reaches of the Xijiang River, represented by the Liujiang and Guijiang Rivers, are characterized by abundant zircon, rutile and leucoxene, reflecting high stability. The Beijiang and Dongjiang Rivers are characterized by the presence of epidote, amphibole, tourmaline, garnet, sillimanite and andalusite, typical of a low stability assemblage and is possibly consistent with erosion of acid magmatic rocks and contact metamorphic rocks. These mineral characteristics are consistent with the geochemical data in reflecting different parent rock types over different drainage areas.

Based on the REE data and heavy mineral analyses, the Pearl River drainage can be recognized as consisting of three regions from east to west (Fig. 7b). Region I is typified by granite intrusions along the SE coast which are drained by the Dongjiang and Beijiang Rivers. This area is characterized by Mesozoic granite and acid extrusive rocks, with

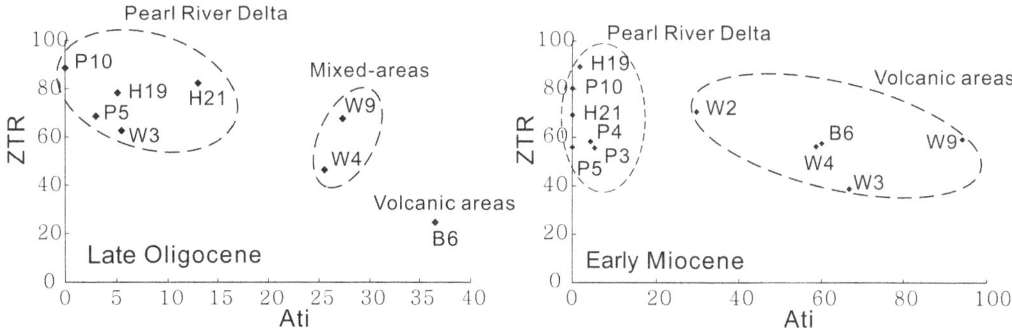

Fig. 4. Ati–ZTR plots of the heavy minerals for the Late Oligocene and Early Miocene from different northern South China Sea wells.

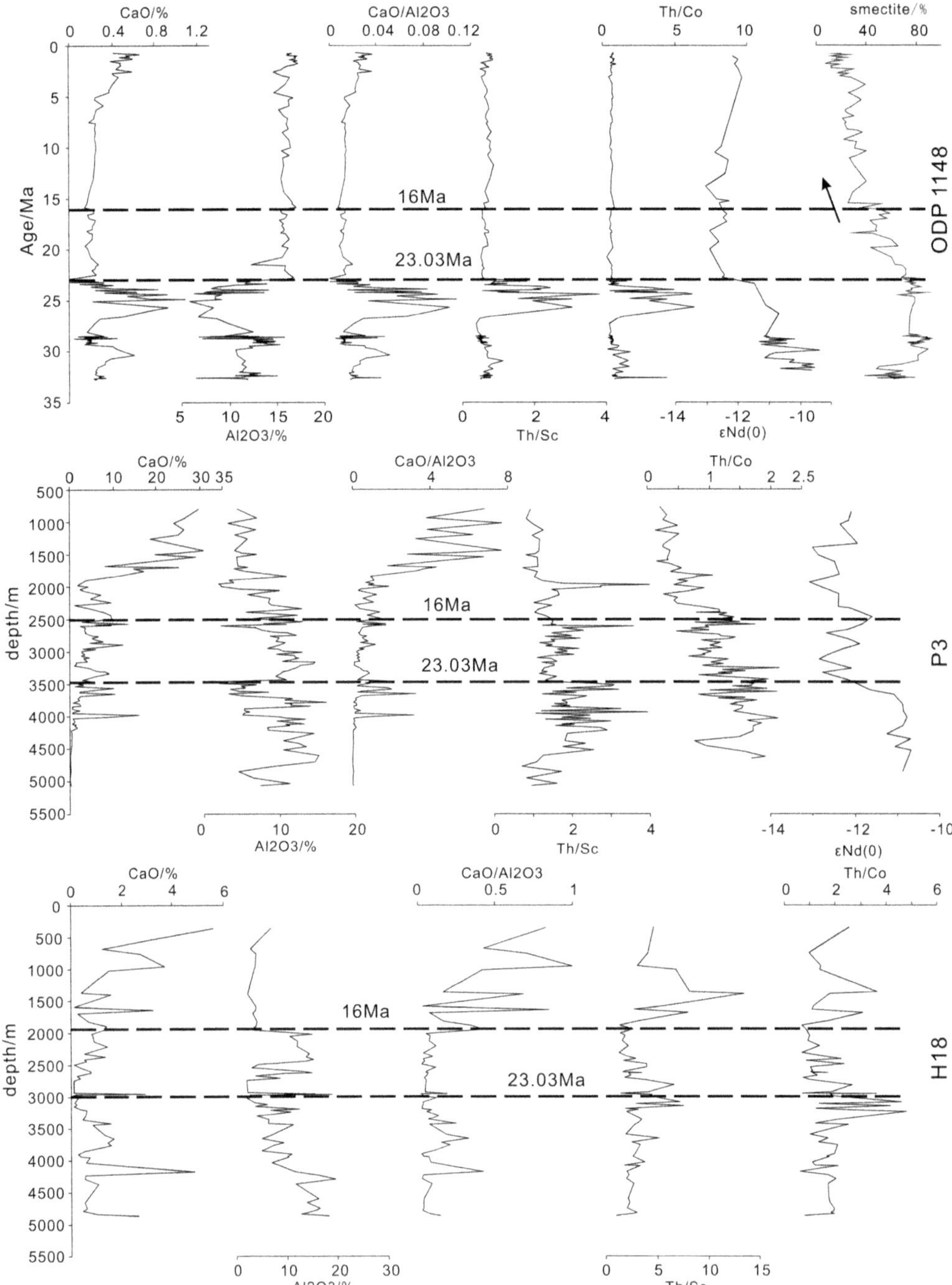

Fig. 5. Percentage abundance of elements in sediments from H18 and P3 compared with the data from ODP Site 1148, showing similar trends (arrow) between these shelves and lower slope sites.

some sedimentary rocks, but is largely free of carbonate rock. Sediments eroded from Region I contain high percentages of epidote, garnet and tourmaline and are generally rich in Al_2O_3 and SiO_2 but poor in CaO. Region II includes the mid-stream of the Xijiang River (represented by the

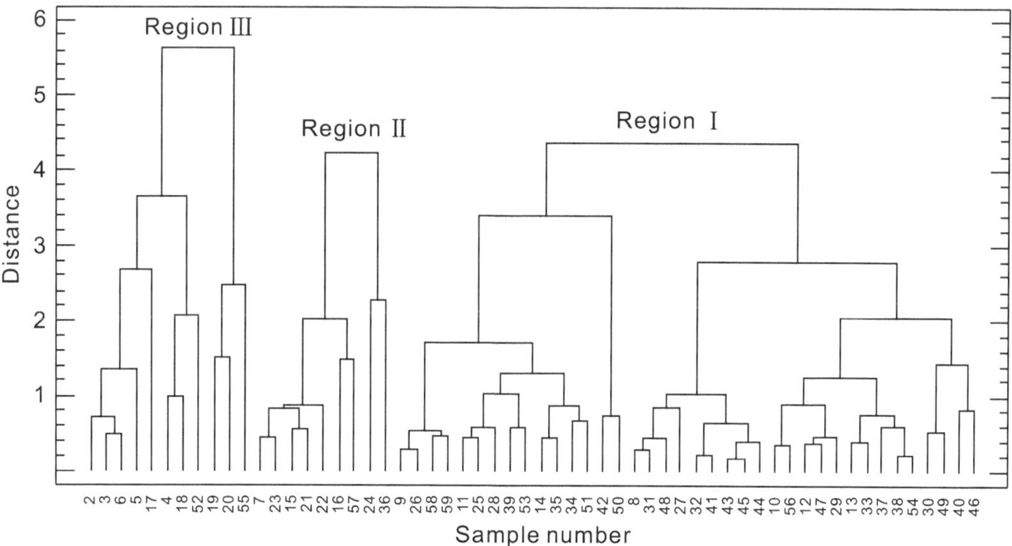

Fig. 6. Cluster analysis of REE in sediments from the Pearl River categorizes three main groups, corresponding to three regional areas.

Liujiang and Guijiang Rivers). This region is mainly covered by sedimentary rocks, particularly Palaeozoic carbonates. The sediment is characterized by fewer heavy mineral types and has high zircon and leucoxene content, reflecting a stable, mature assemblage, caused by repeated erosion, transportation and sedimentation of Palaeozoic sedimentary source rocks, and by high siliciclastic CaO content and low Al_2O_3 and SiO_2 content reflecting the abundance of carbonate rocks. Region III occupies the upper reaches of the Xijiang River (represented by the Nanpanjiang and Beipanjiang Rivers), and is covered mostly by Mesozoic clastic sedimentary rocks and Palaeozoic carbonate rocks. Many types of unstable heavy minerals from near-field sources are present in the sediment due to intense erosion and the short transport distance, although most of them are absent from the lower reaches as the transportation distance increases. Region III has little impact in terms of heavy minerals on the sediment transported to the northern South China Sea, although heavy disruption by farming in the middle and lower reaches during the recent past may have biased part of the data (Hu *et al.* 2013). However, high CaO and low Al_2O_3 and SiO_2 characterize sediments not only from Region III, but also from Region II, due to the similar sedimentary parent rocks found in each.

Sediments in the northern South China Sea

All the Lower Oligocene sediments from ODP Site 1148, as well as Wells H18 and P3, show relative low CaO content, but high Al_2O_3 and SiO_2, while unstable minerals such as apatite and epidote (mainly of metamorphic origin) are enriched in samples from Well H10 (Fig. 3). These characteristics imply near-source deposition for these localities, with sediments mostly eroded from granite and metamorphic rocks in adjacent areas.

Since the Late Oligocene, mineralogical, geochemical, as well as seismic evidence indicates that the palaeo-Pearl River started to provide sediments to the northern South China Sea. With more than 80% smectite and relative high GZi, ZTR and Al_2O_3 but low Ati and CaO (Figs 4 & 5), the Upper Oligocene sediments may have been derived from a source region dominated by silicate parent rocks, particularly magmatic rocks. Meanwhile, granitic parent rocks supplied high amounts of zircon, tourmaline and garnet but no epidote (Fig. 3). Together, these features imply that in the Late Oligocene, the northern South China Sea no longer received sediments from nearby metamorphic rocks exposed in uplifted basement highs within the basin, but instead was supplied from the Mesozoic granite of the South China coast, indicating that the palaeo-Pearl River drainage was limited to the lower reaches of the modern Pearl River in Region I, including the Beijiang and Dongjiang Rivers (Fig. 7b).

We have reconstructed a drastic decrease in smectite during the Miocene, coupled with a fall in Al_2O_3 but an increase in CaO that may reflect enhanced supply of carbonate-sourced sediments. Meanwhile, stable heavy minerals, such as zircon

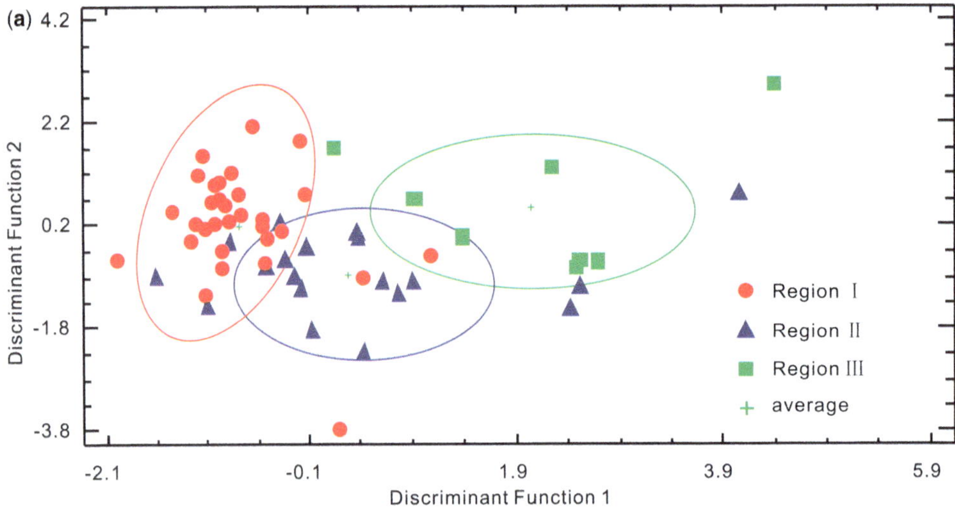

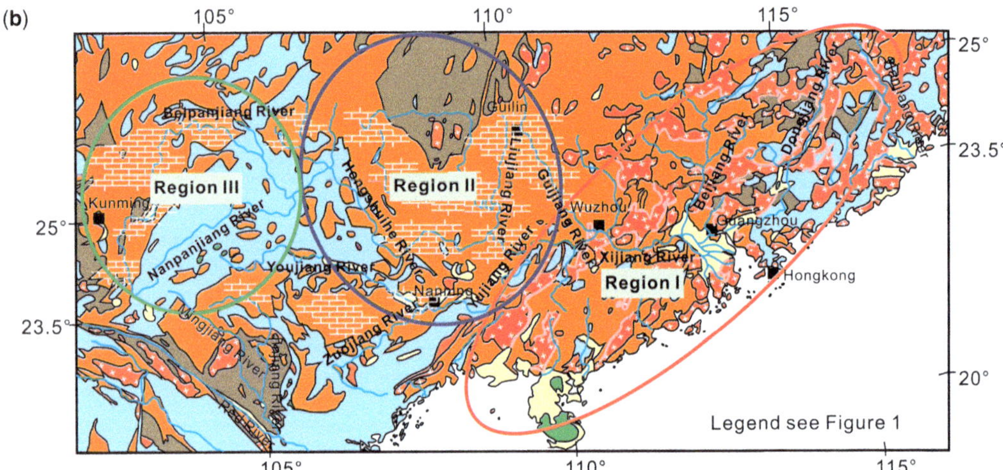

Fig. 7. (**a**) Discriminate analysis of REE in sediments from three regions of the Pearl River as identified by cluster analysis (accuracy: 81%). (**b**) Respectively, Regions I, II and III represent the lower, middle and upper reaches of the river.

Table 1. *Classification of discriminant analyses based on trace elements of sediments from the Pearl River*

	Number of analyses	Region I	Region II	Region III
Region I	31	28	3	0
(Corrected percentages)		(90.3%)	(9.7%)	(0.0%)
Region II	20	4	13	3
(Corrected percentages)		(20.0%)	(65.0%)	(15.0%)
Region III	8	1	0	7
(Corrected percentages)		(12.5%)	(0.0%)	(87.5%)
Percent of cases correctly classified: 81%				

and leucoxene increased their proportion of the heavy mineral suite to 65% and 18% respectively. Unstable minerals have extremely low concentrations, suggesting a lengthening of transport distance, which allowed these minerals to be destroyed (Fig. 3). Other factors may have also acted on unstable minerals, but their decline cannot be attributed only to post-depositional diagenesis. The decline in garnet may also reflect a shift of the dominant source from granites into carbonate rocks. We suggest that flux from carbonate source areas increased greatly after the palaeo-Pearl River expanded its drainage westwards into Region II, i.e. to include the middle reaches of the Xijiang River, including the Guijiang, Liujiang and Youjiang Rivers, or even to Region III, (Nanpanjiang and Beipanjiang Rivers) (Fig. 7b). However, the present data do not allow us to constrain the time when this second drainage expansion happened. Further studies are also required to clarify how climate change, in concert with provenance change, have contributed to the mineral and geochemical changes revealed in this study.

The Late Oligocene/Miocene provenance shift may not have been caused merely by the headwater expansion of the Pearl River, but also by intense tectonic activities associating with volcanic eruptions (Wang *et al.* 2003; Li *et al.* 2005; A. Li *et al.* 2011; Y. Li *et al.* 2011; Wu *et al.* 2012), as implied by elemental and heavy mineral data (Figs 3–5). Major tectonic events in the Early Miocene include spreading of the South China Sea, which could have influenced the sedimentation process and sources, coupled with regional topography changes due to their interrelated nature within the Earth system. Although the palaeo-Pearl River continued to supply sediments to most parts of the northern South China Sea during the Early Miocene, the southern part of the Baiyun Sag appeared to have been affected more by basic volcanic sources, as shown by data from Well W9 (Fig. 2c).

Conclusions

Mineralogical and geochemical analyses of sediment samples from different sites in the northern South China Sea provide new information on the evolving sediment provenance in the region since the Oligocene, which can be summarized as follows.

The chondrite-normalized REE results from palaeo-Pearl River sediments show similar patterns to PAAS, with a negative Eu anomaly, while sediments from basic volcanic parent rocks are characterized by a positive Eu anomaly. In the Early Oligocene, smectite varied considerably between 80% and 39%, while the sediment chemistry was typified by low CaO and high Al_2O_3 and SiO_2. The heavy mineral assemblage was characterized by epidote (up to 42%), zircon, garnet, apatite and minor tourmaline, leucoxene and rutile. These features indicate silicate source rocks of metamorphic and granitic types that were not experiencing intense weathering, although moderate weathering in local areas may have occurred, as indicated by content of 40–60% smectite at ODP Site 1148 (Fig. 5).

In the Late Oligocene, smectite remained stable at a very high level (about 80%), and both zircon and garnet content moderately increased. Epidote and apatite decreased to near zero, although the bulk geochemistry was similar to Early Oligocene samples, with low CaO and high Al_2O_3 and SiO_2. Together, these features reflect the absence of metamorphic rocks from the source areas and longer transport distances, probably coupled with intense erosion.

In the Miocene, smectite content sharply decreased from 80% to 20% and the heavy mineral assemblage was dominated by stable minerals (*c.* 90%), with negligible garnet and low tourmaline. Together with high CaO and low Al_2O_3 and SiO_2 content, this mineral assemblage is interpreted to reflect a shift of source rocks from silicate into carbonate types after the Early Miocene.

REE and heavy mineral data help to separate the Pearl River drainage into three regions from east to west: Region I is the lower reaches along the southeastern coastal area; Region II is the middle reaches of the Xijiang River; and Region III is the head waters of the Xijiang River. They represent three different source regions, each having different parent rock types.

By comparing the inland source data with the offshore sediment record, the evolution of palaeo-Pearl River drainage can be reconstructed. In the Early Oligocene, sediments with abundant epidote, eroded mainly from granite and metamorphic rocks exposed with uplifted basement blocks, contributed to near-source deposition in the northern South China Sea, at a time when the palaeo-Pearl River was probably a young, short river. From the Late Oligocene, however, the palaeo-Pearl River became the dominant source for the northern South China Sea, and its drainage evolution started to affect the sediment provenance on the continental margin. The Oligocene Pearl River extended only into the lower reaches of the modern Pearl River (Beijiang and Dongjiang Rivers) and was mainly derived from granite sources along the SE coast. In the Miocene, the drainage of the palaeo-Pearl River extended inland and reached the midstream of the Xijiang River, as well as the Guijiang, Liujiang and Youjiang Rivers, so that it encompassed a large area of Palaeozoic to Mesozoic clastic sedimentary rocks and carbonate rocks. A change in the parent rock type from silicate rocks to carbonate rocks was accompanied by longer transport distances for the fluvial sediment deposited in most parts of the shallow northern South China Sea. The deep-water northern South China Sea area was not supplied by the palaeo-Pearl River, but was dominated by basic volcanic materials eroded from volcanic centres within the basin from the Oligocene to the Miocene.

We thank Dr S.Y. Foong (Universiti Sains Malaysia) for continual encouragement in writing up this paper. This work was supported by the National Natural Science Foundation of China (Grant Nos. 91128207, 40976023, 91228203), the National Science and Technology Major Projects (Grant No. 2011ZX05025-006-02), and the National Basic Research Program of China (2009CB 219400). We thank the reviewers and the Editor Dr P. Clift for their valuable comments, which greatly improved this presentation.

References

Briais, A., Patriat, P. & Tapponnier, P. 1993. Updated interpretation of magnetic anomalies and seafloor spreading stages in the South China Sea: implications for the Tertiary tectonics of Southeast Asia. *Journal of Geophysical Research: Solid Earth*, **98**, 6299–6328.

Cai, G. 1981. Geological structure of Pearl River and its evolution history. *Pearl River*, **2**, 12–17 [in Chinese].

Chen, S. 2012. Analysis and application of REE patterns and heavy mineral assemblage characteristics in Fanyu Lower Uplift, Pearl River Mouth Basin. *Marine Origin Petroleum Geology*, **17**, 47–54 [in Chinese with English abstract].

Chinese Academy of Geological Science 1975. *Asia Geological Map*. Chinese Academy of Geological Sciences, Beijing [in Chinese].

Chinese Academy of Geological Science 1977. *Identification Manual of Sand Minerals*. Geological Public House, Beijing [in Chinese].

Clift, P. D. 2006. Controls on the erosion of Cenozoic Asia and the flux of clastic sediment to the ocean. *Earth and Planetary Science Letters*, **241**, 571–580.

Clift, P. D. & Lin, J. 2001. Preferential mantle lithospheric extension under the South China margin. *Marine and Petroleum Geology*, **18**, 929–945.

Clift, P. D., Lee, J. I., Clark, M. K. & Blusztajn, J. 2002. Erosional response of South China to arc rifting and monsoonal strengthening: a record from the South China Sea. *Marine Geology*, **184**, 207–226.

CNPC (China National Petroleum Corporation) 1998. *The Method of Separation and Examination of Heavy Minerals for Sedimentary Rock*. Petroleum Industry Press, Beijing.

Davis, J. C. 1986. *Statistics and Data Analysis in Geology*. Wiley, New York.

Gromet, L. P. & Silver, L. T. 1983. Rare-earth element distribution among minerals in a granodiorite and their petrogenetic implications. *Geochimica et Cosmochimica Acta*, **47**, 925–940.

Guangdong Bureau of Geology and Mineral Resources 1988. *Guangdong Province Regional Geology*. Geological Publishing House, Beijing [in Chinese].

Hamilton, P. J., O'Nions, R. K., Bridgewater, D. & Nutman, A. P. 1983. Sm-Nd studies of Archean metasediments and metavolcanics from west Greenland and their implication for the earth's early history. *Earth and Planetary Science Letters*, **62**, 263–272.

Hao, Y. C., Xu, Y. L. & Xu, S. C. 1996. *Research on Micropalaeontology and Palaeoceanography in the Pearl River Mouth Basin, South China Sea*. China University of Geosciences Press, Beijing [in Chinese with English abstract].

He, Z. H., Liu, Z. J. & Zhang, F. 2001. Latest progress of heavy mineral research in the basin analysis. *Geological Science and Technology Information*, **20**, 29–32 [in Chinese with English abstract].

Hu, D., Clift, P. D. *et al.* 2013. Holocene evolution in weathering and erosion patterns in the Pearl River delta. *Geochemistry Geophysics Geosystems*, **14**, 2349–2368, http://doi.org/10.1002/ggge.20166

Huang, C. Y., Yuan, P. B. & Tsao, S. H. 2006. Temporal and spatial records of active arc-continent collision in Taiwan: a synthesis. *Geological Society of America Bulletin*, **118**, 274–288.

Jennrich, R. I. 1977. Stepwise discriminant analysis. *In*: Enslein, K., Ralston, A. & Wilf, H. S. (eds) *Statistical Methods for Digital Computers*. Wiley, New York, 454.

Li, A., Huang, J. & Jiang, H. 2011. Sedimentary evolution in the northern slope of South China Sea since Oligocene and its responses to tectonics. *Geophysics*, **54**, 3233–3245 [in Chinese with English abstract].

Li, Q., Jian, Z. & Su, X. 2005. Late Oligocene rapid transformations in the South China Sea. *Marine Micropaleontology*, **54**, 5–25.

Li, X., Liu, Y. & Tu, X. 2002. Precise determination of chemical compositions in silicate rocks using ICP-AES and ICP-MS: a comparative study of sample digestion techniques of alkali fusion and acid dissolution. *Geochimica*, **31**, 289–294 [in Chinese with English abstract].

Li, X., Wei, G. *et al.* 2003. Geochemical and Nd isotopic variations in sediments of the South China Sea: a response to Cenozoic tectonism in SE Asia. *Earth and Planetary Science Letters*, **211**, 207–220.

Li, X., Zheng, R. & Wei, Q. 2007. Provenance analysis of the Palaeogene strata in the Huizhou depression, Zhujiangkou Basin, Guangdong. *Sedimentary Geology and Tethyan Geology*, **27**, 33–38 [in Chinese with English abstract].

Li, Y., Zheng, R., Gao, B., Hu, X. & Dai, Z. 2011. Characteristics of the detrital response to Oligocene/Miocene geological events in Baiyun Sag, Pearl River Mouth Basin. *Geoscience*, **25**, 476–481 [in Chinese with English abstract].

Liang, B., Zhu, S. & Wu, H. 1989. The characteristics of heavy minerals in the coastal surficial sediments of Guangdong province in relation to sedimentary environments. *Acta Petrologica et Mineralogica*, **8**, 180–188 [in Chinese with English abstract].

Liu, B., Shen, J., Pang, X., He, M., Lian, S. & Qu, L. 2007. Characteristics of continental delta deposits in

Zhuhai formation of Baiyun depression in Pearl River Mouth Basin. *Acta Petrolei Sinica*, **28**, 49–56 [in Chinese with English abstract].

Liu, J., Clift, P. D., Yan, W., Chen, Z., Chen, H., Xiang, R. & Wang, D. 2014. Modern transport and deposition of settling particles in the northern South China Sea: sediment trap evidence adjacent to Xisha Trough. *Deep-Sea Research I*, **93**, 145–155, http://doi.org/10.1016/j.dsr.2014.08.005

Liu, Y., Liu, H. C. & Li, X. H. 1996. Simultaneous and precise determination of 40 trace elements in rock samples using ICP-MS. *Geochimica*, **25**, 552–558 [in Chinese with English abstract].

Liu, Z., Zhao, H., Fan, S. & Chen, Q. 2002. *Geology of South China Sea*. Science Press, Beijing [in Chinese].

Liu, Z., Zhang, G., Lv, R., Shen, H., Guo, R. & Tian, J. 2010. Depression in the deep-water region, the northern South China Sea. *Geoscience*, **24**, 900–909 [in Chinese with English abstract].

McLennan, S. M. 1989. Rare earth elements in sedimentary rocks: influence of provenance and sedimentary processes. *In*: Lipin, B. R. & McKay, G. A. (eds) *Geochemistry and Mineralogy of Rare Earth Elements*. The Mineralogical Society of America, Washington, D.C., 169–200.

McLennan, S. M., Hemming, S., McDaniel, D. K. & Hanson, G. N. 1993. Geochemical approaches to sedimentation, provenance and tectonics. *In*: Johnsson, M. J. & Basu, A. (eds) *Processes Controlling the Composition of Clastic Sediments*. Geological Society of America Special Paper, **284**, 21–40.

Mi, L., Zhang, G. et al. 2011. *Investigation and Evaluation of Oil and Gas Resources Strategy*. Geological Publishing House, Beijing [in Chinese].

Miao, W., Shao, L., Pang, X., Lei, Y., Qiao, P., Li, A. & Wu, M. 2008. REE geochemical characteristics in the northern South China Sea since the Oligocene. *Marine Geology & Quaternary Geology*, **28**, 71–78 [in Chinese with English abstract].

Morton, A. C. & Hallsworth, C. R. 1994. Identifying provenance: specific features of detrital heavy mineral assemblages in sandstones. *Sedimentary Geology*, **90**, 241–256.

Morton, A. C. & Hallsworth, C. R. 1999. Processes controlling the composition of heavy mineral assemblages in sandstones. *Sedimentary Geology*, **124**, 3–29.

Pang, X., Chen, C., Shi, H., Shu, Y., Shao, L., He, M. & Shen, J. 2005. Response between relative sea-level change and the Pearl River deep-water fan system in the South China Sea. *Earth Science Frontiers*, **12**, 167–177 [in Chinese with English abstract].

Pang, X., Chen, C., Wu, M., He, M. & Wu, X. 2006. The Pearl River deep-water fan systems and significant geological events. *Advances in Earth Science*, **21**, 793–799 [in Chinese with English abstract].

Pang, X., Chen, C. et al. 2007*a*. Baiyun Movement, a great tectonic event on the Oligocene-Miocene boundary in the northern South China Sea and its implications. *Geological Review*, **53**, 145–151 [in Chinese with English abstract].

Pang, X., Chen, C. et al. 2007*b*. *The Pearl River Deep-Water Fan System and Petroleum Resource in the South China Sea*. China Science Publishing, 101–104 [in Chinese].

Peng, D., Chen, C., Pang, X., Zhu, M. & Yang, F. 2004. Discovery of a deep-water fan system in South China Sea. *Acta Petrolei Sinica*, **25**, 17–23 [in Chinese with English abstract].

Peng, D., Pang, X. et al. 2005. From shallow-water shelf to deep-water slope: study on deep-water fan systems in South China Sea. *Acta Sedimentologica Sinica*, **23**, 1–11 [in Chinese with English abstract].

Rollinson, H. R. 1993. *Using Geochemical Data: Evaluation, Presentation, Interpretation*. Longman Scientific & Technical, Singapore.

Ru, K. & Pigott, J. D. 1986. Episodic rifting and subsidence in the South China Sea. *AAPG Bulletin*, **70**, 1136–1155.

Shao, L., Li, X., Wang, P., Jian, Z., Wei, G., Pang, X. & Liu, Y. 2004. Sedimentary record of the tectonic evolution of the South China Sea since the Oligocene: evidence from deep sea sediments of ODP Site 1148. *Advances in Earth Science*, **19**, 539–544 [in Chinese with English abstract].

Shao, L., Pang, X., Chen, C., Shi, H., Li, Q. & Qiao, P. 2007. Terminal Oligocene sedimentary environments and abrupt provenance change event in the northern South China Sea. *Geology in China*, **34**, 1022–1031 [in Chinese with English abstract].

Shao, L., Pang, X., Qiao, P., Chen, C., Li, Q. & Miao, W. 2008. Sedimentary filling of the Pearl River mouth basin and its response to the evolution of the Pearl River. *Acta Sedimentologica Sinica*, **26**, 179–185 [in Chinese with English abstract].

Shao, L., Pang, X., Qiao, P., Li, Q., Wei, G. & Wu, M. 2009. Late Oligocene tectonic event in the northern South China Sea and its implications. *Earth Science – Journal of China University of Geoscience*, **34**, 717–724 [in Chinese with English abstract].

Shao, L., Du, J., Li, Q., Pang, X. & Qiao, P. 2015. *Sedimentary Evolution and Sea Level Cycles of Marine Sequences from the Eastern Pearl River Mouth Basin*. CNOOC Internal Report, 2015–8, 1–194.

Stattegger, K. 1991. The Permian–Triassic of the Gartnerkofel-1 core (Carnic Alps, Austria): statistical analysis of the geochemical data. *In*: Holser, W. T. & Schoenlaub, H. P. (eds) *The Permian-Triassic Boundary in the Carnic Alps of Austria (Gartnerkopfel Region)*. Abhandlungen der Geologischen Bundesanstalt, Wien, **45**, 175–192.

Tang, S., Shao, L. & Zhao, Q. 2004. Characteristics of clay minerals in South China Sea since Oligocene and its significance. *Acta Sedimentologica Sinica*, **22**, 337–342 [in Chinese with English abstract].

Taylor, S. R. & McLennan, S. M. 1985. *The Continental Crust: Its Composition and Evolution. An Examination of the Geochemical Record Preserved in Sedimentary Rocks*. Blackwell Scientific Publications, Oxford.

Taylor, S. R. & McLennan, S. M. 1995. The geochemical evolution of the continental crust. *Reviews of Geophysics*, **33**, 241–265.

Wang, C., Zheng, R., Gao, B., Zhu, G., Hu, X. & Li, Y. 2010. Deepwater fan sedimentary characteristics of Zhujiang Formation in Liwan area of Zhujiang River mouth basin. *Geology in China*, **37**, 1628–1637 [in Chinese with English abstract].

WANG, J., ZHANG, X. ET AL. 2002. Integrated geophysical researches on base texture of Zhujiang River Mouth Basin. *Journal of Tropical Oceanography*, **21**, 13–22 [in Chinese with English abstract].

WANG, P., PRELL, W. L. & BLUM, P., 2000. *Proceedings of Ocean Drilling Program Initial Reports 184 [CD-ROM]*. Ocean Drilling Program, Texas A&M University, College Station TX77845-9547, USA.

WANG, P., JIAN, Z. ET AL. 2003. Evolution of the South China Sea and monsoon history revealed in deep-sea records. *Chinese Science Bulletin*, **48**, 2549–2561.

WANG, Y., XU, Q. ET AL. 2011. The Red River submarine fan of Late Miocene in northwestern South China Sea. *Chinese Science Bulletin*, **56**, 1488–1494.

WEI, G., LI, X., LIU, Y., SHAO, L. & LIANG, X. 2006. Geochemical record of chemical weathering and monsoon climate change since the early Miocene in the South China Sea. *Paleoceanography*, **21**, http://doi.org/10.1029/2006PA001300

WEI, G., LIU, Y., MA, J., XIE, L., CHEN, J., DENG, W. & TANG, S. 2012. Nd, Sr isotopes and elemental geochemistry of surface sediments from the South China Sea: implications for Provenance Tracing. *Marine Geology*, **319–322**, 21–34.

WORONOW, A. 1997. Regression and discriminant analysis using raw compositional data–is it really a problem? *In*: PAWLOWSKY-GLAHN, V. (ed.) *IAMG'97 Proceedings of the Third Annual Conference of the International Association for Mathematical Geology*. CIMNE, Barcelona, Part I, 157–162.

WU, M., SHAO, L., PANG, X., QIAO, P., XIANG, X. & ZHAO, M. 2012. REE geochemical characteristics of sediments and their implications in the deepwater area of the northern South China Sea. *Acta Sedimentologica Sinica*, **30**, 672–678 [in Chinese with English abstract].

WU, W. & ZHAO, H. 1982. On silt sources of Lingdingyang of the Zhujiang (Pearl River) estuary by means of mineralogical analysis of the sediments. *Journal of Tropical Oceanography*, **1**, 97–110 [in Chinese with English abstract].

XIANG, X., SHAO, L., QIAO, P. & ZHAO, M. 2011. Characteristics of heavy minerals in Pearl River sediments and their implications for provenance. *Marine Geology & Quaternary Geology*, **31**, 27–35 [in Chinese with English abstract].

YU, Y., ZHANG, C., ZHANG, S., SHI, H. & DU, J. 2012. Research on source direction of Neogene Zhujiang Formation in Huizhou Depression. *Fault-Block Oil & Gas Field*, **19**, 17–21 [in Chinese with English abstract].

ZHANG, X., CHEN, L., SHE, Q., ZHANG, S. & QIAO, P. 2012. Provenance evolution of the paleo-Hanjiang River in the north South China Sea. *Marine Geology & Quaternary Geology*, **32**, 41–48 [in Chinese with English abstract].

ZHU, W., ZHANG, G. ET AL. 2007. *Gas Geology in the Northern Continental Margin Basin of South China Sea*. Petroleum Industry Press, Beijing [in Chinese].

ZHU, W., MI, L. ET AL. 2010. *Atlas of Oil and Gas Basins, China Sea*. Petroleum Industry Press, Beijing [in Chinese].

Testing chemical weathering proxies in Miocene–Recent fluvial-derived sediments in the South China Sea

DENGKE HU[1]*, PETER D. CLIFT[2], SHIMING WAN[3], PHILIPP BÖNING[4], ROBYN HANNIGAN[5], STEPHEN HILLIER[6] & JERZY BLUSZTAJN[7]

[1]*Key Laboratory of Marginal Sea Geology, Chinese Academy of Sciences, Guangzhou 510301, China*

[2]*Department of Geology and Geophysics and Coastal Studies Institute, Louisiana State University, Baton Rouge, LA 70803, USA*

[3]*Key Laboratory of Marine Geology and Environment, Institute of Oceanology, Chinese Academy of Sciences, Qingdao 266071, China*

[4]*Institute of Chemistry and Biology of the Marine Environment (ICBM), Carl von Ossietzky University of Oldenburg, P.O. Box 2503, D-26111 Oldenburg, Germany*

[5]*Department of Environmental, Earth and Ocean Sciences, University of Massachusetts-Boston, Boston, MA 02125, USA*

[6]*The James Hutton Institute, Craigiebuckler, Aberdeen AB15 8QH, UK*

[7]*Department of Geology and Geophysics, Woods Hole Oceanographic Institution, Woods Hole, MA 02543, USA*

**Corresponding author (e-mail: dengke.hu@yahoo.com)*

Abstract: Reconstructing variations in the intensity of chemical weathering in river basins is crucial if we are to understand how climate change impacts environment and whether there are feedbacks between climate and weathering processes. Quantifying chemical weathering is, however, a complicated process, involving a number of competing proxies. We compare weathering records from the Pearl River delta of southern China and Ocean Drilling Program (ODP) Sites 1144 and 1146 on the northeastern slope of the South China Sea in order to test which proxies are the most widely applicable and robust. Comparison with speleothem rainfall records indicates that K/Al tracks precipitation variations most closely and out-performs the widely used Chemical Index of Alteration (CIA). Correlation of K/Al and kaolinite/illite indicates that this clay ratio is also an effective proxy of weathering intensity across all sites and timescales. Kaolinite/smectite, and to a lesser extent smectite/(illite + chlorite), are also indicative of weathering intensity, but show more scatter between sites that may be linked to provenance effects. Mg/Al is relatively immune to grain-size effects, but does not correlate well with other proxies. K/Rb is a reasonably reliable indicator of chemical weathering intensity and may be more sensitive than CIA or K/Al to weathering changes over short timescales and when weathering is not too intense. $^{87}Sr/^{86}Sr$ can be useful but can be influenced by both grain size and provenance effects. In general marine archives of fluvial sediment may record variations in weathering linked to climate, but these are increasingly signals of reworking going further offshore.

Chemical weathering is one of the most important processes that affects the surface of the Earth and one that has potentially large feedbacks on the evolution of global climate through the drawdown of CO_2 during alteration of silicate minerals (Raymo *et al.* 1988; Berner & Berner 1997). If we are to understand the impact that changing climate or tectonic activity has on continental environments it is essential that we are able to reconstruct the intensity of chemical weathering in continental drainage basins over various periods of geological time. The marine sediments preserved in East Asian marginal basins and delivered by the large rivers of the continent potentially provide a long-term record that would allow us to test what controls are most important in governing this process.

From: Clift, P. D., Harff, J., Wu, J. & Qui, Y. (eds) 2016. *River-Dominated Shelf Sediments of East Asian Seas*. Geological Society, London, Special Publications, **429**, 45–72.
First published online December 11, 2015, updated May 26, 2016, http://doi.org/10.1144/SP429.5

Chemical weathering and the East Asia monsoon

A number of attempts have been made to reconstruct the climate of Asia on orbital timescales using techniques such as ice cores from glaciers, biogenic productivity records built from peat, pollen, diatoms, planktonic and benthic foraminifera fossils, as well as variations in loess and clay mineralogy (Clift & Plumb 2008). Over millennial timescales, spanning the past few hundred thousand years, monsoonal rainfall records have been reconstructed in detail using oxygen isotopes in speleothem deposits that show comparable variations to lower resolution records and further provide details of changing climates over centennial to decadal timescales, although recently questions have been raised about the influence of winter temperatures in controlling these archives (Clemens *et al.* 2010). In this study we benefit from palaeoclimate records from the Dongge Cave in SW China (Dykoski *et al.* 2005), which lies within the Pearl River drainage. The oxygen isotopes of these speleothems are interpreted to record the intensity of rainfall at the time of deposition, with enrichment in lighter isotopes associated with heavier rainfall. This approach provides us with a generally accepted independent rainfall proxy to compare with our core data. Although changes in water sources may also affect the isotopic composition of the speleothems they are not thought to be important on the timescales considered, especially as the trend in monsoon intensity reconstructed is also mirrored by other proxies, such as upwelling (Gupta *et al.* 2003).

In this paper we hypothesize that stronger summer monsoon rains result in greater run-off increasing offshore sedimentation while at the same time warmer and wetter conditions inland result in faster chemical weathering of the bedrock and erosion of existing soils on flood plains. However, testing this model is only possible if we can measure the intensity of alteration, and presently there is no consensus about which proxies are best for achieving that goal. We choose to study two contrasting sedimentary systems supplying the South China Sea because they share the same monsoonal climate, but have different tectonic settings and weathering records that span a range of time durations. In order to reconstruct the history of physical erosion and chemical weathering in East Asia we here synthesize existing analyses of sediments taken from the Pearl River delta (which drains tectonically quiescent southern China (Hu *et al.* 2013)) and the NE South China Sea (which is dominated by erosion from tectonically active Taiwan (Hu *et al.* 2012)). We attempt to assess which of these proxies are most suitable for measuring chemical weathering intensity over different geological timescales, irrespective of the provenance or tectonic setting of the sedimentary record involved.

Chemical weathering proxies

Chemical weathering has been monitored by a number of chemical and mineral proxies in the past; however, many of the methods chosen have proven to be applicable either only locally, or only over certain durations of time. There is disagreement about which methods are the most reliable and whether any proxy is applicable universally. Geochemical proxies are based on the principle that silicate weathering affects the major-element chemistry of siliciclastic sediments (Nesbitt & Young 1982; McLennan 1993). Larger cations (e.g. Al, Ba, Rb) remain fixed in the weathering profile, whereas smaller cations (e.g. Ca, K, Na, Sr) are selectively leached during alteration (Nesbitt *et al.* 1980). Alkali metals such as K and Na are lost from rocks during chemical alteration, so that they become less abundant compared to immobile Al or Ti in the residual rocks. Thus ratios like K/Al and K/Rb have been used as tracers of chemical weathering, the assumption being that the soils left by chemical weathering are subsequently eroded and transported to the continental margin where they are preserved as sediments (Wan *et al.* 2010*a*; Hu *et al.* 2013) and that transport itself does not perturb the chemical signal.

A particularly popular proxy is the Chemical Index of Alteration (CIA) that employs both K and Na, as well as non-biogenic Ca, compared to immobile Al (Nesbitt & Young 1982). It should be recognized, however, that this proxy was developed for soils, not marine sediments, and does not account for changes that might occur during sediment transport. The CIA is calculated based on the molecular weights of the oxides, dividing Al_2O_3 by $Al_2O_3 + CaO + Na_2O + K_2O$. In marine deposits this proxy may vary with burial diagenesis because loss of $CaO + Na_2O + K_2O$ may not be restricted to the continental weathering phase, so that the proxy can be misinterpreted when applied just to the continental environment of erosion. Other studies have employed Mg, which is released from mafic minerals during their alteration and is dissolved in the run-off (Duzgoren-Aydin *et al.* 2002; Price & Velbel 2003; Ma *et al.* 2007).

While this relative leaching behaviour is observed in continental soil profiles, sediments in the deep sea may also have experienced hydraulic sorting after the weathered products were delivered to the sea, and it is generally recognized that finer grained sediments will be more altered than coarser grained ones of the same age (Clift *et al.* 2008; Lupker *et al.* 2013). As a result application of these proxies to the marine realm can be complicated and prone to misintepretation.

Isotope systems provide an alternative to major-element chemistry in measuring weathering intensity and are based on the concept that alteration processes fractionate different isotopes of water-mobile elements. Care needs to be taken because fractionation can also occur when different minerals are preferentially weathered (Garçon *et al.* 2014). In a study by Derry & France-Lanord (1996), $^{87}Sr/^{86}Sr$ was applied to the clay mineral fraction to track chemical weathering intensity in Lower Miocene–Recent sediments from the Bengal Fan, but these researchers were unable to resolve whether this weathering occurred in the continental environment or in the rivers during transport. In this case the $^{87}Sr/^{86}Sr$ value increased in the clastic sedimentary fraction as weathering intensified. This proxy, however, is susceptible to the influence of biogenic carbonate in the sediment, and can also change as a result of provenance variations. This proxy is only effective if provenance is known to be stable.

As well as geochemical proxies geologists have attempted to reconstruct chemical weathering using mineralogy, and especially ratios of clay minerals, because these are often generated by the breakdown of other mineral species during weathering processes (Thiry 2000). Some studies have employed the total proportion of clays in sediment to track the degree of alteration in drainage basins, but this proxy is unreliable because clay may be concentrated by hydrodynamic sorting, for example, by currents during transport, or by wave reworking on the shelf. Instead, ratios of different clay minerals are more reliable because these are not dependent on concentration, and certain clay minerals are derived by chemical alteration (e.g. smectite, kaolinite), while others are mainly formed by physical erosion (e.g. illite, chlorite). Ratios such as smectite/(illite + chlorite) have been employed to track the relative intensity of chemical weathering compared to physical erosion in continental margin sediments over a number of timescales (Clift *et al.* 2002; Wan *et al.* 2007; Bouquillon 2010; Colin *et al.* 2010). Application of these proxies can be problematic. Studies of clays around the Mekong delta show that these differ from east to west in the modern system, partly reflecting preferential settling of some clays close to the river mouth (Xue *et al.* 2014). Sediment from the east, South China Sea, side of the delta look like the Mekong River itself while those to the west do not, in having much less illite and chlorite and implying reduced fine particle inputs from the Mekong River towards the Gulf of Thailand.

Nonetheless, the kaolinite/chlorite ratio in marine sediments is often used as an indicator of chemical hydrolysis v. physical processes in continental weathering profiles (Chamley 1989). Kaolinite/chlorite and kaolinite/illite have been used as proxies of humidity, with high values being indicative of enhanced humidity which favours kaolinite formation (Thamban *et al.* 2002, 2005). Comparison of different alteration-derived clay minerals can also image the style of chemical weathering. The kaolinite/smectite ratio can be used as a weathering proxy that reflects variations between humid-warm/tropical (kaolinite-rich) and more dry-seasonal (smectite-rich) climatic conditions (Adatte *et al.* 2002; Alizai *et al.* 2012).

Problems with existing proxies

Theoretically, if all these geochemical and mineralogical proxies were valid and equally tracking chemical weathering, then we would expect them to vary in the same way over similar timescales, potentially in different basins with similar environmental conditions. However, in reality not all sedimentary parameters can be expected to respond in the same linear fashion to changes in weathering intensity. A 24 Ma sequence from ODP Site 1148 in the South China Sea shows that while K/Al and CIA have declined steadily since that time, smectite/illite and $^{87}Sr/^{86}Sr$ values peaked around 15 Ma and then declined, especially after 8 Ma (Clift *et al.* 2014). Different weathering proxies are affected differently by seasonality and temperatures as well as average humidity, resulting in a mixed signal.

Hu *et al.* (2013) have shown that since 9.5 ka K/Al values have been low in sediments from the delta of the Pearl River during periods of strong summer monsoon, as might be expected for a period of increased humidity and strong monsoon (West *et al.* 2005) and consistent with the long-term trend seen at ODP Site 1148 (Clift *et al.* 2014). Similarly, the warm, wet conditions of the Middle Miocene climatic optimum are associated with deposition at ODP Site 1146 of sediment with low values of K/Al and high values of CIA (Wan *et al.* 2009). As might be expected, kaolinite/illite and kaolinite/smectite values in the Pearl River delta also change in phase with the changing climate, with more kaolinite, a chemical weathering product during periods of stronger monsoon.

In contrast, on the nearby NE continental margin of the South China Sea, there is a poor correlation between K/Al and monsoon strength since 14.5 ka at ODP Site 1144, and no link at all with CIA, despite the fact that the drainage systems in Taiwan that feed that site share the same East Asian monsoon climate as the Pearl River basin (Hu *et al.* 2013). Instead, in this location K/Rb and kaolinite/(illite + chlorite) correlate much more closely with monsoon strength, as reconstructed from speleothem records in SW China (Dykoski *et al.* 2005).

Thus, the weathering records from a series of boreholes all positioned relatively close together in SE Asia seem to show contrasting trends in weathering proxies depending on the provenance, topography of the source drainage and timescales of the records. If progress is to be made in understanding what controls chemical weathering we need to test whether the proxies used to track weathering intensity are reliable, or how they are affected by grain size, provenance or other issues that compromise their ability to act in a robust fashion.

Proxies in the Pearl River delta

In order to see which weathering properties are the most reliable it is first necessary to decide which ones are best coupled to the climate. Sediments from the Pearl River delta were chosen to assess any correlation because the cores were taken directly in the river mouth and have a simple provenance (Hu *et al.* 2013) (Fig. 1), which contrasts with the higher degrees of reworking that buffer sediment flux to the continental slope, for example, at ODP Site 1144 (Hu *et al.* 2012). We compare weathering proxies to the regional monsoon rainfall record derived from the stalagmite $\delta^{18}O$ from the Dongge Cave (Dykoski *et al.* 2005). Notably during the last 2.5 kyr the sediments of the Pearl River delta have shown a dramatic shift towards stronger chemical weathering and that trend is interpreted as a reflection of human settlement in the Pearl River basin during that time (Hu *et al.* 2013). This trend, which has recently also been seen in the Red River catchment (Wan *et al.* 2015), has been caused by ploughing during the establishment of widespread agriculture that drives the mobilization of the deeply weathered soils of the area; the soils can then be washed away into the rivers at much faster rates than has typically been observed in a natural setting (Syvitski *et al.* 2005). As a result data younger than 2.5 ka were not considered in our analysis to avoid this process. K/Al and $\delta^{18}O$ records are both measured in high resolution and K/Al is shown to have a close correlation with Dongge Cave $\delta^{18}O$ (Fig. 2). Although ideally we would like to compare monsoon strength and weathering over several full glacial cycles the data to do this do not yet exist. The relationship between K/Al and rainfall intensity appears to be linear, but this may only be because of the limited climate variations seen in the Holocene. The weathering response to greater changes may well be different. Very high rainfall might result in rapid transport times because of large discharge and in turn cause less chemical weathering.

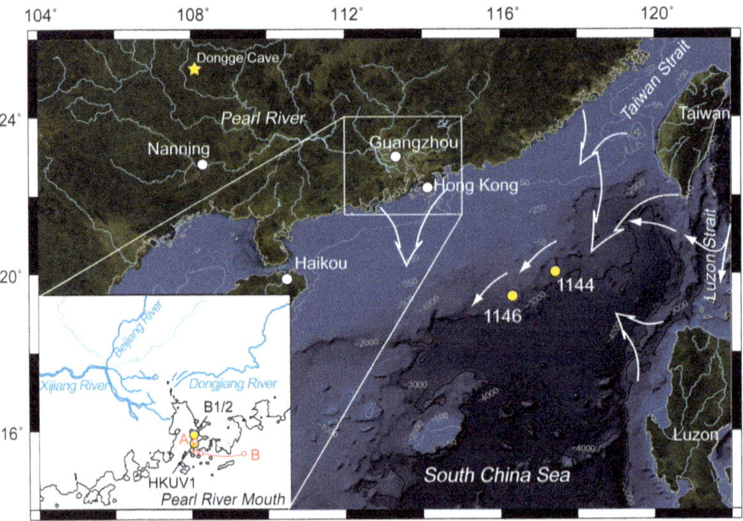

Fig. 1. Satellite image of the Pearl River delta. Map shows the study sites (ODP Sites 1144 and 1146 in the main map, and HKUV1, B2/1 in the inset map). Dongge Cave is the location of speleothem climate studies (Dykoski *et al.* 2005). ODP Sites 1144 and 1146 preserve sediments mainly from Taiwan (broad open arrow) in the last glacial cycle (Wan *et al.* 2010a), but with the potential of modulation by bottom currents (Shao *et al.* 2007) flowing through the eastern gateway of the South China Sea, the Luzon Strait, by the North Pacific Deep Water (NPDW) (Lüdmann *et al.* 2005) (solid line arrows). Isotope studies show that ODP Site 1144 is primarily supplied with sediment eroded from the Taiwan Strait when sea level is lower during glacial times and that the supply was truncated during higher sea levels in the early Holocene.

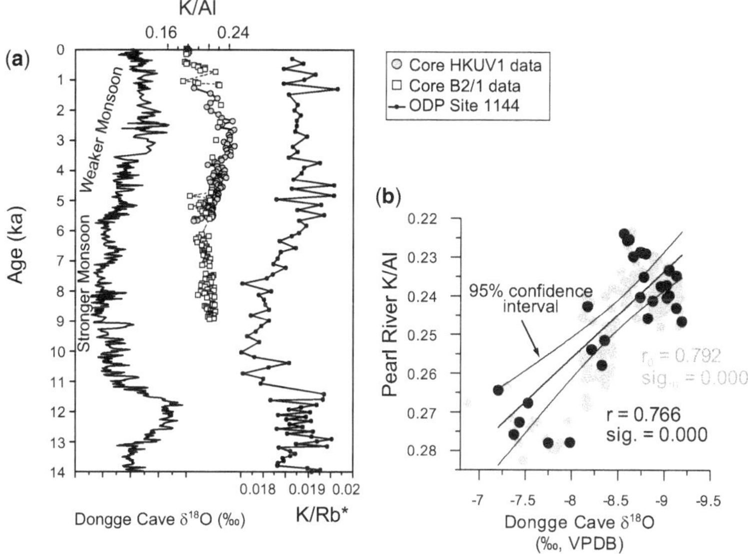

Fig. 2. (**a**) Variations in K/Al in Holocene Pearl River delta sediment and in K/Rb for ODP Site 1144 with respect to $\delta^{18}O$ from Dongge Cave, which is considered a rainfall intensity proxy in South China (Dykoski *et al.* 2005). (**b**) Cross-plot of K/Al against $\delta^{18}O$ showing the correlation. VPDB, Vienna Pee Dee Belemnite.

We further compare all other chemical weathering proxies in the Pearl River delta to the Dongge Cave $\delta^{18}O$ record (Fig. 3a–c, d–f). Because speleothem $\delta^{18}O$ shows rapid changes, comparison with a smoothed value is more appropriate given uncertainties in the core age model, especially as weathering itself tends to take time to react to changing environmental conditions. While flora might change on the scale of years to tens of years, soil chemistry and mineralogy may take hundreds or thousands of years to respond to changes in climatic conditions. Several of the proxies correlate well with the $\delta^{18}O$ values, and K/Al is the most closely correlated with rainfall intensity of any of the weathering proxies. We now investigate this proxy with other measures of chemical weathering in order to assess how well coupled the different weathering proxies are in different drainage basins and over different timescales.

Sampling strategy

Sedimentary records in passive continental margin settings may experience relatively continuous deposition and preservation of material delivered

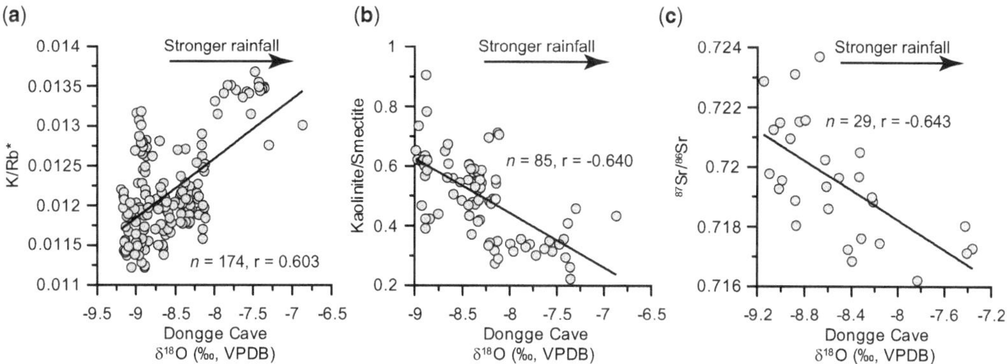

Fig. 3. (**a**) K/Rb*, (**b**) kaolinite/smectite, (**c**) $^{87}Sr/^{86}Sr$ records from Pearl River delta sediments plotted against stalagmite $\delta^{18}O$ records recovered from Dongge Cave (Dykoski *et al.* 2005). K/Rb* was calculate in wt%/($\mu g\ g^{-1}$).

by the rivers draining the adjacent continent. However, continental shelves are susceptible to erosion during periods of low sea level when they are exposed and eroded (Vail *et al.* 1977; Posamentier *et al.* 1988). As a result scientific ocean drilling has tended to concentrate sampling in the deeper water on continental slopes. Slope sequences potentially provide us with the possibility of reconstructing evolving continental weathering conditions over long periods of geological time, even though this runs the risk that the sedimentary record may be at least partly contaminated by reworking on the shelf (Hu *et al.* 2012; Limmer *et al.* 2012).

We examine three drill sites, one located in the delta of the Pearl River in southern China (a composite section composed of two neighbouring Cores HKUV1 and B2/1), and two on the passive margin of the NE South China Sea to the SW of Taiwan (ODP Sites 1144 and 1146). The analysed section from ODP Site 1144 covers only the last 14.5 kyr (Bühring *et al.* 2004), but that at ODP Site 1146 spans 4 myr (Shipboard Scientific Party 2000*b*) (Fig. 1). Both these latter sites have received the vast majority of their sediment from Taiwan (Wan *et al.* 2010*b*). ODP Site 1144 is located on a sediment drift on the continental slope, while ODP Site 1146 forms part of the clastic apron SW of Taiwan. Wan *et al.* (2010*b*) suggest that ODP Site 1146 has increasingly been the recipient of sediment from Taiwan since 3 Ma, following supply from the Pearl River and other adjacent systems draining SE China before that time. We employed data from this site extending back to 4 Ma in order to assess the behaviour of proxies over longer time periods (Fig. 4).

Preservation affects a sediment section's potential as a weathering archive. Delta sediments are susceptible to erosion when sea level falls and consequently are discontinuously preserved (Vail *et al.* 1977). Even after a rise in sea level the accumulation is not uniform. Reconstructions of the Pearl River delta show a steady progradation from north to south since the Early Holocene so that not all parts of the shelf preserve a complete record (Zong *et al.* 2009). Stratigraphy recovered from drilled boreholes and seismic data reveals a *c.* 10-m-thick Holocene sequence (Hang Hau Formation) across the river mouth towards the open shelf (Fig. 4), demonstrating that since the post-glacial sea-level rise there has existed a largely continuous sedimentary record. Radiocarbon dating proves that the core sites from the SW delta preserve a continuous sedimentary archive of discharge by the Pearl River drainage system since 9.5 ka in the Hang Hau Formation (Hu *et al.* 2013). The data we consider here are restricted to the Hang Hau Formation and to the weathered top of the underlying Sham Wat Formation.

Analytical methods

This study mostly represents analysis of geochemical data collected by earlier studies (Wan *et al.* 2010*a*; Hu *et al.* 2012, 2013). However, biogenic carbonate contents can exceed 15% in dry bulk

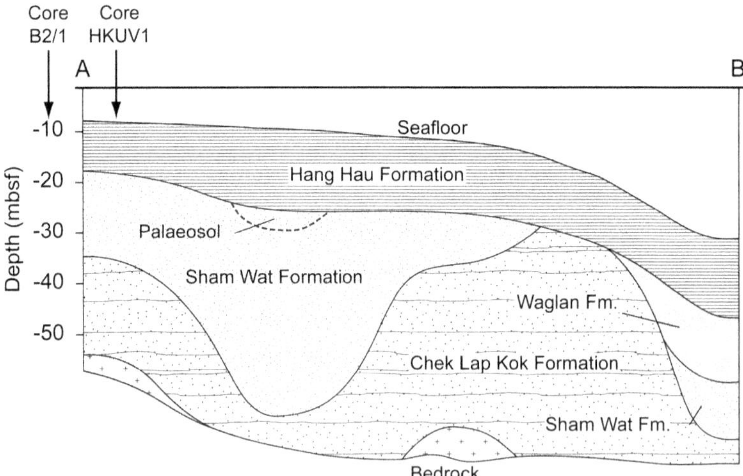

Fig. 4. A stratigraphic profile (position shown in the inset map of Fig. 1) recovered from boreholes drilled, and seismic data obtained inside and outside the Pearl River, by the Hong Kong Geologic Survey (Fyfe 2000) with positions of Cores HKUV1 and B2/1. Core HKUV1 reaches the bottom of Hang Hau Formation at *c.* 10.2 mbsf (metres below sea floor) with a fitted age of 5.97 ka while Core B2/1 goes deeper to 11.71 mbsf with a fitted age of 9.49 ka prior to the bottom of the formation.

Table 1. *Bulk sediment Sr isotopes from ODP Site 1144*

ODP Sample ID	Depth (mcd*)	Age (ka)	$^{87}Sr/^{86}Sr$	Error (2σ)
184-1144C-1H-1-W, 45–47 cm	0.470	0.81	0.711291	3.3E-06
184-1144A-1H-1-W, 49–51 cm	0.950	1.64	0.710990	3.9E-06
184-1144A-1H-1-W, 129–131 cm	1.740	3.02	0.711442	4.0E-06
184-1144A-1H-2-W, 59–61 cm	2.575	4.48	0.711219	4.2E-06
184-1144A-1H-2-W, 139–141 cm	3.340	5.81	0.711525	4.5E-06
184-1144A-1H-3-W, 69–71 cm	4.140	7.21	0.711406	4.4E-06
184-1144A-1H-3-W, 109–111 cm	4.585	7.99	0.711709	3.7E-06
184-1144A-1H-3-W, 119–121 cm	4.655	8.11	0.711625	3.8E-06
184-1144A-1H-3-W, 129–131 cm	4.745	8.27	0.712063	2.6E-06
184-1144A-1H-4-W, 19–21 cm	5.145	8.96	0.712517	4.2E-06
184-1144A-1H-4-W, 79–81 cm	5.750	10.02	0.713034	4.3E-06
184-1144A-1H-4-W, 99–101 cm	5.950	10.37	0.712605	4.6E-06
184-1144A-1H-4-W, 139–141 cm	6.340	11.05	0.711435	4.1E-06
184-1144A-1H-5-W, 9–11 cm	6.640	11.57	0.711272	4.4E-06
184-1144B-2H-3-W, 27–29 cm	6.760	11.75	0.711343	4.0E-06
184-1144B-2H-3-W, 87–89 cm	7.355	12.23	0.711257	3.4E-06
184-1144B-2H-4-W, 17–19 cm	8.160	12.89	0.711219	3.4E-06
184-1144B-2H-4-W, 97–99 cm	8.945	13.52	0.711330	3.4E-06

*mcd = metres common depth.

sediments at ODP Site 1144 (Shipboard Scientific Party 2000*a*) and has affected the $^{87}Sr/^{86}Sr$ values of the sediment reported by Hu *et al.* (2012). In this study sediments were treated with 10% acetic acid to remove carbonate and dissolved in a mixture of HF and $HClO_4$. In order to quantify this impact both bulk sediment and decarbonated fractions from ODP Site 1144 (both organic-free) were measured by Neptune multicollector inductively coupled plasma mass spectrometry (MC-ICP-MS) methods at the Woods Hole Oceanographic Institution (USA) and the Open University (UK). Sample measurements for Sr were normalized to $^{86}Sr/^{88}Sr = 0.1194$ and referenced to a value of 0.710250 for NBS987. The bulk analyses are newly presented in this study and are shown in Table 1.

Samples of sediment from the Sham Wat Formation underlying the Holocene in the Pearl River delta at Core HKUV-1 (>30 ka) were also analysed for major and trace elements to assess the effect of weathering during exposure during sea-level lowstands. Major and trace element analyses were made by conventional X-ray fluorescence (XRF) using a SPECTRO XEPOS Benchtop Energy Dispersive XRF (ED-XRF) at the University of Massachusetts, Boston. Samples were powdered in an agate mill before oxide analysis. Powdered samples (3–5 g dry weight) were measured under a He purge. We used US Geological Survey (USGS) Standard Devonian Shale (SDO-1) as the calibration standard. USGS Green River Shale (SGR-1) was analysed as an unknown to monitor accuracy of the major-element analyses. Measured values for SGR-1 were better than 95% of the known values. Reported errors represent the propagated error of repeat/replicate measures and the measured certified (MC) ratio for each element, with uncertainties of <5%. Results are provided in Table 2.

Clay mineralogy of the >30 ka sediment from the Pearl River was determined using a Siemens X-ray diffractometer (XRD) at the James Hutton Institute, Aberdeen (UK), with Co Kα radiation selected by a diffracted beam monochromator. The clay-sized fraction (<2 μm) was prepared for semi-quantitative analysis by separating it from the rest of sample by application of Stokes's law through settling. The filter transfer method was used to orient clay minerals on a glass slide prior to XRD analysis following the method of Moore & Reynolds (1989). A set of three diffraction patterns (i.e. air-dried, glycolated and heated to 300°C for one hour) were produced for each sample and used for mineral quantification. Estimates of clay mineral abundances were made by fitting profiles and calculating basal areas of the main clay minerals using MDI Jade 6 (Materials Data Inc.) software on glycolated slides. The fitting approach involves all relevant peaks of clay minerals simultaneously for each attempt and is made continuously with a unique configuration and procedure to reduce systematic error. For clay minerals present in amounts >10 wt% uncertainty is estimated to be better than ±5 wt% at the 95% confidence level. Uncertainty of peak area measurement based on repeated measurements is typically <5%, with the smallest peaks having the highest uncertainties (Hillier 2003). Multiple attempts for each slide were made to constrain the randomness of fitting. Results are shown in Table 2.

Table 2. *Major and trace element analyses of sediments from the Sham Wat Formation in Core HKUV-1, together with the associated clay mineral assemblages, normalized to 100%*

Sample ID (Core-section-W/A, depth below section top)	Depth (mbsf*)	Fitted age (Cal. kyr)	MgO	Al_2O_3	SiO_2	P_2O_5	K_2O	CaO	TiO_2	MnO	Fe_2O_3	V	Cr	Co	Ni	Cu
HKUV1-(9.5–10.5 m)-W, 80 cm	10.3	>30	2.28	18.58	52.13	0.11	2.58	1.53	0.90	0.08	6.93	71.0	68.9	19.2	37.6	33.7
HKUV1-(9.5–10.5 m)-W, 90 cm	10.4	>30	2.03	18.25	51.48	0.11	2.52	1.72	0.88	0.08	6.88	69.7	92.0	16.6	37.3	30.0
HKUV1-(10.5-11.5 m)-W, 0 cm	10.5	>30	2.08	16.92	47.89	0.11	2.36	1.46	0.82	0.07	6.49	66.6	87.8	25.1	33.8	29.3
HKUV1-(10.5–11.5 m)-W, 10 cm	10.6	>30	2.52	18.23	50.43	0.10	2.46	1.26	0.88	0.08	6.47	76.6	98.9	21.6	37.2	28.9
HKUV1-(10.5–11.5 m)-W, 20 cm	10.7	>30	2.58	17.98	50.97	0.11	2.46	1.90	0.89	0.09	6.50	83.1	94.8	20.6	38.5	29.3
HKUV1-(10.5–11.5 m)-W, 30 cm	10.8	>30	2.74	17.72	50.57	0.11	2.47	2.02	0.89	0.09	6.72	66.6	90.6	15.2	36.8	26.3
HKUV1-(10.5–11.5 m)-W, 40 cm	10.9	>30	2.05	17.61	49.39	0.11	2.43	1.52	0.87	0.08	6.76	77.1	93.4	20.4	38.2	29.6
HKUV1-(10.5–11.5 m)-W, 50 cm	11	>30	2.14	17.69	49.99	0.11	2.50	1.43	0.90	0.08	7.01	76.3	104.5	18.9	38.9	30.2
HKUV1-(10.5–11.5 m)-W, 60 cm	11.1	>30	2.10	17.76	51.05	0.11	2.55	1.60	0.91	0.07	6.88	83.9	80.9	16.6	38.2	29.8
HKUV1-(10.5–11.5 m)-W, 70 cm	11.2	>30	2.33	17.28	50.23	0.11	2.43	1.67	0.87	0.08	6.55	62.9	84.1	15.8	38.8	28.8
HKUV1-(10.5–11.5 m)-W, 80 cm	11.3	>30	2.47	18.12	51.77	0.11	2.55	1.28	0.91	0.07	6.50	69.2	76.7	18.0	38.6	31.4
HKUV1-(10.5–11.5 m)-W, 90 cm	11.4	>30	2.28	18.05	50.87	0.11	2.55	1.41	0.90	0.07	7.20	83.7	87.8	25.4	37.8	31.5
HKUV1-(11.5–12.x m)-W, 0 cm	11.5	>30	1.86	17.51	50.16	0.11	2.49	1.43	0.90	0.08	6.86	67.9	79.0	19.9	38.2	32.0
HKUV1-(11.5–12.x m)-W, 10 cm	11.6	>30	2.56	17.95	50.60	0.11	2.52	1.31	0.92	0.08	6.90	81.0	100.3	20.2	38.7	30.1
HKUV1-(11.5–12.x m)-W, 20 cm	11.7	>30	2.06	17.81	50.65	0.11	2.49	1.39	0.91	0.08	6.79	69.7	105.4	20.4	36.5	30.4
HKUV1-(11.5–12.x m)-W, 30 cm	11.8	>30	2.20	17.50	50.33	0.11	2.45	1.47	0.87	0.09	6.93	71.0	112.3	19.1	37.0	29.0
HKUV1-(11.5–12.x m)-W, 40 cm	11.9	>30	2.16	17.14	48.90	0.11	2.44	2.63	0.88	0.08	6.70	74.7	87.4	19.8	35.3	28.7
HKUV1-(11.5–12.x m)-W, 50 cm	12	>30	1.97	17.35	49.94	0.11	2.47	1.83	0.88	0.07	6.71	74.5	87.4	19.4	39.4	29.0
HKUV1-(11.5–12.x m)-W, 60 cm	12.1	>30	2.56	19.21	53.43	0.11	2.72	1.51	0.96	0.10	6.89	80.2	70.3	18.3	43.3	35.4

Sample ID (Core-section-W/A, depth below section top)	Zn	Ga	As	Rb	Sr	Y	Ba	Pb	Th	U	Smectite (%)	Illite (%)	Kaolinite (%)	Chlorite (%)
HKUV1-(9.5–10.5 m)-W, 80 cm	114.4	25.0	18.7	178.7	117.4	40.0	358	38.3	20.3	15.6	16	45	31	8
HKUV1-(9.5–10.5 m)-W, 90 cm	117.5	24.6	15.1	177.4	122.6	40.1	452	42.8	19.8	17.2	19	42	31	8
HKUV1-(10.5–11.5 m)-W, 0 cm	112.0	25.0	13.0	180.7	112.7	39.0	404	38.0	20.0	15.0	16	43	39	2
HKUV1-(10.5–11.5 m)-W, 10 cm	116.0	26.1	13.4	182.6	96.9	40.2	457	38.4	20.3	16.1	19	39	41	1
HKUV1-(10.5–11.5 m)-W, 20 cm	115.9	22.8	15.1	175.3	137.9	39.5	367	38.0	19.7	16.7	21	38	39	2
HKUV1-(10.5–11.5 m)-W, 30 cm	114.8	23.9	15.7	171.4	134.9	38.5	404	41.7	20.5	15.4	17	46	29	8
HKUV1-(10.5–11.5 m)-W, 40 cm	116.4	25.1	17.2	179.1	109.7	39.2	371	38.2	20.5	17.9	19	45	34	3
HKUV1-(10.5–11.5 m)-W, 50 cm	113.8	26.8	13.9	175.8	109.2	40.7	341	40.7	19.6	18.0	34	39	22	5
HKUV1-(10.5–11.5 m)-W, 60 cm	113.4	24.8	14.5	175.2	124.0	39.8	336	39.9	20.4	18.1	22	44	29	5
HKUV1-(10.5–11.5 m)-W, 70 cm	106.9	24.0	12.7	171.2	111.5	40.1	363	38.0	19.0	19.1	16	49	27	7
HKUV1-(10.5–11.5 m)-W, 80 cm	117.9	26.5	15.5	179.5	98.2	41.3	377	39.1	20.0	18.4	19	45	26	10
HKUV1-(10.5–11.5 m)-W, 90 cm	118.0	26.1	16.7	175.6	118.6	39.6	412	39.8	19.8	18.7	19	44	34	3
HKUV1-(11.5–12.x m)-W, 0 cm	118.7	24.2	15.5	176.5	100.9	40.7	374	39.5	20.9	17.5	18	45	26	12
HKÙV1-(11.5–12.x m)-W, 10 cm	118.8	26.0	14.1	175.6	102.1	41.4	380	41.1	20.8	18.9	15	46	36	2
HKUV1-(11.5–12.x m)-W, 20 cm	115.4	25.2	11.2	175.2	101.8	40.1	384	40.5	19.3	18.1	19	45	16	19
HKUV1-(11.5–12.x m)-W, 30 cm	112.8	24.6	15.8	176.3	106.1	40.0	294	37.9	19.4	17.5	17	48	34	2
HKUV1-(11.5–12.x m)-W, 40 cm	112.3	25.0	14.4	172.7	143.3	39.7	359	38.5	20.1	16.5	14	48	22	16
HKUV1-(11.5–12.x m)-W, 50 cm	109.2	24.5	13.8	174.1	115.9	40.3	313	37.7	19.3	19.6	15	49	26	10
HKUV1-(11.5–12.x m)-W, 60 cm	138.1	27.4	15.7	176.4	102.8	38.8	335	46.4	20.1	17.3	15	49	28	8

*mbsf = metres below sea floor

Results

Two different sedimentary systems: Pearl River and Taiwan

Sediments deposited on the slope and in the delta can be compared using both major elements and isotopic methods. These highlight the contrasting sources of the sediment. Slope sediments from ODP Sites 1144 and 1146 sit in a group on a triangular plot of K, Ti and Al where they have consistently higher K content compared to another array of values from the Pearl River boreholes (Fig. 5a). The diagram shows that there is no overlap in terms of major composition, suggesting that the two groups represent separate sedimentary systems. The sediments from ODP Site 1144 form a more restricted range than those at ODP Site 1146, but this is not surprising given that their sedimentation spans a much shorter time interval. Sr and Nd isotopes can be used together to further investigate the relationship between the two areas and potential sources, bearing in mind that Sr is water-mobile and potentially affected by weathering as well as provenance, while Nd isotopes are generally considered to reflect only provenance (Goldstein & Jacobsen 1988). Although it has recently been noted that finer grained suspended sediments preferentially sample basaltic sources compared with coarser grained bed-load transported sediments (Garçon & Chauvel 2014), this is not considered critical in this study because none of the source areas is rich in basalt and grain-size variations are modest.

As anticipated the silicate fraction at ODP Site 1144 has higher $^{87}Sr/^{86}Sr$ values than the bulk sediment, reflecting the presence of some biogenic carbonate in the bulk samples (Fig. 5b). The decarbonated samples plot in almost the same part of the figure as the samples from the Pearl River delta. When compared to other sediment isotope analyses from the region, both datasets plot closest to sediments from rivers draining Taiwan, but with slightly higher ε_{Nd} values. We note that the Holocene Pearl River delta sediments plot at lower $^{87}Sr/^{86}Sr$ values than the modern Pearl River, albeit with the same ε_{Nd} values, a discrepancy that Hu *et al.* (2013) interpreted to reflect stronger alteration of the modern river load compared to the recent geologic past. None of the sediments show significant displacement towards the high ε_{Nd} values associated with the Luzon Arc.

Further comparison of the sediment provenance is possible by examining clay mineralogy (Fig. 5c). We employ a standard triangular diagram to compare the cored sediment mineralogy with modern river systems. As with the isotopes, the Pearl River Holocene sediments of the Hang Hau Formation do not show overlap with the modern Pearl River, as measured by Liu *et al.* (2007), although delta sediments older than 30 ka (top of the Sham Wat Formation), as well as those deposited since 2.5 ka, do show a closer affinity to the smectite-poor composition of the modern river. There is a progressive increase in kaolinite (i.e. indicating more weathering) in the pre-2.5 ka Holocene and the >30 ka sequence, which was deposited during the last sea-level highstand and then exposed and weathered during the Last Glacial Maximum (Fig. 5c). The diagram again shows a clear separation between the sediments from the ODP core sites and both the Pearl River delta and the modern Pearl River. ODP Sites 1144 and 1146 (<4 Ma) show very similar clay assemblages and some overlap with compositions from modern Taiwanese rivers, albeit with slightly more smectite, probably caused by moderate alteration since deposition. Burial depths at ODP Site 1144 are <10 m, but reach 270 m at ODP Site 1146 (Shipboard Scientific Party 2000*b*), although burial itself is not the critical issue here but rather it is the time since erosion, which provides the chance for alteration to occur in a fluid-rich environment (Jeandel & Oelkers 2015).

None of the sediments younger than 4 Ma lie close to compositions detected on Luzon. However, sediment at ODP Site 1146 older than 4 Ma varies in a much broader array between modern Taiwan and Luzon rivers compared to the younger sediment, showing a constantly higher level of kaolinite and wider range of smectite, indicative of more intense alteration compared to the post-4 Ma material.

There is a clear separation of sediment between the ODP core sites and the Pearl River delta. The slope sites show closest comparison with modern Taiwanese rivers and we concur with earlier studies of the cores that this island is the dominant source of sediment to the NE part of the South China Sea (Shao *et al.* 2001; Liu *et al.* 2010; Wan *et al.* 2010*b*), with little influence from either Luzon or from rivers draining southern China. The sediments of the Pearl River delta must have been sourced from the Pearl River itself, yet the sediments show significant differences between the Holocene delta and the modern river, reflecting anthropogenic disturbance of the Pearl River basin (Hu *et al.* 2013).

The closer geochemical correspondence between the >30 ka Pearl River sediments and the modern river system compared to the Holocene is especially instructive, because the >30 ka sediments have been exposed and weathered during the Last Glacial sea-level lowstand, while the pre-2.5 ka Holocene represents fresher material that has experienced only one cycle of erosion and weathering. The modern river is now carrying material whose alteration state is more typical of older deposits, weathered during at least one sea-level cycle, rather than what is seen through much of the Holocene. This comparison

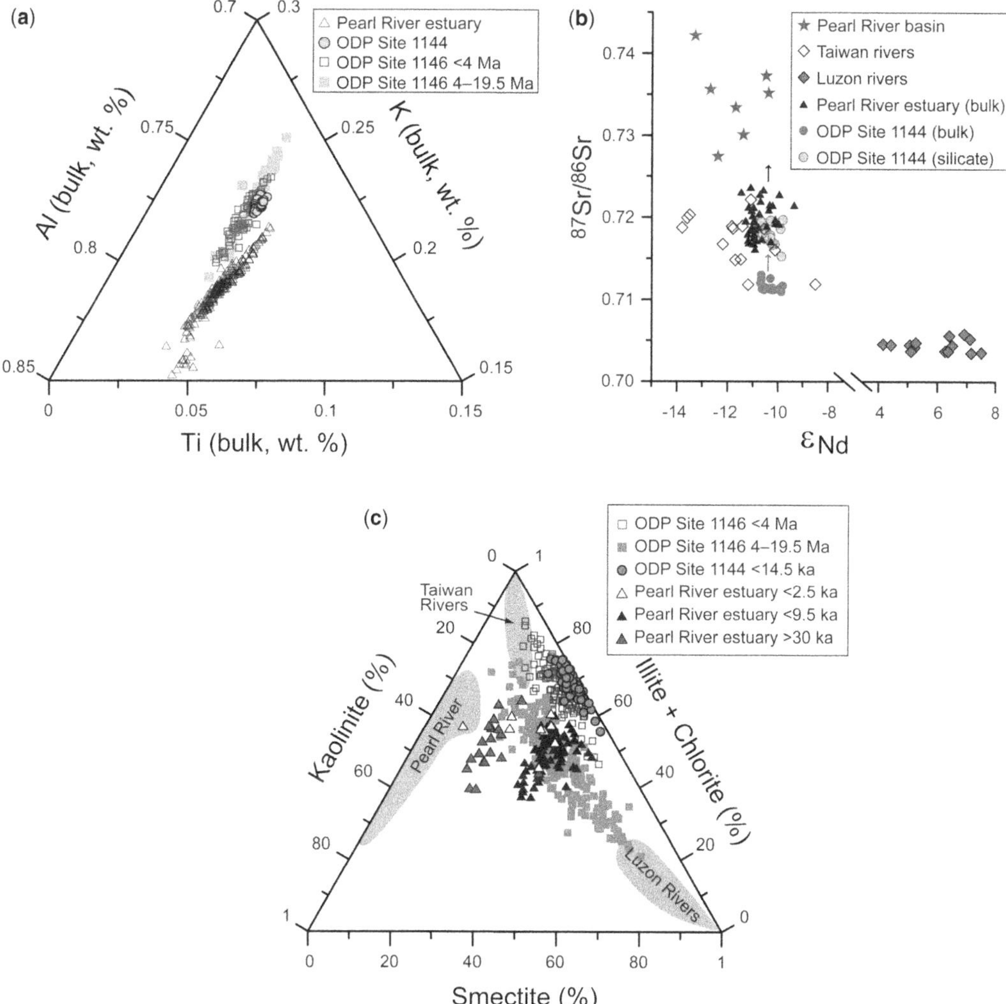

Fig. 5. (**a**) Ternary diagram shows different component variations of major elements (Ti, K, Al) from three core sites. Data from Hu *et al.* (2012, 2013). ODP Site 1144 since 14.5 ka has relatively constant geochemical content compared to ODP Site 1146 and the Pearl River delta, which show broader arrays spanning 19.5 Ma and 9.5 ka respectively. (**b**) ε_{Nd} against $^{87}Sr/^{86}Sr$ plots of ODP Site 1144 and Pearl River delta sediment compared to their potential source rivers. Data are from Hu *et al.* (2012, 2013), with bulk Sr data from ODP Site 1144 presented in this study. Modern Pearl River data are from Liu *et al.* (2007); data of Taiwan rivers are from Chen & Lee (1990) and Lan *et al.* (2002); data from Luzon rivers are from Goldstein & Jacobsen (1987, 1988) and GEOROC (http://georoc.mpch-mainz.gwdg.de). Decarbonated samples of ODP Site 1144 show a shift of $^{87}Sr/^{86}Sr$ values compared to the bulk sediment (shown as a grey arrow). (**c**) Ternary diagram showing clay mineral components of studied core sites and potential source rivers. Data from ODP Site 1144 and Site HKUV1, B2/1 (i.e. from the Pearl River) are from Hu *et al.* (2012, 2013). Data of ODP Site 1146 are from Wan *et al.* (2010*a*). Grey areas represent components of modern rivers studied by Liu *et al.* (2010). Site HKUV1 and B2/1 are further categorized by sedimentary ages according to the delta stratigraphy.

further highlights the major problems in using shelf sediments as proxies for reconstructing weathering histories because their repeated exposure during lowstands results in more alteration that is not related to the environmental conditions at the time of sedimentation. We presume that the fresher 2.5–9.5 ka sediments from the delta will alter to a state similar to that seen in the >30 ka material after the next glacial period, following prolonged exposure and in spite of the fact that weathering rates are slower during what would be a drier, cooler period. Post-depositional alteration is a more serious issue for shelf sediments compared to those on the slope, because shelf sediments are exposed during sea-level lowstands when they may be intensively weathered, although slope sediments are not totally immune to post-depositional alteration during early diagenesis (Jeandel *et al.* 2011; Jeandel & Oelkers 2015).

Competing weathering proxies

Having determined the source of the sediments and highlighted the primary compositional differences between them we now look at how a number of commonly used chemical weathering proxies compare both within individual sites and more generally across the study region. We examine K/Al, K/Rb* (K/Rb calculated from % and ppm respectively), $^{87}Sr/^{86}Sr$ and Mg/Al as geochemical ratios that reduce as weathering proceeds, together with CIA that rises as alteration increases. As discussed above we consider K/Al to be potentially the most reliable because of its closer coupling to climatic records, at least since 9.5 ka. Thus, proxies that correlate with K/Al may also be controlled by climatically modulated processes. We further investigate ratios of clay minerals as proxies for the relative strength of chemical weathering compared to physical erosion. As well as testing how well these proxies relate to one another we investigate how each might be affected by grain size because this has been linked to the degree of weathering in sediments within the marginal seas close to China (Zhao & Yan 1993).

We calculate Pearson's correlation coefficients to score the correlations between proxies while using the significance of linear fitting to evaluate the reliability of the coefficients. Results are provided in Tables 3 and 4 for the Pearl River sediment spanning time since 9.5 ka and ODP Site 1144 since 14.5 ka respectively. As for ODP Site 1146, we calculate results for sediment <4 Ma and 4–19.5 Ma separately based on the previous provenance study at this site (Wan *et al.* 2010*b*) and the differences between various properties of the two sediment age groups. Results from ODP Site 1146 prior to and after 4 Ma are provided in Tables 5 and 6, respectively. We also show results of this test graphically to see how these proxies co-vary, at each core site and between different sites.

K/Al and clay mineral proxies. Combined data from all three sites show correlations between clay mineral proxies and the K/Al ratios, although the relationship is less well defined using kaolinite/smectite or smectite/(illite + chlorite) and is best with kaolinite/illite (Fig. 6a–c). Samples from ODP Site 1144 (<14.5 ka) show no correlation, but this mostly reflects the limited variability there. The delta sediment and post-4 Ma ODP Site 1146 show better correlations between K/Al and kaolinite/illite. Pearl River delta sediment has lower values of K/Al with greater values of kaolinite/illite, kaolinite/smectite and smectite/(illite + chlorite) compared to sediments from ODP Sites 1144 and 1146. This implies that ODP Sites 1144 and 1146 sediments are less weathered than material from the Pearl River delta, which is unsurprising given the rapid erosion of their Taiwanese sources. Climate is secondary to tectonic setting in controlling degrees of sediment weathering.

Plots of individual sites from the Pearl River delta (Fig. 6d–f) and ODP Site 1146 (Fig. 6g–i) show correlation between the kaolinite/illite ratio and K/Al. Only weak correlations are seen between K/Al and kaolinite/smectite in the Pearl River delta (Fig. 6b). Weak correlation is also seen between n K/Al and smectite/illite + chlorite in sediment <4 Ma from Site 1146 (Fig. 6c). ODP Site 1146 sediments dated at 4–19.5 Ma, however, show generally better correlation between K/Al and kaolinite/illite and smectite/(illite + chlorite). It is noteworthy that 4–19.5 Ma sediments from ODP Site 1146 have higher smectite content, so that when we plot smectite/(illite + chlorite) against K/Al, these sediments span a much wider range compared to the younger sediments from all sites (Fig. 6c). Kaolinite/smectite has no correlation with K/Al either pre-dating, or since, 4 Ma at ODP Site 1146.

K/Al and K/Rb. K/Rb has been successfully applied as a monitor of chemical weathering at ODP Site 1144, where K/Al shows little coherent variation at least since 14 ka (Hu *et al.* 2012). Figures 7a–c plot this proxy against K/Al, while Figures 7d–f compare it against kaolinite/illite, with reasonably good correlation observed between K/Rb and K/Al when all three datasets are plotted together. However, there are poor correlations between K/Rb and K/Al within the Pearl River delta and at ODP Site 1146 when each site is considered separately (Figs 7b, c). There is a modest correlation between kaolinite/illite and K/Rb in data from all three sites spanning the last 4 myr but samples dating 4–19.5 Ma from ODP Site 1146 do not follow this relationship (Fig. 7d). At each site there is no clear relationship between kaolinite/illite and K/Rb (Fig. 7e, f).

Sr isotopes. $^{87}Sr/^{86}Sr$ measurements are only available over millennial timescales in the Pearl River delta and at ODP Site 1144. There are no Sr isotope data from ODP Site 1146 spanning longer timescales. Because $^{87}Sr/^{86}Sr$ values increase in the presence of carbonate we only compare carbonate-free samples with other weathering proxies

Table 3. *Pearson's correlation coefficient (Corr.) and significance of linear fitting (Sig.) calculated from Pearl River delta sediment <9.5 ka*

	K/Al	K/Rb*	Clay mineralogy			$^{87}Sr/^{86}Sr$	Mg/Al	Grain size		
			Kao/Ill	Kao/Sm	Sm/Ill + Ch			Mean	Si/Al	
K/Al	1	**0.482**	− **0.459**	− **0.446**	0.266	− **0.713**	**0.394**	− 0.122	**0.914**	Corr.
	–	0.000	0.000	0.000	0.010	0.000	0.000	0.087	0.000	Sig.
K/Rb*	**0.482**	1	− 0.099	**0.359**	− 0.037	− **0.335**	0.278	0.085	**0.353**	Corr.
	0.000	–	0.349	0.000	0.725	0.057	0.000	0.236	0.000	Sig.
Kaolinite/illite	− **0.459**	− 0.099	1			0.302	− **0.383**	− 0.006	− **0.540**	Corr.
	0.000	0.349	–			0.223	0.000	0.958	0.000	Sig.
Kaolinite/smectite	− **0.446**	**0.359**		1		0.393	−0.271	−0.038	−**0.513**	Corr.
	0.000	0.000		–		0.106	0.009	0.724	0.000	Sig.
Smectite/illite + chlorite	0.266	−0.037			1	−0.374	0.093	−0.035	0.290	Corr.
	0.010	0.725			–	0.126	0.380	0.751	0.005	Sig.
$^{87}Sr/^{86}Sr$	−**0.713**	−**0.335**	0.302	0.393	−0.374	1	−**0.589**	0.339	−**0.728**	Corr.
	0.000	0.057	0.223	0.106	0.126	–	0.000	0.066	0.000	Sig.
Mg/Al	**0.394**	0.278	−**0.383**	−0.271	0.093	−**0.589**	1	−0.207	**0.528**	Corr.
	0.000	0.000	0.000	0.009	0.380	0.000	–	0.004	0.000	Sig.
Mean	−0.122	0.085	−0.006	−0.038	−0.035	0.339	−0.207	1	−0.123	Corr.
	0.087	0.236	0.958	0.724	0.751	0.066	0.004	–	0.085	Sig.
Si/Al	**0.914**	**0.353**	−**0.540**	−**0.513**	0.290	−**0.728**	**0.528**	−0.123	1	Corr.
	0.000	0.000	0.000	0.000	0.005	0.000	0.000	0.085	–	Sig.
Number of samples (*n*)	203	202	92	92	92	33	203	197	203	

Acceptable correlations are in bold throughout.

Table 4. *Pearson's correlation coefficient (Corr.) and significance of linear fitting (Sig.) calculated from ODP Site 1144 <14.5 ka*

	K/Al	K/Rb*	Clay mineralogy			$^{87}Sr/^{86}Sr$	Mg/Al	CIA	Grain size		
			Kao/Ill	Kao/Sm	Sm/Ill + Ch				Mean	Si/Al	
K/Al	1	−0.101	0.212	0.176	0.035	0.322	**0.450**	−0.184	0.231	0.231	Corr.
	–	0.327	0.037	0.085	0.736	0.242	0.000	0.152	0.026	0.023	Sig.
K/Rb*	−0.101	1	−0.197	−0.170	0.212	−**0.805**	**0.444**	−0.127	**0.322**	**0.563**	Corr.
	0.327	–	0.053	0.095	0.037	0.000	0.000	0.326	0.002	0.000	Sig.
Kaolinite/illite	0.212	−0.197	1			**0.431**	0.044	0.064	0.064	−0.157	Corr.
	0.037	0.053	–			0.084	0.666	0.621	0.542	0.125	Sig.
Kaolinite/smectite	0.176	−0.170		1		**0.438**	−0.024	0.082	0.108	−0.170	Corr.
	0.085	0.095		–		0.079	0.815	0.527	0.303	0.095	Sig.
Smectite/illite + chlorite	0.035	0.212			1	−0.203	0.291	−0.060	−0.140	0.105	Corr.
	0.736	0.037			–	0.435	0.004	0.642	0.181	0.307	Sig.
$^{87}Sr/^{86}Sr$	0.322	−**0.805**	**0.431**	**0.438**	−0.203	1	−**0.468**	0.209	−0.178	−**0.694**	Corr.
	0.242	0.000	0.084	0.079	0.435	–	0.079	0.454	0.526	0.004	Sig.
Mg/Al	**0.450**	**0.444**	0.044	−0.024	0.291	−**0.468**	1	−0.263	**0.417**	**0.614**	Corr.
	0.000	0.000	0.666	0.815	0.004	0.079	–	0.039	0.000	0.000	Sig.
CIA	−0.184	−0.127	0.064	0.082	−0.060	0.209	−0.263	1	−0.244	−0.133	Corr.
	0.152	0.326	0.621	0.527	0.642	0.454	0.039	–	0.062	0.303	Sig.
Mean	0.231	**0.322**	0.064	0.108	−0.140	−0.178	**0.417**	−0.244	1	**0.547**	Corr.
	0.026	0.002	0.542	0.303	0.181	0.526	0.000	0.062	–	0.000	Sig.
Si/Al	0.231	**0.563**	−0.157	−0.170	0.105	−**0.694**	**0.614**	−0.133	**0.547**	1	Corr.
	0.023	0.000	0.125	0.095	0.307	0.004	0.000	0.303	0.000	–	Sig.
Number of samples (n)	97	97	100	100	100	17	97	62	93	97	

Acceptable correlations are in bold throughout.

Table 5. *Pearson's correlation coefficient (Corr.) and significance of linear fitting (Sig.) calculated from ODP Site 1146 <4 Ma*

	K/Al	K/Rb*	Clay mineralogy			Mg/Al	CIA	Grain size		
			Kao/Ill	Kao/Sm	Sm/Ill + Ch			Mean	Si/Al	
K/Al	1	0.171	**−0.491**	0.108	**−0.510**	0.217	**−0.655**	0.019	0.298	Corr.
	–	0.097	0.000	0.298	0.000	0.033	0.000	0.861	0.044	Sig.
K/Rb*	0.171	1	0.067	−0.016	0.096	0.168	**−0.339**	−0.072	−0.127	Corr.
	0.097	–	0.519	0.878	0.355	0.101	0.001	0.501	0.399	Sig.
Kaolinite/illite	**−0.491**	0.067	1			−0.287	**0.470**	−0.013	**−0.324**	Corr.
	0.000	0.519	–			0.005	0.000	0.904	0.028	Sig.
Kaolinite/smectite	0.108	−0.016		1		−0.318	0.154	0.065	−0.015	Corr.
	0.298	0.878		–		0.002	0.138	0.547	0.920	Sig.
Smectite/illite + chlorite	**−0.510**	0.096			1	0.178	0.134	−0.035	−0.081	Corr.
	0.000	0.355			–	0.086	0.198	0.746	0.592	Sig.
Mg/Al	0.217	0.168	−0.287	−0.318	0.178	1	**−0.553**	−0.309	0.072	Corr.
	0.033	0.101	0.005	0.002	0.086	–	0.000	0.003	0.635	Sig.
CIA	**−0.655**	**−0.339**	**0.470**	0.154	0.134	**−0.553**	1	−0.055	**−0.430**	Corr.
	0.000	0.001	0.000	0.138	0.198	0.000	–	0.609	0.003	Sig.
Mean	0.019	−0.072	−0.013	0.065	−0.035	−0.309	−0.055	1	**0.537**	Corr.
	0.861	0.501	0.904	0.547	0.746	0.003	0.609	–	0.000	Sig.
Si/Al	0.298	−0.127	**−0.324**	−0.015	−0.081	0.072	**−0.430**	**0.537**	1	Corr.
	0.044	0.399	0.028	0.920	0.592	0.635	0.003	0.000	–	Sig.
Number of samples (*n*)	96	96	95	95	95	96	96	90	47	

Acceptable correlations are in bold throughout.

Table 6. *Pearson's correlation coefficient (Corr.) and significance of linear fitting (Sig.) calculated from ODP Site 1146, 4–19.5 Ma*

	K/Al	K/Rb*	Clay mineralogy			Mg/Al	CIA	Grain size		
			Kao/Ill	Kao/Sm	Sm/Ill + Ch			Mean	Si/Al	
K/Al	1	**0.543**	**−0.629**	**0.439**	**−0.735**	**0.325**	**−0.868**	**0.500**	**−0.367**	Corr.
	–	0.000	0.000	0.000	0.000	0.000	0.000	0.000	0.002	Sig.
K/Rb*	**0.543**	1	−0.207	0.138	**−0.403**	0.215	**−0.382**	0.074	**−0.499**	Corr.
	0.000	–	0.005	0.067	0.000	0.004	0.000	0.328	0.000	Sig.
Kaolinite/illite	**−0.629**	−0.207	1			**−0.509**	**0.673**	**−0.483**	0.004	Corr.
	0.000	0.005	–			0.000	0.000	0.000	0.976	Sig.
Kaolinite/smectite	**0.439**	0.138		1		−0.314	**−0.431**	**0.457**	**−0.463**	Corr.
	0.000	0.067		–		0.000	0.000	0.000	0.000	Sig.
Smectite/illite + chlorite	**−0.735**	**−0.403**			1	0.087	**0.724**	**−0.607**	**0.595**	Corr.
	0.000	0.000			–	0.247	0.000	0.000	0.000	Sig.
Mg/Al	**0.325**	0.215	**−0.509**	−0.314	0.087	1	−0.338	0.030	0.272	Corr.
	0.000	0.004	0.000	0.000	0.247	–	0.000	0.688	0.022	Sig.
CIA	**−0.868**	**−0.382**	**0.673**	**−0.431**	**0.724**	−0.338	1	**−0.771**	0.177	Corr.
	0.000	0.000	0.000	0.000	0.000	0.000	–	0.000	0.139	Sig.
Mean	**0.500**	0.074	**−0.483**	**0.457**	**−0.607**	0.030	**−0.771**	1	0.021	Corr.
	0.000	0.328	0.000	0.000	0.000	0.688	0.000	–	0.861	Sig.
Si/Al	**−0.367**	**−0.499**	0.004	**−0.463**	**0.595**	0.272	0.177	0.021	1	Corr.
	0.002	0.000	0.976	0.000	0.000	0.022	0.139	0.861	–	Sig.
Number of samples (*n*)	178	178	178	178	178	178	178	177	71	

Acceptable correlations are in bold throughout.

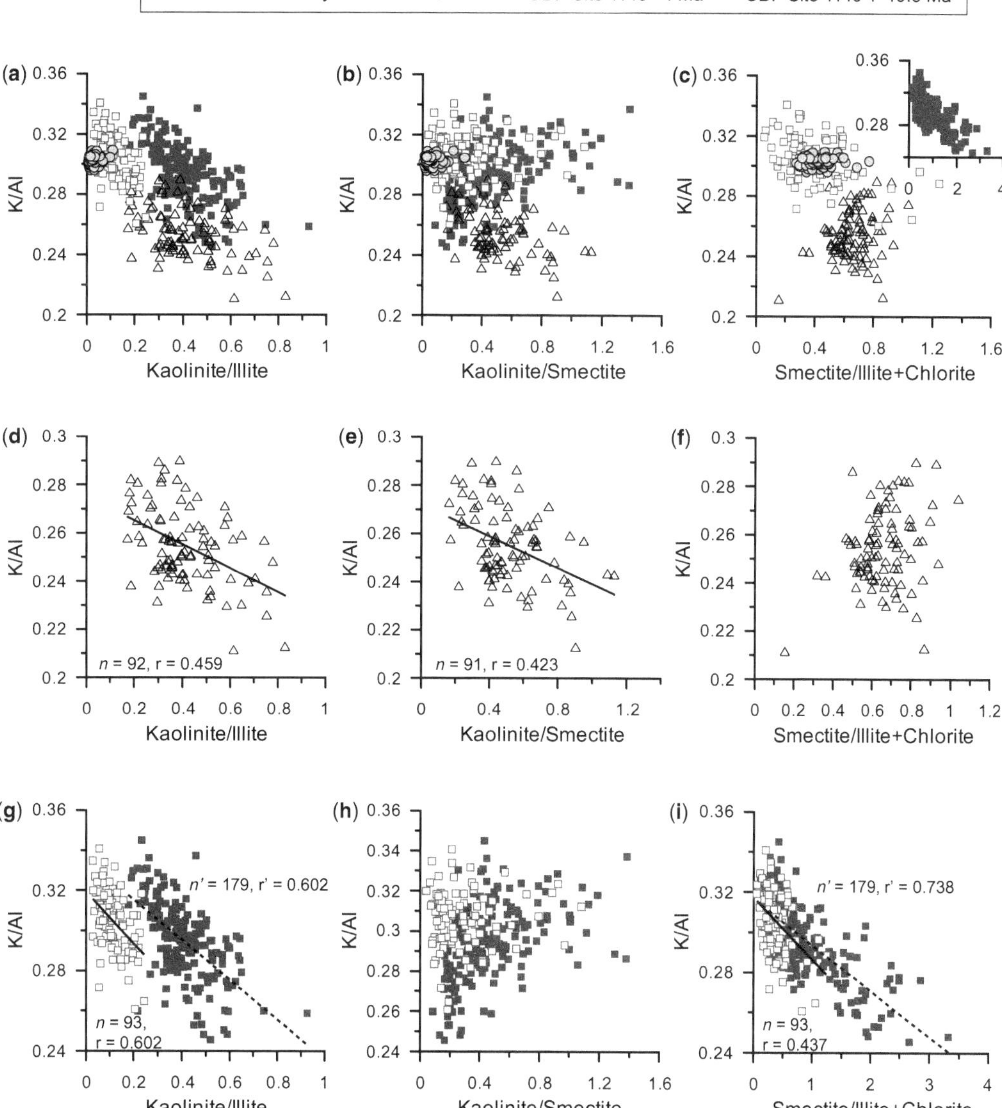

Fig. 6. Cross-plots of K/Al against ratios of clay mineral content as kaolinite/illite, kaolinite/smectite and smectite/(illite + chlorite) from (**a–c**) combined data of all three core sites over all timescales, (**d–f**) from data of the Pearl River delta spanning 9.5 ka, (Hu *et al.* 2012, 2013) and (**g–i**) from data of ODP Site 1146 spanning post-4 Ma and 4–19.5 Ma (Wan *et al.* 2010*a*). Solid lines are least-squares trend lines where added. Linear trend line using least square method is added throughout, when applicable ($R^2 > 0.1$).

(Fig. 8). $^{87}Sr/^{86}Sr$ correlates reasonably well with K/Al (Fig. 8a) in the Pearl River delta and a little less well with K/Rb (Fig. 8b) or with Mg/Al (Fig. 8c). At ODP Site 1144 $^{87}Sr/^{86}Sr$ correlates best with K/Rb, albeit on a different trend compared to that defined by the Pearl River delta. K/Al and Mg/Al show no clear coupling to $^{87}Sr/^{86}Sr$ values at ODP Site 1144. Plots of K/Al and K/Rb v. $^{87}Sr/^{86}Sr$ show that the two areas define separate trends, with ODP Site 1144 sediments defining higher K/Al and K/Rb arrays. In contrast, Mg/Al v. $^{87}Sr/^{86}Sr$ shows an overlapping array with a negative correlation. Kaolinite/illite values from both sites only show weak correlations with $^{87}Sr/^{86}Sr$ (Fig. 8d), again with the two studied regions defining different positively correlated arrays.

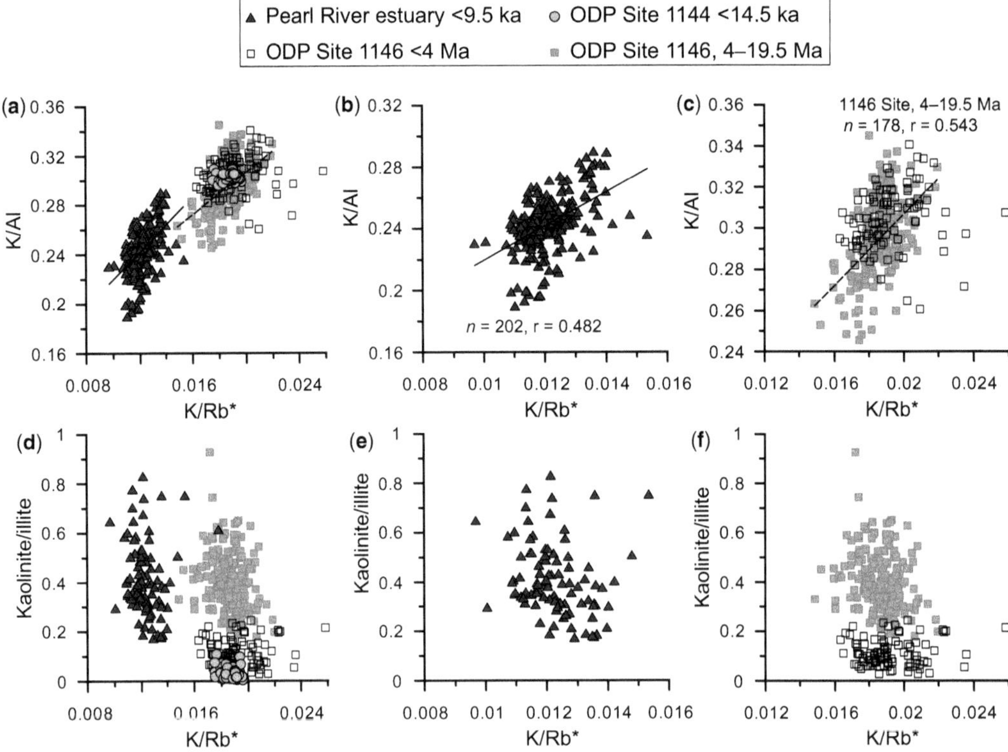

Fig. 7. Cross-plots of K/Rb against K/Al in (**a**) for all datasets, (**b**) from the Pearl River delta Holocene, (**c**) from ODP Site 1146. Cross-plots of K/Rb against kaolinite/illite (**d**) for all datasets, (**e**) from the Pearl River delta Holocene, (**f**) from ODP Site 1146. Data are from Hu *et al.* (2012, 2013) and Wan *et al.* (2010*a*).

Mg/Al. Mg/Al is predicted to act as a proxy for the mean state of sediment alteration because this ratio measures the loss of Mg^{2+} from Mg-rich feldspar, or secondary chlorite during the process of hydrolysis during which Al is concentrated. Mg is also preferentially lost during breakdown of mafic minerals (Dessert *et al.* 2003). Figure 9 plots Mg/Al in relation to other proxies both at individual sites and with all sites synthesized together. These diagrams show only modest correlation at best between Mg/Al and K/Al, K/Rb or kaolinite/illite for each site. Data from ODP Site 1146 yield the best correlation between Mg/Al and kaolinite/illite over 19.5 Ma (Fig. 9c), while a poorer correlation is seen when considering only those sediments dating 4–19.5 Ma. There is no reasonable correlation seen for <4 Ma sediment at this site. No correlation is seen when combining data together across all the sites.

Chemical Index of Alteration (CIA). CIA could not be calculated from the Pearl River delta because of a lack of Na data from those samples. $^{87}Sr/^{86}Sr$ does not correlate with CIA data at ODP Site 1144, which is the only site where such data are available. At ODP Site 1146 CIA correlates with K/Al (Fig. 10a) and kaolinite/illite (Fig. 10c) within the 0–4 and 4–19.5 Ma groups, although these fall along two separate trends that parallel each other. The 4–19.5 Ma sediment has consistently higher kaolinite/illite values and relatively lower values of K/Al compared to the younger sediment group and is inferred to be more altered. K/Rb and Mg/Al have a similar relationship with CIA, showing very weak, poorly defined negative correlations when combining all data together and showing no significant difference between the 0–4 and 4–19.5 Ma sediments. At ODP Site 1144 there is no coherent correlation between CIA and K/Al, K/Rb, kaolinite/illite or Mg/Al, but this probably reflects the small variation in these proxies at that site, at least since 14.5 ka. All ODP Site 1144 samples plot within the range of values seen at ODP Site 1146 for sediment younger than 4 Ma, except in terms of Mg/Al and CIA (Fig. 10d). For those proxies the ODP Site 1144 sediments show consistently higher values of Mg/Al (less altered) than ODP Site 1146, but still lie within the range of CIA of

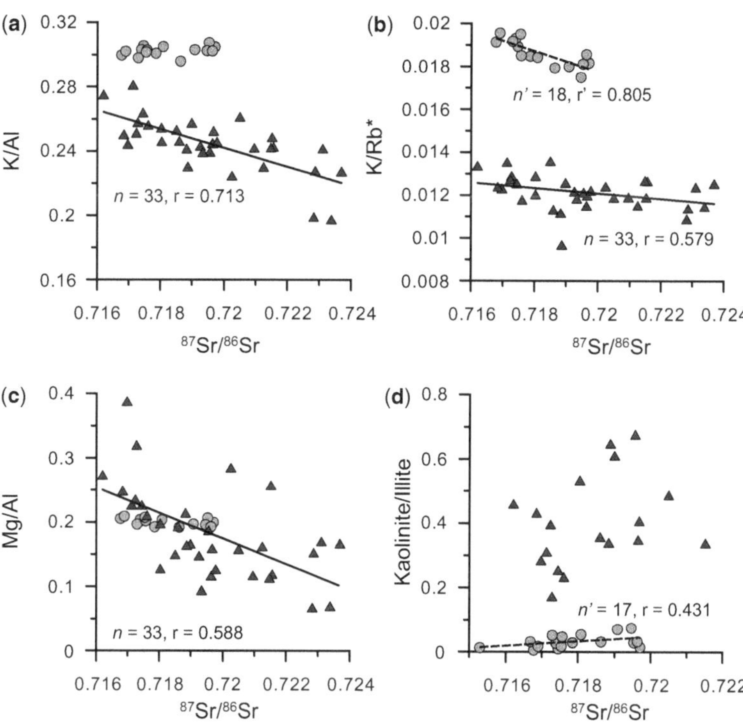

Fig. 8. Sr isotopes of the detrital clastic fraction plotted with chemical weathering proxies (**a**) K/Al, (**b**) K/Rb, (**c**) Mg/Al and (**d**) kaolinite/illite at Sites HKUV1 and B2/1 and ODP Site 1144. Least-squares trend line added where applicable. Data are from Hu *et al.* (2012, 2013).

sediment deposited since 4 Ma at this site. Overall 4–19.5 Ma sediments at ODP Site 1146 yield higher values of CIA than younger sediment from either sites.

Grain-size effects

Despite the widely recognized link between finer grained sediment and more alteration (Lupker *et al.* 2013), grain size shows little correlation with weathering proxies (K/Al, K/Rb, Mg/Al, $^{87}Sr/^{86}Sr$, CIA) at any of the three core sites, despite the fact that the Pearl River delta cores are more silty and consistently coarser than the other materials considered (Fig. 11a–c; K/Rb, Mg/Al not shown). The Pearl River sediments are in general more weathered than those from the other core sites despite their coarser grain size, indicating that provenance/tectonic setting is a more important influence. The only exception is at ODP Site 1146 when plotting grain size against K/Al or CIA for the entire 19.5 myr time period (Fig. 11a, c). However, it is noteworthy that the highest values of K/Al and K/Rb are from slope sediment associated with the finest grain size, while the coarsest material in the Pearl River delta has the lowest (most weathered) proxy values. Although $^{87}Sr/^{86}Sr$ values at ODP Site 1144 are higher for finer sediment, the Holocene Pearl River delta sediment does not show this trend (Fig. 11b). The contrasting grain sizes between the two different sedimentary systems are shown by the fact that all the coarsest materials were derived from the Pearl River delta. In the Pearl River delta, sediments of all grain sizes have very similar geochemical proxy ratios, although it is recognized that the overall range is not very high.

We also plot proxy ratios against Si/Al because Si tends to increase in sandier sediments, while Al is higher in muddy, clay-rich deposits (Figs 11d–f). K/Al shows good positive correlation with Si/Al within the Pearl River delta. There is less alteration in sandier materials that reveals strong grain-size control over degrees of alteration at that site.

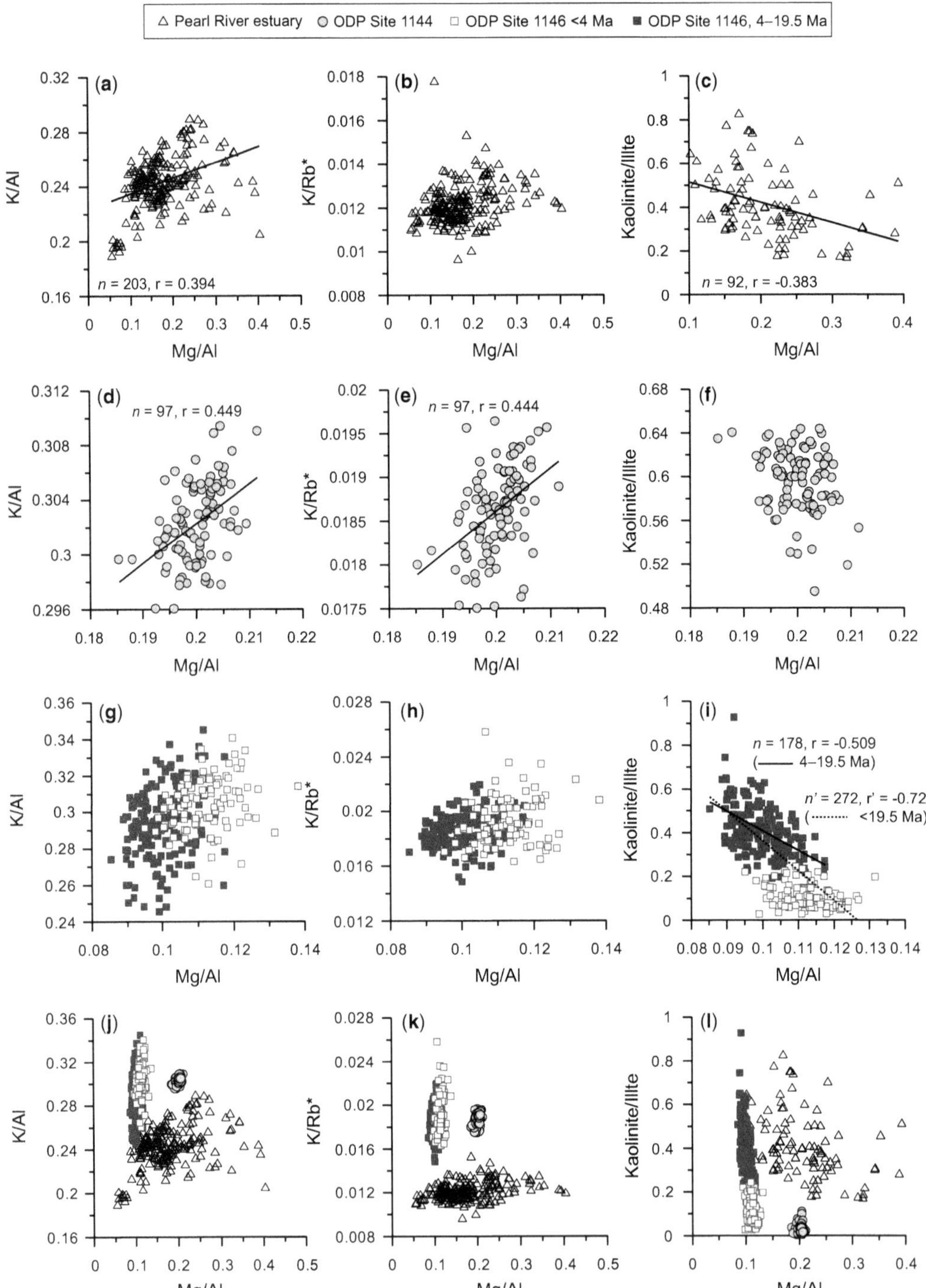

Fig. 9. Mg/Al in relation to K/Al, K/Rb and kaolinite/illite shows very modest correlations at all three core sites over different timescales. (**a–c**) plots of Pearl River estuary over 9.5 ka, (**d–f**) plots of ODP Site 1144 over 14.5 ka, (**g–i**) plots of ODP Site 1146 over 4 Ma. Plots (**j–l**) are combined results of all core sites over all time spans showing no correlation between Mg/Al with other proxies of chemical weathering intensity.

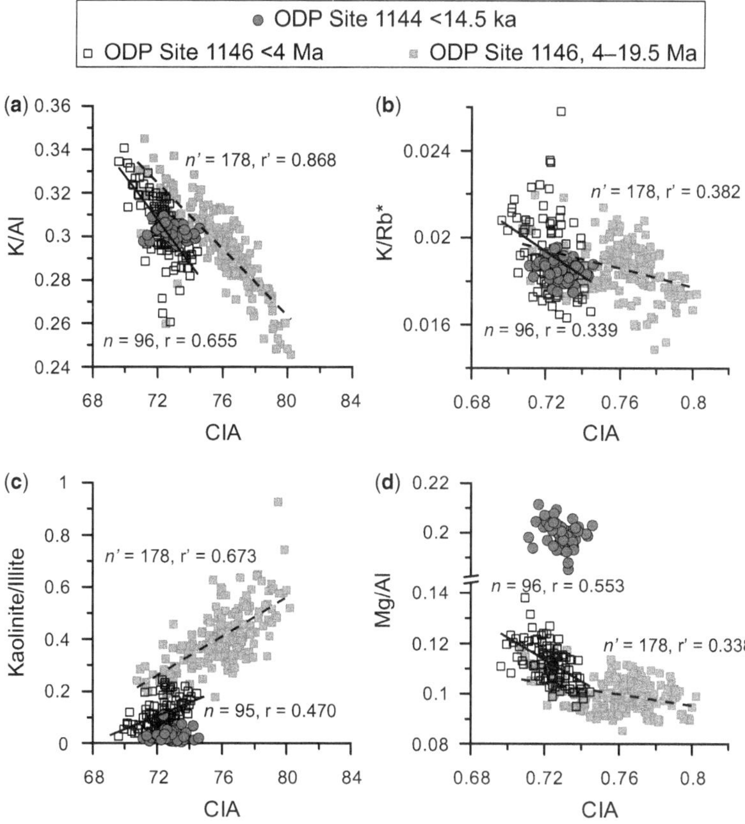

Fig. 10. Cross-plots of CIA against (**a**) K/Al, (**b**) K/Rb, (**c**) kaolinite/illite and (**d**) Mg/Al from ODP Sites 1144 and 1146.

However, when plotted together with the ODP Sites 1144 and 1146 material there is no coherent trend and these other sites show limited variation within individual sites (Fig. 11f). A similar relationship is seen between K/Rb and Si/Al. Both sediments from ODP Site 1144 and the Pearl River delta display reasonable correlation between $^{87}Sr/^{86}Sr$ and Si/Al, which is consistent with more alteration associated with muddier sediments (Fig. 11e), although we cannot discount mineralogical differences in the finer sediments being the primary cause for the change in Sr isotope in the finer grain size fraction (Garçon *et al.* 2014). CIA only shows a very weak correlation with Si/Al in ODP Site 1146 0–4 Ma sediment (Fig. 11f).

Discussion

Alternative proxies for weathering intensity

We use the results presented above to determine which proxies are likely to make the most reliable measures of weathering intensity, across different time periods and contrasting tectonic settings. We make this judgement based on intercorrelations and our observation that K/Al tracks climate change, especially rainfall, quite closely in the simple system of the Pearl River Holocene. Some of the discrepancies between different proxies may be related to different degrees of mobility in the key elements, such as K, Mg and Sr. During interglacial times, faster weathering might be expected to be linked to the strong summer monsoon, because more rainfall results in higher humidity, while temperatures were also higher. However, faster alteration of weathering-susceptible minerals does not always result in production of more weathered material because the short duration of the interglacial period is also significant in limiting the total degree of alteration. Moreover, greater runoff/discharge is likely to result in shorter sediment transport times, especially in mountainous catchments (i.e. Taiwan), thus resulting in less time for chemical alteration to occur, despite rapid alteration

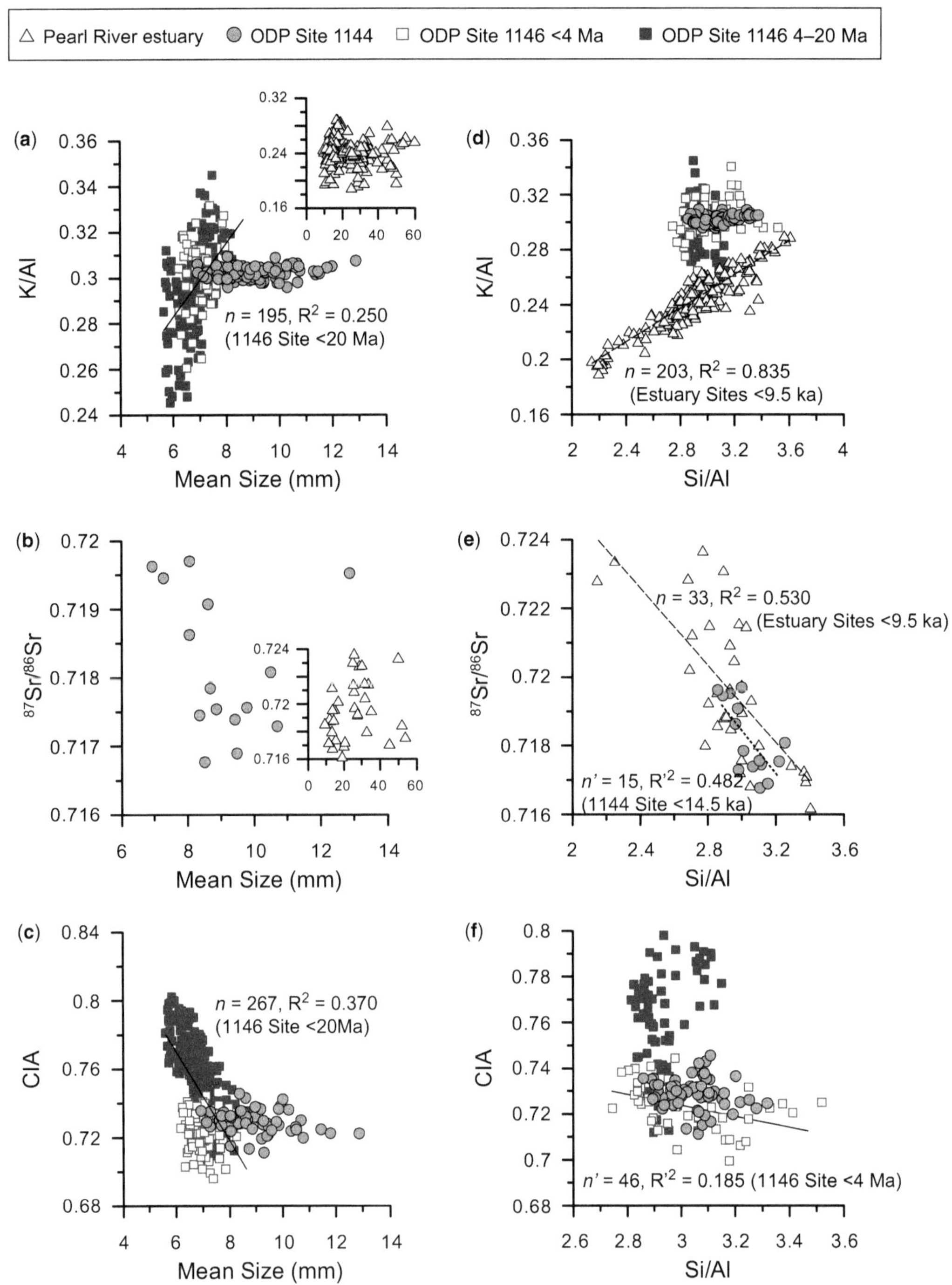

Fig. 11. Grain size, as measured by particle size analyser (**a–c**) and proxied by Si/Al (**d–f**) in relation to K/Al (a, d), $^{87}Sr/^{86}Sr$ (b, e) and CIA (c, f), plotted from combined data of all core sites shows no correlation between this factor to other prospective proxies of chemical weathering intensity.

rates. In turn this can result in the deposition of less altered material, even though weathering is faster and despite the evidence from the Pearl River that periods of stronger monsoons in that system are associated with sedimentation of more altered material, at least during the Holocene (Fig. 2a).

Alternatively, sedimentation of more weathered material on continental margins during wet, interglacial periods may at least partly reflect enhanced erosion of stored sediment from fluvial terraces that has had time to alter, potentially over a number of climatic cycles, rather than being an immediate response to changing rates of weathering. Estimates from the Indus River basin suggest that such recycling is around half the sediment budget in the Holocene in that system (Clift & Giosan 2014), although in the case of Taiwan those rivers lack the extensive flood plains that allow sediment storage and weathering.

During glacial times, when humidity was reduced and temperatures were cooler, rates of alteration would have been slower (West *et al.* 2005). In tropical latitudes glacial-era temperatures were not much cooler than today (Annan & Hargreaves 2013), so the effect is probably more related to precipitation. At the same time sea level was lower, and continental shelves were exposed for lengthy periods of time (tens of thousands of years), so that unstable minerals may have been altered more completely despite the drier, colder climate. Prior to 4 Ma, before high-amplitude sea-level and climate fluctuations occurred (Miller *et al.* 2005), the situation was different because of the greater climatic stability, at least over millennial timescales. This generally more stable and warmer climate is reflected in the more weathered character of the 4–19.5 Ma sediment recovered at ODP Site 1146 compared to the <4 Ma material, although these sediments also reflect some alteration during burial diagenesis (Jeandel & Oelkers 2015). Stable sea levels also would slow the flux of material from source to sink by removing the pumping action of rising and falling sea levels to transport sediment across the shelf break. If we are to understand how chemical alteration is linked to climate and sea level, it is critical to have at least one proxy that is reliable in all settings and over a range of timescales. Our analysis suggests that K/Al is the best proxy considered here.

A series of proxies, whether they are geochemical, isotopic or based on clay mineralogy, should in theory have behaved similarly under the same climatic stimulus, and at least some of them should correlate with one another in a coherent fashion. For example, alkali and alkali earth metals (K^+, Na^+, Ca^{2+}, Mg^{2+}) are vulnerable to leaching from eroded rock compared with water-immobile metals, such as Al, so that ratios like K/Al, kaolinite/illite and the CIA of weathered rocks should follow the same or at least similar trends through time. When they do not this may indicate that one element is not responsive because of the finite amount of time needed to release it by alteration, compared to other elements that are more mobile, although this is probably modulated by different minerals containing more or less Mg compared to K which are weathered preferentially. This may be why K/Al, and to a lesser extent CIA, respond coherently over timescales of >10 kyr but Mg/Al does not.

Ideally good proxies should not be strongly affected by provenance and be applicable over millennial and longer geological timescales. The low degree of chemical variation seen at ODP Site 1144 suggests that some proxies have limits to their application on millennial timescales (e.g. CIA and Mg/Al), presumably because they are not very sensitive to the limited variation in alteration that can occur over such short time periods (<10 kyr). Previous studies of the Pearl River and Taiwan sedimentary systems addressed here allow us to test which of the targeted proxies were linked to variations in the intensity of the East Asian monsoon. While each showed some linkage between at least one weathering proxy and monsoon strength (Figs 2 & 3) it was less clear whether these records reflected a synchronous weathering response to climate change. We believe it is more likely that the response in the marine weathering record is a function of reworking of previously weathered material being more eroded and transported during time of strong rainfall. This material may come from fluvial terraces and flood plains, as in the case of the Pearl River delta or from the exposed shelf in the case of the slope sites.

Clay minerals and K/Al. Assemblages of clay minerals are believed to be a complex of both neoformation from primary minerals and transformation of pre-existing clay minerals, in response to a wide range of processes including weathering, transport, sedimentation, diagenesis and authigenesis (Hillier 2006; Fagel 2007). However, regardless of which pathway clay minerals are formed under, the general trend is consistent in that smectite and kaolinite tend to accumulate in more weathered residual soil/mud, whereas kaolinite is likely to be the ultimate product during the clay cycle under the influence of warm and wet environments (Barshad 1966). How long it takes for clay mineralogy, regardless of its transport or depositional history, to respond to changes in the climatic forcing is an outstanding question. Recent work on the flood plains and delta of the Indus River suggest that these lags may not be much longer than the uncertainties in the age control in Quaternary sediments, i.e. *c.* 1–2 kyr (Alizai *et al.* 2012). Data from the Pearl River delta showing that clay ratios change with speleothem climate records could reflect degrees of reworking from flood plains rather than an immediate response to climate change (Fig. 3b), so that it is possible that weathering responses could be slower (*c.* 4–5 kyr).

Similarly, changes in clay mineralogy since 14.5 ka at ODP Site 1144 show only weak response to monsoon changes compared to the stalagmite record from Dongge Cave (Dykoski *et al.* 2005), a feature that Hu *et al.* (2012) attribute to reworking of older sediments by intensifying monsoon run-off or reworking from the adjacent shelf. This site does not allow the response times of the clay mineralogy to climatic forcing to be constrained beyond being less than 30 ka.

Over timescales of 1 myr clay mineralogy at ODP Site 1146 shows coherent links with monsoon strength indicators (Wan *et al.* 2007), but whether this is a direct response or one reflecting changing degrees of recycling from onshore flood plains or the continental shelf is unclear. An outstanding issue is whether the higher degrees of alteration seen in the 4–19.5 Ma sediment at ODP Site 1146 reflects more weathering under warmer, wetter conditions prior to the onset of Northern Hemispheric Glaciation (NHG), or whether some of this alteration has occurred after burial, as a result of fluid flux through the continental margin sedimentary apron.

Sediment from all core sites reveals a coherently negative correlation between clay ratios kaolinite/illite, kaolinite/smectite, as well as smectite/(illite + chlorite) and K/Al (Figs 6a–c). This suggests that all these proxies are behaving coherently to the process of alteration, but that kaolinite/illite has the clearest response. In contrast, smectite/(illite + chlorite) and kaolinite/smectite do not show good correlations with K/Al at the Pearl River delta or at ODP Site 1146 respectively, indicating a more complicated and potentially less robust response. At ODP Site 1144 none of the clay ratios show coherence with K/Al. Overall, we conclude that kaolinite/illite is probably the most effective clay proxy of weathering intensity across all sites and timescales.

Mg/Al and K/Rb. Kaolinite/illite was plotted against Mg/Al, based on the fact that Mg acts as a water-mobile element during mineral alteration. As shown in Figures 9f, i, l all three sites show very weak correlation between Mg/Al and kaolinite/illite for sediment younger than 4 Ma. Considering data spanning all timescales results in no clear trend (Fig. 9c). Together with the fact that Mg/Al does not correlate well with either K/Al or K/Rb we conclude that although Mg/Al is relatively immune to grain-size effects this is not a good universal proxy of weathering intensity, and should only be applied with other proxies under certain circumstances.

Over millennial scales in the Pearl River delta both K/Al and K/Rb values show well-defined, coherent variations with the speleothem-based monsoon record (Figs 2 & 3). However, K/Al, unlike K/Rb, shows no such correlation since 14.5 ka with other weathering proxies at ODP Site 1144 (Hu *et al.* 2012). The trend of chemical alteration at ODP Sites 1144 and 1146 can be observed using the Al_2O_3–($CaO + Na_2O$)–K_2O (A–CN–K) ternary diagram (Nesbitt & Young 1984, 1989). ODP Sites 1144 and 1146 plot as a coherent single array on the predicted weathering trend for average upper continental crust (Nesbitt & Young 1984) parallel to the A–CN line (Fig. 12). This implies that the two sites have similar average compositions to their source rock and that siliciclastic weathering has occurred mainly by the breakdown of micas and plagioclase, but not yet K-feldspar.

ODP Site 1144 (<14.5 ka) sediment shows even less alteration from the bedrock source composition, while the modern Pearl River drainage system reaches the highest level of chemical weathering seen in any material considered here, followed by ODP 1146 spanning the last 19.5 myr. K/Al shows no trend in weathering intensity at ODP Site 1144 since 14.5 ka. However, K/Rb represents leaching of K, because isomorphous Rb is retained in the lattices, so that K/Rb is a more sensitive proxy than K/Al. This former ratio demonstrates good correlation with K/Al over various timescales except in <14.5 ka ODP Site 1144 sediment, and also shows links with clay mineral ratios, leading us to conclude that it is a reasonably reliable indicator of chemical weathering intensity. It may be more sensitive than K/Al to weathering changes over short timescales and when considering less altered materials.

Strontium isotopes. $^{87}Sr/^{86}Sr$ displays a good correlation with K/Al, K/Rb and Mg/Al in the Pearl River delta, and a less well-defined, but reasonable correlation with K/Rb, but not K/Al, in sediment >14.5 ka from ODP Site 1144 (Fig. 8). The offset between trends at the different sites probably reflects the fact that $^{87}Sr/^{86}Sr$ is sensitive to both chemical weathering and provenance (Derry & France-Lanord 1996). In our compiled datasets provenance can be controlled by Nd isotopes so that changes in $^{87}Sr/^{86}Sr$ values can often be interpreted to reflect climatically modulated weathering, although without supporting datasets its ambiguous nature will always result in it being a less reliable alteration measure.

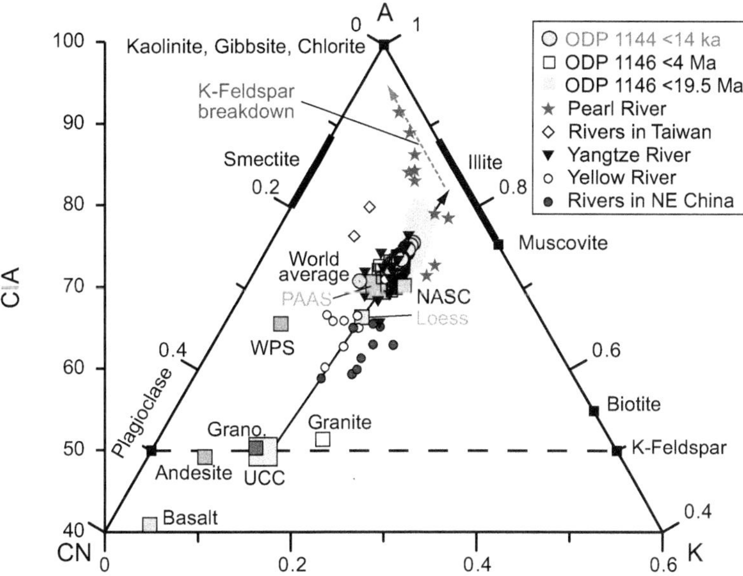

Fig. 12. Al_2O_3–($CaO + Na_2O$)–K_2O (A–CN–K) ternary diagram of siliciclastic sediments after Wan *et al.* (2010*a*) and Shao *et al.* (2012). We show sediment from ODP Sites 1144 and 1146 compared to that from other Asian rivers and to major upper crustal rock compositions. UCC, upper continental crust; WPS, West Philippine Sea; NASC, North American shale composite; Grano., average composition of granodiorite; PAAS, Average Post-Archaean Australian Shale.

Clay mineralogical proxies in the northern South China Sea

Kaolinite/illite appears to correlate well with K/Al across sites and timescales, but other clay ratios are less coherent in their response. Smectite/(illite + chlorite) correlates with other proxies and is believed to have recorded chemical weathering intensity changes at ODP Site 1146 since 4 Ma. However, the modern rivers draining into the South China Sea (i.e. Pearl River and smaller rivers on Taiwan and Luzon) comprise radically different clay mineral assemblages so that provenance variations can also have potentially impacted clay assemblages (Liu *et al.* 2007). In the northern South China Sea, sediment from ODP Sites 1144 and 1146 contains relatively more smectite than seen in the modern Pearl River. These sites may receive smectite from Luzon carried by bottom currents linked to the Kuroshio Current (Shao *et al.* 2007), although isotopic evidence argues that this flux must be modest in volume. Indeed, we now know that the rivers are not stable end-members with regard to their clay minerals and that the Pearl River used to be a more significant source of smectite prior to the onset of agriculture at *c.* 3 ka (Hu *et al.* 2013). The higher smectite at the drill sites compared to the modern rivers may in part reflect weathering of the sediment on the shelf during sea-level lowstands prior to resedimentation on to the slope. The overall coherency of clay minerals and K/Al suggests that provenance has not been dominant in governing clay ratios in this region at least over the time periods studied.

Conclusions

Comparison of variations in geochemical and clay mineral weathering indicators from three core sites reveal two contrasting sedimentary systems spanning *c.* 10 kyr to 10 myr. While sites on the shelf near the Pearl River delta represent a direct record of continental weathering and recycling onshore, the exposure and alteration of such sequences during sea-level lowstands mean that the original weathering record has a low preservation potential. Sedimentary archives on the slope are typically linked to climatic cycles and have better preservation potential, but are often influenced by reworking and erosion of pre-existing deposits on the shelf. Samples pre-dating 4 Ma are generally more weathered than younger material because of burial diagenesis and/or cycles of repeated weathering when the sediment is preserved on the shelf and subject to exposure during sea-level lowstands. Warmer, wetter conditions before the Pliocene, probably also played a part in increase alteration. The Pearl River delta comprises sediment that is more weathered than that shed from Taiwan. The greater weathering of the older samples may indicate warmer, wetter conditions prior to the onset of the NHG, as well as post-depositional alteration. Rapid erosion and sediment transport from Taiwan explains the low degree of sediment alteration observed at ODP Sites 1144 and 1146 since 4 Ma.

While several weathering proxies have shown local application we conclude that clay mineral ratio kaolinite/illite, as well as K/Al are the most consistently reliable proxies of chemical weathering intensity over wide ranges of geological settings and time. K/Al shows close correspondence to changes in the $\delta^{18}O$ monsoon proxy from speleothem records. Kaolinite/smectite, and to a lesser extent smectite/(illite + chlorite), are also indicative of weathering conditions, but show more scatter between sites. K/Rb is ideal to be used as an alternative to K/Al over periods of 1–10 kyr, especially when alteration is modest, but it does not track other proxies well over longer time periods and when weathering is intense. $^{87}Sr/^{86}Sr$ values appear to be reliable weathering proxies provided the sediment provenance is stable and the degree of detrital carbonate is minor or can be removed. Mg/Al is the proxy least affected by grain-size issues, but shows poor coupling to other weathering proxies and is hard to interpret as a robust measure of weathering in the continental basin.

We acknowledge financial support from the Swire Educational Trust and South China Sea Institute of Oceanology PhD Funding (Grant No. MSGL09-06). We thank Rod Sewell and colleagues at the Hong Kong Geological Survey for access to Core B2/1 and Wyss Yim at the University of Hong Kong (HKU) for access to core HKUV1. ODP Kochi Core Centre staff are acknowledged for access to ODP Site 1144 cores. PC thanks the Charles T. McCord Chair at Louisiana State University for financial support.

References

ADATTE, T., KELLER, G. & STINNESBECK, W. 2002. Late Cretaceous to early Paleocene climate and sea-level fluctuations: the Tunisian record. *Palaeogeography, Palaeoclimatology, Palaeoecology*, **178**, 165–196, http://doi.org/10.1016/s0031-0182(01)00395-9

ALIZAI, A., HILLIER, S., CLIFT, P. D., GIOSAN, L., HURST, A., VANLANINGHAM, S. & MACKLIN, M. 2012. Clay mineral variations in Holocene terrestrial sediments from the Indus Basin. *Quaternary Research*, **77**, 368–381, http://doi.org/10.1016/j.yqres.2012.01.008

ANNAN, J. D. & HARGREAVES, J. C. 2013. A new global reconstruction of temperature changes at the Last Glacial Maximum. *Climate of the Past*, **9**, 367–376, http://doi.org/10.5194/cp-9-367-2013

BARSHAD, I. 1966. The effect of a variation in precipitation on the nature of clay mineral formation in soils from acid and basic igneous rocks. *In*: HELLER, L. & WEISS, A. (eds) *International Clay Conference*. Israel Program for Scientific Translation, Jerusalem, 167–173.

BERNER, R. A. & BERNER, E. K. 1997. Silicate weathering and climate. *In*: RUDDIMAN, W. F. (ed.) *Tectonic Uplift and Climate Change*. Springer, New York, 353–365.

BOUQUILLON, A. 2010. History of ceramics. *In*: BOCH, P. & NIEPCE, J.-C. (eds) *Ceramic Materials: Processes, Properties and Applications*. ISTE, London, 29–53.

BÜHRING, C., SARNTHEIN, M. & ERLENKEUSER, H. 2004. Toward a high-resolution stable isotope stratigraphy of the last 1.1 m.y.: Site 1144, South China Sea. *In*: PRELL, W. L., WANG, P., BLUM, P., REA, D. K. & CLEMENS, S. C. (eds) *Proceedings of the Ocean Drilling Program (ODP), Scientific Results*. Ocean Drilling Program, College Station, TX, **184**, 1–29.

CHAMLEY, H. 1989. *Clay Sedimentology*. Springer-Verlag, Berlin.

CHEN, C. H. & LEE, T. 1990. A Nd-Sr isotopic study on river sediments of Taiwan. *Proceedings of the Geological Society of China*, **33**, 339–350.

CLEMENS, S. C., PRELL, W. L. & SUN, Y. 2010. Orbital-scale timing and mechanisms driving Late Pleistocene Indo-Asian summer monsoons: reinterpreting cave speleothem $\partial^{18}O$. *Paleoceanography*, **25**, PA4207, http://doi.org/10.1029/2010PA001926

CLIFT, P., LEE, J., CLARK, M. & BLUSZTAJN, J. 2002. Erosional response of South China to arc rifting and monsoonal strengthening; a record from the South China Sea. *Marine Geology*, **184**, 207–226.

CLIFT, P. D. & GIOSAN, L. 2014. Sediment fluxes and buffering in the post-glacial Indus Basin. *Basin Research*, **25**, 1–18, http://doi.org/10.1111/bre.12038

CLIFT, P. D. & PLUMB, R. A. 2008. *The Asian Monsoon: Causes, History and Effects*. Cambridge University Press, Cambridge.

CLIFT, P. D., LONG, H. V. ET AL. 2008. Evolving east Asian river systems reconstructed by trace element and Pb and Nd isotope variations in modern and ancient Red River-Song Hong sediments. *Geochemistry Geophysics Geosystems*, **9**, Q04039, http://doi.org/10.1029/2007gc001867

CLIFT, P. D., WAN, S. & BLUSZTAJN, J. 2014. Reconstructing chemical weathering, physical erosion and monsoon intensity since 25 Ma in the northern South China Sea: a review of competing proxies. *Earth-Science Reviews*, **130**, 86–102, http://doi.org/10.1016/j.earscirev.2014.01.002

COLIN, C., SIANI, G., SICRE, M. A. & LIU, Z. 2010. Impact of the East Asian monsoon rainfall changes on the erosion of the Mekong River basin over the past 25,000 yr. *Marine Geology*, **271**, 84–92, http://doi.org/10.1016/j.margeo.2010.01.013

DERRY, L. A. & FRANCE-LANORD, C. 1996. Neogene Himalayan weathering history and river $^{87}Sr/^{86}Sr$; impact on the marine Sr record. *Earth and Planetary Science Letters*, **142**, 59–74.

DESSERT, C., DUPRÉ, B., GAILLARDET, J., FRANCOIS, L. M. & ALLEGRE, C. 2003. Basalt weathering laws and the impact of basalt weathering on the global carbon cycle. *Chemical Geology*, **202**, 257–273.

DUZGOREN-AYDIN, N. S., AYDIN, A. & MALPAS, J. 2002. Re-assessment of chemical weathering indices: case study of pyroclastic rocks of Hong Kong. *Engineering Geology*, **63**, 99–119.

DYKOSKI, C. A., EDWARDS, R. L. ET AL. 2005. A high-resolution, absolute-dated Holocene and deglacial Asian monsoon record from Dongge Cave, China. *Earth and Planetary Science Letters*, **233**, 71–86.

FAGEL, N. 2007. Clay minerals, deep circulation and climate. *In*: CLAUDE, H. M. & ANNE DE, V. (eds) *Developments in Marine Geology*. Elsevier, Amsterdam, 139–184.

FYFE, J. 2000. *The Quaternary Geology of Hong Kong*. Hong Kong Geological Survey, Geotechnical Engineering Office, Civil Engineering Department, The Government of the Hong Kong SAR.

GARÇON, M. & CHAUVEL, C. 2014. Where is basalt in river sediments, and why does it matter? *Earth and Planetary Science Letters*, **407**, 61–69, http://doi.org/10.1016/j.epsl.2014.09.033

GARÇON, M., CHAUVEL, C., FRANCE-LANORD, C., LIMONTA, M. & GARZANTI, E. 2014. Which minerals control the Nd–Hf–Sr–Pb isotopic compositions of river sediments? *Chemical Geology*, **364**, 42–55, http://doi.org/10.1016/j.chemgeo.2013.11.018

GOLDSTEIN, S. J. & JACOBSEN, S. B. 1987. The Nd and Sr isotopic systematics of river-water dissolved material: implications for the sources of Nd and Sr in seawater. *Chemical Geology: Isotope Geoscience Section*, **66**, 245–272.

GOLDSTEIN, S. J. & JACOBSEN, S. B. 1988. Nd and Sr isotopic systematics of river water suspended material: implications for crustal evolution. *Earth and Planetary Science Letters*, **87**, 249–265, http://doi.org/10.1016/0012-821x(88)90013-1

GUPTA, A. K., ANDERSON, D. M. & OVERPECK, J. T. 2003. Abrupt changes in the Asian southwest monsoon during the Holocene and their links to the North Atlantic Ocean. *Nature*, **421**, 354–356.

HILLIER, S. 2003. Quantitative analysis of clay and other minerals in sandstones by X-ray powder diffraction (XRPD). *In*: WORDEN, R. H. & MORAD, S. (eds) *Clay Mineral Cements in Sandstones*. International Association of Sedimentologists, Special Publication, Wiley-Blackwell, Oxford, **34**, 213–251, http://doi.org/10.1144/GSL.ENG.2006.021.01.03

HILLIER, S. 2006. Formation and alteration of clay materials. *In*: REEVES, G. M. & CRIPPS, J. C. (eds) *Clays Materials Used in Construction*. Geological Society, London, Engineering Geology Special Publications, **21**, 29–71.

HU, D., BÖNING, P. ET AL. 2012. Deep sea records of the continental weathering and erosion response to East Asian monsoon intensification since 14 ka in the South China Sea. *Chemical Geology*, **326–327**, 1–18, http://doi.org/10.1016/j.chemgeo.2012.07.024

HU, D., CLIFT, P. D. ET AL. 2013. Holocene evolution in weathering and erosion patterns in the Pearl River Delta. *Geochemistry, Geophysics, Geosystems*, **14, 2349–2368**, http://doi.org/10.1002/ggge.20166

JEANDEL, C. & OELKERS, E. H. 2015. The influence of terrigenous particulate matter dissolution on ocean chemistry and global element cycles. *Chemical Geology*, **395**, 50–66.

JEANDEL, C., PEUCKER-EHRENBRINK, B. ET AL. 2011. Ocean margins: the missing term in oceanic element budgets? *Eos*, **92**, 217–224.

LAN, C. Y., LEE, C. S., SHEN, J., LU, C. Y., MERTZMAN, S. A. & WU, T. W. 2002. Nd-Sr isotopic composition and geochemistry of sediments from Taiwan and their implications. *Western Pacific Earth Sciences*, **2**, 205–222.

LIMMER, D. R., BOENING, P. ET AL. 2012. Geochemical record of Holocene to recent sedimentation on the Western Indus continental shelf, Arabian Sea. *Geochemistry, Geophysics, Geosystems*, **13**, Q01008, http://doi.org/10.1029/2011GC003845

LIU, Z., COLIN, C., HUANG, W., LE, K., TONG, S., CHEN, Z. & TRENTESAUX, A. 2007. Climatic and tectonic controls on weathering in south China and Indochina Peninsula: clay mineralogical and geochemical investigations from the Pearl, Red, and Mekong drainage basins. *Geochemistry, Geophysics, Geosystems*, **8**, Q05005, http://doi.org/10.1029/2006GC0 01490

LIU, Z., COLIN, C. ET AL. 2010. Clay mineral distribution in surface sediments of the northeastern South China Sea and surrounding fluvial drainage basins: source and transport. *Marine Geology*, **277**, 48–60, http://doi.org/10.1016/j.margeo.2010.08.010

LÜDMANN, T., WONG, H. & BERGLAR, K. 2005. Upward flow of North Pacific Deep Water in the northern South China Sea as deduced from the occurrence of drift sediments. *Geophysical Research Letters*, **32**, L05614, http://doi.org/10.1029/2004GL021967

LUPKER, M., FRANCE-LANORD, C., GALY, V., LAVÉ, J. & KUDRASS, H. 2013. Increasing chemical weathering in the Himalayan system since the Last Glacial Maximum. *Earth and Planetary Science Letters*, **365**, 243–252.

MA, J., WEI, G., XU, Y., LONG, W. & SUN, W. 2007. Mobilization and re-distribution of major and trace elements during extreme weathering of basalt in Hainan Island, South China. *Geochimica et Cosmochimica Acta*, **71**, 3223–3237.

MCLENNAN, S. M. 1993. Weathering and global denudation. *The Journal of Geology*, **101**, 295–303.

MILLER, K. G., KOMINZ, M. A. ET AL. 2005. The Phanerozoic record of global sea-level change. *Science*, **312**, 1293–1298.

MOORE, D. & REYNOLDS, R. 1989. *X-ray Diffraction and the Identification and Analysis of Clay Minerals*. Oxford University Press, Oxford.

NESBITT, H. W. & YOUNG, G. M. 1982. Early Proterozoic climates and plate motions inferred from major element chemistry of lutites. *Nature*, **299**, 715–717.

NESBITT, H. W. & YOUNG, G. M. 1984. Prediction of some weathering trends of plutonic and volcanic rocks based on thermodynamic and kinetic considerations. *Geochimica et Cosmochimica Acta*, **48**, 1523–1534.

NESBITT, H. W. & YOUNG, G. M. 1989. Formation and diagenesis of weathering profiles. *The Journal of Geology*, **97**, 129–147.

NESBITT, H. W., MARKOVICS, G. & PRICE, R. C. 1980. Chemical processes affecting alkalis and alkaline-earths during continental weathering. *Geochimica et Cosmochimica Acta*, **44**, 1659–1666.

POSAMENTIER, H. W., JERVEY, M. T. & VAIL, P. R. 1988. Eustatic controls on clastic deposition – conceptual framework. *In*: Wilgus, C. K., Hastings, B. S., Kendall, C. G. S. C., Posamentier, H. W., Ross, C. A. & Wagoner, J. C. V. (eds) *Sea Level Change: An Integrated Approach.* GCS-SEPM, Tulsa, OK, Special Publication, **42**, 110–124.

Price, J. R. & Velbel, M. A. 2003. Chemical weathering indices applied to weathering profiles developed on heterogeneous felsic metamorphic parent rocks. *Chemical Geology*, **202**, 397–416.

Raymo, M. E., Ruddiman, W. F. & Froelich, P. N. 1988. Influence of Late Cenozoic mountain building on ocean geochemical cycles. *Geology*, **16**, 649–653.

Shao, J., Yang, S. & Li, C. 2012. Chemical indices (CIA and WIP) as proxies for integrated chemical weathering in China: inferences from analysis of fluvial sediments. *Sedimentary Geology*, **265–266**, 110–120, http://doi.org/10.1016/j.sedgeo.2012.03.020

Shao, L., Li, X., Wei, G., Liu, Y. & Fang, D. 2001. Provenance of a prominent sediment drift on the northern slope of the South China Sea. *Science in China Series D: Earth Sciences*, **44**, 919–925.

Shao, L., Li, X. et al. 2007. Deep water bottom current deposition in the northern South China Sea. *Science in China Series D: Earth Sciences*, **50**, 1060–1066.

Shipboard Scientific Party 2000*a*. Site 1144. *In*: Wang, P., Prell, W. L. & Blum, P. et al. (eds) *Proceedings of the Ocean Drilling Program, Initial Reports. Chapter 5.* Ocean Drilling Program, College Station, TX, **184**, 1–97, http://www-odp.tamu.edu/publications/184_IR/VOLUME/CHAPTERS/IR184_05.PDF

Shipboard Scientific Party 2000*b*. Site 1146. *In*: Wang, P., Prell, W. L. & Blum, P. et al. (eds) *Proceedings of the Ocean Drilling Program, Initial Reports. Chapter 7*, Ocean Drilling Program, College Station, TX, **184**, 1–97.

Syvitski, J. P. M., Vorosmarty, C. J., Kettner, A. J. & Green, P. 2005. Impact of humans on the flux of terrestrial sediment to the global coastal ocean. *Science*, **308**, 376–380.

Thamban, M., Purnachandra Rao, V. & Schneider, R. R. 2002. Reconstruction of late Quaternary monsoon oscillations based on clay mineral proxies using sediment cores from the western margin of India. *Marine Geology*, **186**, 527–539, http://doi.org/10.1016/s0025-3227(02)00268-2

Thamban, M., Naik, S. et al. 2005. Changes in the source and transport mechanism of terrigenous input to the Indian sector of Southern Ocean during the late Quaternary and its palaeoceanographic implications. *Journal of Earth System Science*, **114**, 443–452, http://doi.org/10.1007/bf02702021

Thiry, M. 2000. Palaeoclimatic interpretation of clay minerals in marine deposits: an outlook from the continental origin. *Earth-Science Reviews*, **49**, 201–221.

Vail, P. R., Mitchum, R. M. et al. 1977. Seismic stratigraphy and global changes of sea level, part 3: relative changes of sea level from coastal onlap. *In*: Payton, C. E. (ed.) *Seismic Stratigraphy: Applications to Hydrocarbon Exploration.* American Association of Petroleum Geologists Memoir, **26**, 63–81.

Wan, S., Li, A., Clift, P. & Stuut, J. 2007. Development of the East Asian monsoon: mineralogical and sedimentologic records in the northern South China Sea since 20 Ma. *Palaeogeography, Palaeoclimatology, Palaeoecology*, **254**, 561–582.

Wan, S., Kürschner, W. M., Clift, P. D., Li, A. & Li, T. 2009. Extreme weathering/erosion during the Miocene climatic optimum: evidence from sediment record in the South China Sea. *Geophysical Research Letters*, **36**, 1–5, http://doi.org/10.1029/2009gl040279

Wan, S., Clift, P. D., Li, A., Li, T. & Yin, X. 2010*a*. Geochemical records in the South China Sea: implications for East Asian summer monsoon evolution over the last 20 Ma. *In*: Clift, P. D., Tada, R. & Zheng, H. (eds) *Monsoon Evolution and Tectonic-Climate Linkage in Asia.* Geological Society, London, Special Publications, **342**, 245–263.

Wan, S., Li, A., Clift, P. D., Wu, S., Xu, K. & Li, T. 2010*b*. Increased contribution of terrigenous supply from Taiwan to the northern South China Sea since 3 Ma. *Marine Geology*, **278**, 115–121, http://doi.org/10.1016/j.margeo.2010.09.008

Wan, S., Toucanne, S. et al. 2015. Human impact overwhelms long-term climate control of weathering and erosion in southwest China. *Geology*, **43**, 439–442, http://doi.org/10.1130/G36570.1

West, A. J., Galy, A. & Bickle, M. J. 2005. Tectonic and climatic controls on silicate weathering. *Earth and Planetary Science Letters*, **235**, 211–228, http://doi.org/10.1016/j.epsl.2005.03.020

Xue, Z., Liu, J. P. et al. 2014. Sedimentary processes on the Mekong subaqueous delta: clay mineral and geochemical analysis. *Journal of Asian Earth Sciences*, **79A**, 520–528, http://doi.org/10.1016/j.jseaes.2012.07.012

Zhao, Y. Y. & Yan, M. C. 1993. Geochemical record of the climate effect in sediments of the China Shelf Sea. *Chemical Geology*, **107**, 267–269, http://doi.org/10.1016/0009-2541(93)90188-o

Zong, Y., Huang, G., Switzer, A., Yu, F. & Yim, W. W. S. 2009. An evolutionary model for the Holocene formation of the Pearl River delta, China. *The Holocene*, **19**, 129.

Geochemical characteristics and palaeoenvironmental reconstruction of the sediments from the Gulf of Tonkin, South China Sea

ZHENANG CUI[1,2]*, YUEMING HOU[2,3], ZHEN XIA[1], KAI LIANG[1], QIAO XUE[1] & LIANG ZHANG[1]

[1]*Guangzhou Marine Geological Survey, Guangzhou 510760, China*

[2]*School of Marine Sciences, Guangxi University, Nanning 530004, China*

[3]*CNOOC China Ltd Shenzhen, Guangzhou 510240, China*

**Corresponding author (e-mail: cuizhenang@163.com)*

Abstract: Geochemical investigations of sediments from core GC22 from the Tonkin Gulf, South China Sea, have been carried out in order to reconstruct the palaeoenvironmental evolution of the area during the Holocene. Vertical variations in Al/Ti, K/Al and Mg/Al clearly indicate the degree of chemical weathering in the source area. Zr/Ti and SiO_2/Al_2O_3 recorded the history of current velocity changes, and La/Co v. La/Sc, combined with distribution patterns of rare earth elements, suggested that Hainan Island was the main source of the sediments during the Holocene. Based on the results of the analysis, the evolution of palaeoenvironments in the Tonkin Gulf can be divided into four stages: (1) 10.12–6.46 ka BP, the regional climate got warmer, and the sea level of the gulf rose rapidly, which is indicated by rapid declines in Sr/Ba and CaO. (2) 6.46–4.3 ka BP, the gulf had a stable depositional environment, and the local climate became cold and dry. (3) 4.3–3.55 ka BP, the currents and sedimenary provenance in the gulf were significantly influenced by the opening of the Qiongzhou Strait. (4) 3.55 ka BP–present, the regional sea level remained roughly stable.

Holocene environmental fluctuation has been a research hotspot in past global change studies because it is crucial in understanding the present climatic system and predicting future global changes (Feng *et al.* 2006; Dusar *et al.* 2011; Oswald & Foster 2011). In recent years, numerous palaeoclimatic and palaeoenvironmental studies have been carried out on the marine sediment record of the South China Sea. A strong link between sediment supply from the largest rivers of Asia (e.g. the Pearl and Mekong rivers) and East Asian monsoon activity during the Quaternary has been established on the basis of palaeoceanographic, mineralogical and geochemical studies on South China Sea sediments (Wei *et al.* 2004; Yang *et al.* 2008; Li *et al.* 2010; Liu *et al.* 2010). However, only a few such studies exist for the Gulf of Tonkin (Li *et al.* 2002, 2006, 2010; Wan *et al.* 2015). Thus far, the transport process and flux variability of the surrounding river sediments into the Gulf of Tonkin during the Quaternary remain unclear, and the changes in regional palaeoclimate and environment during the Holocene have been debated.

Previous palaeoclimate and sediment provenance studies in the Gulf of Tonkin concentrated on faunal assemblages, heavy minerals, clay minerals and the stable isotope record, but did not investigate the major and trace element chemistry (Fang *et al.* 1992; Li *et al.* 2010; Liu *et al.* 2010; Tanaka *et al.* 2011). Many of the major and trace elements are associated with the terrigenous materials in marine sediments, which provide important information about their origin, and can be used to reconstruct the chemical and physical conditions of the palaeoenvironment in the catchment areas.

In view of these earlier studies, the purpose of the present work is to document the major and trace element composition (including rare earth elements (REEs)) of the sediments in order to decipher the sedimentary provenance and source-area palaeoweathering using gravity core GC22, located at 107° 57′ 52″ E, 18° 28′ 36″ N in the Gulf of Tonkin (Fig. 1), and hence to define Holocene environmental changes in the regions.

Geographical and geological setting

The Gulf of Tonkin is a semi-closed gulf located in the NW of the South China Sea, and connected with the South China Sea through the southern end of the gulf and the Qiongzhou Strait, which is located between Hainan Island and the Leizhou Peninsula (Fig. 1). The seafloor of the gulf is flat, and slopes from the NW to the SE. The maximum depth of the water in the gulf is 100 m (Yu & Mu 2006). The Red River (Song Hong River) provides the major

From: Clift, P. D., Harff, J., Wu, J. & Qui, Y. (eds) 2016. *River-Dominated Shelf Sediments of East Asian Seas*. Geological Society, London, Special Publications, **429**, 73–85.
First published online December 9, 2015, updated March 3, 2016 and May 26, 2016, http://doi.org/10.1144/SP429.12

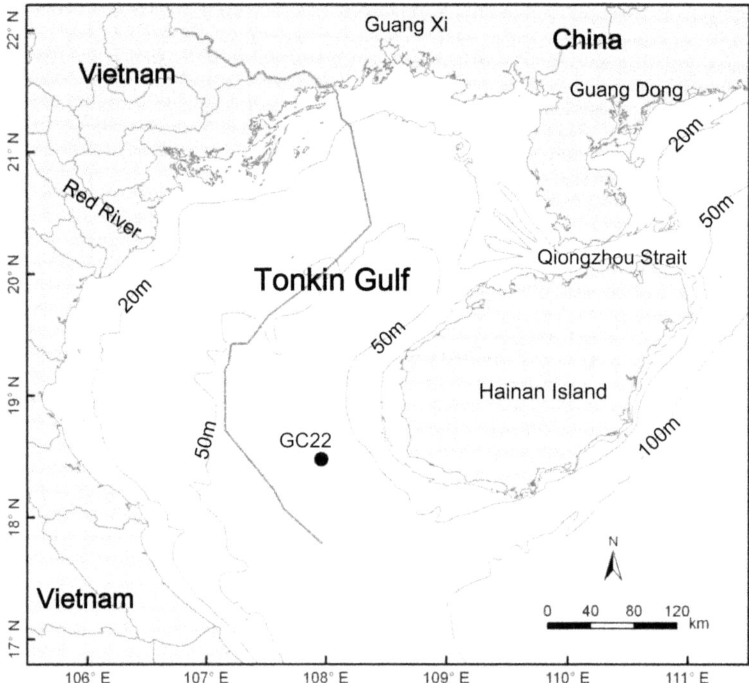

Fig. 1. Geographical setting of the Tonkin Gulf and the location of gravity core GC22. The grey thin line in the Tonkin Gulf is a bathymetric curve, and the broken line indicates the boundary between Vietnam and The People's Republic of China.

riverine discharge into the gulf, along with some smaller coastal rivers. Discharge from the Pearl River, about 400 km to the NE, may reach the gulf through the Qiongzhou Strait (Tang *et al.* 2003). In addition, coastal erosion is a significant source of sediments.

Located in the northern tropics, the Gulf of Tonkin region experiences a monsoonal climate driven by the Southwest Monsoon in summer and the Northeast Monsoon in winter. The seasonally reversing monsoon winds play an important role in the hydrology and general circulation in the Gulf of Tonkin. The beginning of the Northeast Monsoon (winter season) is in September, and the first appearance of the Southwest Monsoon is in May, which expands to cover the entire gulf during July and August (Shaw & Chao 1994). The average water temperature over the year is about 24°C, with a maximum surface temperature of 37°C and a bottom minimum temperature of 2°C; the annual average rainfall is about 1670 mm (Yu & Mu 2006).

Geologically, Cambrian–Recent strata outcrop on the land around the Gulf of Tonkin. Granites are the dominant rocks and occupy more than 60% of western Hainan. Basic igneous rocks of Tertiary and Quaternary age compose the basement of the Leizhou Peninsula, while Palaeozoic sandstones and shales are widely exposed in the coastal areas of Guangxi, China. Over most of the land area around the Gulf of Tonkin, the solid bedrock is unevenly mantled by Quaternary alluvium, regolith or the deeply weathered layers. The Red River drainage system has considerable differences in its lithologies, including limestone, and metamorphic, granitoid, felsic and ultramafic rocks. The sediments derived from these different rocks are intensively mixed within the estuary of the Red River.

The sedimentary evolution of the Gulf of Tonkin has been analysed in several studies using sediment cores (Shi *et al.* 2007; Yao *et al.* 2009). According to these studies, the oldest marine sediments deposited in the gulf since the Last Glacial Maximum (LGM) are dated at 12.0–11.5 cal ka BP, when the basin was transgressed during the last post-glacial transgression, during a continuous rise in sea level over a long period of time. As sea level continued to rise, the Qiongzhou Strait completely opened up around 8.5 ka and the sea level reached its present level at approximately 6.5 ka.

Material and methods

Our investigations concentrate on gravity core GC22, which was collected in the Gulf of Tonkin by the Chinese research vessel *Fendou 5* in 2009. Core GC22 is 280 cm long and was acquired in the southern part of the Gulf of Tonkin in water depths of 75 m, about 110 km east of Hainan Island (Fig. 1). Onboard, the core was cut into two halves, photographed, described, and then subsampled for sedimentological and geochemical analysis. The core was sampled at 5 cm intervals, and the outer rims of the samples were removed to avoid contamination, yielding a total of 56 subsamples (Cui *et al.* in press). These samples were stored in polyethylene bags and kept frozen at $-20^{\circ}C$ until chemical analysis was performed.

The grain size of the bulk samples was measured using a laser particle size analyser (Mastersizer 2000), after removing organic matter and biogenic carbonate from the samples with 10% H_2O_2 and 1 N HCl, respectively. Prior to geochemical analysis, the 56 bulk samples were desalinated with distilled water and dried at $60^{\circ}C$ in a clean oven, and then ground to a 200 mesh or less in an agate mortar. Ten major elements (Si, Al, Fe, Mn, Ca, Na, K, Ti and P) were determined by X-ray fluorescence spectrometry (Philips PW-2404). Thirteen trace elements (Sc, V, Cr, Co, Ni, Cu, Zn, Mg, Ga, Sr, Ba, Pb, Zr and Y) and REEs were measured usning an inductively coupled plasma mass spectrometer (ICP-MS, X Series 2).

The precision and accuracy of concentrations were monitored by repeated analyses of international geostandard MAG-1. The results showed that major and trace element concentrations were accurate to within ± 5 and $\pm 8\%$, respectively. The analysis of grain size and all of the geochemistry was carried out in the laboratory of the Guangzhou Marine Geology Survey (GMGS), China.

^{14}C age measurements from core GC22 were made on handpicked mono-specific planktonic foraminiferal species *Globigerinoides sacculifer* (250–500 μm size fraction). Accelerated mass spectrometer (AMS) ^{14}C age measurements were made at different intervals in the core at the State Key Laboratory of Nuclear Physics and Technology, Peking University, using 8–10 mg of ultrasonically cleaned samples of *Globigerinoides sacculifer*. Data were calibrated using CALIB version 7.02 calibration dataset.

Results

Lithological stratigraphy and chronology

Sediments from core GC22 were divided into four sedimentary units: units 1–4 in descending order, based on lithology, colour, sedimentary structures, texture, contact character, succession character and fossil components (Fig. 2).

Unit 1 consists of two subunits. The upper subunit (1A) (0–10 cm) is mostly composed of very soft to soft, greyish olive, homogenous silty clay. The mean grain size (MZ) is 6.25Φ. The lower subunit (1B) is 138 cm long (10–148 cm) and composed of dark greenish grey, plastic, homogenous silty clay with scattered shell fragments, needles of Echinoidea and some sandy nests (quartz, shell fragments and foraminifera).

Unit 2 consists of 32 cm (148–180 cm) of dark greenish grey, plastic, homogenous silty clay that includes a layer of accumulated bioclasts with shell fragments of mussels and snails. Volcanic debris was occasionally detected in this unit. Unit 3 (180–230 cm) is composed of dark greenish grey, plastic, homogenous silty sand with some bioclasts, foraminifera, needles and shells of Echinoidea, mussels and snail shells, as well as volcanic debris. At a depth of 195 cm, we identified a sharp, very thin layer (<1 cm thick) containing fine sand with almost homogenous grain size (Fig. 2). Unit 4 (230–280 cm) is made up of olive grey, plastic, homogenous silty clay with abundant shell fragments and sparse volcanic debris. A very large oyster shell, 10 cm in diameter, was found at the bottom of Unit 4.

The AMS ^{14}C dating results measured on benthic foraminifera (G. Sacculifer) yielded calibrated ages of 1300 ± 20 years BP at 63–65 cm, 4800 ± 27 years BP at 203–205 cm and $10\,122 \pm 32$ years BP at 278–280 cm, respectively. The age–depth model was based on linear interpolation between the mean calibrated ^{14}C ages so that average sedimentation rates could be obtained. Overall, sedimentation rates increased from the early Holocene to the late Holocene, with 0.16 mm a^{-1} for 10.122–4.8 ka BP, 0.40 mm a^{-1} for 4.8–1.3 ka BP and 0.54 mm a^{-1} for 1.3–0 ka BP.

Major and trace element compositions

The major and trace element compositions for the core sediments are given in Table 1, together with loss on ignition (LOI) and the mean grain size (MZ). Overall, the vertical distribution of most major and trace elements within the studied sedimentary sequence are closely related to lithology, and are different in the different units (Figs 3 & 4). The mean grain size ranges from 5.52 to 6.61Φ, with an average of 6.25Φ, indicating that silt dominates the core sediments. Among the major elements measured for the entire profile, SiO_2 is the dominant component, averaging 59.6%, followed by Al_2O_3, averaging 13.0%, and CaO, averaging 5.9%. The three elements amount to an average of more than 75%. The peak CaO content occurs in Unit 4, with an average value 7.4%, which can be attributed to the presence of shells or shell fragments (Cui *et al.* in press).

The relationships between the studied variables are presented by Pearson correlation coefficients

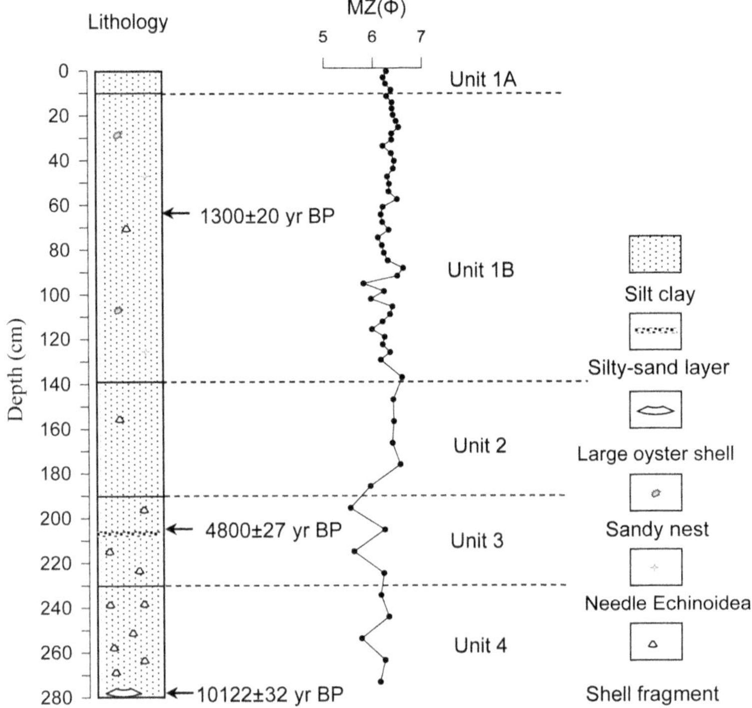

Fig. 2. Geological columns of the core GC22 and the downcore distribution of average grain size (MZ). Dashed horizontal lines indicate the boundaries between depositional units (units 1A, 1B, 2, 3 and 4). The three ages at depths of 64, 204 and 279 cm, which are marked with asterisks, were obtained by AMS ^{14}C dating.

(Table 2; Cui *et al.* in press). As expected, there is a negative correlation between SiO_2 and Al_2O_3 (−0.19). Similarly, Fe_2O_3, TiO_2, K_2O, Na_2O, MgO, P_2O_3 and most trace elements (with the exception of Sr) have strong or moderate positive correlations with Al_2O_3, indicating that these major elements are associated with aluminous clay minerals. The negative correlation of CaO with SiO_2 and Al_2O_3 suggests that the high content of CaO in the sediments is caused by a high concentration of calcium carbonate from biogenic shells. The significant relationship between CaO and Sr suggests that they are derived from the same source. Overall, there is no clear relationship between grain-size compositions and elemental concentrations in the sediments (Table 2), although some trace elements (e.g. Zn, Pb, Rb, Co, V, Fe and Mn) have a weak positive relationship with the mean grain size, which indicates that grain size may not be the dominant controlling factor over sediment geochemistry.

Rare earth element (REE)

Concentrations of REEs, as well as light/heavy REEs (LREE/HREE), La/Sm, Gd/Yb, δEu and δCe in GC22 sediments are presented in Table 3. The ratio of LREEs (La to Sm) to HREEs (Gd to Lu) is used to indicate the fractionation of total REEs between light and heavy REEs. La/Sm and Gd/Yb ratios indicate fractionation within LREE and HREEs, respectively. The Eu (δEu) and Ce (δCe) anomalies are defined as $Eu_N/[(Sm_N)(Gd_N)]^{1/2}$ and $Ce_N/[(La_N)(Pr_N)]^{1/2}$, respectively, where N indicates chondrite normalization.

In chondrite-normalized REE plots (Fig. 5) (data from Masuda *et al.* 1973), the four units of core GC22 show very similar patterns, including their abundances, with a higher concentration of LREE than HREE and a distinct negative Eu anomaly. The uniformity in the REE patterns suggests that the sediments may derive from a homogenous source (McLennan 1989). In general, the REE patterns of the GC22 sediments are very similar to those of chondrite-normalized upper continental crust (UCC). Compared with the UCC, the lower LREE/HREE and less prominent Eu anomaly of the core sediment suggests a less-fractionated sediment provenance than in average upper-crustal sources.

As for the major and trace elements, the downcore distributions of the REEs and REE fractionation parameters for core GC22 match well

Table 1. *Concentrations of major (%) and trace (ppm) elements, LOI (%) and mean grain size (Φ) in GC22 sediments*

		MZ	SiO_2	Al_2O_3	Fe_2O_3	CaO	MgO	K_2O	Na_2O	MnO	P_2O_5	TiO_2	LOI	Co	Cu	Ni	Pb	Cr	Sr	Zn	Zr	Sc	V	Ga	Ba	Y	La
GC22	Min	5.52	56.0	11.3	4.11	5.26	2.03	2.17	0.76	0.07	0.07	0.53	8.70	9.10	9.37	26.7	18.7	54.1	202.0	68.2	56.0	9.7	58.8	13.3	282.0	21.9	29.3
samples	Max	6.61	61.8	14.4	5.24	9.35	2.67	2.76	2.18	0.08	0.12	0.71	11.8	13.5	15.4	33.1	27.4	79.5	320.0	85.2	79.8	12.7	86.5	18.9	371.0	28.5	39.9
(n = 56)	Mean	6.25	59.6	13.0	4.71	5.86	2.32	2.46	1.69	0.08	0.10	0.61	9.6	11.8	11.6	29.4	22.3	65.8	236.2	76.7	71.9	11.3	74.1	16.4	331.5	24.1	36.9
	CV	0.04	0.01	0.1	0.04	0.13	0.05	0.06	0.10	0.04	0.11	0.05	0.07	0.09	0.13	0.05	0.09	0.12	0.08	0.05	0.08	0.06	0.10	0.08	0.07	0.05	0.05
Unit1	Min	6.10	57.3	12.6	4.50	5.30	2.22	2.33	0.76	0.07	0.09	0.57	8.70	11.4	9.37	26.7	20.5	54.5	217	73.9	69.4	10.4	69.7	15.1	322.5	22.3	35.5
(n = 30)	Max	6.61	61.3	14.4	5.24	5.93	2.67	2.76	2.18	0.08	0.12	0.71	10.2	13.5	15.4	33.1	27.4	77.0	251	85.2	79.8	12.7	86.5	18.9	371.0	28.5	39.9
	Mean	6.33	59.5	13.4	4.79	5.62	2.38	2.51	1.70	0.07	0.10	0.61	9.53	12.3	12.1	29.1	23.7	65.8	234	78.3	75.5	11.5	78.8	17.1	345.6	24.6	37.3
	CV	0.02	0.01	0.03	0.03	0.03	0.04	0.04	0.12	0.03	0.07	0.05	0.05	0.05	0.15	0.04	0.07	0.13	0.04	0.03	0.04	0.05	0.06	0.06	0.03	0.05	0.03
Unit 2	Min	5.80	58.2	12.1	4.44	5.26	2.25	2.52	1.66	0.07	0.09	0.56	8.83	11.1	9.66	27.6	20.2	54.1	224	73.9	66.3	10.7	69.2	15.1	327.0	21.9	33.5
(n = 6)	Max	6.39	61.0	13.2	4.78	6.29	2.46	2.75	1.78	0.08	0.10	0.63	9.78	12.9	10.5	29.2	22.5	59.8	245	78.0	76.4	[illegible]	76.6	16.9	347.0	24.5	36.9
	Mean	6.15	60.0	12.6	4.56	5.73	2.35	2.62	1.72	0.07	0.10	0.59	9.18	12.1	10.1	28.4	21.6	56.7	234	75.8	72.9	10.9	73.3	15.8	337.3	23.1	35.7
	CV	0.04	0.02	0.03	0.03	0.06	0.04	0.03	0.02	0.03	0.06	0.05	0.05	0.06	0.03	0.02	0.05	0.04	0.04	0.02	0.06	0.02	0.04	0.04	0.03	0.05	0.04
Unit 3	Min	5.97	58.8	11.4	4.11	5.34	2.04	2.18	1.65	0.07	0.09	0.56	9.00	9.10	9.94	27.4	18.7	55.4	216	68.3	63.3	9.7	59.8	13.3	291.0	22.5	29.3
(n = 10)	Max	6.58	60.6	13.6	4.94	6.88	2.46	2.58	1.76	0.08	0.10	0.67	9.99	13.0	12.1	32.9	22.7	79.5	249	81.6	76.5	12.1	75.1	17.4	356.0	24.7	39.2
	Mean	6.32	59.6	12.8	4.65	5.72	2.30	2.38	1.71	0.08	0.09	0.62	9.64	11.2	11.2	30.4	20.2	70.2	231	77.4	70.0	11.3	70.2	16.2	320.4	23.5	36.7
	CV	0.03	0.01	0.05	0.05	0.08	0.05	0.05	0.02	0.04	0.04	0.05	0.04	0.10	0.07	0.06	0.07	0.12	0.05	0.05	0.05	0.06	0.06	0.07	0.06	0.03	0.08
Unit 4	Min	5.52	56.0	11.3	4.19	5.44	2.03	2.17	1.45	0.07	0.07	0.53	9.11	9.50	10.4	27.7	19.2	60.7	202	68.2	56.0	10.0	58.8	13.5	282.0	22.2	34.8
(n = 10)	Max	6.30	61.8	13.0	4.92	9.35	2.29	2.40	1.84	0.08	0.11	0.63	11.83	11.8	12.3	32.3	21.9	74.1	320	76.0	67.2	11.5	69.2	16.0	309.0	24.7	39.9
	Mean	6.00	59.3	12.1	4.60	6.81	2.16	2.31	1.58	0.08	0.09	0.59	10.17	10.6	11.3	30.0	20.7	67.2	249	71.9	62.6	10.7	64.2	14.8	296.9	23.7	36.9
	CV	0.05	0.04	0.05	0.05	0.21	0.04	0.04	0.08	0.04	0.14	0.04	0.11	0.08	0.06	0.06	0.05	0.06	0.17	0.04	0.06	0.05	0.05	0.06	0.03	0.03	0.04

CV, coefficient of variation; n, number of samples, LOI, loss on ignition.

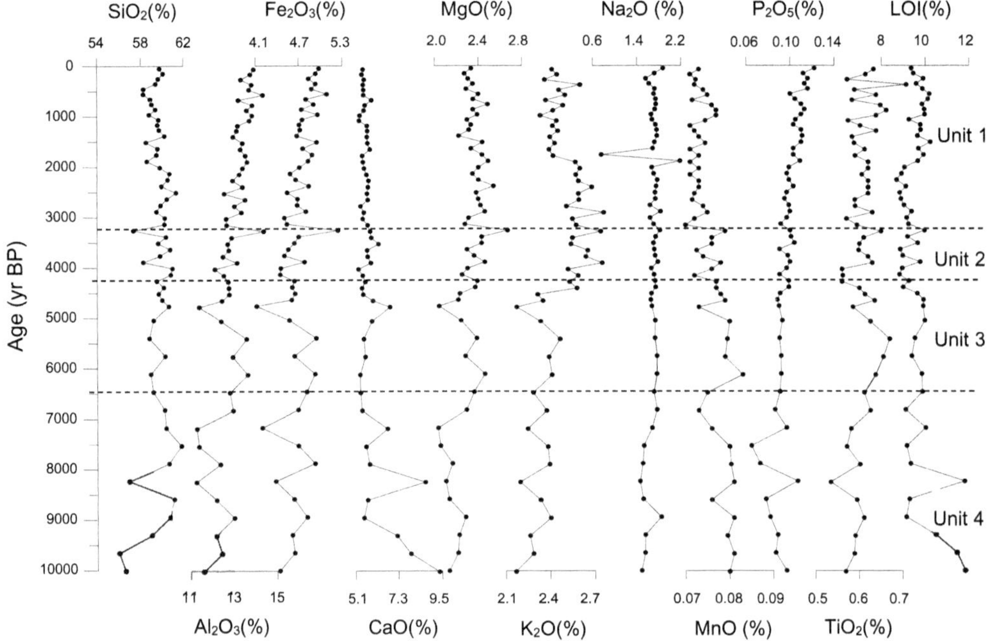

Fig. 3. Downcore variations of major elements in sediments recovered by core GC22.

with the lithological units (Fig. 6). The REE fractionation parameters (LREE/HREE, La/Sm, Gd/Yb, δEu and δCe) show extremely poor correlations with mean grain size, suggesting that REE fractionation is not controlled by the sediment grain size (Cui *et al.* in press).

Discussion

Analysis of geochemical controls

The chemical composition of clastic sediment is the net result of a number of geological factors. These include source-rock composition, the intensity of

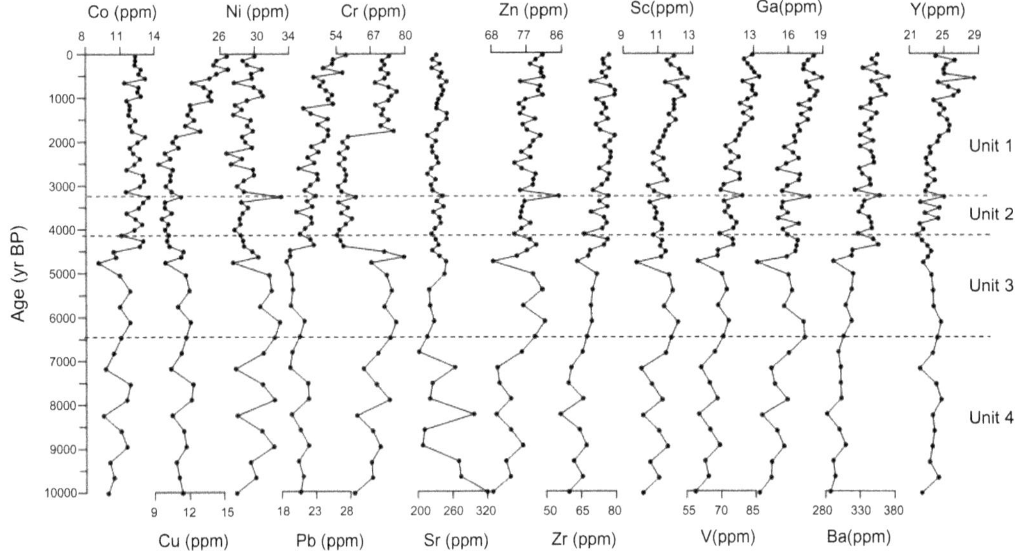

Fig. 4. Downcore variations of trace elements in sediments recovered in core GC22.

Table 2. *Pearson's correlation matrix for major and trace elements, MZ in GC22 sediments (Cui* et al. *in press)*

	MZ	SiO_2	Al_2O_3	Fe_2O_3	CaO	MgO	K_2O	Na_2O	MnO	P_2O_5	TiO_2	LOI	Co	Cu	Ni	Pb	Cr	Sr	Zn	Zr	Sc	V	Ga	Ba	Y	REEs	la/Sm	Gd/Yb	δEu	δCe
SiO_2	−0.20																													
Al_2O_3	**0.52**	−0.19																												
Fe_2O_3	0.41	−0.28	**0.84**																											
CaO	−0.29	**−0.55**	**−0.59**	−0.45																										
MgO	0.42	−0.20	**0.78**	**0.64**	−0.47																									
K_2O	0.17	0.20	0.48	0.30	**−0.51**	**0.71**																								
Na_2O	0.18	0.05	0.27	0.11	−0.27	0.30	0.35																							
MnO	−0.08	−0.35	−0.30	0.04	0.38	−0.20	−0.35	−0.15																						
P_2O_5	0.30	−0.33	**0.55**	0.30	−0.10	0.42	0.16	0.18	−0.43																					
TiO_2	0.33	−0.07	**0.58**	**0.53**	−0.36	0.45	0.30	0.31	0.05	0.16																				
LOI	0.02	**−0.83**	−0.21	−0.02	**0.78**	−0.27	**−0.62**	−0.29	0.44	0.16	−0.14																			
Co	0.32	0.02	**0.70**	**0.63**	**−0.60**	**0.79**	**0.75**	0.28	−0.26	0.43	0.35	−0.42																		
Cu	0.30	−0.21	**0.55**	**0.62**	−0.23	0.01	−0.24	−0.12	−0.06	0.44	0.33	0.21	0.18																	
Ni	0.22	−0.13	0.30	**0.55**	−0.24	0.15	−0.12	−0.04	**0.52**	−0.32	0.47	0.04	0.05	0.38																
Pb	0.26	−0.12	**0.64**	**0.58**	−0.37	0.44	0.28	0.07	−0.41	**0.66**	0.11	−0.10	**0.66**	**0.58**	−0.09															
Cr	0.20	−0.19	0.25	0.42	−0.12	−0.19	**−0.57**	−0.20	0.27	0.09	0.34	0.35	−0.21	**0.75**	**0.55**	0.14														
Sr	−0.13	**−0.64**	−0.38	−0.30	**0.87**	−0.30	−0.44	−0.26	0.32	0.23	−0.33	**0.82**	−0.39	−0.03	−0.32	−0.09	0.00													
Zn	**0.54**	−0.18	**0.89**	**0.77**	**−0.65**	**0.80**	**0.52**	0.34	−0.14	0.48	**0.58**	−0.26	**0.77**	0.41	0.41	**0.53**	0.17	−0.43												
Zr	0.38	0.03	**0.75**	**0.50**	**−0.61**	**0.79**	**0.66**	0.35	−0.44	**0.62**	0.45	−0.41	**0.85**	0.19	−0.08	**0.65**	−0.12	−0.35	**0.78**											
Sc	**0.50**	−0.21	**0.81**	**0.79**	**−0.56**	0.48	0.14	0.19	−0.05	0.44	**0.54**	−0.03	**0.51**	**0.73**	0.47	**0.54**	**0.60**	−0.31	**0.77**	**0.57**										
V	0.45	−0.09	**0.85**	**0.70**	**−0.61**	**0.65**	0.44	0.22	−0.42	**0.72**	0.40	−0.21	**0.77**	**0.57**	0.03	**0.79**	0.23	−0.26	**0.78**	**0.84**	**0.82**									
Ga	0.45	−0.12	**0.85**	**0.77**	**−0.65**	**0.60**	0.32	0.22	−0.26	**0.58**	0.45	−0.17	**0.66**	**0.63**	0.23	**0.68**	0.39	−0.33	**0.82**	**0.71**	**0.88**	**0.92**								
Ba	0.40	0.04	**0.72**	**0.50**	**−0.60**	**0.67**	**0.59**	0.22	−0.43	**0.65**	0.25	−0.36	**0.82**	0.29	−0.18	**0.72**	−0.07	−0.22	**0.70**	**0.88**	**0.57**	**0.89**	**0.78**							
Y	0.28	−0.29	**0.60**	**0.66**	−0.24	0.25	−0.08	−0.08	−0.07	0.37	0.38	0.23	0.35	**0.69**	0.33	**0.51**	**0.60**	−0.04	0.46	0.35	**0.71**	**0.62**	**0.62**	0.41						
REEs	0.39	−0.15	**0.56**	**0.59**	−0.30	0.31	0.04	0.09	−0.05	0.12	**0.58**	0.03	0.17	0.44	**0.54**	0.17	**0.50**	−0.26	0.48	0.23	**0.55**	0.37	0.43	0.15	**0.68**					
la/Sm	−0.08	0.05	−0.38	−0.31	0.19	−0.49	−0.41	0.00	0.23	**−0.53**	−0.14	0.16	**−0.66**	−0.13	0.25	**−0.62**	0.19	−0.05	−0.37	**−0.64**	−0.19	**−0.56**	−0.36	**−0.64**	−0.19	0.16				
Gd/Yb	0.20	−0.15	0.17	0.10	−0.03	0.37	0.36	0.47	0.02	0.24	0.39	−0.04	0.29	−0.17	−0.08	0.08	−0.19	0.05	0.27	0.35	0.08	0.21	0.13	0.23	−0.04	0.16	−0.32			
δEu	−0.06	0.25	−0.21	−0.21	−0.09	−0.21	−0.06	−0.16	−0.16	−0.07	−0.46	−0.12	0.00	−0.02	−0.18	−0.08	−0.13	−0.07	−0.16	−0.10	−0.10	−0.02	−0.05	0.07	−0.10	−0.39	0.13	−0.40		
δCe	−0.13	0.21	−0.39	−0.27	0.05	**−0.50**	**−0.52**	−0.11	0.25	−0.38	−0.06	0.04	**−0.56**	0.01	0.30	−0.47	0.34	−0.16	−0.31	**−0.56**	−0.17	−0.46	−0.32	**−0.60**	−0.18	0.01	0.46	−0.31	−0.04	
LREE/HREE	0.13	0.04	−0.03	−0.07	0.09	−0.36	−0.37	−0.01	−0.11	−0.08	0.06	0.14	−0.49	0.28	0.16	−0.18	0.37	−0.02	−0.19	−0.35	0.06	−0.16	−0.06	−0.34	0.09	0.48	**0.69**	−0.20	−0.09	0.33

The significant coefficients are given in bold. The results are based on 56 samples ($n = 56$). LOI, loss on ignition.

Table 3. *Concentrations of REEs (ppm) and LREE/HREE, δEu, $(La/Sm)_{NASC}$ and $(Gd/Yb)_{NASC}$ ratios in GC22 sediments*

		La	Ce	Pr	Nd	Sm	Eu	Gd	Tb	Dy	Ho	Er	Tm	Yb	Lu	LREE/ HREE	δEu	δCe	La/Sm	Gd/Yb
All Samples ($n = 56$)	Min	29.3	61.0	6.91	26.0	5.05	1.06	4.48	0.72	4.24	0.78	2.24	0.36	2.32	0.34	7.08	0.61	0.91	0.87	0.98
	Max	39.9	79.4	9.31	32.8	6.43	1.29	5.69	0.89	5.03	0.98	2.90	0.44	3.03	0.47	9.15	0.70	0.98	1.07	1.30
	Mean	36.9	74.5	8.38	30.6	5.81	1.16	5.12	0.80	4.61	0.88	2.56	0.40	2.56	0.40	8.46	0.66	0.94	0.95	1.16
	CV	0.05	0.04	0.05	0.04	0.05	0.04	0.05	0.04	0.04	0.04	0.05	0.05	0.04	0.06	0.04	0.03	0.02	0.04	0.05
Unit 1 ($n = 30$)	Min	35.5	70.6	8.18	29.3	5.60	1.13	4.79	0.76	4.40	0.83	2.44	0.37	2.42	0.38	8.11	0.61	0.91	0.89	0.98
	Max	39.9	78.9	9.31	32.8	6.43	1.29	5.69	0.89	5.03	0.98	2.90	0.43	3.03	0.47	8.88	0.69	0.96	0.99	1.30
	Mean	37.3	75.4	8.64	30.8	5.98	1.18	5.17	0.81	4.63	0.89	2.61	0.40	2.58	0.41	8.48	0.65	0.94	0.94	1.16
	CV	0.03	0.03	0.03	0.03	0.03	0.03	0.04	0.03	0.03	0.04	0.04	0.04	0.05	0.05	0.03	0.02	0.01	0.03	0.05
Unit 2 ($n = 6$)	Min	33.5	67.3	7.75	28.5	5.36	1.10	4.66	0.72	4.24	0.80	2.32	0.36	2.32	0.38	7.86	0.63	0.92	0.90	1.09
	Max	36.9	76.1	8.73	31.7	6.09	1.22	5.69	0.83	4.96	0.93	2.81	0.42	2.64	0.44	8.80	0.70	0.95	1.00	1.27
	Mean	35.7	72.2	8.25	29.9	5.63	1.16	5.16	0.78	4.57	0.85	2.57	0.39	2.51	0.40	8.26	0.66	0.94	0.95	1.19
	CV	0.04	0.04	0.05	0.04	0.05	0.04	0.07	0.05	0.06	0.06	0.06	0.06	0.05	0.06	0.04	0.04	0.01	0.04	0.06
Unit 3 ($n = 10$)	Min	29.3	61.0	6.91	26.0	5.05	1.08	4.69	0.74	4.31	0.84	2.35	0.37	2.43	0.34	7.08	0.63	0.91	0.87	1.03
	Max	39.2	79.4	8.53	32.0	5.98	1.21	5.50	0.82	4.80	0.89	2.65	0.44	2.73	0.43	9.15	0.69	0.98	1.07	1.27
	Mean	36.7	74.0	8.06	30.5	5.66	1.15	5.20	0.79	4.60	0.87	2.54	0.41	2.57	0.40	8.37	0.65	0.96	0.97	1.17
	CV	0.08	0.08	0.06	0.06	0.05	0.04	0.06	0.03	0.04	0.02	0.04	0.06	0.04	0.06	0.06	0.03	0.02	0.06	0.06
Unit 4 ($n = 10$)	Min	34.8	68.7	7.41	28.5	5.24	1.06	4.48	0.72	4.24	0.78	2.24	0.37	2.38	0.36	8.17	0.64	0.92	0.93	1.05
	Max	39.9	79.1	8.55	31.1	5.81	1.19	5.26	0.84	4.86	0.92	2.60	0.41	2.68	0.43	9.08	0.70	0.98	1.05	1.18
	Mean	36.9	73.8	8.01	30.1	5.58	1.11	4.86	0.78	4.55	0.86	2.43	0.39	2.53	0.39	8.63	0.66	0.96	0.99	1.11
	CV	0.04	0.04	0.04	0.03	0.03	0.04	0.05	0.05	0.04	0.05	0.05	0.04	0.04	0.06	0.03	0.03	0.02	0.03	0.04

CV, coefficient of variation; n, number of samples.

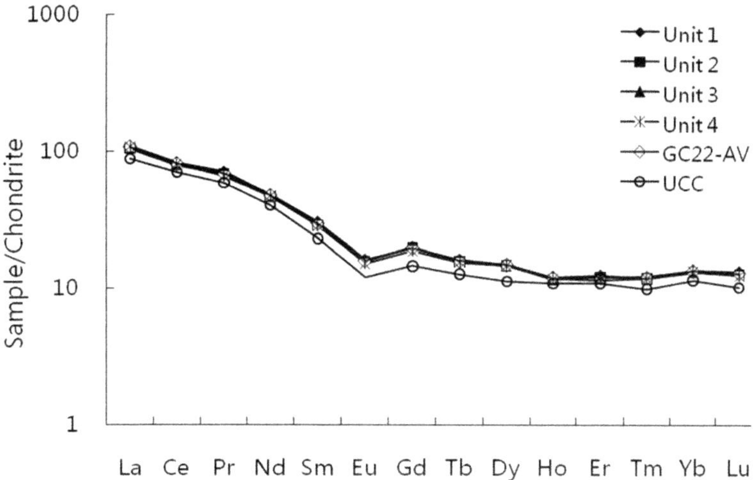

Fig. 5. Chondrite-normalized REE distribution patterns of the sediments from three units at GC22 and the normalized average over the core (GC22-AV) show a similar trend, and, compared with UCC, the core sediments plot closer to UCC and therefore indicate a cratonic source.

weathering, the rate of sediment supply, sorting during transportation and deposition (McLennan 1989).

During chemical weathering, Al is resistant to leaching and is enriched in weathering products (Nesbitt & Young 1982; Nesbitt & Markovics 1997). K and Mg are easily removed from primary minerals during chemical weathering. However, they tend to be combined into the weathering products and are generally enriched in weathering profiles (Weaver 1967; Nesbitt *et al.* 1980). Enhanced chemical weathering may result in lower Al/K and Al/Mg in weathering products if the degree of chemical weathering is not extremely high. Sediments from core GC22 show higher K/Al and Mg/Al values in units 1 and 2, and relatively low K/Al and Mg/Al ratios in units 3 and 4 (Fig. 7). These patterns

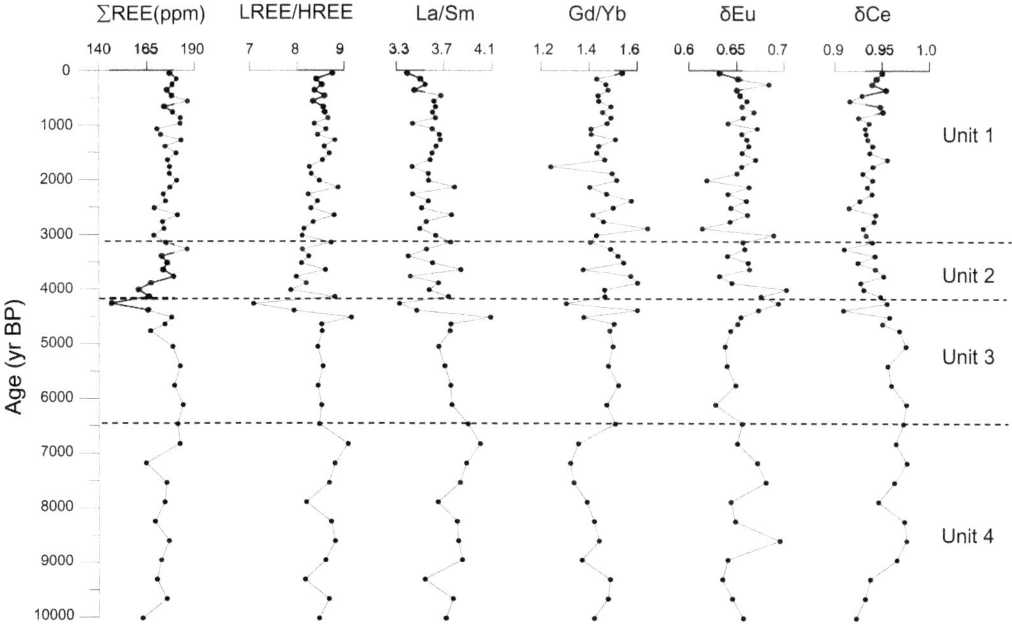

Fig. 6. Downcore variations of rare earth elements in sediments recovered in core GC22.

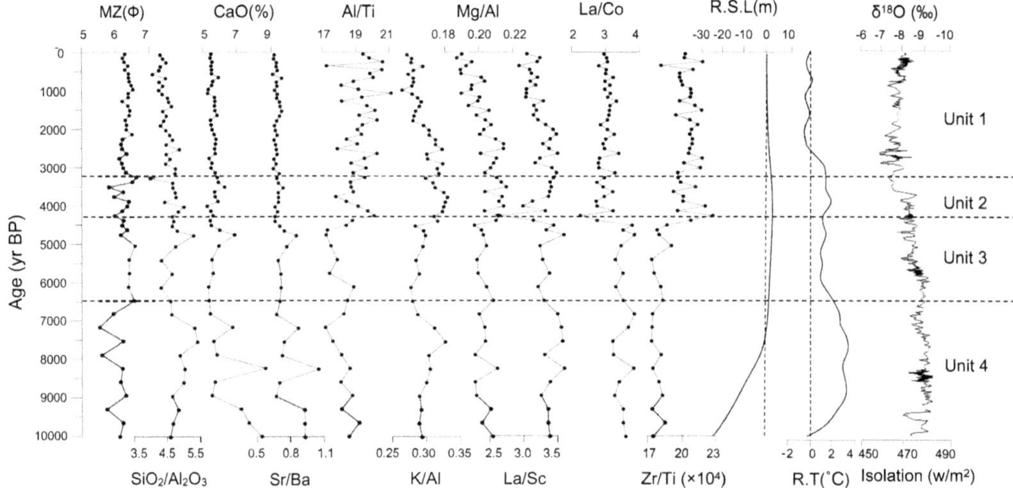

Fig. 7. Downcore variations of select indexes in sediments recovered in core GC22. The curves of relative sea level (Tanabe *et al.* 2003) and relative temperature (Li *et al.* 1997) indicate the sea-level and climate changes in the northern South China Sea during the Holocene. The curves of oxygen isotope and average summer insolation were from stalagmite D4 in Dongge cave, China (Carolyn *et al.* 2005).

may indicate that chemical weathering was stronger in the early Holocene than during the middle and late Holocene.

Al_2O_3, TiO_2 and Zr are relatively immobile during chemical weathering. They are, however, influenced by sorting during sedimentary processes, and mineral fractionation is an important factor controlling the composition of finer sediments (Taylor & McLennan 1985; Condie 1993; Dellwig *et al.* 2000; Wei *et al.* 2004; Roddaz *et al.* 2006). Heavy mineral enrichments are often associated with particle sorting effects caused by wave action and high current velocities. Thus, the observed high Zr levels provide an indication of a possible rise in energetic conditions from the Holocene to the present system.

The Zr/Ti ratio is known to be a variable related to grain-size mineral fractionation, where Ti is concentrated in finer minerals such as rutile, while Zr is relatively enriched in coarser flux such as zircon (Taylor & McLennan 1985; Garcia *et al.* 1994; Ishiga *et al.* 2000). The Zr/Ti ratio displays a systematic variation with an obvious shift to higher values above a depth of 248 cm (Fig. 7). The upper part of the section has higher Zr/Ti values than the lower part, suggesting the upper sediments were the result of sedimentation under relatively higher energy conditions. The SiO_2/Al_2O_3 ratio provides an estimate of the sediment textural maturity. This is because of the relative enrichment of Al-rich phyllosilicates at the expense of Si-rich phases in the fine-grained sediments (Weltje & Eynatten 2004; López-González *et al.* 2006; Roddaz *et al.* 2006). Thus, a decrease in the SiO_2/Al_2O_3 ratio is generally related to a decrease in the textural maturity. SiO_2/Al_2O_3 ratios for core GC22 sediments vary from 4.0 to 5.4, with the mean value of 4.6, which is higher than 3.3 found in post-Archaean Australian Shale (Brijraj *et al.* 2006). It is noteworthy that there is also a slight decrease in SiO_2/Al_2O_3 at a depth of 248 cm, and a continuous decrease is also visible in the upper part, indicating the gradually increasing sediment cycle, which is in agreement with previous results of Zr/Ti.

Although factors such as chemical weathering and hydrodynamic sorting can affect the sediment composition, source composition also plays a major role in controlling the composition of sediments. To investigate the origin of core GC22 sediments, Cui *et al.* (in press) used La/Co and La/Sc values combined with distribution patterns of REEs. They suggested that western Hainan Island was the main source of sediment. Moreover, the eastern Qiongzhou Strait is also a very important source of sediments to the core location. Contributions from this source varied spatially and temporally with strait evolution. With the opening of the Qiongzhou Strait, the provenance and hydrodynamic conditions of the Tonkin Gulf underwent a significant change. At around 4.5 ka, significant volumes of material from the eastern area of the Qiongzhou Strait were transported into the study area. The Red River, however, contributed only slightly, if at all, to the core sediments during the entire Holocene.

As mentioned above, chemical weathering, hydraulic sorting and provenance changes had changed elemental concentration distribution, which provided useful information for palaeoenvironmental research.

Depositional history and environmental changes

The variations in the geochemistry of core GC22 sediments can be used as a proxy for the changes in the depositional environment. We now combine geochemical proxies with the lithologies in GC22 to reconstruct a history of depositional environments in the Gulf of Tonkin (Fig. 7).

At the beginning of Holocene (10.12–6.46 ka BP), when the climate warmed and sea level rose rapidly, core GC22 was close to the land and in a shallow-water environment, as confirmed by the abundant fossil remains and the large oyster shell found in Unit 4. The relatively large variability and large grain size of sediments compared to the other units suggests dynamic and high-energy conditions. The sediments in Unit 4 are also distinguished by their high absolute concentrations of CaO and high Sr/Ba values, which suggest that the core site was in an offshore marine area and closer to a land area and/or estuary than other units. The lower K/Al and Mg/Al are interpreted to indicate that the sediments experienced relatively strong chemical weathering, indicating a warm and humid climate, which is consistent with the palaeoclimate records on shores such as those from the Dongge cave (Carolyn *et al.* 2005). The short transportation distance from the source areas may lead to the poorly sorted sediments in this unit, which was characterized by lower Zr/Ti and higher SiO_2/Al_2O_3 values.

Unit 3 (230–180 cm) is distinguished from Unit 4 by a sharp increase in grain size. Deposition started in 6462 years BP and ended at 4300 years BP (interpolated ages). Sediments of this unit have a lower abundance of shell remains and concentrations of CaO compared to Unit 4. Moreover, Sr/Ba values also display a sharp decrease in Unit 3 compared with Unit 4. Thus, the core location is inferred to have been located far from the land area in Unit 3 times compared to Unit 4, reflecting the higher sea level. Numerous studies have suggested that there was a period of higher sea level at 6–4 ka BP in the Gulf of Tonkin (Tanabe *et al.* 2003; Huang & Wang 2007; Sun *et al.* 2009; Yao *et al.* 2009), which is consistent with our interpretation. SiO_2/Al_2O_3, a proxy for the textual maturity of sediments, also shows a decrease at the boundary between units 3 and 4. This may result from the longer transport distance from the source area caused by a rise in sea level, but is not caused by greater hydraulic sorting, given that the sediments have similar ratios of Zr/Ti. The similar values of K/Al and Mg/Al in Unit 3 indicate similar weathering conditions compared to Unit 4. Meanwhile, the high sea level reworked loose coastal sediments and soils in the inundated zone, and these were transported out to sea and dispersed with the tide into other coastal areas of the Gulf of Tonkin. These materials may also have been transferred into the central areas of the Gulf of Tonkin and deposited in low-energy conditions. This may explain the increasing portion of terrestrial matter. We note that there is a 1 cm-thick silty-sand layer at a depth of 205 cm (*c.* 4.9 ka BP). The uniform grain size of the sediments, and the sharp and erosional contact with the lower sediments, together with the higher Sr/Ba and CaO values, are indicative of a deposit generated under high-energy conditions. Therefore, we can infer that the area might have experienced bottom current activity at around 4.9 ka BP.

The sharp decrease in K/Al, Mg/Al, La/Sc, La/Co and Zr/Ti between units 2 and 3 may indicate a significant change of palaeoenvironment. The opening of the Qiongzhou Strait greatly influenced the current in the Tonkin Gulf (Cui *et al.* in press), which might have resulted in a sharp increase in depositional energy, causing an enrichment of heavy minerals (e.g. zircon) in sediments, as indicated by a sharp increase in Zr/Ti values. Meanwhile, the materials from the eastern Qiongzhou Strait (including the material from Pearl River) could have been transported into the Tonkin Gulf. This would represent a significant change in the supply of sediment. In addition, values of K/Al and Mg/Al in the core sediments would have been influenced by climate fluctuation impacting on the chemical weathering process.

The geochemical characteristics of Unit 1 are similar to those of Unit 2 (Fig. 7). There is only a small decrease in grain size from 6.1Φ in Unit 2 to 6.3Φ in Unit 1. This may indicate a similar depositional environment, and the major sediment flux still being derived from western Hainan. In view of the cold and dry climate of this period (Li *et al.* 1997), the downward trend in K/Al and Mg/Al values may be attributed to changes in sediment provenance.

Conclusions

The major, trace and rare earth element composition of sediments from core GC22 recorded Holocene environmental conditions around the Gulf of Tonkin, South China Sea. Based on the geochemical data, in conjunction with the lithological stratigraphy, the main conclusions are as follows:

- The major, trace and rare earth element chemistry of core GC22 sediments reveal that they were derived mostly from Hainan Island during the whole of the Holocene. The effect of weathering and physical sorting on the composition of the sediments is negligible.

- At 10.12–6.46 ka BP, site GC22 was in a shallow-water environment with a warm and humid climate; at 6.46–4.3 ka BP, with a high and stable sea level, the location had a stable deposition environment; at 4.3–3.55 ka BP, the Qiongzhou Strait significantly influenced the sediment provenance and the strength of currents in the Gulf of Tonkin; and at 3.55 ka BP–present, the sediment provenance may have changed slightly under the influence of basic stability of the regional sea level.

This work was supported by both China Geological Survey (CGS) research program (Grant No. 1212010914027) and BMBF (Grant No. 03F0607A) in Germany. The authors would like to express sincere thanks to HIGS (Hainan Institute of Geological Survey, China) for providing some geochemical data of Hainan west soils. We are especially grateful to Profs P. D. Clift and Q. Yan, and the two anonymous reviewers for their thorough and incisive comments that substantially improved the manuscript. Although Detlef E. Schulz-Bull and Joanna J. Waniek were not authors on this paper, we would like to express thanks for their useful work.

References

Brijraj, K., Das, A. S. *et al.* 2006. Geochemistry of Mansar Lake sediments, Jammu, India: implication for source-area weathering, provenance, and tectonic setting. *Journal of Asian Earth Sciences*, **26**, 649–668.

Carolyn, A., Dykoski, R. *et al.* 2005. A high-resolution, absolute-dated Holocene and deglacial Asian monsoon record from Dongge Cave, China. *Earth and Planetary Science Letters*, **233**, 71–86.

Condie, K. C. 1993. Chemical composition and evolution of the upper continental crust: contrasting results from surface samples and shales. *Chemical Geology*, **104**, 1–37.

Cui, Z. A., Schulz-Bull, D. E. *et al.* In press. Geochemical characteristics and provenance of Holocene sediments (Core STAT22) in the Beibu Gulf, South China Sea. *Journal of Coastal Research*, http://doi.org/10.2112/JCOASTRES-D-14-00238.1

Dellwig, O., Hinrichs, J. *et al.* 2000. Changing sedimentation in tidal flat sediments of the southern North Sea from the Holocene to the present: a geochemical approach. *Journal of Sea Research*, **44**, 195–208.

Dusar, B., Verstraeten, G. *et al.* 2011. Holocene environmental change and its impact on sediment dynamics in the Eastern Mediterranean. *Earth-Science Reviews*, **108**, 137–157.

Fang, Z., Zhao, J. X. & Mcculloch, M. T. 1992. Geochemical and Nd isotopic study of Paleozoic bimodal volcanics in Hainan Island, South China – Implications for rifting tectonics and mantle reservoirs. *Lithos*, **29**, 127–113.

Feng, Z. D., Tang, L. Y. *et al.* 2006. Holocene vegetation variations and the associated environmental changes in the western part of the Chinese Loess Plateau. *Palaeogeography, Palaeoclimatology, Palaeoecology*, **241**, 440–456.

Garcia, D., Fonteiller, S. M. & Moutte, J. 1994. Sedimentary fractionations between Al, Ti, and Zr and genesis of strongly peraluminous granites. *Journal of Geology*, **102**, 411–422.

Huang, W. & Wang, P. X. 2007. Accumulation rate characterstics of deep-water sedimentation in the South China Sea during the last glaciation and the Holocenen. *Acta Oceanologica Sinica*, **29**, 69–73 [in Chinese].

Ishiga, H., Nakamura, T. *et al.* 2000. Geochemical record of the Holocene Jomon transgression and human activity in coastal lagoon sediments of the San'in district, SW Japan. *Global and Planetary Change*, **25**, 223–237.

Li, X. J., Han, J. X. *et al.* 1997. Holocence palaeclimatic changes in the northern South China Sea. *Geological Research of South China Sea*, **9**, 86–114.

Li, G. Z., Liang, W. & Liu, J. 2002. Heavy mineral assemblies and silt sources in the sediment of Lianzhou Bay. *Guangxi Sciences*, **9**, 119–123.

Li, Z., Saito, Y. *et al.* 2006. Palynological record of climate change during the last deglaciation from the Song Hong (Red River) delta, Vietnam. *Palaeogeography, Palaeoclimatology, Palaeoecology*, **235**, 406–430.

Li, Z., Zhang, Y. L. *et al.* 2010. Palynological records of Holocene monsoon change from the Gulf of Tonkin (Beibuwan). *Northwestern South China Sea Quaternary Research*, **74**, 8–14.

Liu, Z. F., Colin, C. *et al.* 2010. Clay mineral distribution in surface sediments of the northeastern South China Sea and surrounding fluvial drainage basins: source and transport. *Marine Geology*, **277**, 48–60.

López-González, N., Borrego, J. *et al.* 2006. Geochemical variations in estuarine sediments: provenance and environmental changes (Southern Spain). *Estuarine, Coastal and Shelf Science*, **67**, 313–320.

Masuda, A., Nakamura, N. & Tanaka, T. 1973. Fine structures of mutually normalized rare-earth patterns of chondrites. *Geochimica et Cosmochimica Acta*, **37**, 239–244.

Mclennan, S. R. 1989. Rare earth elements in sedimentary rocks: influences of provenance and sedimentary processes. *Reviews in Mineralogy*, **21**, 169–200.

Nesbitt, H. W. & Markovics, G. 1997. Weathering of grandioritic crust, long-term storage of elements in weathering profiles, and petrogenesis of siliciclastic sediments. *Geochimica et Cosmochimica Acta*, **61**, 1653–1670.

Nesbitt, H. W. & Young, G. M. 1982. Early Proterozoic climate and plate motions inferred from major element chemistry of lutites. *Nature*, **299**, 715–717.

Nesbitt, H. W., Markovics, G. & Price, R. C. 1980. Chemical processes affecting alkalis and alkaline earths during continental weathering. *Geochimica et Cosmochimica Acta*, **44**, 1659–1666.

Oswald, W. W. & Foster, D. R. 2011. A record of late-Holocene environmental change from southern New England, USA. *Quaternary Research*, **76**, 314–318.

RODDAZ, M., VIERS, J. ET AL. 2006. Controls on weathering and provenance in the Amazonian foreland basin: insights from major and trace element geochemistry of Neogene Amazonian sediments. *Chemical Geology*, **226**, 31–65.

SHAW, P. T. & CHAO, S. Y. 1994. Surface circulation in the South China Sea. *Deep Sea Research Part I: Oceanographic Research Papers*, **41**, 1663–1683.

SHI, X. J., YU, K. F. & CHEN, G. T. 2007. Progress in researches on sea-level changes in South China Sea since mid-Holocene. *Marine Geology and Quaternary Geology*, **27**, 121–132.

SUN, G. H., QIU, Y. & ZHU, B. D. 2009. The tectonic implications of Holocene sea-level relics in the South China Sea and adjacent areas. *Acta Oceanologica Sinica*, **31**, 58–68.

TANG, D., KAWAMURA, H. ET AL. 2003. Seasonal and spatial distribution of chlorophyll-α concentrations and water conditions in the Gulf of Tonkin, South China Sea. *Remote Sensing of Environment*, **85**, 475–483.

TANABE, S., HORI, K. ET AL. 2003. Song Hong (Red River) delta evolution related to millennium-scale Holocene sea-level changes. *Quaternary Science Reviews*, **22**, 2345–2361.

TANAKA, G., KOMATSU, T. ET AL. 2011. Temporal changes in ostracod assemblages during the past 10 000 years associated with the evolution of the Red River delta system, northeastern Vietnam. *Marine Micropaleontology*, **81**, 77–87.

TAYLOR, S. R. & MCLENNAN, S. R. 1985. *The Continental Crust: Its Composition and Evolution*. Blackwell, London.

WAN, S., TOUCANNE, S. ET AL. 2015. Human impact overwhelms long-term climate control of weathering and erosion in southwest China. *Geology*, http://doi.org/10.1130/G36570.1

WEAVER, C. E. 1967. Potassium, illite and the ocean. *Geochimica et Cosmochimica Acta*, **31**, 2181–2196.

WEI, G. J., LIU, Y. ET AL. 2004. Major and trace element variations of the sediments at ODP Site 1144, South China Sea, during the last 230 ka and their paleoclimate implications. *Palaeogeography, Palaeoclimatology, Palaeoecology*, **212**, 331–342.

WELTJE, G. J. & EYNATTEN, H. V. 2004. Quantitative provenance analysis of sediments: review and outlook. *Sedimentary Geology*, **171**, 1–11.

YANG, S. Y., YIM, W. W. & HUANG, G. Q. 2008. Geochemical composition of inner shelf Quaternary sediments in the northern South China Sea with implications for provenance discrimination and paleoenvironmental reconstruction. *Global and Planetary Change*, **60**, 207–221.

YAO, Y. T., HARFF, J. ET AL. 2009. Reconstruction of paleocoastlines for the northwestern South China Sea since the Last Glacial Maximum. *Science in China, Series D: Earth Science*, **52**, 1127–1136.

YU, Y. J. & MU, Y. T. 2006. The new institutional arrangements for fisheries management in Gulf of Tonkin. *Marine Policy*, **30**, 249–260.

Post-glacial mud depocentre in the southern Beibu Gulf: acoustic features and sedimentary environment evolution

Y. NI[1,2]*, J. HARFF[3], Z. XIA[1,2], J. J. WANIEK[4], M. ENDLER[4] & D. E. SCHULZ-BULL[4]

[1]*Guangzhou Marine Geological Survey, China Geological Survey, Guangzhou 510760, China*

[2]*Key Laboratory of Marine Mineral Resources, Ministry of Land and Resources, Guangzhou 510760, China*

[3]*Institute of Marine Sciences, University of Szczecin, Mickiewicza 18, 70–383, Szczecin, Poland*

[4]*Leibniz-Institute for Baltic Sea Research, Warnemünde, Seestrasse 15, 18119 Rostock, Germany*

**Corresponding author (e-mail: niyugen@163.com)*

Abstract: This study documents the Southern Beibu Gulf Mud Depocentre (SBGMD) for the first time using high-resolution sub-bottom seismic reflection profiles. The SBGMD is located near the southern mouth of the Beibu Gulf in water depths of 50–80 m, and covers an area of more than 11 000 km^2. The acoustic architecture of the SBGMD is characterized by a homogenous seismic unit surrounded by an erosive area where gullies and sand/mud waves occur. The SBGMD unconformably overlies a parallel-bedded seismic unit.

Based on ^{14}C-dating, together with acoustic, lithological and geochemical analysis of gravity core SO31, which was sampled within the high-sedimentation-rate area of the SBGMD, the post-glacial sedimentary history may be summarized as follows: (1) from 12.8 to 11.6 cal ka BP, the SBGMD was deposited in a low-energy, brackish/freshwater and intensely terrestrial-influenced environment; (2) from 11.6 to 8.4 cal ka BP, the SBGMD underwent marine transgression; and (3) from 8.4 cal ka BP to present, a continuous shallow-marine environment has prevailed. Comparing the age and depth of the SBGMD base with the established sea-level curve, lacustrine sedimentation is predicted to have occurred in the SBGMD area before 12.8 cal ka BP.

Post-glacial shelf mud deposits are encountered worldwide and have received considerable attention, with the main focus being on their formation mechanisms (Dronkers & Miltenburg 1996; Lesueur *et al.* 1996; Hanebuth & Lantzsch 2008; Vinzon *et al.* 2009), sedimentary evolution (Garnaud *et al.* 2003; Grossman *et al.* 2006; Lantzsch *et al.* 2009), and implications for palaeoenvironmental and palaeoclimatic change (Polyakova & Stein 2004; Morigi *et al.* 2005; Nizou *et al.* 2010; Virtasalo *et al.* 2014).

On the Chinese continental shelf and adjacent areas, several mud deposits/mud depocentres have also been the subject of detailed research: for example, Bohai Sea Mud, North Yellow Sea Mud, Shandong Coastal Mud, Central Yellow Sea Mud, SW Yellow Sea Mud, SE Yellow Sea Mud, SW Cheju Island Mud, Yangtze River Proximal Mud, Zhejiang–Fujian Coastal Mud, Pearl River Proximal Mud, Western Guangdong Coastal Mud (Pearl River Distal Mud), and the Eastern and Western Mouth Tidal Delta of the Qiongzhou Strait (the 'butterfly delta' system of the Qiongzhou Strait) (Fig. 1) (Yang *et al.* 2003; Liu *et al.* 2004, 2007; Liu *et al.* 2014; Ni *et al.* 2014). However, previous studies on the mud depocentre in the southern Beibu Gulf (Fig. 2) are limited.

The Beibu Gulf is located on the continental shelf of the NW South China Sea. Harff *et al.* (2013) devoted an anthology to the depositional environment of the South China Sea's northern shelf, including the Beibu Gulf, while Ma *et al.* (2010) investigated the grain-size distribution of surface sediments in the eastern Beibu Gulf. Mud depocentres have been described in the NE Beibu Gulf (i.e. the Western Mouth Tidal Delta (WMTD) of the Qiongzhou Strait) (Fig. 2), comprising mainly the tidal delta slope and the tidal delta toe area of the WMTD (Ni *et al.* 2014), and in the NW Beibu Gulf, the Ba Lat Mouth Depocentre (BLMD), also known as the Ba Lat pro-delta (van den Bergh *et al.* 2007*a*, *b*). These features were imaged by high-resolution acoustic profiling and geological sampling, but no information about the morphology and acoustic facies of a mud depocentre in the southern Beibu Gulf has previously been published.

The purpose of this paper is: (1) to report the acoustic features of the Southern Beibu Gulf Mud

From: Clift, P. D., Harff, J., Wu, J. & Qui, Y. (eds) 2016. *River-Dominated Shelf Sediments of East Asian Seas*. Geological Society, London, Special Publications, **429**, 87–98.
First published online March 14, 2016, updated May 26, 2016, http://doi.org/10.1144/SP429.13

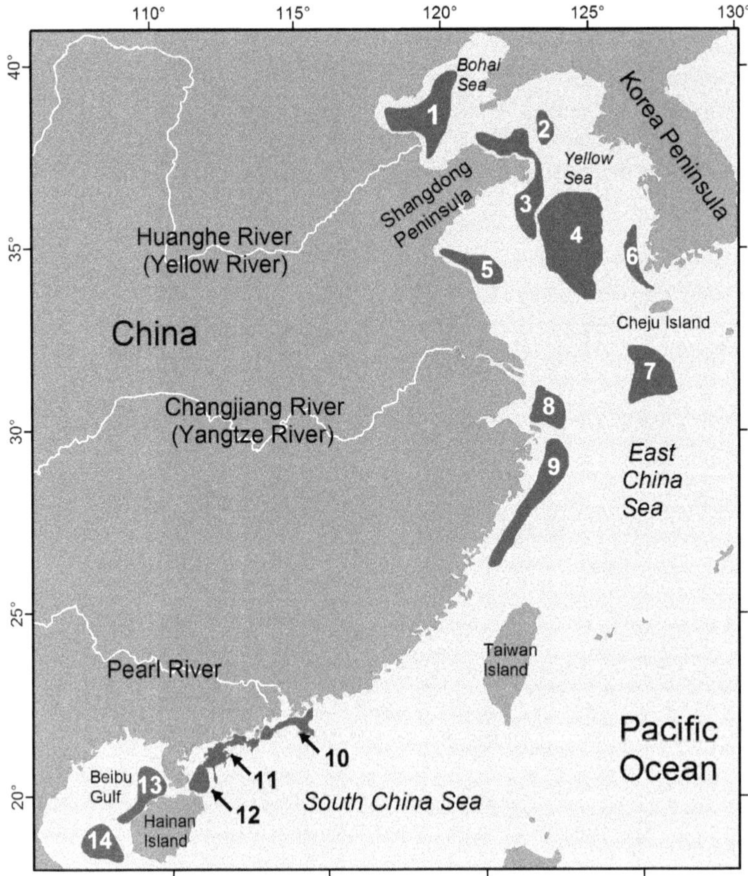

Fig. 1. The mud deposits/mud depocentres on the Chinese continental shelf and the adjacent areas (dark grey area with number) (modified from Yang *et al.* 2003; Liu *et al.* 2004, 2007; Liu *et al.* 2014; Ni *et al.* 2014). 1, Bohai Sea Mud; 2, North Yellow Sea Mud; 3, Shandong Coastal Mud; 4, Central Yellow Sea Mud; 5, SW Yellow Sea Mud; 6, SE Yellow Sea Mud; 7, SW Cheju Island Mud; 8, Yangtze River Proximal Mud; 9, Zhejiang–Fujian Coastal Mud; 10, Pearl River Proximal Mud; 11, Western Guangdong Coastal Mud (Pearl River Distal Mud); 12, Eastern Mouth Tidal Delta of the Qiongzhou Strait; 13, Western Mouth Tidal Delta of the Qiongzhou Strait; 14, the Southern Beibu Gulf Mud Depocentre (from this study).

Depocentre (SBGMD) based on high-resolution sub-bottom acoustic profiling; and (2) to present the post-glacial sedimentary history of the SBGMD based on accelerator mass spectrometry (AMS) ^{14}C dating, as well as on lithological and geochemical analysis of gravity core SO31, which was collected within the SBGMD.

Geological setting

The Beibu Gulf is a semi-closed shallow gulf located in the NW South China Sea, with maximum water depths of about 100 m and a mean water depth of about 40 m (Fig. 2). The circulation system is controlled by marine currents, tides and wind. The gulf is connected to the South China Sea by its wide opening to the south and the narrow Qiongzhou Strait to the east.

The tectonic setting is dominated by the Yinggehai–Song Hong Basin, the subsidence of which is influenced by the deep-seated Red River Fault Zone (Tanabe *et al.* 2003; Clift & Sun 2006; Xie *et al.* 2008). During the Pleistocene, this basin was mainly filled with sediments supplied by the Red River. At the sea-level lowstand during the last glaciation and, in particular, during the Last Glacial Maximum (*c.* 20 ka), the shelf was exposed, and so erosion and fluvial sedimentation dominated the environment. A sea-level reconstruction of the western margin of the South China Sea since 20 ka, compiled by Tanabe *et al.* (2003), was based on data

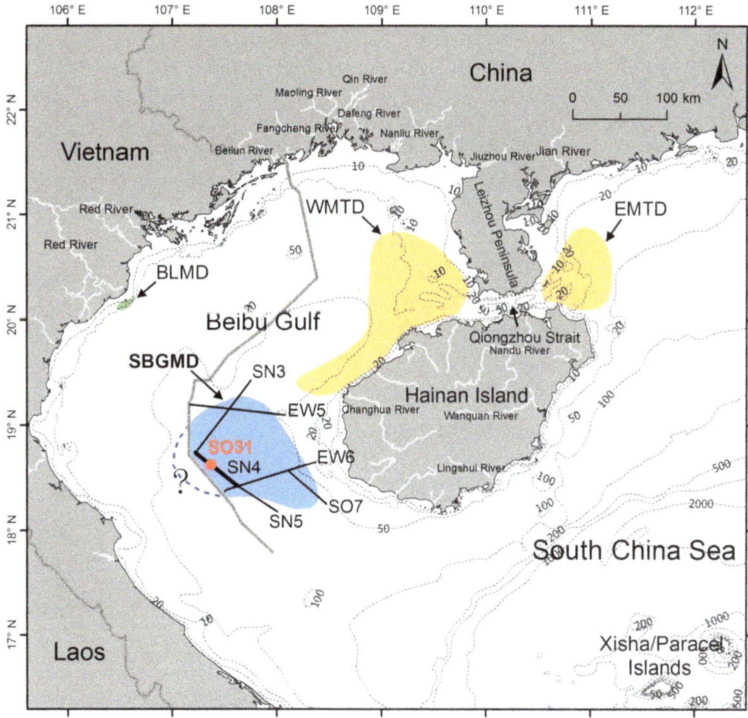

Fig. 2. The locations of the post-glacial mud depocentre in the southern Beibu Gulf, sub-bottom profiles and gravity core SO31. SBGMD, Southern Beibu Gulf Mud Depocentre; WMTD, Western Mouth Tidal Delta of the Qiongzhou Strait (after Ni *et al.* 2014); EMTD, Eastern Mouth Tidal Delta of the Qiongzhou Strait (after Ni *et al.* 2014); BLMD, Ba Lat Mouth Depocentre (after van den Bergh *et al.* 2007*a*, *b*); the blue area indicates the SBGMD and the blue dash line with question mark indicates the unknown extension of the SBGMD within the Vietnamese territorial water in the Beibu Gulf; the yellow area indicates the WMTD and EMTD; the green area indicates the BLMD; the red dot marks the position of gravity core SO31; the black solid lines mark the sub-bottom survey lines and the thicker one marks the sub-bottom profile where gravity core SO31 is located; the grey dash lines mark the isobaths.

from the Sunda Shelf (Biswas 1973; Hanebuth *et al.* 2000), the Malacca Strait (Geyh *et al.* 1979) and the Song Hong (Red River) Delta (Tran & Ngo 2000; Lam & Boyd 2001). This showed that the sea level at 20 ka was about 120 m lower than present, before the sea level started to rise during the Late Pleistocene and Holocene. Yao *et al.* (2009) studied in detail the effect of the post-glacial sea-level change on the northern margin of the South China Sea based on a recent digital elevation model (ETOPO2v2 database) that was superimposed on relative sea-level curves (Chen *et al.* 1991; Tanabe *et al.* 2003; Zong 2004) and sediment thickness maps. In the Beibu Gulf area, the rising post-glacial sea level caused a continuous transgression that submerged the shelf, which had been exposed to subaerial weathering during the Last Glacial Maximum.

According to Tanabe *et al.* (2003) and Funabiki *et al.* (2007), the Red River Delta formed in response to the Holocene sea-level rise. Between 9 and 6 cal ka BP, the Red River Delta prograded into the Yinggehai–Song Hong Basin and formed wide tidal flats as a result of early Holocene inundation. The modern subaqueous muddy pro-delta of the Red River is limited in the nearshore zone. Previous investigations of the Ba Lat pro-delta showed that most suspended sediment from the river plume settled within 10 km of the river mouth (van Maren & Hoekstra 2004, 2005). The pro-delta was confined to a range in water depth of 30 m, with the depocentre (covering an area of *c.* 160 km^2 and 2–5 m in thickness) located 9–18 km SW of the mouth of the Ba Lat river (van den Bergh *et al.* 2007*a*, *b*). In the east, an additional sediment source appeared when the Qiongzhou Strait, an 80 km-long and 30 km-wide water channel located between Hainan Island and Leizhou Peninsula, connected the Beibu Gulf with the NW South China Sea during the early Holocene (Fig. 2). The hydrodynamic regime of this strait is dominated by strong reciprocal tidal currents and is characterized by a year-round westwards mean flow, which transports large volumes of water and sediment from the South China Shelf into the Beibu Gulf (Shi *et al.* 2002), leading to the formation of the tidal 'butterfly delta' system at both mouths of the Qiongzhou Strait (Fig. 2) (Ni *et al.* 2014). This tidal delta system started to develop at about 8.7 ka BP when Qiongzhou Strait was completely opened, as measured by ^{14}C dating. The western mouth tidal delta of Qiongzhou Strait covers an area of about 11 000 km^2 (Fig. 2), with a maximum thickness of >23 m (Ni *et al.* 2014), making this muddy deposit the most important mud depocentre in the NE Beibu Gulf.

Dataset and methods

Four research cruises (FENDOU5 in 2009, SONNE 219 in 2011, HAIYANG4 in 2012 and FENDOU5 in 2012) were conducted in the Beibu Gulf and adjacent areas, collecting a total of 4100 km of high-resolution seismic profiles (including 2820 km of 'SES' sub-bottom profiles and 1280 km of 'Parasound' sub-bottom profiles), and 87 hydrographic and sediment sampling stations occupied. The FENDOU5 cruise in 2009 and the SONNE 219 cruise in 2011 were Sino-German cruises within the framework of the bilateral BEIBU research project ('Holocene environmental evolution and anthropogenic impact of Beibu Gulf, South China Sea'), in a collaboration between the Guangzhou Marine Geological Survey (GMGS), China, and the Leibniz Institute for Baltic Sea Research (IOW), Germany. The HAIYANG4 cruise in 2012 and the FENDOU5 cruise in 2012 were conducted by the Guangzhou Marine Geological Survey.

We selected six sub-bottom profiles (lines EW5, EW6, SN3, SN4, SN5 and SO7) and one gravity core (SO31) for this study (Fig. 2). The 'SES' profiles EW5, EW6, SN3, SN4 and SN5 (Fig. 3) were obtained using a SES-96 parametric sub-bottom profiler (with a primary frequency of 100 kHz and a secondary frequency of 4–12 kHz) during the FENDOU5 cruise in 2009. The 'Parasound' profile SO7 was measured using a Parasound P70 parametric sub-bottom profiler (with a primary frequency of 18–33 kHz and a secondary frequency of 0.5–6 kHz) during the SONNE 219 cruise in 2011. Gravity core SO31 was collected by RV SONNE at 107°22′8″E, 18°37′26″N at a water depth of 65 m (Fig. 2). With a length of 738 cm, the core was subsampled at 2 cm intervals for the grain-size analysis and at 3 cm intervals for the geochemical analysis.

The 'SES' sub-bottom profiles were processed using ISE2.92 software at the Guangzhou Marine Geological Survey, and the 'Parasound' sub-bottom profiles were processed using the Acoustic & Sediment Physics Matlab Toolbox at Leibniz Institute for Baltic Sea Research (cf. Endler *et al.* 2015).

Grain-size analysis was carried out using a laser diffraction particle size analyser (a Mastersizer 2000, measuring from 0.02 to 2000 μm) at the Guangzhou Marine Geological Survey. The carbonate fraction was not removed before grain-size analysis was performed because this was an insignificant component. Ten major elements (Si, Al, Fe, Ca, Mg, K, Na, Mn, P and Ti) and two trace elements (Cr and Zr) were determined using a X-ray fluorescence spectrometer (Axios), seven trace elements (Co, Cu, Ni, Sr, Zn, V and Ba) were measured using an inductively coupled plasma-optical emission spectrometer (ICP-OE Opima 4300DV) and three trace elements (Sc, Ga and Pb) were analysed using an inductively coupled plasma-mass spectrometer (ICP-MS X Series2) at the Guangzhou Marine Geological Survey. The measurement accuracy of the X-ray fluorescence spectrometer (Axios), ICP-OE Opima 4300DV and ICP-MS X Series2 was better than 5, 4 and 3%, respectively. The total organic carbon (TOC) was measured based on the Dichromate Titration Method at the Guangzhou Marine Geological Survey. Major element concentrations were measured as oxides.

Ten AMS ^{14}C dates of mixed foraminifers from core SO31 were measured by the Beta Analytic Radiocarbon Dating Laboratory and calibrated using the Calib 7.0.1 program (Stuiver *et al.* 2014) (Table 1). The Marine Reservoir Correction ($\Delta R =$ 18, ΔR uncertainly = 37) was made according to the data (Nos 406, 407 and 408) from the nearby Xisha/Paracel Islands (Southon *et al.* 2002). The age–depth model was calculated based on the calibrated ^{14}C ages via linear interpolation (the 277 cm-depth age was abandoned as an isolated abnormal value) (Fig. 4). The picking of the foraminifers, including the dating samples and the 1 cm interval sampling work, was carried out at Nanjing University.

Results

The Southern Beibu Gulf Mud Depocentre

The Southern Beibu Gulf Mud Depocentre (SBGMD) is located near the southern opening of the Beibu Gulf, and covers an area of about 11 000 km^2 in Chinese waters according to our study (Fig. 2). However, how the SBGMD extends into Vietnamese waters is still unknown owing to a lack of published data.

The SBGMD can be recognized based on high-resolution sub-bottom acoustic/seismic profiles (Fig. 3). The SBGMD is situated at an average water depth of 65 m (range 50–80 m) with a relatively smooth seafloor, surrounded by an erosive area where gullies and sand waves/mud waves are observed (Fig. 3). Acoustically, the SBGMD is characterized by a homogenous seismic unit

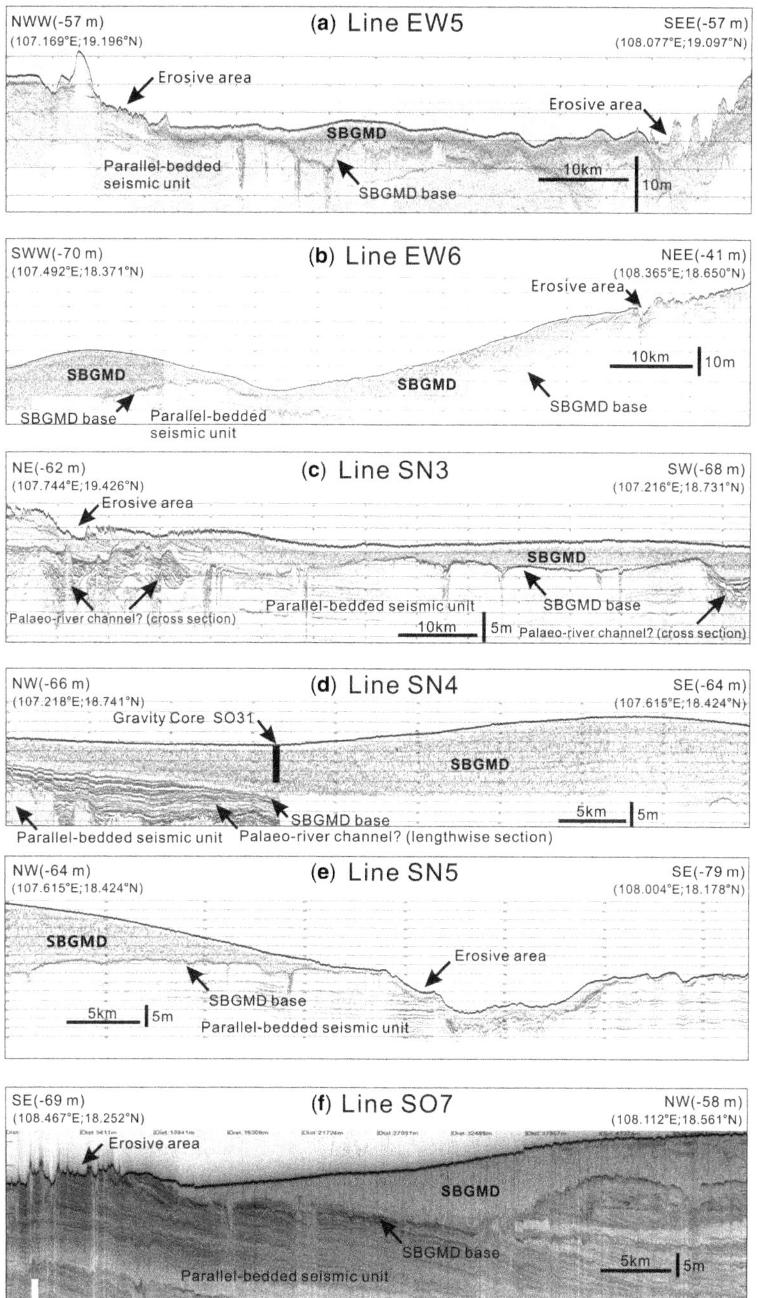

Fig. 3. Sub-bottom profiles of the SBGMD. (**a**) The seismic 'SES' profile line EW5; (**b**) the seismic 'SES' profile line EW6; (**c**) the seismic 'SES' profile line SN3; (**d**) the seismic 'SES' profile line SN4; (**e**) the seismic 'SES' profile line SN5; and (**f**) the seismic 'Parasound' profile line SO7.

unconformably resting on a parallel-bedded seismic unit (Fig. 3). The thickness of the SBGMD mostly exceeds 5 m, with a maximum thickness of >18 m in the SW part, where the average thickness also exceeds 10 m and the seafloor is characterized by a gentle relief (Fig. 3).

Table 1. *List of AMS ^{14}C ages for selected depths of core SO31 in the SBGMD*

Depth (cm)	Material	Conventional radiocarbon age (^{14}C years BP)	2σ calibrated result (cal years BP)	Mean calibrated age (cal years BP)
73	Foraminifera	940 ± 30	464–621	530
142	Foraminifera	1680 ± 30	1106–1310	1220
205	Foraminifera	2460 ± 30	1951–2257	2090
277	Foraminifera	970 ± 30	486–634	550
337	Foraminifera	3510 ± 30	3237–3496	3380
403	Foraminifera	4370 ± 30	4353–4648	4490
469	Foraminifera	4860 ± 30	4981–5281	5140
538	Foraminifera	9710 ± 40	10 436–10 736	10 590
601	Foraminifera	10 060 ± 50	10 823–11 187	11 040
733	Foraminifera	11 350 ± 50	12 656–12 944	12 800

Note: dates were measured by the Beta Analytic Radiocarbon Dating Laboratory, Miami, FL, USA.

Line EW5 (Fig. 3a) is 82 km long and crosses the northern part of the SBGMD from west to east. The boundary of the SBGMD in this line can be easily identified by erosion: that is, four erosive gullies in the west and the sand/mud wave area in the east of the line. The seabed of SBGMD is undulating, especially at the western end of this line, and the homogenous seismic unit is relatively thin, with an average thickness of about 5 m. The base of the SBGMD, which is unconformable, is clearly visible in the western part of the SBGMD, but is difficult to identify in the east where the underlying parallel-bedded unit is diffuse but still recognizable.

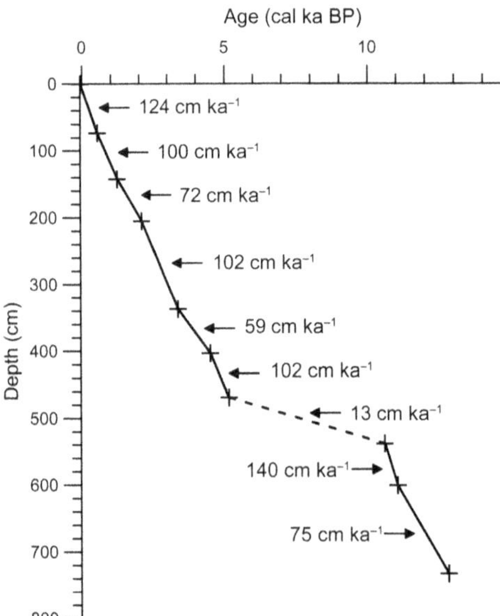

Fig. 4. The age–depth model of gravity core SO31, for the age deviations see Table 1. The dashed line marks the dramatic low sedimentation rate interval.

Line EW6 (Fig. 3b) is 88 km long and cuts through the southern part of the SBGMD from NE to SW. At the SW end of this line, the SBGMD shows its greatest thickness (>10 m) where the seafloor is characterized by low relief. The homogenous seismic unit is <5 m thick, but then increases to around 10 m in the SW–middle part of this line, where the seafloor is characterized by a low depression. At the NE end of this line, the SBGMD wedges out where it meets the erosive area characterized by small gullies and sand/mud waves. The unconformable base of the SBGMD is clearly developed at the SW and NE ends of this line, although it is not recorded in the most SW part. The unconformable base is not recognizable in the middle part of this line, probably because of the presence of overlying gas-bearing sediment. Although the underlying parallel-bedded seismic unit is obscure along the whole of Line EW6, parallel bedding is still visible at the SW part of this line.

With a length of 94 km, Line SN3 (Fig. 3c) crosses the NW part of the SBGMD from NE to SW. The thickness of the SBGMD along this line is mostly <5 m. The unconformable base is distinct, and the underlying parallel-bedded seismic unit is recognizable, although not very clear. Interestingly, there are palaeo-river channel deposits directly resting on top of the parallel-bedded seismic unit at both ends of this line, which are characterized by a high-amplitude acoustic facies. At the NE end of this line, the profile shows that the SBGMD ends in a sand/mud wave field, and that the homogenous layer also turns into a bedded layer with a truncated surface (wavy seabed).

Line SN4 (Fig. 3d) is entirely located inside the SW part of the SBGMD, with a length of 55 km and striking NW–SE. The seafloor is characterized by low relief at the SE end of this line, where the average thickness of the homogenous seismic unit is >10 m and the maximum thickness is >18 m.

Accurate thickness data cannot be given here because the lower boundary of the SBGMD was not recorded. At the NW end of this line, the SBGMD is thinner and directly overlies the high-amplitude palaeo-river channel acoustic unit, which discontinuously overlies the lower seismic unit. Because the NW end of Line SN4 is connected to the SW end of Line SN3, the lower seismic unit can be correlated to the parallel-bedded seismic units in the other profiles. It is also worthwhile mentioning that the palaeo-river channel deposits are cross-cut by Line SN3 at its SW end, so that the channel axis coincides with Line SN4.

Line SN5 (Fig. 3e) is located at the southern part of the SBGMD. This line is 50 km long and has a strike of NW–SE. The NW end of this line connects to the SE end of Line SN4. At the NW end of this line, the seafloor slopes as the SBGMD gradually thins to the SE and pinches out where it encounters an erosive valley. Although there is also a thin homogenous seismic unit covering the erosive valley, this is not considered to belong to the SBGMD. The unconformable base of the SBGMD is clear, as well as the underlying parallel seismic unit.

Line SO7 (Fig. 3f), which is 52 km long and strikes NW–SE, is located in the SE part of the SBGMD. Most of this line clearly shows that the homogenous SBGMD unconformably rests on the parallel-bedded seismic unit. The SBGMD is relatively thick in the NW–middle part of this line, where it has an average thickness of 10 m, but thins out at the SE end of this line, where the parallel-bedded seismic unit is directly exposed on the seafloor and sand/mud waves occur.

Lithological and geochemical record of core SO31

Gravity core SO31 is located on Line SN4 in the SW part of the SBGMD, which was taken in the homogenous seismic unit but did not penetrate it (Figs 2 & 3d). Based on the AMS ^{14}C ages, the age–depth model of core SO31 shows that the mean deposition rate was 57 cm ka^{-1}, reaching a maximum value of 140 cm ka^{-1}, while the minimum is just 13 cm ka^{-1} (Fig. 4). Moreover, three sedimentation rate steps in core SO31 are distinct: (1) 733–538 cm depth, from 12.8 to 10.6 cal ka BP, with an average sedimentation rate of 88 cm ka^{-1}; (2) 538–469 cm depth, from 10.6 to 5.1 cal ka BP, with an average sedimentation rate of only 13 cm ka^{-1}; and (3) 469–0 cm depth, from 5.1 cal ka BP to present, with an average sedimentation rate of 90 cm ka^{-1} (Fig. 4). We note that these steps show sedimentation rate differences based on the available ages, but may not represent the sedimentation variations in core SO31 in detail.

Gravity core SO31 is homogenous at the macroscopic level without distinct layers, although the dominant grey sediment colour gradually becomes brownish grey in the lowermost 2 m.

Despite this, gravity core SO31 can clearly be divided into three layers: (1) Unit 1: the lower layer (738–640 cm); (2) Unit 2: the intermediate layer (640–510 cm); and (3) Unit 3: the upper layer (510–0 cm), based on the changing concentrations of major (Fig. 5) and trace (Fig. 6) elements, particle size distribution, median grain size (Φ50), total organic carbon (TOC), and elemental ratios (i.e. Sr/Ba and Ti/Ca) (Fig. 7). The 640 cm boundary can be easily defined by the sudden change in major element concentrations (SiO_2, Al_2O_3, Fe_2O_3, Na_2O, K_2O and TiO_2) (Fig. 5), trace element concentrations (Cu, Ni, Sr, Ba and Zr) (Fig. 6), TOC (Fig. 7d) and Ti/Ca (Fig. 7f). The boundary at 510 cm is also clearly determined as the inflection point of major element concentrations (SiO_2, Al_2O_3, Fe_2O_3, MgO, CaO, Na_2O, MnO and P_2O_5) (Fig. 5), trace element concentrations (Cu, Sr, Zn, Ba and Zr) (Fig. 6), TOC (Fig. 7d), Sr/Ba (Fig. 7e) and Ti/Ca (Fig. 7f). The ages at depths of 640 and 510 cm in core SO31 are 11.6 and 8.4 cal ka BP, respectively, based on the age–depth model of this core (Fig. 4) and assuming linear interpolation.

Grain size is a solid indicator mirroring the hydrodynamic conditions during sedimentation. The TOC may be used to reflect the redox environment and the supply of organic material: for example, the high TOC content in sediments from Lake Huguang Maar indicates an anoxic environment (Yancheva *et al.* 2007). Sr/Ba values can be used as an indicator of salinity and have been successfully applied in the Yangtze Delta, where Sr/Ba values are in direct proportion to salinity (Chen *et al.* 1997). Sr/Ba values of >0.35 indicate salt-water environments, <0.2 indicate freshwater environments, while values in-between show brackish-water environments (Chen *et al.* 1997). The Ti/Ca ratio can be a useful proxy of clastic input and has been used as a proxy for physical weathering intensity in the northern South China Sea at the Ma timescale (Clift *et al.* 2014). In this study, Ti/Ca is interpreted to indicate a change in the sedimentary environment linked to sea-level rise because Ti represents the terrestrial influence, including fluvial run-off, whereas Ca represents a marine influence from the skeletons and shells of marine organisms.

The particle-size distribution in core SO31 (Fig. 7b) shows that the sediments are clayey silt, and are made up of a mixture of clay (30%), silt (67%) and sand (3%) (bulk average values), with 32% clay, 67% silt and 1% sand in Unit 1; 28% clay, 67% silt and 5% sand in Unit 2; and 30% clay, 67% silt and 3% sand in Unit 3. The silt content is quite stable (*c.* 67%) through the whole core,

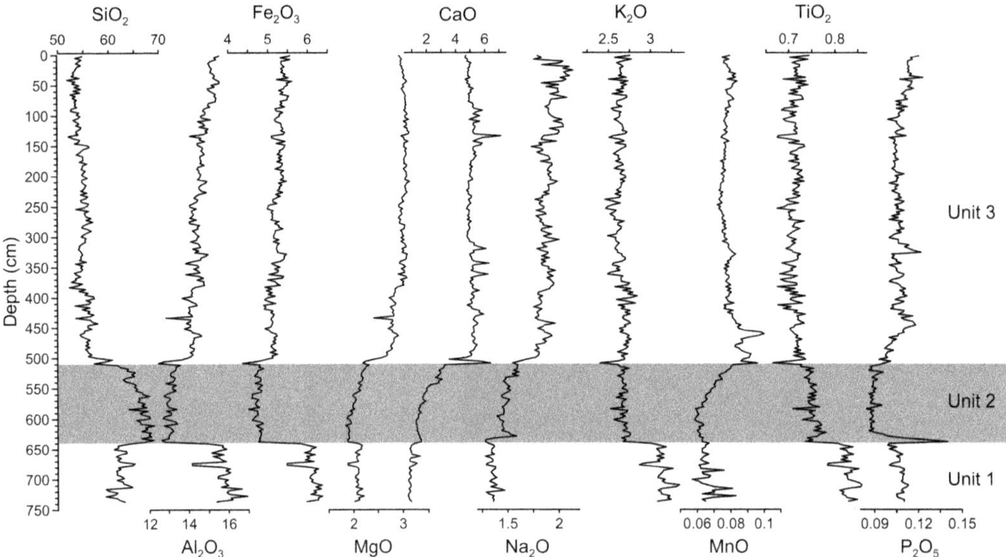

Fig. 5. The vertical distribution of major elements in gravity core SO31. The major elements are measured as oxides; the grey shaded area marks Unit 2 (640–510 cm) of core SO31; The abscissa axis element content is given in per cent.

while the clay and sand content shows a reciprocal relationship (Fig. 7b). Grain size varies vertically, with the coarsest sediment in Unit 2 and the finest sediment in Unit 1 (Fig. 7b). Such a variation is also revealed by the median grain size (Φ50), which has a layer average value of 7.3 in Unit 1, 6.8 in Unit 2 and 7.2 in Unit 3 (Fig. 7c).

In Unit 1 of core SO31, the TOC content is relatively steady, with an average value of 0.7%. The TOC content sharply declines from 0.7 to 0.4% above a depth of 640 cm. In Unit 2, the TOC content is the lowest, with an average value of 0.4%. From a depth of 510 to 480 cm, the TOC content rapidly increases from 0.5 to 0.8%, and remains at a high

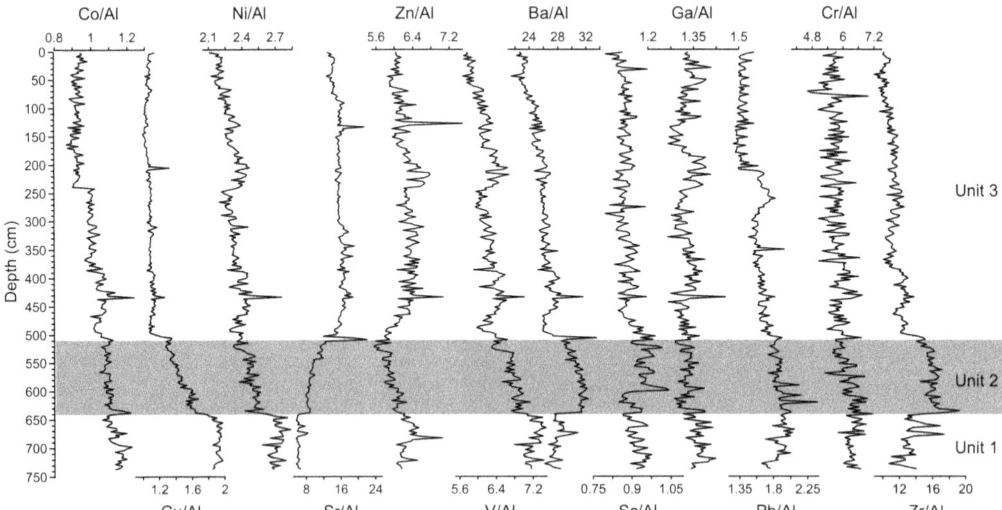

Fig. 6. The vertical distribution of trace elements in gravity core SO31. The trace elements are standardized by Al to eliminate the grain-size effect; The grey shaded area marks Unit 2 (640–510 cm) of core SO31; the abscissa axis element content is given in 10^{-4}.

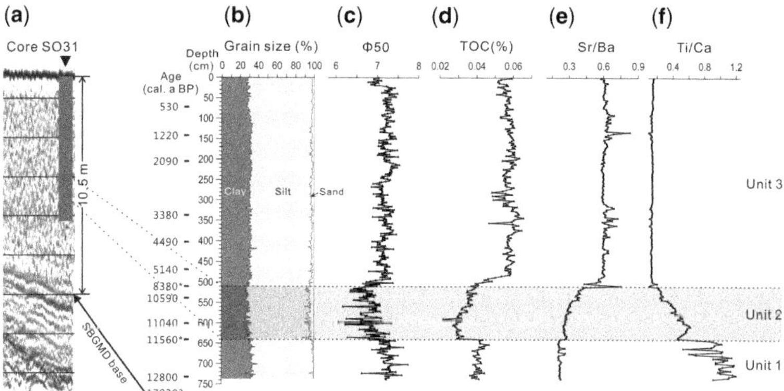

Fig. 7. The ^{14}C-dated acoustical, lithological and geochemical records of gravity core SO31 in the SBGMD. (**a**) The sub-bottom profile where gravity core SO31 is located: the distance of 10.5 m between the seafloor and the SBGMD base was estimated based on the sub-bottom profile record using a mean sound velocity of 1500 m s^{-1}; (**b**) vertical grain-size distribution; (**c**) vertical median grain-size (Φ50) distribution; (**d**) vertical total organic carbon (TOC) content distribution; (**e**) vertical Sr/Ba ratio distribution; and (**f**) vertical Ti/Ca ratio distribution, given as an oxides ratio. The grey shaded area marks Unit 2 (640–510 cm). Ages with asterisks were calculated based on the age–depth model (Fig. 4) via linear interpolation. The age with a question mark was calculated based on the age–depth model (Fig. 4) via linear extrapolation.

level above that level. Unit 3 has the highest TOC content, with an average value of 0.8% (Fig. 7d).

Sr/Ba values are steady and low in Unit 1 (average = 0.2) (Fig. 7e), while there is a constant high value in Unit 3 (average = 0.6). From a depth of 640 to 510 cm (Unit 2), Sr/Ba gradually increases from 0.2 to 0.4, but with an isolated peak of 0.8 at a depth of about 510 cm. Similar peaks also occur in most major and trace elements (Figs 5 & 6).

Ti/Ca values (Fig. 7f) show intense fluctuations, but are generally high in Unit 1, with an average value of 1.0, whereas there is a steady low value in Unit 3, with an average value of 0.1. Above 640 cm, Ti/Ca falls sharply from 1.0 to 0.5. Overall, Ti/Ca gradually decreases from 0.5 to 0.1 upwards in Unit 2 (640–510 cm), although there is a short interval of increasing values from 640 to 600 cm.

Post-glacial sedimentary environments

As noted above, the SBGMD is located near the southern opening of the Beibu Gulf (Fig. 2). When considering the post-glacial transgression from south to north, we conclude that the SBGMD was deposited as early as seawater began intruding into the gulf, so that it may record the entire history of sedimentary environments and sea-level change since that time. In this study, gravity core SO31 was used to analyse the sedimentary history of the SBGMD, which was taken from the homogeneous seismic unit of the SBGMD, and recovered a 738 cm-long homogenous muddy core (Figs 3d & 7a). However, core SO31 did not penetrate the SBGMD base and, thus, the AMS ^{14}C-dated lithological and geochemical record from this core only covers the sedimentary history of the SBGMD since 12.8 cal ka BP.

From 12.8 to 11.6 cal ka BP (i.e. from 733 to 640 cm in Core SO31 (Unit 1)), the lowest sand content (1%) and the highest value of Φ50 (7.3) indicate the weakest hydrodynamic conditions. The relatively high TOC content (0.7%) reflects a relatively anoxic environment, while the lowest Sr/Ba value (0.2) would imply a brackish/freshwater environment. The highest Ti/Ca value (1.0) represents a dominant terrestrial influence (Fig. 7). In addition, benthic foraminifers are present, but there are no planktonic foraminifers in Unit 1. Based on the above proxies, we conclude that, from 12.8 to 11.6 cal ka BP, the SBGMD was inferred to have been deposited in a low-energy, reductive, brackish/freshwater and strongly terrestrial-influenced environment (e.g. nearshore water, lagoon or lacustrine environment).

From 11.6 to 8.4 cal ka BP (i.e. from 640 to 510 cm in core SO31 (Unit 2)), the SBGMD indicates a marine transgressive environment that is characterized by the highest sand content (5%), the smallest Φ50 (6.8), the lowest TOC content (0.4%) and a gradually upwards-increasing Sr/Ba ratio from 0.2 to 0.4. Ti/Ca gradually decreases upsection from 0.5 to 0.1 (Fig. 7). Compared to the steady low Sr/Ba values (0.2) in Unit 1 and the steady high Sr/Ba ratio (0.6) in Unit 3, the gradual upwards increase of Sr/Ba values in Unit 2 implies a gradual upwards increase of salinity and, thus, indicates that salty water gradually intruded into the area of the SBGMD during 11.6–8.4 cal ka BP.

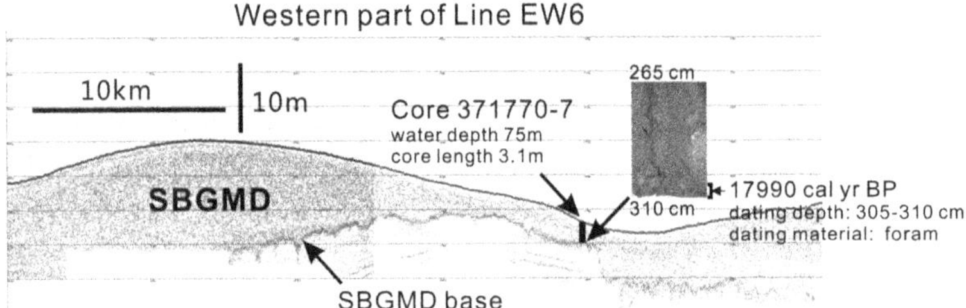

Fig. 8. The western part of sub-bottom profile Line EW6 and gravity core 371770-7. AMS ^{14}C dating of the bottom of core 371770-7 shows the age of the SBGMD base to be 17.99 cal ka BP.

This assumption is also supported by the gradual upwards decline in the Ti/Ca ratio from 0.5 to 0.1 in Unit 2, which represents a gradually increasing marine influence and a gradually decreasing terrestrial influence. Moreover, this marine transgression is also reflected by the relatively strong hydrodynamic conditions indicated by the coarsest sediment and relatively high levels of bottom water oxygen (the lowest TOC content). Moreover, it is noteworthy that at the 11.6 cal ka BP interface, there is a sudden change in the vertical distribution of the TOC content, Ti/Ca, and most of the major and trace elements (Figs 5, 6 & 7) that is likely to reflect the marine transgression and the change in the sedimentary environment. 11.6 cal ka BP coincides with the end of Younger Dryas (Fairbanks 1989). The 8.4 cal ka BP interface corresponds to the complete opening time of the Qiongzhou Strait (Yao *et al.* 2009; Ni *et al.* 2014). An isolated peak occurs in the Sr/Ba and Ti/Ca values, as well as in most of the major and trace elements (Figs 5, 6 & 7). This is probably a result of the addition of a new sediment input through the Qiongzhou Strait.

From 8.4 cal ka BP to present (i.e <510 cm in Core SO31 (Unit 3)), the SBGMD was deposited in a continuous shallow-marine environment that was characterized by homogenous fine sediments (30% clay, 67% silt and 3% sand; $\Phi50 = 7.2$), a steady highest TOC content (0.8%), persistent highest Sr/Ba values (0.6) and the very stable lowest Ti/Ca ratio (0.1) (Fig. 7). The relatively consistent grain size, TOC, Sr/Ba, Ti/Ca, and major and trace element composition (Figs 5, 6 & 7) reflect the sedimentary environment in this upper layer of the core, which is a relatively stable marine environment (i.e. salty, low-energy, reductive, shallow sea), similar to that seen in the present day. In addition, planktonic foraminifers are abundant in this layer.

Discussion

Although gravity core SO31 just records part of the sedimentary history of the SBGMD (since 12.8 cal ka BP), the high-resolution sub-bottom profiles and other gravity cores within the SBGMD still allow us to discuss its sedimentary environment before 12.8 cal ka BP.

Line SN4 indicates that the base of the SBGMD, where core SO31 is located, is 10.5 m below the seafloor, assuming a mean sound velocity of 1500 m s^{-1} (Fig. 7a). Following the age–depth model of core SO31 (Fig. 4), the SBGMD base would be 17.0 cal ka BP assuming a simple linear extrapolation. Certainly, this age is a very rough estimate, but this estimate is supported by the ^{14}C dating of correlated sediments in gravity core 371770-7, which is located on the western part of Line EW6 and reaches the base of the SBGMD (Fig. 8). Core 371770-7 was taken at a water depth of 75 m and has a length of 310 cm (107°47′52″E; 18°28′36″N). The AMS ^{14}C dating age at 305–310 cm (core bottom) is 17.99 cal ka BP (Fig. 8). The dating material was foraminifers, and the dating sample preparation was performed at the Guangzhou Institute of Geochemistry, Chinese Academy of Science, and the dating measurement carried out at Peking University.

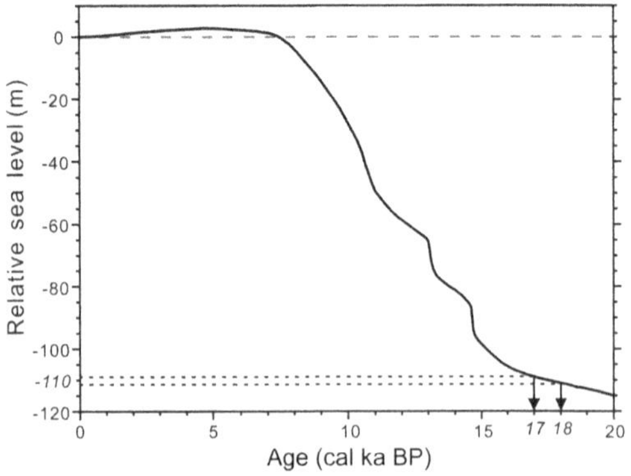

Fig. 9. Sea-level curve of the western margin of the South China Sea since 20 ka (modified from Tanabe *et al.* 2003).

Comparing the age of the SBGMD base to the established regional sea-level curve (Tanabe *et al.* 2003), it indicates that the sea level at 17.0 or 17.99 cal ka BP was about 110 m below modern sea level (Fig. 9). However, the depth of the SBGMD base where core SO31 was located is 75.5 m (water depth of 65 m; SBGMD thickness of 10.5 m) below modern sea level, and the depth of the SBGMD base where core 371770-7 was located is 78.1 m (water depth of 75 m; SBGMD thickness of 3.1 m) below modern sea level. How could the SBGMD begin to deposit at approximately 35 m above sea level? If our timescale is right, this means that the SBGMD must have been subjected to a lacustrine environment at this early time. In other words, there may have been lacustrine deposition occurring in the SBGMD area before 12.8 cal ka BP. Alternatively, the sedimentation rate may not be linear at the base of the SBGMD and the age at the base is actually much younger than we infer here, and the age of 17.99 cal ka BP at the base of core 371770-7 may suggest that the foraminifers used to derive this age are at least partly reworked if the regional sealevel reconstructions are close to being correct and assuming that the AMS ^{14}C analysis was accurate.

Conclusions

The main conclusions of this study are as follows:

- The Southern Beibu Gulf Mud Depocentre (SBGMD) is located near the southern opening of the Beibu Gulf in water depths of 50–80 m and covering an area of more than 11 000 km^2.
- The SBGMD is surrounded by an erosive area where gullies and sand/mud waves occur. It is acoustically characterized by a homogenous seismic unit unconformably overlying a parallel-bedded seismic unit.
- From 12.8 to 11.6 cal ka BP, the SBGMD was deposited in a low-energy, brackish/freshwater and intensely terrestrial-influenced environment, characterized by the finest sediment, the lowest Sr/Ba ratio, the highest Ti/Ca values and relatively high TOC content.
- From 11.6 to 8.4 cal ka BP, the SBGMD experienced a marine transgression, as indicated by the coarsest sediment, the gradual increase in Sr/Ba values, the gradual decrease in Ti/Ca values and the lowest TOC content.
- From 8.4 cal ka BP to present, the SBGMD has remained in a continuous shallow-marine environment similar to at present, resulting in deposition of fine sediment, with the highest Sr/Ba and lowest Ti/Ca values, together with the highest TOC content.

This study was funded by projects from the China Geological Survey (No. 1212010914027), BMBF, Germany (No. 03F0607A), and the open fund of the Key Laboratory of Marine Mineral Resources, Ministry of Land and Resources, China (No. KLMMR-2014-A-06). The authors thank all the crews and scientists of the FENDOU5 cruise in 2009, the SONNE 219 cruise in 2011, the HAIYANG4 cruise in 2012 and the FENDOU5 cruise in 2012 for their hard work in collecting the seismic data and sediment samples. The authors also thank colleagues from the Guangzhou Marine Geological Survey (GMGS) who helped with the laboratory measurements. The first author thanks Peihong Jia from Nanjing University for the foraminifer picking work and helpful discussions. The authors are grateful to Peter D. Clift and Paul J. Liu for their helpful comments and suggestions that improved the manuscript significantly.

References

Biswas, B. 1973. Quaternary changes in sea-level in the South China Sea. *Geological Society of Malaysia Bulletin*, **6**, 229–256.

Chen, J.R., Chen, X.S., Zhao, X.T., Zheng, F. & Sun, S.X. 1991. Study of sea level in Holocene change at Luhuitou area of Hainan Province. *Geological South China Sea*, **3**, 77–86 [in Chinese].

Chen, Z., Chen, Z. & Zhang, W. 1997. Quaternary stratigraphy and trace-element indices of the Yangtze Delta, Eastern China, with special reference to marine transgressions. *Quaternary Research*, **47**, 181–191.

Clift, P. & Sun, Z. 2006. The sedimentary and tectonic evolution of the Yinggehai-Song Hong basin and the southern Hainan margin, South China Sea: implications for Tibetan uplift and monsoon intensification. *Journal of Geophysical Research*, **111**, B06405, http://doi.org/10.1029/2005JB004048

Clift, P.D., Wan, S. & Blusztajn, J. 2014. Reconstructing chemical weathering, physical erosion and monsoon intensity since 25 Ma in the northern South China Sea: a review of competing proxies. *Earth-Science Reviews*, **130**, 86–102.

Dronkers, J. & Miltenburg, A.G. 1996. Fine sediment deposits in shelf seas. *Journal of Marine Systems*, **7**, 119–131.

Endler, M., Endler, R., Bobertz, B., Leipe, T. & Arz, H. 2015. Linkage between acoustic parameters and seabed sediment properties in the south-western Baltic Sea. *Geo-Marine Letters*, **35**, 145–160, http://doi.org/10.1007/s00367-015-0397-3

Fairbanks, R.G. 1989. A 17,000-year glacio-eustatic sea level record: influence of glacial melting rates on the Younger Dryas event and deep-ocean circulation. *Nature*, **342**, 637–642.

Funabiki, A., Haruyama, S., Quy, N.V., Hai, P.V. & Thai, D.H. 2007. Holocene delta plain development in the Song Hong (Red River) delta, Vietnam. *Journal of Asian Earth Sciences*, **30**, 518–529.

Garnaud, S., Lesueur, P., Clet, M., Lesourd, S., Garlan, T., Lafite, R. & Brun-Cottan, J.C. 2003. Holocene to modern fine-grained sedimentation on a macrotidal shoreface-to-inner-shelf setting (eastern Bay of the Seine, France). *Marine Geology*, **202**, 33–54.

GEYH, M.A., KUDRASS, H.R. & STREIF, H. 1979. Sea-level changes during the late Pleistocene and Holocene in the Strait of Malacca. *Nature*, **278**, 441–443.

GROSSMAN, E.E., EITTREIM, S.L., FIELD, M.E. & WONG, F.L. 2006. Shallow stratigraphy and sedimentation history during high-frequency sea-level changes on the central California shelf. *Continental Shelf Research*, **26**, 1217–1239.

HANEBUTH, T., STATTEGGER, K. & GROOTES, P.M. 2000. Rapid flooding of the Sunda Shelf: a late-glacial sea-level record. *Science*, **288**, 1033–1035.

HANEBUTH, T.J.J. & LANTZSCH, H. 2008. A Late Quaternary sedimentary shelf system under hyperarid conditions: unravelling climatic, oceanographic and sea-level controls (Golfe d'Arguin, Mauritania, NW Africa). *Marine Geology*, **256**, 77–89.

HARFF, J., LEIPE, T., WANIEK, J. & ZHOU, D. (eds) 2013. Depositional Environments and Multiple Forcing Factors at the South China Sea's Northern Shelf. *Journal of Coastal Research*, **SI 66**, 90.

LAM, D.D. & BOYD, W.E. 2001. Some facts of sea-level fluctuation during the late Pleistocene–Holocene in Ha Long Bay and Ninh Binh area. *Journal of Sciences of the Earth*, **23**, 86–91 [in Vietnamese with English abstract].

LANTZSCH, H., HANEBUTH, T.J.J. & BENDER, V.B. 2009. Holocene evolution of mud depocentres on a high-energy, low-accumulation shelf (NW Iberia). *Quaternary Research*, **72**, 325–336.

LESUEUR, P., TASTET, J.P. & MARAMBAT, L. 1996. Shelf mud fields formation within historical times: examples from offshore the Gironde estuary, France. *Continental Shelf Research*, **16**, 1849–1870.

LIU, J.P., MILLIMAN, J.D., GAO, S. & CHEN, P. 2004. Holocene development of the Yellow River's subaqueous delta, North Yellow Sea. *Marine Geology*, **209**, 45–67.

LIU, J.P., XU, K.H., LI, A.C., MILLIMAN, J.D., VELOZZI, D.M., XIAO, S.B. & YANG, Z.S. 2007. Flux and fate of Yangtze River sediment delivered to the East China Sea. *Geomorphology*, **85**, 208–224.

LIU, Y., GAO, S., WANG, Y., YANG, Y., LONG, J., ZHANG, Y. & WU, X. 2014. Distal mud deposits associated with the Pearl River over the northwestern continental shelf of the South China Sea. *Marine Geology*, **347**, 43–57.

MA, F., WANG, Y., LI, Y., YE, C., XU, Z. & ZHANG, F. 2010. The application of geostatistics in grain size trend analysis: a case study of eastern Beibu Gulf. *Journal of Geographical Sciences*, **20**, 77–90.

MORIGI, C., JORISSEN, F.J. *ET AL.* 2005. Benthic foraminiferal evidence for the formation of the Holocene mud-belt and bathymetrical evolution in the central Adriatic Sea. *Marine Micropaleontology*, **57**, 25–49.

NI, Y., ENDLER, R. *ET AL.* 2014. The 'butterfly delta' system of Qiongzhou Strait: morphology, seismic stratigraphy and sedimentation. *Marine Geology*, **355**, 361–368.

NIZOU, J., HANEBUTH, T.J.J., HESLOP, D., SCHWENK, T., PALAMENGHI, L., STUUT, J.B. & HENRICH, R. 2010. The Senegal River mud belt: a high-resolution archive of paleoclimatic change and coastal evolution. *Marine Geology*, **278**, 150–164.

POLYAKOVA, Y.I. & STEIN, R. 2004. Holocene paleoenvironmental implications of diatom and organic carbon records from the southeastern Kara Sea (Siberian Margin). *Quaternary Research*, **62**, 256–266.

SHI, M., CHEN, C. *ET AL.* 2002. The role of Qiongzhou Strait in the seasonal variation of the South China Sea circulation. *Journal of Physical Oceanography*, **32**, 103–121.

SOUTHON, J., KASHGARIAN, M., FONTUGNE, M., METIVIER, B. & YIM, W.W.-S. 2002. Marine reservoir corrections for the Indian Ocean and Southeast Asia. *Radiocarbon*, **44**, 167–180.

STUIVER, M., REIMER, P.J. & REIMER, R. 2014. *CALIB 7.0.1*, http://calib.qub.ac.uk/calib/

TANABE, S., HORI, K., SAITO, Y., HARUYAMA, S., VU, V.P. & KITAMURA, A. 2003. Song Hong (Red River) delta evolution related to millennium-scale Holocene sea-level changes. *Quaternary Science Reviews*, **22**, 2345–2361.

TRAN, N. & NGO, Q.T. 2000. Development history of deposits in the Quaternary of Vietnam. *In*: NGUYEN, T.V. (ed.) *The Weathering Crust and Quaternary Sediments in Vietnam*. Department of Geology and Minerals of Vietnam, Hanoi, 177–192 [in Vietnamese].

VAN DEN BERGH, G.D., BOER, W., SCHAAPVELD, M.A.S., DUC, D.M. & VAN WEERING, Tj.C.E., 2007*a*. Recent sedimentation and sediment accumulation rates of the Ba Lat prodelta (Red River, Vietnam). *Journal of Asian Earth Sciences*, **29**, 545–557.

VAN DEN BERGH, G.D., VAN WEERING, Tj.C.E., BOELS, J.F., DUC, D.M. & NHUAN, M.T. 2007*b*. Acoustical facies analysis at the Ba Lat delta front (Red River Delta, North Vietnam). *Journal of Asian Earth Sciences*, **29**, 532–544.

VAN MAREN, D.S. & HOEKSTRA, P. 2004. Seasonal variation of hydrodynamics and sediment dynamics in a shallow subtropical estuary: the Ba Lat River, Vietnam. *Estuarine, Coastal and Shelf Science*, **60**, 529–540.

VAN MAREN, D.S. & HOEKSTRA, P. 2005. Dispersal of suspended sediments in the turbid and highly stratified Red River plume. *Continental Shelf Research*, **25**, 503–519.

VINZON, S.B., WINTERWERP, J.C., NOGUEIRA, R. & DE BOER, G.J. 2009. Mud deposit formation on the open coast of the larger Patos Lagoon–Cassino Beach system. *Continental Shelf Research*, **29**, 572–588.

VIRTASALO, J.J., RYABCHUK, D., KOTILAINEN, A.T., ZHAMOIDA, V., GRIGORIEV, A., SIVKOV, V. & DOROKHOVA, E. 2014. Middle Holocene to present sedimentary environment in the easternmost Gulf of Finland (Baltic Sea) and the birth of the Neva River. *Marine Geology*, **350**, 84–96.

XIE, X., MÜLLER, R.D., REN, J., JIANG, T. & ZHANG, C. 2008. Stratigraphic architecture and evolution of the continental slope system in offshore Hainan, northern South China Sea. *Marine Geology*, **247**, 129–144.

YANCHEVA, G., NOWACZYK, N.R. *ET AL.* 2007. Influence of the intertropical convergence zone on the East Asian monsoon. *Nature*, **445**, 74–77.

YANG, S.Y., JUNG, H.S., LIM, D.I. & LI, C.X. 2003. A review on the provenance discrimination of sediments in the Yellow Sea. *Earth Science Reviews*, **63**, 93–120.

YAO, Y., HARFF, J., MEYER, M. & ZHAN, W. 2009. Reconstruction of paleocoastlines for the northwestern South China Sea since the Last Glacial Maximum. *Science in China Series D: Earth Sciences*, **52**, 1127–1136.

ZONG, Y.Q. 2004. Mid-Holocene sea-level highstand along the Southeast Coast of China. *Quaternary International*, **117**, 55–67.

Last Glacial Cycle and seismic stratigraphic sequences offshore western Hainan Island, NW South China Sea

HONGJUN CHEN[1,2,3,4]*, JAN HARFF[3,5], YAN QIU[1,3], ANDRZEJ OSADCZUK[5], JINPENG ZHANG[1,3], MICHAL TOMCZAK[5], ZHAOGUO CUI[1,3], GUANGQIANG CAI[1,3], MINGMING WEN[1,3] & LIQING LI[1,3]

[1]*Guangzhou Marine Geological Survey, China Geological Survey, Guangzhou 510075, China*

[2]*CAS Key Laboratory of Marginal Sea Geology, South China Sea Institute of Oceanology, Chinese Academy of Sciences, Guangzhou 510301, China*

[3]*Key Laboratory of Marine Mineral Resources, Ministry of Land and Resources, Guangzhou 510075, China*

[4]*University of Chinese Academy of Sciences, Beijing 100049, China*

[5]*Institute of Marine and Coastal Sciences, University of Szczecin, Szczecin 70383, Poland*

**Corresponding author (e-mail: chhju@163.com)*

Abstract: In order to investigate the relationship between sea-level changes during the Last Glacial Cycle (LGC) and seismic stratigraphic sequences on the NW continental shelf of the South China Sea, a sparker single-channel high-resolution seismic profiling was correlated with a sediment core taken offshore western Hainan Island. Interpretation of the seismic stratigraphy in relation to local and global sea-level changes could be refined.

According to the age model, developed from AMS (accelerator mass spectrometry)-^{14}C and OSL (optically stimulated luminescence) dates, and $\delta^{18}O$ (oxygen isotope) stratigraphy, the 88.3 m sediment core reflects environmental change since 110 ka. Correlation with a sediment core in the adjacent basin based on stable oxygen isotope records allows a zonation of the top 15 m of the sediment core and the identification of marine isotope stages (MIS) 1–5e.

Seven seismic reflectors interpreted as unconformities were identified. These erosional surfaces have been dated by interpolation using an age–depth model compared to global eustatic curves. The results indicate that seven sea-level cycles can be distinguished in the study area during the LGC, and these are correlated with regional and global sea-level change models. Further research confirms that seismic stratigraphy in the NW South China Sea can be intimately related to LGC sea-level changes associated with regional surface uplift and sediment supply.

Understanding the current and future climate and environmental system requires insight into the development of the global climate during the Quaternary and the Last Glacial Cycle (LGC) in particular. In contrast to deep marine basins with low sedimentation rates, high-frequency palaeoceanographic and climatic changes are recorded on continental margins, and especially by shelf sediments. Because the South China Sea is not strongly affected by glacio-isostatic adjustment of the Earth's crust to loading and unloading by ice sheets, or by destructive advance of ice sheets during glacial periods, the wide shelves record the effects of climate-induced sea-level cycles, so that the change between marine transgression and regression can be reconstructed in high resolution in this region (Hanebuth *et al.* 2003, 2006, 2009, 2011; Schimanski & Stattegger 2005; Harff *et al.* 2013).

Basic concepts in sequence stratigraphy allow the sedimentary architecture of continental margins to be interpreted as a record of climatically controlled eustatic sea-level changes, vertical movements of the crust (in low latitudes, mainly tectonic effects) and riverine sediment supply. Atmospheric temperature reconstructions are mainly based on oxygen isotope proxies from Antarctic and Greenland ice cores, whereas sea-level changes can be reconstructed by radiometrically dated corals (Lambeck & Chappell 2001; Waelbroeck *et al.* 2002) or by changes in the oxygen isotopic signature of planktonic foraminifera shells in semi-enclosed basins (e.g. the Red Sea and the Mediterranean Sea) that reflect palaeo-salinities resulting from the basin isolation effect (Siddall *et al.* 2004; Rohling *et al.* 2009; Grant *et al.* 2012). Recently, Lobo & Ridente (2013) showed the effect of Milankovitch cycles in

From: Clift, P. D., Harff, J., Wu, J. & Qui, Y. (eds) 2016. *River-Dominated Shelf Sediments of East Asian Seas*. Geological Society, London, Special Publications, **429**, 99–121.
First published online December 1, 2015, updated January 28, 2016 and May 26, 2016, http://doi.org/10.1144/SP429.9

insolation on the architecture of sediments covering continental shelves. This influence was proven even for high-frequency sequences of fourth and fifth order with time intervals of 100–200 and 10–40 kyr. Lobo & Ridente (2013) compared their study of globally distributed seismic profiles on continental shelves with a composite of sea-level data from different sources, as well as a stochastically interpreted sea-level dataset published by Grant *et al.* (2012). Comparable regional investigations have been carried out by Hanebuth *et al.* (2011) in the South China Sea by comparing seismic profiles on the Sunda Shelf with a sea-level curve derived from numerical simulation corrected for hydrostatic effects by Nakada & Lambeck (1989).

Comparable studies are scarce for the northern margin of the South China Sea. Seismostratigraphic analyses were performed by Palamenghi *et al.* (2015) for the Late Cenozoic of the Pearl River Mouth Basin. Clift & Sun (2006) studied the Neogene sedimentary history of the Yinggehai–Song Hong Basin, and Xie *et al.* (2008) compared the sequence stratigraphic inventories of the same basin SW of Hainan and the Qiongdongnan Basin (SE of Hainan: Fig. 1). The post-glacial sea level related to the palaeogeographical history of the Beibu Gulf Basin and adjacent areas was relatively well investigated by the comprehensive study of Yao *et al.* (2009), who compiled relative sea-level curves from the NW coast of the South China Sea and reconstructed the post-glacial marine transgression. Their palaeogeographical model of the opening of the Qiongzhou Strait (Fig. 1) was confirmed by seismostratigraphic analyses by Ni *et al.* (2014). However, scientific investigations of the Late Pleistocene sea-level history comparable to earlier studies of the Sunda Shelf are extremely rare for the NW margin of the South China Sea.

In 2008, the Guangzhou Marine Geological Survey (GMGS) gathered seismic reflection data from west of Hainan, as well as taking a sediment core at Site HDQ2 (Fig. 1) that penetrated the late Pleistocene sedimentary record over a 88.3 m core length. Here, the results of a comprehensive study of the seismic records and the sediments from Core HDQ2 are presented in terms of the seismic stratigraphic architecture of the study area and its interpretation in light of the regional sea-level history

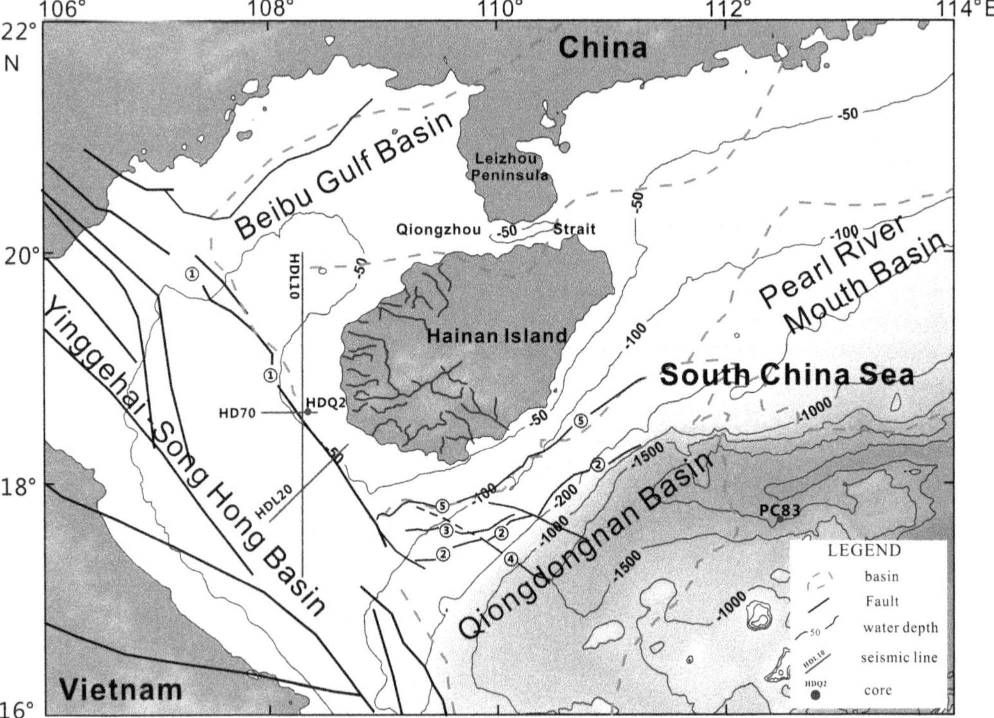

Fig. 1. Tectonic and bathymetric map of the NW South China Sea showing the location of the Yinggehai–Song Hong Basin and its adjacent basins (modified from Sun *et al.* 2009; Wang *et al.* 2011). The source of the bathymetry is from NOAA GEODAS. The red solid lines are seismic profiles (HDL10, HD70 and HDL20) in the Yinggehai–Song Hong Basin. The red circles are Core HDQ2 to the west of offshore Hainan Island and Core PC83 in deep water.

during the Late Pleistocene. The study took place within the framework of a bilateral Chinese–Polish collaborated project 'Sedimentary Environment and Climate Evolution since the Late Pleistocene in the Beibu Gulf and its Adjacent Area' (SECEB), jointly conducted by the Guangzhou Marine Geological Survey and the University of Szczecin.

Geological setting

The South China Sea is one of the largest marginal seas of the western Pacific Ocean. It is located within the influence of the East Asian monsoon and experiences the earliest onset of the Asian summer monsoon (Zhao & Wang 1999). The study area is located on the NW continental shelf, within the Beibu Gulf, west of Hainan Island (Fig. 1). The Beibu Gulf is one of the largest embayments of the South China Sea and forms a shelf basin with an average water depth of 45 m that reaches maximum values of approximately 100 m close to the shelf edge. The monsoon periodically changes the wind, tides and marine currents that control the circulation system within the gulf. Water exchange with the South China Sea is governed by the wide opening to the south and the narrow Qiongzhou Strait to the east, separating Hainan Island and the Leizhou Peninsula (Bauer *et al.* 2013).

The study area covers a corner on the shoulder of two basins: the NW-striking Yinggehai Basin and the NE-striking Qiongdongnan Basin (Fig. 1). The position in the upper NW edge of the Qiongdongnan Basin is bounded to extensional NE-trending faults that were first formed during the Oligo-Miocene during the opening stages of the South China Sea. To the west, the Qiongdongnan Basin, Hainan Island and the Beibu Gulf Basin are bounded by the NW-trending Yinggehai–Song Hong Basin, which is one of the world's largest pull-apart basins (Fig. 1). This basin is genetically connected to the Red River Fault Zone (Tanabe *et al.* 2003; Clift & Sun 2006; Xie *et al.* 2008), having been filled by sediments supplied by the Red River eroded from the SE flank of the Tibetan Plateau.

Because of the tectonic setting, the slopes of the Yinggehai-Song Hong and Qiongdongnan basins are quite different in their structure. Whereas the former is characterized by progradational slope clinoforms (owing to the high rate of sediment supply), the Qiongdongnan slope shows vertical stacking, gravitational faults, rollover and slump deposits. The study area on the shelf between the Yinggehai-Song Hong and Qiongdongnan basins is separated from the sediment source of the Red River, so that the sediments that have accumulated on the shelf corner are derived from Hainan and partly from the landmass of South China (Xie *et al.* 2008). Cenozoic coastal development along the NW margin of the South China Sea was determined by rift tectonics and sediment supply by river sources. Shi *et al.* (2011) noted that a marked reduction in sediment supply from Hainan Island into the neighbouring basins has occurred since the Miocene and concluded that the main source of fill for the Yinggehai–Song Hong Basin from the Miocene to Recent has been the Red River, which delivers the erosional products driven by the Tibetan Plateau and Vietnamese coastal uplift. Furthermore, Van Hoang *et al.* (2010) commented on the influence of tectonic uplift in conjunction with an intensification of the summer monsoon on an increase in physical erosion since the end of the Oligocene. Despite the dominance of the Red River as a supplier of sediments for the Yinggehai–Song Hong Basin, near-source deltas around Hainan at different periods led Shi *et al.* (2011) to the interpretation that Hainan Island has, for long time, been a major source of sediment to both the Yinggehai–Song Hong and the Qiongdongnan basins. During the Quaternary, the climatically controlled sea level (together with river discharge) took over the role of controlling the coastal geomorphology. Because the Chinese marginal seas, such as the Yellow Sea, the East China Sea and the South China Sea, were continuously connected to the world ocean during the Quaternary, global ice–water volume effects must also be reflected in the sediment architecture of the South China Sea margins. Wang & Li (2009) reported a late Pleistocene sea-level curve for the South China Sea margin that covered the time span of marine isotope stages (MIS) 3–MIS 1 (Huang *et al.* 1986). However, sediments deposited during lowstand periods are extremely scarce on the shelf, whereas highstand sediments have been preserved in basins adjacent to the major river deltas (Hanebuth *et al.* 2006). Hanebuth *et al.* (2006) investigated late Pleistocene highstand sediments (apparently MIS 3) directly below the Red River delta, and proposed that a direct correlation between the palaeo-sea level in the South China Sea margins and sites in the Pacific or Atlantic oceans is complicated owing to differences in the isostatic-tectonic history of the continental margins. Differences in the hydro-isostatic adjustment and its influence on sea-level curves from different parts of the South China Sea can be compensated for by modelling approaches, such as that used in Nakada & Lambeck (1989). For a comparison of South China Sea palaeo-sea-level data with global eustatic curves, the main focus should lie with the identification of sea-level cyclicity reflecting high-frequency Milankovitch cycles rather than the absolute correlation of sea-level data.

The effect of post-glacial sea-level change on the northern margin of the South China Sea has been studied in detail by Yao *et al.* (2009). In the Beibu Gulf area, the rising post-glacial sea level caused a continuous transgression of the shelf, which was exposed to subaerial weathering during the Last Glacial Maximum (LGM). Marine transgressive sediments can be found in seismic records as a marker horizon with basin-wide correlation. Transgression submerged the Beibu Gulf before the Qiongzhou Strait opened at approximately 9 ka (Yao *et al.* 2009; Ni *et al.* 2014), and served as a permanent connection between the Beibu Gulf and the NE South China Shelf. In the NW, the Red River delta dominates the coast of the Beibu Gulf. According to Tanabe *et al.* (2003), this delta formed in response to the Holocene palaeo-sea-level change. Between 9 and 6 ka, the Red River delta formed by progradation into the Yinggehai–Song Hong Basin and was drowned as a result of early Holocene inundation that formed wide tidal flats. At 6 ka, the sea level was around 3 m higher than present (Tanabe *et al.* 2003; Hanebuth *et al.* 2011); however, the flats emerged after 6 ka, when the sea level dropped to the modern level, and the flats have since been converted to marine terraces. Within the research area, marine sediments are represented by a fine-grained mud layer, a few metres thick, which is replaced in nearshore areas by coarser-grained sediments.

Materials and methods

Seismic data profiling and processed

During 2006–08, the study area was investigated three times. Two cruises with the Chinese R/V *Fendou No. 5* collected seismic profiles and, during a third cruise, a drilling campaign was performed. A combination of digital GPS (DGPS) and radar systems, with a positional accuracy of around 5 m, was used for navigation for the R/V *Fendou No. 5*.

High-resolution two-dimensional (2D) single-channel sparker seismic profiles (source of 4000–8000 J) were acquired during the cruises. These have a vertical resolution of about 1–5 m, with a maximum acoustic penetration of up to 400 ms (two-way travel time). The acquired data were converted to SEG-Y format and further processed using Geovation 2D software at the data-processing centre of the GMGS. All of the seismic profiles were interpreted on a workstation using Discovery commercial software, which was developed by Halliburton. Nearly 470 km of seismic profiles have been collected. We identified seismic units, major erosional discontinuities and depositional sequences based on reflector terminations (onlap, downlap and toplap), as well as the seismic facies. Subsequently, we tied the 2D seismic interpretation to Core HDQ2 in order to achieve age control.

Core HDQ2 (18° 37.98′ N, 108° 20.58′ E) is located on the continental shelf of the NW South China Sea (Fig. 1), approximately 36.9 km from the western coast of Hainan in a water depth of 44 m. The core recovered a total of 88.3 m of sediment. The chronostratigraphic framework of the core is based on AMS (accelerator mass spectrometry)-^{14}C and optically stimulated luminescence (OSL) dates, and oxygen isotope stratigraphy.

Age dating methods

AMS-^{14}C dating. Seven bulk sediment samples were pre-processed at the Guangzhou Institute of Geochemistry, Chinese Academy of Sciences (GIGCAS). Accelerator mass spectrometry (AMS) analyses were performed at the State Key Laboratory of Nuclear Physics and Technology (Peking University). In addition, three AMS-^{14}C dates were analysed at the radiocarbon laboratory of Beta Analytic (USA) using foraminifera assemblages. Sediment samples were sieved (mesh size 125 μm) and approximately 2000 foraminifera shells (a single specimen with minimal abrasion) were picked for each sample (samples 372150, 372151 and 372152). Primary ^{14}C data were calibrated using the CALIB 7.1 (with calibration curve Marine13 and Delta $R = 0$).

OSL dating. The analyses were carried out by the Guangzhou Institute of GIGCAS. The OSL samples were treated with 0.1 N HCl, 0.01 N $Na_2C_2O_4$ and 30% H_2O_2 to remove carbonate, clay coatings and organic matter. Subsequently, the polymineral fine-grain fraction (4–11 μm) was extracted from the sediments. 34% fluorosilicic acid (H_2SiF_6) was used for 5 days to remove feldspar grains from the polymineral fine grains in order to obtain a pure quartz fraction from the three samples. The purity of the quartz extract was evaluated by petrographical inspection. The OSL single-aliquot regeneration (SAR) protocol was applied to fine-grained quartz to determine the equivalent dose (D_e) of the three samples. The SAR analyses were conducted using an automated Daybreak 2200 TL/OSL Reader at the Experimental and Testing Centre of Marine Geology, Ministry of Land and Resources of China (KLMMR). The U, Th and K_2O content was determined for the bulk sediment to in order to estimate the exposure of quartz grains to ionizing radiation, and the moisture content of a sample during burial was also determined. The D_e values of the samples were calculated by integrating the 0–1 s region of the OSL decay curve, and the final 5 s of stimulation was subtracted as background. All dose–response curves were fitted using a saturating exponential function.

Foraminifera analysis

Samples for foraminiferal analysis were oven-dried at 60°C, weighed and disaggregated by soaking in tap water. Samples were wet-sieved over a 63 μm sieve, dried at 60°C and weighed in order to calculate the weight per cent of the coarse fraction (>63 μm). Samples were further sieved through a 154 μm sieve, and each sample was placed on a piece of glass, mixed using a blade or pin and then an aliquot was split from the mixed sample before picking.

Planktonic foraminifera were picked from the >154 μm fraction, containing 200–400 specimens. Complete and broken planktonic foraminifera tests were counted, and a planktonic fragment ratio was calculated according to the method of Le & Shackleton (1992). Benthic foraminifera were picked from the >154 μm fraction until 100 or more specimens had been isolated (maximum number 415; average 168; <120 in 10 samples).

Oxygen stable isotopic composition of foraminifera

The analyses were carried out by the Key Laboratory of Marine Mineral Resources, Ministry of Land and Resources of China (KLMMR). Only 32 samples were suitable to use for analysing the calcareous planktonic foraminifera (*Globigerinoides ruber*). For stable isotope analyses in Core HDQ2, only those samples that had no sign of chemical or physical alteration were analysed. The size of the foraminifera *Globigerinoides ruber* chosen for isotope analysis was within the range 0.25–0.35 mm. The detailed method is given in Xu *et al.* (2005). The $\delta^{18}O$ results are shown in Table 1.

Mineralogical, geochemical ($CaCO_3$) analyses

The clay fraction (<2 μm) was separated according to Stoke's settling velocity principle after removing carbonate and organic matter with 10% H_2O_2 and 0.5 N HCl, respectively. Clay mineralogy determinations were performed by standard X-ray diffraction (XRD) on a D/Max 2500PC diffractometer with Cu Ka radiation (40 kV, 200 mA) at KLMMR.

Three XRD runs were performed, following air-drying, ethylene glycol solvation for 7 h at 50°C and heating at 490°C for 2 h. Identification of clay minerals was made mainly using the position of the (001) series of basal reflections on the three XRD diagrams. Mixed-layer clays, mainly of smectite–illite, were included in 'smectite', and mixed layers, mainly of chlorite–illite, with very minor

Table 1. *Isotope characteristics of the foraminifera of Core HDQ2 at different layers*

Station	ID	Depth (cm)	Foraminifera species (P)	$\delta^{18}O_{(V\text{-}PDB)}$ (‰)
HDQ2	08026C3281	0	*G. ruber*	−3.25
	08026C3282	20	*G. ruber*	−3.64
	08026C3283	40	*G. ruber*	−3.80
	08026C3284	60	*G. ruber*	−3.79
	08026C3285	640	*G. ruber*	−2.71
	08026C3286	760	*G. ruber*	−3.19
	08026C3287	780	*G. ruber*	−3.65
	08026C3288	820	*G. ruber*	−3.51
	08026C3289	840	*G. ruber*	−3.25
	08026C3291	900	*G. ruber*	−3.47
	08026C3292	920	*G. ruber*	−3.71
	08026C3293	940	*G. ruber*	−1.34
	08026C3294	960	*G. ruber*	−3.15
	08026C3295	980	*G. ruber*	−3.42
	08026C3296	1000	*G. ruber*	−3.37
	08026C3297	1020	*G. ruber*	−2.75
	08026C3298	1040	*G. ruber*	−2.91
	08026C3299	1060	*G. ruber*	−3.45
	08026C3300	1080	*G. ruber*	−2.06
	08026C3301	1100	*G. ruber*	−2.10
	08026C3302	1120	*G. ruber*	−3.57
	08026C3303	1140	*G. ruber*	−3.91
	08026C3304	1160	*G. ruber*	−3.38
	08026C3305	1180	*G. ruber*	−2.93
	08026C3306	1200	*G. ruber*	−3.00
	08026C3307	1220	*G. ruber*	−2.98
	08026C3308	1240	*G. ruber*	−3.21
	08026C3309	1260	*G. ruber*	−2.89
	08026C3310	1280	*G. ruber*	−2.39
	08026C3311	1300	*G. ruber*	−2.67
	08026C3312	1360	*G. ruber*	−3.60
	08026C3313	1380	*G. ruber*	−2.90

abundance were not calculated. The relative percentages of the four main clay mineral groups were estimated by weighting integrated peak areas of characteristic basal reflections (smectite 17 Å; illite 10 Å; kaolinite/chlorite 7 Å) in the glycolated curve using MDI Jade 6.5 software. Relative proportions of kaolinite and chlorite were determined using the ratio of 3.57 Å/3.54 Å peak areas. The illite chemistry index is inferred from the ratio of 5 Å/10 Å peak areas on the XRD diagrams generated after ethylene glycol treatment. The measurement of illite crystallinity is made by computing the integrated breadth of the glycolated 10 Å peak. Based on the XRD method, the semi-quantitative evaluation of each clay mineral has an accuracy of about 5%.

For each sediment sample, 7 g of was dried, weighed and washed, with grain sizes of between 0.25 and 0.063 mm being taken for analysis. The light minerals, heavy minerals and electromagnetic minerals were separated according their physical characteristics. The volume percentage of each was then calculated under a stereo-microscope, and then transferred into the sediment weight percentage.

For the calcium carbonate analyses, 8 ml of acetic acid (10%) was added to each sample (0.5000 g) in a boron silicate tube. All the tubes were heated in a boiling water bath for 30 min. The resulting solutions were analysed for calcium carbonate using the EDTA complexometry method (Skoog & West 1969).

Grain-size analyses

In total, 375 mineralogical samples were measured for particle-size-distribution analysis, which was performed at a resolution of 20 cm along the core, or more frequently if lithological properties changed. Measurements were conducted in the laboratory of the Palaeogeography Unit, Faculty of Geosciences (University of Szczecin, Poland). Samples were freeze-dried and burned at 550°C in order to remove organic matter. The mineral content was sieved through a Retsch laboratory mesh size of 258 μm and then determined using a particle-size analyser (Mastersizer Micro ver. 2.19 Malvern Instruments Ltd), which enabled measurements to be made of particles with average diameters ranging from 0.3 to 300 μm. The sediment grain-size fraction >258 μm was weighed and analysed in 82 samples using a sieve shaker, with mesh sizes of 355, 500, 710, 1000, 1400 and 2000 μm. The results of the fine and coarse particles were merged together, and are presented graphically as the percentage content of each class.

Results

Age dating

The purpose of determining the age of the sediment material by radiocarbon or OSL methods is to identify the time of deposition of each stratigraphic unit. The age data show inverted or confused time series in the upper part of the core. These results are potentially affected by admixtures of younger or older organic matter during transport and deposition. It is, therefore, essential to confirm the reliability of these dates (Hanebuth *et al.* 2006). We note that the OSL age at a depth of 88 m is younger than the ages at a depth of 78 m (Table 2). This could be because of the younger sediment from the upper part mixing during the drilling of the core. According to these data, Core HDQ2 contains a late Pleistocene (*c.* 100 ka BP) to Holocene (*c.* 11 ka BP) succession, covering a time span from MIS 5.4 to MIS 1.

Stratigraphic framework

A sediment piston core (PC83) was sampled in the SE part of the Qiongdongnan Basin (17° 39.52′ N,

Table 2. *The AMS-^{14}C, OSL and calibrated ages of Core HDQ2*

No.	Depth (m)	Age (ka)	Error	Calibrated age (cal ka BP)	SD	Dating method
1	0.01	4.300	±30	4320–4520*		AMS-^{14}C†
2	0.63	9.740	±38	11.220–11.169‡		AMS-^{14}C§
3	1.10	23.734	±85	28.435‖	±168	AMS-^{14}C§
4	3.40	28.620	±113	34.014‖	±185	AMS-^{14}C§
5	3.60	35.501	±188	40.813‖	±241	AMS-^{14}C§
6	7.00	33.405	±166	38.796‖	±221	AMS-^{14}C§
7	8.05	35.300	±370	38.620–40.240*		AMS-^{14}C†
8	9.50	25.268	±143	30.421‖	±209	AMS-^{14}C§
9	14.25	31.111	±159	36.473‖	±207	AMS-^{14}C§
10	19.99	43.180	±920	44.400–48.050*		AMS-^{14}C†
11	22.70	63.000			7	OSL¶
12	23.40	67.000			7	OSL¶
13	29.00	70.000			9	OSL¶
14	39.80	78.000			8	OSL¶
15	45.6	83.000			9	OSL¶
16	65.2	97.000			11	OSL¶
17	78.0	107.000			11	OSL¶
18	88.0	67.000			9	OSL¶

*AMS-^{14}C ages have been calibrated using the CALIB 7 (calibration curve Marine13 and Delta $R = 0$).
†Measured at the Beta Analytic radiocarbon dating laboratory (Miami, FL, USA).
‡AMS-^{14}C ages have been calibrated using the program CALIB 5.0.1.
§Measured in the State Key Laboratory of Nuclear Physics and Technology (Peking University) and the Guangzhou Institute of Geochemistry, Chinese Academy of Sciences (GIGCAS).
‖Correction provided by Professor R. G. Fairbanks (http://radiocarbon.ldeo.columbia.edu).
¶Measured in the GIGCAS.

112° 32.62′ E, water depth 1917 m) in order to allow stratigraphic correlation. The stratigraphy of Core PC83 was calibrated using AMS-^{14}C ages and relative geomagnetic palaeo-intensity records, which include continuous mineral magnetic and palaeomagnetic records measured by discrete sediment samples (Yang *et al.* 2009). The palaeo-intensity records were also correlated with the NAPIS-75, S. Atlantic-1089, Sint-200 and NOPAPIS-250 records to determine the chronological relative palaeo-intensity framework for the South China Sea (Yang *et al.* 2009) (Fig. 2). The planktonic foraminiferal oxygen isotopic records of Core PC83 (Cai *et al.* 2012) reveal a transition from MIS 5 to the Holocene (MIS 1). The boundaries of the MIS 1–MIS 5 stages in Core PC83 are at depths of 85, 200, 415 and 560 cm, respectively (Cai *et al.* 2012) (Fig. 2).

However, in Core HDQ2, well-preserved planktonic foraminifera *Globigerinoides ruber* only appeared at depths of 0–60 and 640–1380 cm, respectively (Fig. 3; Table 1). The oxygen isotope curve of the planktonic foraminifera *Globigerinoides ruber* in Core HDQ2 is depicted in Figure 3. From the oxygen isotope curve, it can be seen that the δ^{18}O values at depths of 0–60 cm range from 1.03 to 1.19‰, and from −3.25‰ to −3.80‰, with an average δ^{18}O value of −3.62‰. The δ^{18}O values at depths of 640–1380 cm range from −0.2 to −1.47‰ and from −1.34‰ to −3.91‰. The whole-core δ^{18}O values range from −1.34 to −3.91‰, with an average value of −3.14‰ (Table 2). The lowest δ^{18}O value is at a depth of 1140 cm, which has a negative bias of −3.91‰, while the greatest δ^{18}O value is at a depth of 940 cm, which has a positive bias of −1.34‰. The trend of oxygen isotopic variation in Core HDQ2 over the 0–60 cm interval shows a step-like increase from the lighter value of −3.79‰ to the heavier value of −3.26‰. There are four negative bias layers at depth intervals of 1300–1380, 1120–1260, 960–1020 and 780–880 cm, respectively. There are three positive bias layers at depth intervals of 1260–1300, 1080–1120 and 920–960 cm.

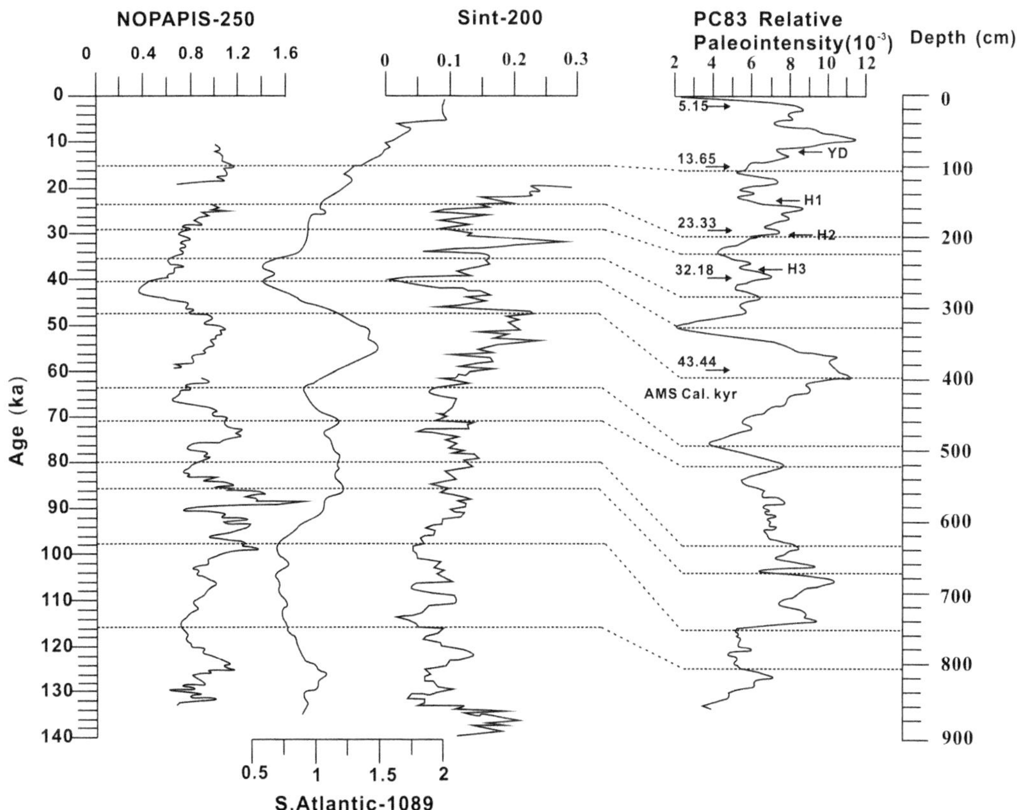

Fig. 2. The stratigraphic framework of Core PC83 in the northern South China Sea correlated to NAPIS (Laj *et al.* 2000), the Sub-Antarctic South Atlantic (Stoner *et al.* 2003), Sint200 (Guyodo & Valet 1996) and NOPAPIS-250 (Yamamoto *et al.* 2007) (after Yang *et al.* 2009).

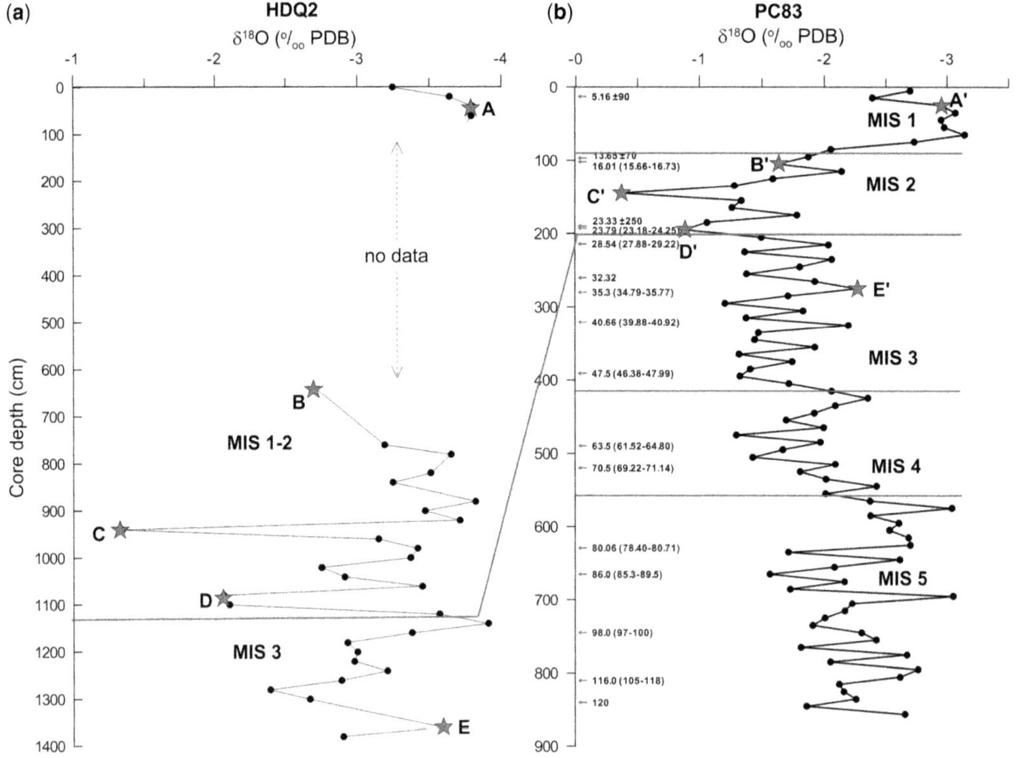

Fig. 3. (**a**) Oxygen isotopic stratigraphy of planktonic foraminifera from Core HDQ2 (0–1380 cm). (**b**) Oxygen isotopic stratigraphy of planktonic foraminifera from Core PC83 in the northern South China Sea. The red stars are the corresponding age control points between Cores HDQ2 and PC83. The dashed lines show the link between age control points. The solid red lines are the boundaries of MIS 1–MIS 5. The red arrows mark the ^{14}C ages in Core PC83 (Yang *et al.* 2009). The purple arrows mark the calculated age based on the age–depth model from Core PC83.

The δ^{18}O curve of the planktonic foraminifera *G. ruber* for Core HDQ2 is generally consistent with that in Core PC83 at the 0–400 cm interval and SO49-8KL at the 0–300 cm interval. Comparison of the δ^{18}O curve between cores HDQ2 and PC83 shows that the upper part of the Core HDQ2 (0–1380 cm) can be divided into MIS 1–MIS 3 stages by correlating identical curve patterns in down-core records (Fig. 3). The boundary of MIS 1–MIS 2 in Core HDQ2 cannot be identified because there are no planktonic foraminifera *G. ruber* in the 60–640 cm interval. Comparison identifies the boundary of MIS 2–MIS 3 in Core HDQ2, located at a depth of 11.40 m. Corresponding points in cores HDQ2 and PC83 are marked by comparison with the δ^{18}O curve (Fig. 3). Age-controlled points in Core PC83 were calculated using this cross-correlated age–depth model (Fig. 4a) (Yang *et al.* 2009), with the calculated results shown in Table 3. The calculated ages of the control points are younger than those of the AMS-^{14}C ages at the 0.63–9.5 m interval in Core HDQ2 (Table 2). This is because the AMS-^{14}C ages are possibly affected by the mixing of older sediments eroded during incision of rivers (discussed below). In Figure 4b, all age data from Core HDQ2 (AMS-^{14}C, δ^{18}O and OSL) are plotted against core depth below seafloor. In the subsection 'Seismo-stratigraphic and sequence-stratigraphic analysis', later in this paper, we describe how the age data from Core HDQ2 are integrated and compared with global eustatic data within the framework of a sequence stratigraphic model.

Late Pleistocene sedimentary inventory of sediment Core HDQ2

According to the correlation with Core PC83, it is assumed that the sediments from Core HDQ2 record the time span from MIS 5.4 to MIS 1 (110 to *c.* 11.0 ka). The material from Core HDQ2 is generally wet and compacted. Sediment colours span a spectrum from dark grey to grey and from brown to yellow from the bottom to the upper part. Characteristic

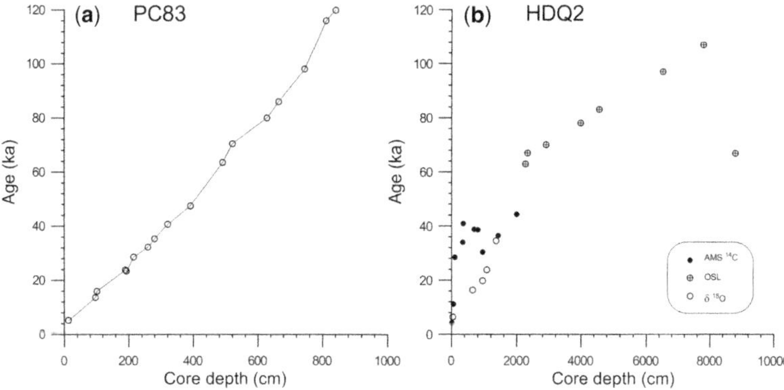

Fig. 4. (**a**) Correlated age–depth model of Core PC83 (Yang *et al.* 2009). (**b**) Age–depth (AMS-^{14}C, δ^{18}O and OSL) data from Core HDQ2.

sedimentary features are confined to individual facies sections. Sediment structures vary from rhythmic lamination and sand or silt layers to irregular sandy patches or clusters and burrows. Mica content ranges from almost 0% to <4.05% through the whole core. Quartz shows a dominance of poorly rounded grains, with content ranging from almost 0% to, at most, 90.5%, which mainly consists of sand. Feldspar content ranges from 0% up to 1% over the whole core. Illite and kaolinite range from 22 to 64%, from and 7 to 23%, respectively. A detailed facies description is given in Table 4 and is discussed below.

Facies interpretation and sedimentary cycles

Core HDQ2 consists of six sedimentary facies units (SFU) that can be further subdivided into sedimentary facies sections (SFS) (Table 4). All units and sections are described and numbered in descending order. A lithological description and facies interpretation based on lithology, grain-size measurements, and mineralogical and palaeontological data interpretation are shown in Table 4.

*Unit SFU 6 (depth in core: 78.70–*c*. 88.30 m).* The basal SFU 6 is characterized by grey silts with abundant mud and sand clusters. Grey to steel-grey clay-rich silts with sandy silts and sand layers or irregular clusters/patches are found in lower part of the unit. The unit shows a general upwards-coarsening succession. Parallel laminations are occasionally recognized in the unit (Fig. 5a). Sand content varies greatly from approximately 2 to 24% throughout. The highest mica content is located at 81.8–83.80 m of the unit, but few planktonic or benthic foraminifera are found (Fig. 6). OSL ages suggest that the unit accumulated rapidly prior to 107 ka (Table 2).

The lack of marine shells and well-preserved foraminifera suggest that tidal influence was relatively high in this unit. The irregular mud and sand clusters/patches were caused by high water energy. SFU 6 was not penetrated by the drilling. It is represented only by the top of the section consisting of silty sediments, which we interpret to indicate a nearshore environment (Table 4).

Table 3. *Calculated ages of corresponding control points between cores HDQ2 and PC83*

Points	Depth at PC83 (cm)	Depth at HDQ2 (cm)	Age (ka)
A–A'	25	40	6.4
B–B'	105	640	16.3
C–C'	145	940	19.8
D–D'	195	1080	23.9
E–E'	275	1360	34.6

*Unit SFU 5 (depth in core: 48.86–*c*. 78.70 m).* SFU 5 consists of three sections (sections 5.3–5.1), with fine-grained sediments in the upper part (Table 5; Fig. 6). The basal section consists of steel-grey silts, sand and sandy mud with scattered shell fragments (Section 5.3) and gravel (clasts 2–4 cm in diameter) at the bottom. Steel-grey to grey silts are interbedded with sand layers with rounded gravel clasts up to 8 cm in diameter at a depth of 59.95 m (Section 5.2) (Fig. 5b). Grey silts are characterized by occasional sand layers, laminations and irregular clusters (Section 5.1) (Fig. 5c).

Sand content ranges from 0.5 to 76% throughout the unit. The lower part of the unit (76–78.7 m) shows a fining-upwards succession. In contrast, no grain-size trend can be found in the upper part of the unit. A higher content of mica, $CaCO_3$, and planktonic and benthic foraminifera are found in

Table 4. *Detailed facies descriptions for sediment core HDQ2 and facies interpretations based on visual lithological core documentation*

Interval (core depth) (m)	Sediment description (lithology, colour, texture, structure)	Sedimentary facies		
		Unit	Section	Environment
0–*c.* 0.63	Dark grey, gravel mud and transition upwards into gravel muddy sand, with complete foraminifers	1	1.1	Coastal/open shelf
0.63–*c.* 1.10	Gravel (5 cm in diameter in the lower part)		1.2	Coastal
1.10–*c.* 3.49	Brown–yellow silt and sandy silt interlayered with thin silt–sand layers, and irregular brown sandy patches. Burrows were found in the upper part	2	2.1	Lake
3.49–*c.* 6.10	Grey silt interlayered with sandy silt lamination and irregular sandy silt cluster		2.2	Lagoonal
6.10–*c.* 11.20	Grey to grey–black sandy silt, with gravel and scattered white shell fragments at the upper part		2.3	Tidal Shelf/ coastal
11.20–*c.* 14.25	Grey to grey–black sandy silt	3	3.1	Shallow shelf
14.25–*c.* 22.48	Homogeneous light-grey sandy silt and grey silt (gradually coarsening upwards) with scattered white shell fragments and abundant irregular sandy patches or clusters		3.2	Coastal/beach
22.48–*c.* 23.50	Grey gravel sand (gradually coarsening downwards) with more white shell fragments		3.3	Coastal/river mouth
23.50–*c.* 24.80	Grey–yellow coarser sand at the top. Grey silt with thin silt and sandy silt layers or irregular patches	4	4.1	Lake
24.80–*c.* 29.00	Grey silt with thin silt and sandy silt layers or irregular patches		4.2	Lagoonal
29.00–*c.* 38.00	Grey sandy silt and sandy mud with abundant sand layers or patches		4.3	Coastal
38.00–*c.* 40.90	Grey clay silts with abundant yellow–brown sandy patches or clusters, gravel mud at the lower part		4.4	River mouth sand
40.90–*c.* 48.86	Grey silts with thin silt sand layers and irregular clusters or patches Grey gravel mud with scattered shell fragments		4.5	Coastal
48.86–*c.* 59.70	Grey silts with scattered sand layers, lamination and irregular clusters	5	5.1	Tidal zone
59.70–*c.* 76.00	Steel-grey to grey silts interbedded by sand layers; a 8 cm rounded gravel at a depth of 59.95 m		5.2	Coastal
76.00–*c.* 78.70	Steel-grey silts sand and sandy mud with scattered shell fragments; gravel (2–4 cm in diameter) at the bottom of 20 cm Grey silts with abundant mud and sand clusters		5.3	Coastal/shallow shelf
78.70–*c.* 88.30	Grey to steel-grey clay silts with sandy silts and sand layers or irregular clusters or patches in the lower part	6	6	Near shore

Section 5.3. OSL dates indicate deposition from 97 to 107 ka.

Rhythmic silt–sand interbedding is common in tidally influenced coastal environments (Nio & Yang 1991; de Boer 1998; Dalrymple *et al.* 2003). Silts interbedded with sand layers are consistent with the unit being deposited in a tidally influenced environment. The lack of mud containing thin peaty layers suggests that the site was not located in a marshy environment. The gravel in the unit suggests that the sediment environment was higher energy in the lower part of sections 5.1 and 5.3 (Fig. 5c). We interpret the facies of Unit 5 to be tidal zone, coastal and shallow-shelf deposits (Table 4).

*Unit SFU 4 (depth in core: 23.50–*c*. 48.86 m).* SFU 4 consists of five sections (sections 4.5–4.1) (Table 4; Fig. 6). Section 4.5 comprises grey silts with thin layers and irregular clusters or patches of silty sand. Grey gravelly mud with scattered shell fragments is also found. Section 4.4 is mainly composed of grey clay-rich silts with abundant yellow–brown sandy patches or clusters, and gravelly mud at the lower part (Fig. 5d). Grey sandy silt and sandy mud with abundant sand layers or patches is shown in Section 4.3, which also contains foraminifera. Section 4.2 is composed of grey silt with thin silt and sandy silt layers or irregular patches. Grey–yellow coarser sand is found at the top of

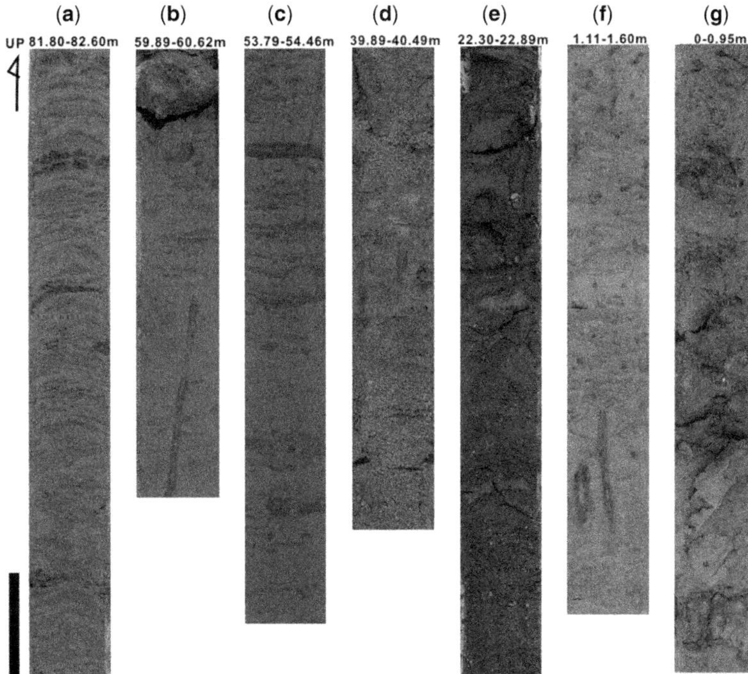

Fig. 5. Photographs of typical sedimentary facies of Core HDQ2. The scale bar is 10 cm. (**a**) SFU 6: depth of 81.80–82.60 m. Grey to steel-grey clay-rich silts with sandy silts and layers or irregular clusters or patches of sand. (**b**) SFU 5: depth of 59.89–60.62 m. Steel-grey to grey silts interbedded with sand; a 8 cm rounded gravel at 59.95 m. (**c**) SFU 5: depth of 53.79–54.46 m. Grey silts with scattered sand layers, laminations and irregular clusters. (**d**) SFU 4: depth of 39.89–40.49 m. Grey clay-rich silts with abundant yellow–brown sandy patches or clusters, with gravelly mud in the lower part. (**e**) SFU 3: depth of 22.30–22.89 m. Grey sandy silt and grey silt (gradually coarsening upwards) with scattered white shell fragments and abundant irregular sandy patches or clusters. (**f**) SFU 2: depth of 1.11–1.60 m. Brown–yellow silt and sandy silt is interbedded with thin silt sand layers, and irregular brown sandy patches. (**g**) SFU depth 1: depth of 0–0.95 m. Dark-grey, gravelly mud transitioning upwards into gravelly muddy sand.

the section. Grey silt with thin silt and sandy silt layers or irregular patches is found in Section 4.1. OSL dates within the unit indicate that this accumulated at around 70.0–83.0 ka.

Sand content ranges from approximately 2 to 82% throughout the unit and shows a fining-upwards succession. A higher content of reworking planktonic and benthic foraminifera are found in Section 4.3, compared to the rest of the unit. There is a low content of foraminifera in Sections 4.1 and 4.2. A few shell fragments are shown at a depth of 45–47 m in the lower part of the unit.

We interpret the facies of Unit 4 to be lake, lagoonal, coastal and river-mouth sand (Table 4).

*Unit SFU 3 (depth in core: 11.20–*c. *23.50 m).* SFU 3 consists of three sections (sections 3.3–3.1) (Table 4; Fig. 6). The basal section consists of grey gravelly sand (gradually fining upwards) with more white shell fragments (Section 3.3) (Fig. 5e). Homogeneous light-grey sandy silt and grey silt (gradually coarsening upwards) with scattered white shell fragments and abundant irregular sandy patches or clusters is shown in Section 3.2. The upper part of Section 3.1 is composed of grey to grey–black sandy silt. Radiocarbon dates within the unit indicate deposition from 36.4 to 67.0 ka (Table 2).

The sand content ranges from 5 to 58%, and has a higher content in the upper and lower part of SFU 3. The reworked planktonic and benthic foraminifera are mainly distributes in Section 3.1. Shell fragments are seen in the upper and lower sections of the unit.

We interpreted the facies of Unit 3 as shallow-shelf, coastal/beach and coastal/river-mouth deposits (Table 4).

*Unit SFU 2 (depth in core: 1.10–*c. *11.20 m).* SFU 2 consists of three sections (sections 2.3–2.1) (Table 4; Fig. 6). Grey to grey–black sandy silt, with gravel and scattered white shell fragments are

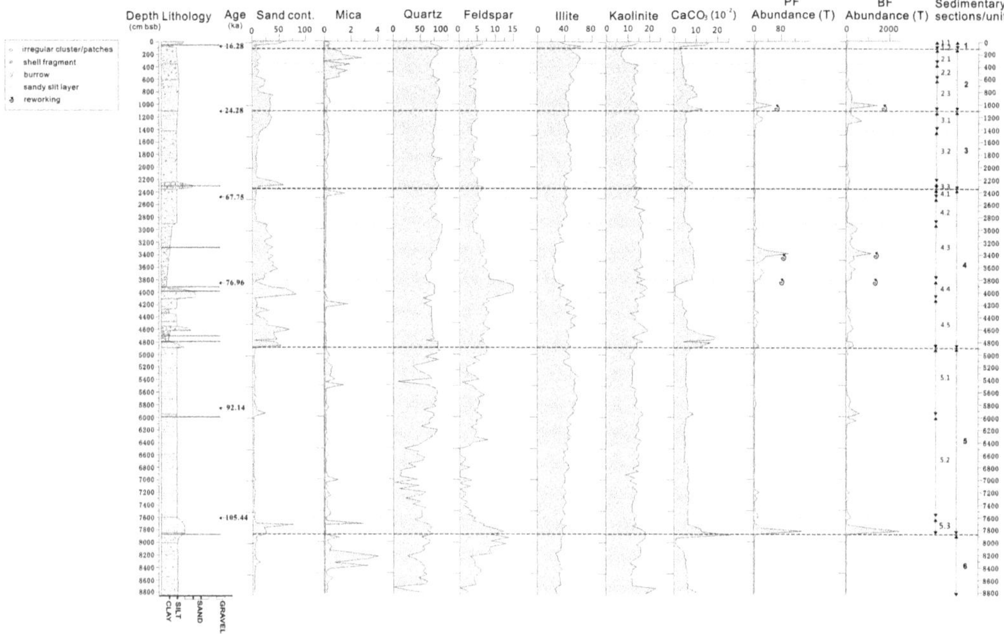

Fig. 6. Comparison of grain-size data, selected age dates, mineral composition, carbonate content, and the distribution of planktonic and benthic foraminifera in Core HDQ2. cm bsb, centimetres below sea bottom.

Table 5. *Summary of seismic facies characterizing the dominant seismo-stratigraphic units (SSUs)*

Graphic scheme	Seismic facies	Seismic reflector geometry	Interpretation	Observed in the SSUs
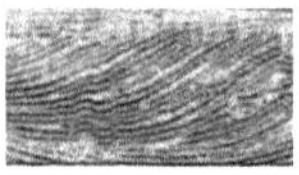	F1	High-angle (3–4°) oblique clinoforms	Sedimentary prisms of the delta front	B
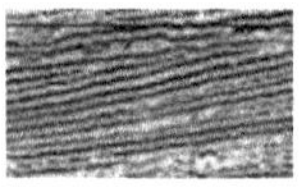	F2	Low-angle (<1°) oblique subparallel clinoforms	Distal portion of subaqueous delta, marine deposits	C
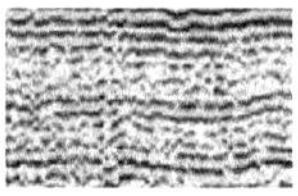	F3	Parallel clinoforms or weak lateral continuity	Sandy and/or silty bodies incorporated in muddy sediments	A, B
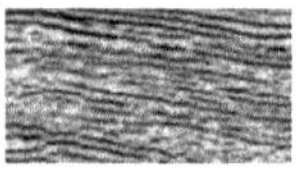	F4	Parallel clinoforms or high lateral continuity	Sandy marine deposits	D, E, F
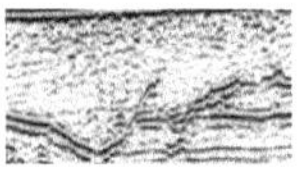	F5	U-shape incised fill	(buried) Channel	A, B, D
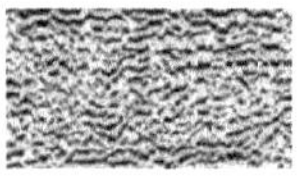	F6	Discontinuous to chaotic	Nearshore deposits	C, E, F, G

shown in Section 2.3. Section 2.3 is composed of grey silt interlayered with laminated sandy silt and irregular sandy silt clusters. Brown–yellow silt and sandy silt is interlayered with thin silty sand layers, and irregular brown sandy patches. Burrows were found in the upper part of Section 2.1 (Fig. 5f). Radiocarbon dates within the unit indicate deposition from around 28.4 to 30.4 cal ka BP (Table 2).

The sand content ranges from 4 to 38% and the succession of SFU 2 generally fines up. Reworked planktonic and benthic foraminifera are mainly distributed in Section 2.3. We identify this unit as representing sedimentation in a lake, lagoon and tidal shelf/coastal from top to bottom.

*Unit SFU 1 (depth in core: 0–*c. *1.10 m).* The uppermost parts of SFU 1 represents a highstand sea-level succession. Gravel (Section 1.2) and dark-grey, gravelly mud transition upwards into gravelly muddy sand, containing complete foraminifera (Section 1.1) (Figs 5g & 6; Table 4). Radiocarbon dates within the unit indicate deposition from 11.2 to 28.4 ka (Table 2). The sand content ranges from 3 to 61% and the succession as a whole fines upwards. This unit is interpreted as coastal and coastal/open-shelf deposits (Table 4; Fig. 6).

Four units (SFU 2–5) can be distinguished with a similar fining-upwards sequence. SFU 1 is interpreted as the first stage in the most recent sea-level rise. The bottom of each complete cycle consists of a coarser-grained section (sand and gravel). These sections have been identified as coastal, coastal beach or coastal/river sediments. These bottom sections are followed by silty sediments interpreted as being deposited in a tidal, lagoonal, lake or shallow-shelf environment. Each of the SFUs can be interpreted as the reflection of sea-level cycles, starting with coastal/river-mouth sediments to shallow-shelf deposits truncated by erosion before a new cycle starts with disconformably overlying coastal sediments. Besides the controlling grain-size parameters, the cyclicity is reflected also by the downcore distribution of $CaCO_3$, which is enriched in deeper-water facies and in the carbonate content (Fig. 6). Carbonate is related to biogenic calcareous debris, which is enriched in coastal sediments. No deeper-water facies sediments are found in the succession. This is because the core is nearshore and relatively far from the shelf edge during lowstand stages of the sea-level cycle.

Seismic architecture of the Late Pleistocene–Holocene succession

Reflection surfaces (R). Seven major bounding reflection surfaces are named as R1–R7 in the seismic profiles (Figs 7–9), which allow the sequence to be assigned to seismo-stratigraphic units (SSUs) A–G from the top to the bottom. The SSUs are bounded by continuous reflectors of strong–medium amplitude underlined by a change in seismic facies. These surfaces are interpreted as unconformities representing Late Pleistocene–Holocene sea-level cyclicity. To calculate the approximate depth to the seismic reflectors, a velocity of $1600\ m\ s^{-1}$ was used.

Seismic facies (F). The term ‘seismic facies’ refers to the characteristic seismic parameters, including configuration-continuity, amplitude, frequency, signature and interval velocity, of a group of adjoining seismic reflections pertaining to a seismic unit or sequence.

Each depositional unit displays an elementary pattern consisting of two main seismic facies. In this study, six different seismic facies types were recognized, each having a particular sedimentological significance (Table 5).

Seismic facies F1 corresponds to relatively high-angle (3°–4°) oblique or sigmoid clinoforms. Seismic facies F2 corresponds to low-angle oblique, subparallel clinoforms ($<1^\circ$) with tangential lower (downlapping) terminations. Seismic facies include prisms that pinch out in landwards and seawards directions, and extend laterally in a strike direction with the same geometry over the shelf area. Seismic facies F1 is interpreted as the sedimentary prism of a delta front, and seismic facies F2 as the distal marine portion of a subaqueous delta. These are observed within SSUs B–C (Table 5).

Seismic facies F3, located on the shelf, is characterized by parallel clinoforms or weak lateral continuity, and with reflections that are laterally discontinuous. This facies is observed within SSUs A and B, and indicates discontinuous sandy and/or silty bodies, irregular in thickness and texture, interbedded with muddy sediments. The depositional energy is variable, but locally could be relatively high (Table 5).

Seismic facies F4 displays parallel clinoforms with a high lateral continuity, and is characterized by a sequence of horizontal or subhorizontal parallel reflections that are laterally continuous, and with variable frequencies and amplitudes. The facies is interpreted as high-energy deposits interbedded with relatively low-energy deposits. These are observed within SSUs D–F, and are interpreted as sandy marine deposits (Table 6).

Seismic facies F5 shows a U-shape and infill of chaotic facies. It is observed within SSUs A, B and D, and is interpreted as channel deposits formed within a high-energy setting (Table 5).

Seismic facies F6 is observed within SSUs C, E, F and G. It is composed of discontinuous to chaotic reflections bounded by medium- to high-amplitude surfaces. On continental shelves, this heterogeneous

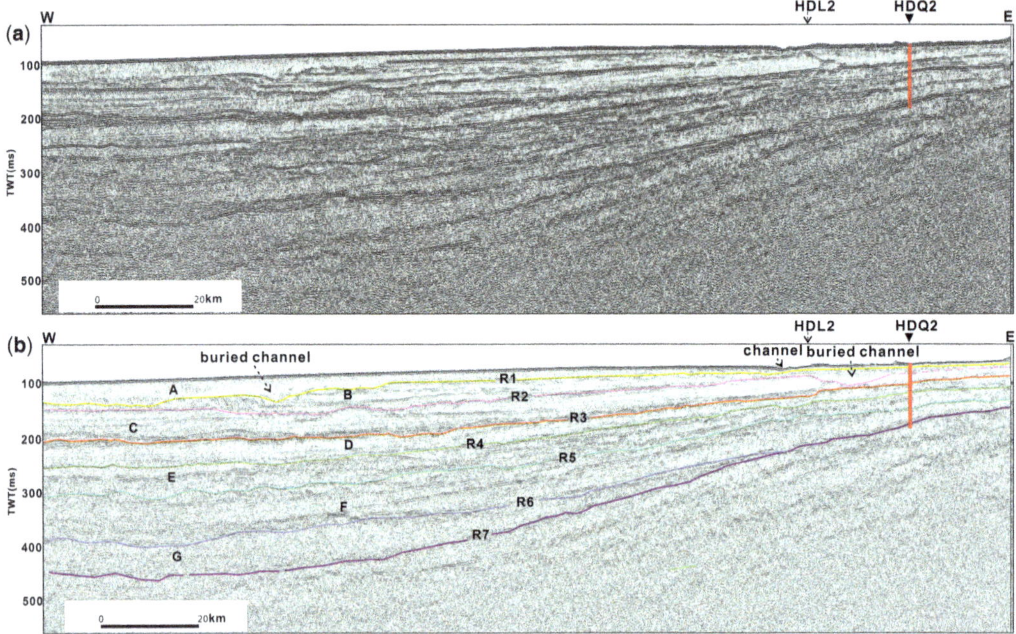

Fig. 7. (**a**) Uninterpreted high-resolution single-channel seismic Profile HD70 west of Hainan (see Fig. 1 for the location). (**b**) Stratigraphic interpretation of seismic Profile HD70. Vertical scale: two-way travel time (TWT) in milliseconds (ms). Reflectors interpreted as erosional surfaces on the shelf are marked by coloured lines. The dashed line marks the inferred position of reflector surface R2. The east–west-orientated Profile HD70 across the continental shelf illustrates the stacking of the uppermost seven SSUs (A–G) bounded by regional unconformities (R1–R7).

reflection pattern is interpreted as indicating the presence of coarse sediments deposited in high-energy settings (Table 6).

From the seismic characteristics and sedimentological significance, the seismic facies recognized on the shelf may be grouped into three associations:

(1) Parallel, high lateral continuity and parallel weak lateral continuity: seismic facies corresponding to predominantly muddy sediments (with more or less important sandy and/or silty intercalations) that accumulated essentially on flat horizontal surfaces, indicating relative quiet depositional environments (highstand systems tract, HST).

(2) Discontinuous to chaotic: seismic facies indicating heterogeneous and mostly coarse (nearshore) sediments (lowstand systems tract, LST).

(3) Parallel oblique, tangential oblique and shingled: prograding seismic facies showing different internal dip patterns and variable angular relationships at their boundaries due to sedimentation during falling sea-level stages (falling-stage systems tract, FSST).

Seismic facies associations (1) and (2) represent essentially the gross granulometry and lithology of the sediments, and are mainly the result of vertical accretion, but not the direction from which the sedimentary input comes. Seismic facies association (3) is caused by lateral accretion, and indicates the direction of sedimentary accretion by granulometric and lithological significance.

Architecture of seismo-stratigraphic units. The basal SSU G displays high-amplitude and chaotic reflection facies in the north–south-trending Profile HDL10, but high-amplitude discontinuous or low continuous reflections in profiles HD70 and HDL20. The upper limit of SSU G was detected at a depth ranging between 64 and 360 m below the sea bottom, and corresponds to an irregular erosional surface. The thickness of wedge-shaped SSU G increases seawards in a west or SW direction (Figs 7–9). The thickness of this unit is between 0 and 120 m, with an average thickness for SSU G of 24 m. Maximum values are found near the north end of Profile HDL10. The unit is eroded and absent close to the shoreline of Hainan.

Seismo-stratigraphic units D–F have similar seismic facies. They display parallel clinoforms or high lateral continuity, quasi-parallel and low continuous or discontinuous internal reflections in the nearshore area, becoming parallel and continuous

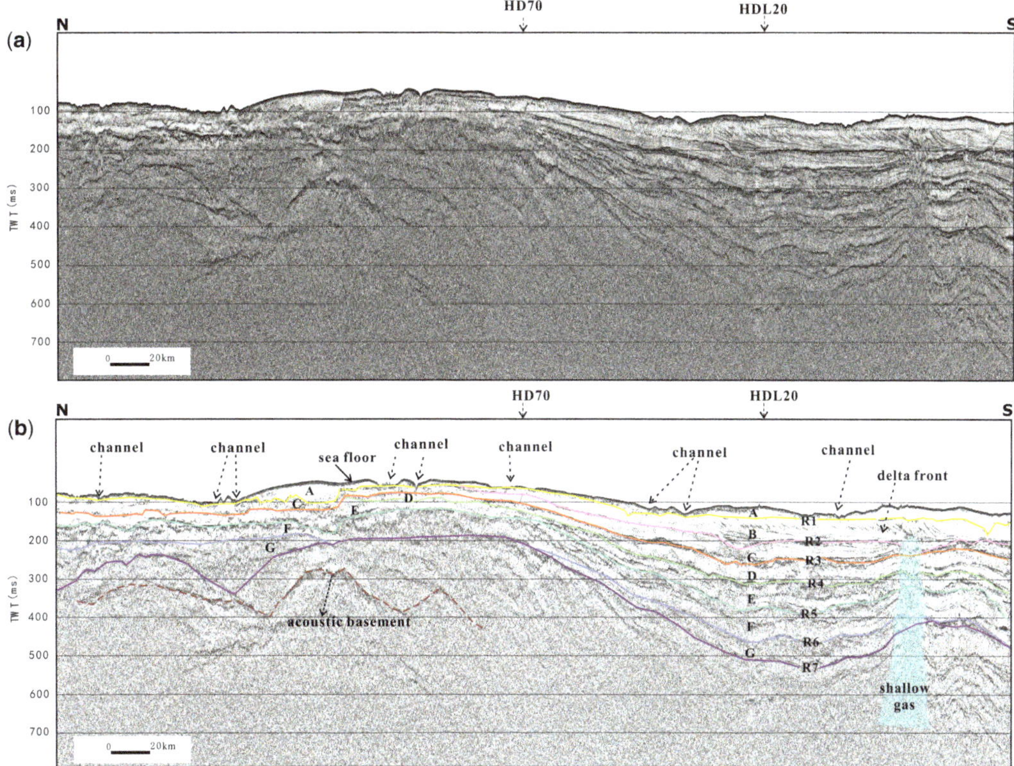

Fig. 8. (**a**) Uninterpreted high-resolution single-channel seismic Profile HDL10 west of Hainan (see Fig. 1 for the location). (**b**) Stratigraphic interpretation of seismic Profile HDL10. Vertical scale: two-way travel time (TWT) in milliseconds (ms). Reflectors interpreted as erosional surfaces on the shelf are marked by thick coloured lines. The north–south-orientated Profile HDL10 across the continental shelf illustrates the stacking of the uppermost seven SSUs (A–G) bounded by regional unconformities (R1–R7).

in a seawards direction. The average thickness of SSUs F, E and D is 4, 8 and 2.4 m in a SW (seawards) direction, respectively (Fig. 6). However, SSU F is thicker than SSUs E and D in the western and southern sectors (Figs 7 & 8). SSU D shows incised valleys at several locations and is absent in the northern part of the shelf area (Fig. 7).

Sequence Boundary SB 2 lies between SSUs D and C. SSU C is characterized by reflectors with high amplitude and good lateral continuity, as well as low-angle subparallel clinoforms ($<1^\circ$) with tangential lower (downlap) terminations in the SW and southern shelf areas (regression). At the top, SSU C is bounded by a rough erosional surface (R6). The thickness of SSU C ranges from 0.4 to 4 m, with an average thickness of 2.4 m (Figs 6 & 7).

SSU B is characterized by high-amplitude, good lateral continuity reflectors and relatively high-angle (from 3° to 4°) or sigmoid clinoforms. U-shaped incised fill facies was locally developed at the base of SSU B. The thickness of SSU B ranges from 0 to 8 m, with an average thickness of 4.8 m. SSU B is absent at the nearshore area (Figs 7–9).

Sequence Boundary SB 1 lies between SSU A and B. SSU A is characterized by parallel clinoforms or by weak lateral continuity. U-shaped incised fill facies are locally developed at its base. The thickness of SSU A ranges from 0.8 to 4 m, with an average thickness of 1.6 m (Figs 6–8). Transgressive marine sediments overlie SSU A. Basinwards, these sediments are developed as fine-grained, muddy facies. Nearshore, the facies is coarser grained because of the higher energy of the coastal zone.

Comparison of seismic and lithological data and coastline shifts

When applying a sequence stratigraphic model, seismic reflectors are interpreted as an expression of (regional/local) unconformities and may be compared with the lithological inventory of the sediment succession in order to identify regularities in the

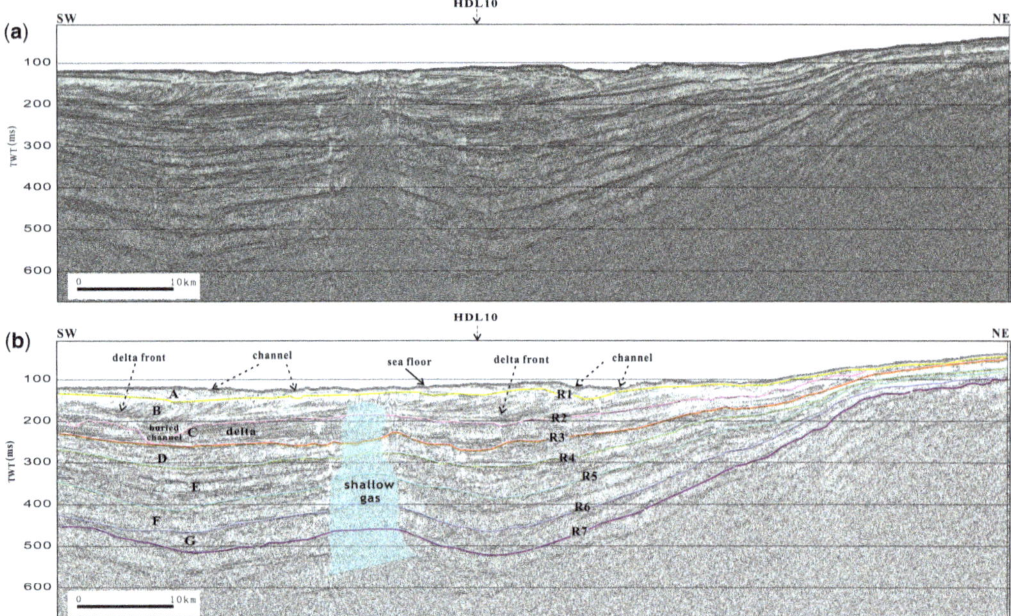

Fig. 9. (**a**) Uninterpreted high-resolution single-channel seismic profile HDL20 west of Hainan (see Fig. 1 for the location). (**b**) Stratigraphic interpretation of seismic Profile HDL20. Vertical scale: two-way travel time (TWT) in milliseconds (ms). Reflectors interpreted as erosional surfaces on the shelf are marked by thick coloured lines. The SW–NE-orientated Profile HDL20 across the continental shelf illustrates the stacking of the uppermost seven SSUs (A–G) bounded by regional unconformities (R1–R7).

accumulation. For this purpose, the seismic reflection Profile HD70 at Site HDQ2 is graphically correlated with grain-size distribution data measured downcore and the SFU (Table 4). In Figure 10, seismic reflector (R) depths (two-way time, TWT) and SSU are plotted against the percentages of gravel, sand, silt and clay within the sediment succession. These grain-size data clearly show cycles coinciding with the SFU and the sedimentary facies layers, which have been determined by lithological core description (Table 4). Each cycle starts with coarser (more coastal) sediments, which rest discontinuously on top of the pre-Quaternary basement interpreted here as transgression markers (related to a landwards shift of the coastline). The depth below seafloor of the bottom boundary of these coarse sedimentary facies layers is used to mark the depths of the reflectors. Upsection, the sediments fine, reflecting a rising sea level connected with a landwards shift of the coastline and a deepening of water depth at the drill site. The pattern of the depositional cycles can be deducted clearly from the sand and gravel content curves.

We highlight the following:

- The cycle of SFU 1 is incomplete because it is Holocene and is still an ongoing sedimentary cycle.
- The transgression and sea-level highstand phases of SFU 2 cannot be simply separated because the unit's upper boundary has been reworked during sea-level regression during the formation of incised river valleys at Site HDQ2, as discussed below.
- According to the grain-size distribution of SFU 4 between 39 and 50 m below sea bottom (bsb)

Table 6. *Characteristics of the seven late seismo-stratigraphic Units (SSUs) identified in seismic profiles, their bounding surfaces and the thickness at the shelf (see the text for details)*

Bounding reflector surface	SSU	Surfaces		SSU thickness (m)
		Base	Top	
R1	A	R1	Seafloor	16
R2	B	R2	R1	48
R3	C	R3	R2	24
R4	D	R4	R3	24
R5	E	R5	R4	80
R6	F	R6	R5	40
R7	G	R7	R6	24

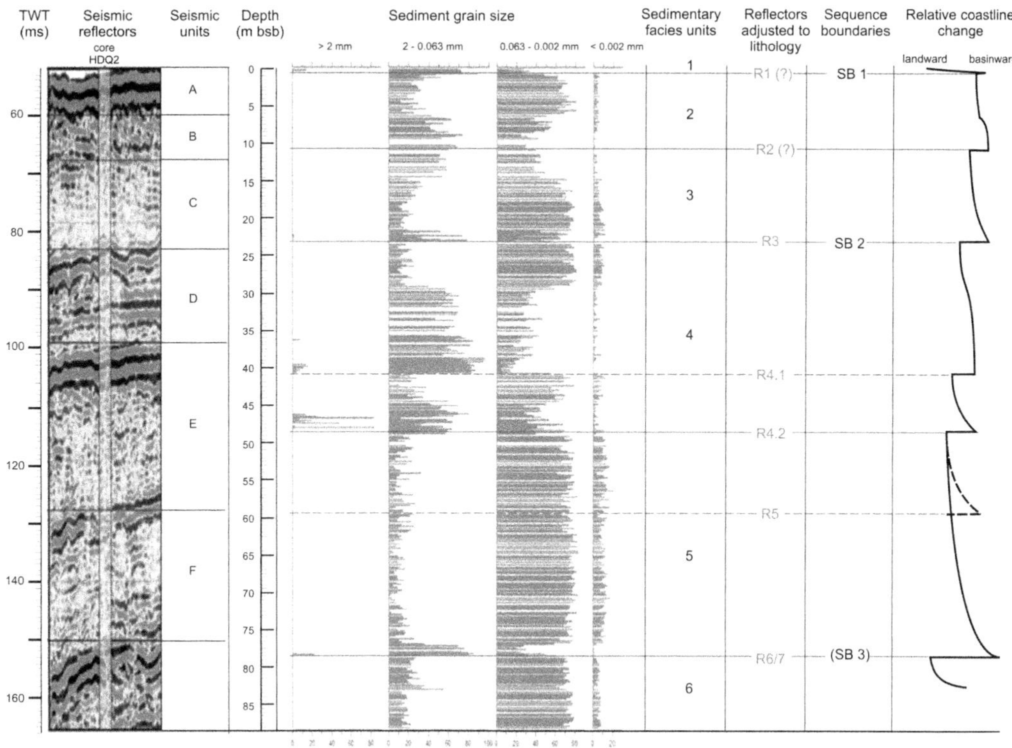

Fig. 10. Seismic reflection data of Profile HD70 at Site HDQ2: grain-size distribution data measured downcore in Core HDQ2, seismic stratigraphic units (SSUs), sedimentary facies units (SFUs) (according to Table 4), reflectors (R), sequence boundaries (SB) and a curve expressing relative (local) coastline shift.

(see Fig. 10), Reflector R4 has to be split into two sub-reflectors: R'42 and R'41. Between these reflectors, a local subcycle of sea-level change has caused some minor coastline shifts.

In SFU 5, Reflector R5 does not show lithological correlation in the grain-size data displayed in Figure 10, which we interpret as a sampling effect. In the first sampling campaign used for the measurement of grain-size data depicted in Figure 5, an increased sand content was measured at a depth of 60 m below seafloor. For this reason, a subcycle of sea-level (and coastline) change (marked by dashed lines in Fig. 10) is interpreted between 49 and 60 m below seafloor in Core HDQ2.

As can see from seismic line HD70 (Fig. 7), SSU A is missing at the location of Core HDQ2. SSU B here rests discontinuously on SFU 6. Movement in the coastline has been constructed as a qualitative expression of the effect of sea-level change on this shift (landward/basinward) expressed as transgression and regression. According to the erosional pattern on the top of SSUs D, E and F, these mark regional sea-level lowstands. Sequence boundaries SB 1 and SB 2 have been defined between SSUs A and B (SB 1), and between SSUs C and D (SB 2). Because Core HDQ2 did not penetrate SFU 6, and SSU G is missing at the drilling site, the boundary between SSU B and SFU 6 is regarded as a quasi Sequence Boundary (SB 3).

Seismo-stratigraphic and sequence-stratigraphic analysis

For a comprehensive view of the area of investigation, and in particular the temporal evolution of sea level, the local data are compared to global sea-level curves. This method was previously applied by Lobo & Ridente (2013) in order to identify high-resolution Milankovitch cycles (fourth–fifth order in terms of sequence stratigraphy hierarchical classification) in the architecture of marine shelf sediments. Comparative studies of the Mediterranean, the Gulf of Mexico and the East China Sea showed that fourth- and fifth-order Milankovitch periodicities can be identified on time intervals of 100–20 ka. For this correlation, Lobo & Ridente (2013) used a composite sea-level curve (master curve) from different data sources published by Waelbroeck *et al.* (2002) that was also used for this study. As mentioned above, because of different vertical crustal movements in different marginal seas, absolute values of sea-level change are hard to compare. However, for studies of sea-level cycles, the variability of the data is of interest rather than a comparison of absolute values. Therefore, we also used oxygen isotopes from Red Sea and Mediterranean Sea foraminifera to trace sea-level changes. Grant *et al.* (2012) developed a probabilistic model of the LGC sea-level change based on a statistical analysis of empirical data from different sources, including an estimate of confidence levels. In Figure 11, the Waelbroeck *et al.* (2002) and Grant *et al.* (2012) sea-level data are used as a reference frame to interpret age data measured for sediments sampled from Core HDQ2. The comparison of the two curves in Figure 11 shows a divergence in the amplitude caused by the different sources and treatment of the primary data. Nonetheless, both curves are almost identical in dating the major anomalies and inflection points. Because of the large volume of the data, the Grant *et al.* (2012) curve takes advantage of a statistically significant higher periodicity of sea-level change. This higher variability is used in this study to correlate the position of reflectors to sea-level lowstand phases. As mentioned above, the age of Core HDQ2 sediments cannot be approximated by best-fit curves as would be normal for age models of continuous sediment successions. In near-coast sediment sequences, hiatuses and secondary thickness reduction by subaerial and subaquatic erosion must be considered. Because regional erosional processes correlate to global sea-level changes, the Waelbroeck *et al.* (2002) and Grant *et al.* (2012) data are used as a framework for the interpretation of our data.

In Figure 11, reflector depth lines are plotted against AMS and OSL age data from Core HDQ2, then superposed with the Waelbroeck *et al.* (2002) and the Grant *et al.* (2012) curve. In addition, the

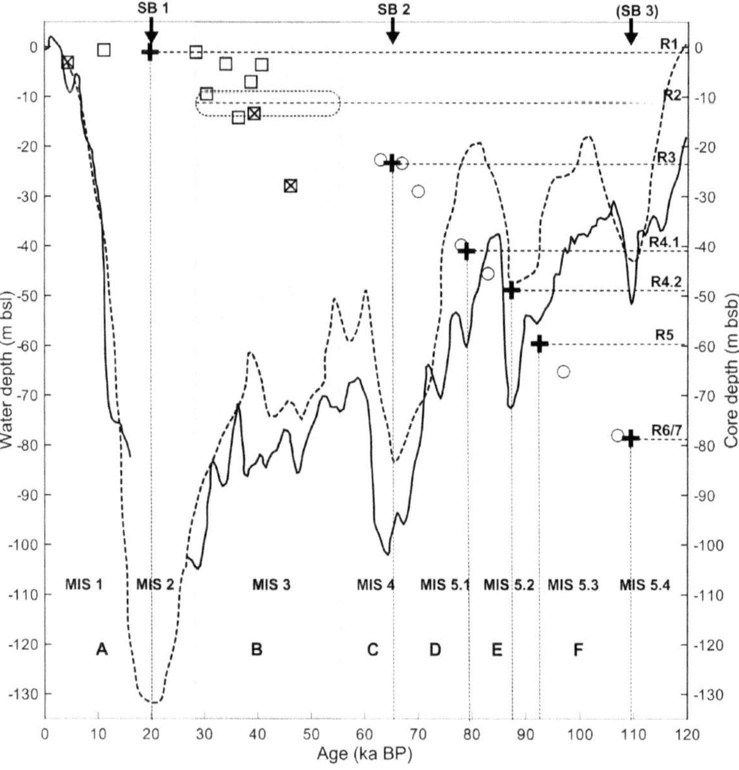

Fig. 11. Age assignment of Reflectors R1–R6/R7 (adjusted to lithology) in Core HDQ2 compared to global sea-level curves: dashed line, Waelbroeck *et al.* (2002); solid line, Grant *et al.* (2012); squares, AMS sediment ages (bulk); crossed out squares, AMS sediment ages (foraminifera); circles, OSL sediment ages; black crosses, age–depth assignment of corrected reflectors; SBX, Sequence Boundary X; A–F, seismo-stratigraphic units (SSUs); m bsl, metres below sea level; m bsb, metres below sea bottom.

diagram shows the AMS and OSL sediment age data.

The OSL data between approximately 110 and 65 ka BP can be clearly approximated by a linear age–depth function (Fig. 11). However, erosional processes and hiatuses during the formation of the sedimentary column do not allow simple linear functions as an age model. Instead, we here use a two-step procedure for the time span MIS 5–MIS 4 to identify the age of seismic reflectors (the coarse transgressive sediments at the base of a sedimentary sequence or its subordinated units). As a first step, the ages of samples above and below a reflector are interpolated in order to assign an age value to the reflector as a first approximation. In a second step, assuming that the transgressive horizon represents the beginning of a sea-level rise after a lowstand phase, the interpolation data have been adjusted to the next sea-level minimum using the average of values from the Grant *et al.* (2012) and Waelbroeck *et al.* (2002) curves (red circles in Fig. 11). In case of reflectors R6 and R7, in the first step the age value was determined by extrapolation of OSL data from sediments on top of the reflector, as there is no OSL data below.

In the case of reflectors R1 and R2, the interpolation method cannot be applied because, according to the interpretation of seismic Profile HD70 (Fig. 7), Core HDQ2 penetrated sediments accumulated during MIS 4–MIS 2, which have been redeposited by fluvial processes. Reflector R2 cannot be dated based on samples from Core HDQ2. Figure 7 shows that incised and infilled valleys have destroyed the primary layering structure of the sediments. As a result, AMS dating from this part of the sediment profile does not show age data ordered downcore in the sediment succession (Fig. 11). For this reason, a transgressive horizon that can be correlated with Reflector R2 is not developed at the site of Core HDQ2. From the lithology of Core HDQ2 and the AMS dates (Fig. 11), it only possible to determine a time interval during which the basinwards Reflector R2 was formed. The time limits for that interval are set here hypothetically to 28 and 56 ka assigned to relative lowstands of the sea-level curve of Grant *et al.* (2012). Furthermore, the boundary between terrestrial, early marine and younger (Holocene) marine sediments is disguised in Core HDQ2 by fluvial and nearshore processes, so that the age of Reflector R1 is inferred here as 20 ka BP – the time of the LGM and the most important sea-level lowstand during the LGC. More precise age dating will be possible if Late Pleistocene–Holocene sediments from an undisturbed shelf area SW of Hainan become available.

In summary, the sediments sampled by Core HDQ2 can be subdivided into two sedimentary sequences (SS 1 and SS 2) bounded by sequence boundaries SB1 (R1), SB2 (R3) and SB3 (R6–R7).

Table 7. *Seismic reflectors of sediment core HDQ2: depth – adjusted to lithology (metres below sea bottom (m bsb)) and age (ka). Because of the approximate character of the age model used here, the age data are rounded correspondingly*

Reflectors	Depth (m bsb)	Age (ka)
R1	1.1	20
R2	11.2	28–56
R3	23.5	65
R41	40.9	79
R42	48.9	87
R5	59.7	92
R6/R7	78.4	110

Sediment sequence SS 2 corresponds to the time span between the sea-level lowstand of MIS 5.4 and MIS 4. The sediment sequence comprises SSU B–D, and represents the change from a transgressive systems tract (TST) to highstand (HST) and falling-stage systems tracts (FSST). SSU C represents a sea-level subcycle in a relatively shallow coastal environment, the high-energy dynamics of which generated coarse-grain-size SFUs that represented a relatively stable coastline during the MIS 5.2 lowstand environment. SSU D developed during a FSST, which was then eroded during the MIS 4 lowstand stage, as shown by locally incised valleys (Fig. 7).

Sediment sequence SS 2 is bounded at the bottom by Reflector R3, and SSU E represents the start of a new sedimentary cycle with typical FSSTs. Prograding delta fronts (Seismic Facies Types F1 and F2), and incised valleys and erosion on the top of units SSU B (and partly SSU C) indicate that the units are parts of a regressive system, that were terminated by the sea-level lowstand (Reflector R1) during the LGM at about 20 ka. Sequence Boundary SB 1 terminates the sedimentary sequence, which comprises sediments from MIS 3 and MIS 2. The sea-level fall during the LGM caused a wide shelf and slope area to be exposed to erosion. On top of this truncation surface (SB 1), Late Pleistocene terrestrial and marine Holocene sediments rest discontinuously.

The age data of the reflectors in Core HDQ2, which represent transgressive sediments resting discontinuously on the basement, are listed in Table 7.

Discussion

The high-resolution sediment records of the western Pacific marginal seas have great potential to improve the understanding of the variations in interactions between land and sea, and deserve more attention (Wang 1999). At the same time, classic shelf systems represent a major sink for both terrestrial and marine materials during their transport from the coast to the continental slope (Lantzsch *et al.* 2009).

Remains of regressive systems are usually scarce on the shelf. Knowledge of the palaeo-environment during the time span between MIS 5 and MIS 2 is poorly understood. Therefore, the search for complete sediment successions on continental shelves and their correlation with possible driving forces, including variations in climate, are at the cutting edge of palaeo-oceanography. The shelves of the South China Sea preserve sediment with palaeo-oceanographic and palaeoclimatological proxy-data that are almost unmasked by signals of glacio-isostaic adjustment. Core HDQ2 drilled west of Hainan and its corresponding seismic profiles are of high scientific value in terms of decoding the sediment proxies in order to reveal signals of sea-level change and their correlation with insolation as high-frequency Milankovitch cycles.

A special problem for sediment data retrieved close to the shore is the development of an appropriate age model. Unconformities cover time spans of different length with no sedimentation or erosion of older sediments. For the identification of unconformities, the joint interpretation of seismic data and sedimentological records is usually effective. Radiocarbon methods can be used for dating transgressive sediments up to an age of approximately 40 ka. Older sandy sediments can be dated by OSL methods. Because of the incomplete records in nearshore positions, age models used for basin sediments that link discrete age data by interpolation methods are not appropriate on the shelf. Here, we recommend superimposing the measured data over global sea-level curves in order to adjust the age data to known global sea-level cycles induced by periodical insolation changes. As Lobo & Ridente (2013) showed, fourth- to fifth-order cycles in sequence stratigraphy classification are found in shelf sediments, covering time intervals of 100–200 and 10–40 ka, respectively. Similar findings have been reported in the current study. For a correct interpretation, it is essential to identify systems tracts in the sedimentary architecture and to allocate them to the phases of sea-level cycle found in the sedimentary record.

The age data show that Core HDQ2 and SSU A–G reflect nearly 110 kyr of sedimentary records from MIS 5e to MIS 1. The characteristics of seismic facies suggest that SSUs D–G, which developed during MIS 5, belong to a highstand muddy transgression system in the southern outer-shelf area, which is characterized by parallel clinoforms and laterally continuous wedge-shaped sedimentation in the landwards direction. SSUs D–G, as shown in Core HDQ2, are interpreted as coastal sand, gravel or silty sand, with a seismic character as chaotic or discontinuous reflectors. The top surface of SSU D shows erosion caused by a falling sea level during MIS 4.

Seismo-stratigraphic units B and C were mainly formed during MIS 4–MIS 2, as a FSST. Buried channels, the result of sea-level lowstands, were also developed at the base of SSUs A and B. SSUs B and C have low- ($<1°$) to high-angle ($3–4°$) oblique subparallel clinoforms, and were interpreted as subaqueous delta and sedimentary prisms within a delta front. Based on the prograding direction of the sediments in SSUs B and C, the source of the sediments can be constrained to be from western Hainan. Research of the slope evolution offshore Hainan also indicates that the Yinggehai–Song Hong Basin is characterized by a series of progradational slope clinoforms with a rapid seawards shift as a result of a large sediment supply from the Red River and Hainan (Xie *et al.* 2008). Vertical crustal movements may have occurred since the late Pleistocene–middle Holocene in this area (Wang & Zhou 1990; Yao *et al.* 2009). This tectonic movement may be the cause of the remarkable terrigenous sediment eroded and discharged through the channel and rivers along the west coast and offshore Hainan.

At the LGM, sea level dropped quickly, and was more than 100 m below that of the present sea level. The shelf area of NW South China Sea was exposed and subjected to erosion and/or terrestrial sedimentation (Li *et al.* 1990; Chen *et al.* 1994; Huang *et al.* 1995; Xie *et al.* 2008), resulting in the formation of palaeo-channels in the shelf area.

Transgressive sediments deposited during the sea-level rise after the LGM are well preserved, particularly in incised valleys of fluvial and estuarine systems (Hanebuth *et al.* 2011). However, major parts of the inner shelf are not covered by younger, late Holocene sediments and older strata are exposed widely on the modern seafloor (Kögler *et al.* 1982). Since the LGM, by about 15 cal ka BP, a fast retreat of the coastline began as the post-glacial sea level rose rapidly, leading to a land loss of around 24×10^4 km^2 over the next 5 kyr in the NW South China Sea. During the early Holocene, the coastline kept retreating and the Qiongzhou Strait completely opened up around 9 cal ka BP (Yao *et al.* 2009). Some tiny channels were distributed at the seafloor in the study area, but are not filled with fluvial or estuarine sediments because, owing the rapid rise in sea level after the LGM, there was not enough sediment to fill the channel. Alternatively, strong monsoon-driven ocean currents may have passed through the offshore of Hainan and eroded any sediment supplied.

Objectives for further studies

Hanebuth *et al.* (2011) found sediments of 'Forced Regressive Systems' on the Sunda Shelf as isolated sediment bodies that formed during intermediate sea-level fluctuations between MIS 4 and MIS 2. It is important to correlate these sediments with those on the northern shelf of the South China Sea. Further studies should focus on: the mass balancing of sediment accumulation; source-to-sink models for 'Forced Regressive Systems'; and the reconstruction of the interrelationship of relative sea-level variation embedded into the driving forces of sediment formation (palaeoclimate analysis, reconstruction of palaeo-river valleys, and analysis of tectonic uplift/subsidence in the source and sink area).

Conclusions

The high-resolution 2D seismic reflection profiles and core data presented here provide a general view of the Late Quaternary stratigraphic architecture in the western offshore of Hainan, NW South China Sea.

Interpretations of nearly 400 km of sparker seismic-reflection profiles provide the basic data for studying the Late Quaternary highstand and lowstand wedges related to the fluctuations in the relative sea level in the study area. A series of sedimentary wedges can be grouped into seven distinct seismo-stratigraphic units (SSUs) (A–G), separated by erosional unconformities. The age of seismic reflectors were calculated according to a special age–depth model, based on the correlation of sediment age data with global sea-level curves. The thickness and seismic characteristics of each SSU were also detailed in this paper. The sequence boundaries, showing a highly irregular surface, appear to be the result of subaerial erosion and marine processes during the LGC. Repeated falls in sea level, coupled with high sediment discharge, as well as the tectonically driven uplift of Hainan, may have resulted in the development of prograding lowstand delta wedges on the shelf margin that took place during MIS 5–MIS 2.

The authors express their sincere thanks to the Guangzhou Marine Geological Survey (GMGS) and the University of Szczecin for the support and active promotion of the bilateral Chinese–Polish collaboration project providing the frame of co-operation.

The authors also thank the crews of R/V Fendou No. 5 for assisting us with data collection. This study was funded by the project of China Geological Survey (SN: GZH201500207; 1212010611302), and the Chinese-Polish collaboration project SECEB. Part of the results presented in this publication were achieved along with a research project funded by the Polish National Science Centre (NCN) allocated on the basis of decision No. DEC-2011/01/N/ST10/07708.

Professor Ryszard K. Borówka, Dean of the Faculty of Geosciences, University of Szczecin, deserves thanks for constructive scientific discussions and access to sedimentological laboratories. Special gratitude is devoted to Aleksandra Truszkowska, who measured the grain size facies of HDQ2 core sediments as part of her MSc; she was technically assisted by Bożena Kosińska and Bartosz Bieniek. We are indebted to Professor Helge Arz from the Leibniz Institute for Baltic Sea Research, Warneműnde (Germany) for helping to manage the AMS-^{14}C dating of three samples that were based on foraminifera assemblages.

References

Bauer, A., Radziejewska, T. *et al.* 2013. Regional differences of hydrographical and sedimentological properties in the Beibu Gulf, South China Sea. *In*: *Depositional Environments and Multiple Forcing Factors at the South China Sea's Northern Shelf.* Journal of Coastal Research, Special Issue 66, 49–71.

Cai, G. Q., Peng, X. C., Chen, H. J., Zhang, J. P. & Zhang, Y. L. 2012. The component characteristics and paleoenvironmnetal sighificances for sediments in core 83PC from the Xisha Trough, South China Sea. *Marine Geology and Quaternary Geology*, **32**, 77–84 [in Chinese].

Chen, M., Zhao, H., Wen, X., Zhang, Q. & Song, C. 1994. Quaternary foraminiferal group and sporopollen zones in cores L2 and L16 in the Lingdingyang estuary. *Marine Geology and Quaternary Geology*, **14**, 11–22 [in Chinese].

Clift, P. & Sun, Z. 2006. The sedimentary and tectonic evolution of the Yinggehai-Song Hong basin and the southern Hainan margin, South China Sea: implications for Tibetan uplift and monsoon intensification. *Journal of Geophysical Research*, **111**, B06405, http://doi.org/10.1029/2005JB004048

Dalrymple, R. W., Baker, E. K., Harris, P. T. & Hughes, M. G. 2003. Sedimentology and stratigraphy of a tide-dominated, forelandbasin delta (Fly river, Papua New Guinea). *In*: Sidi, F. H., Nummendal, D., Imbert, P., Darman, H. & Posamentier, H. W. (eds) *Tropical Deltas of Southeast Asia – Sedimentology, Stratigraphy, and Petroleum Geology*. SEPM (Society for Sedimentary Geology), Special Publications, **76**, 147–173.

de Boer, P. L. 1998. Intertidal sediments: composition and structure. *In*: Eisma, D. (ed.) *Intertidal Deposits: River Mouths, Tidal Flats, and Coastal Lagoons*. CRC Press, Boca Raton, FL, 345–362.

Grant, K. M., Rohling, E. J. *et al.* 2012. Rapid coupling between ice volume and polar temperature over the past 150,000 years. *Nature*, **491**, 744–747.

Guyodo, Y. & Valet, J. P. 1996. Relative variations in geomagnetic intensity from sedimentary records: the past 200,000 years. *Earth and Planetary Science Letters*, **143**, 23–36.

Hanebuth, T. J. J., Stattegger, K., Schimanski, A., Lüdmann, T. & Wong, H. K. 2003. Late Pleistocene forced-regressive deposits on the Sunda Shelf (Southeast Asia). *Marine Geology*, **199**, 139–157.

HANEBUTH, T. J. J., SAITO, Y., TANABE, S., VU, Q. L. & NGO, Q. T. 2006. Sea levels during late marine isotope stage 3 (or older?) reported from the Red River delta (northern Vietnam) and adjacent regions. *Quaternary International*, **145–146**, 119–134.

HANEBUTH, T. J. J., STATTEGGER, K. & BOJANOWSKI, A. 2009. Termination of the last glacial maximum sea-level lowstand: the Sunda-Shelf data revisited. *Global and Planetary Change*, **66**, 76–84.

HANEBUTH, T. J. J., VORIS, H. K. & YOKOYAMA, Y. 2011. Formation and fate of sedimentary depocentres on Southeast Asia's Sunda Shelf over the past sea-level cycle and biogeographic implications. *Earth-Science Reviews*, **104**, 92–110.

HARFF, J., LEIPE, T., WANIEK, J. & ZHOU, D. (eds) 2013. Depositional environments and multiple forcing factors at the South China Sea's northern shelf. *In*: *Depositional Environments and Multiple Forcing Factors at the South China Sea's Northern Shelf. Journal of Coastal Research*, Special Issue 66, 90.

HUANG, Y., ZOU, H. & ZHANG, K. 1995. Quaternary coastline evolution in the north of South China Sea (in Chinese). *Tropical Geomorphology*, **16**, 1–21.

HUANG, Z., LI, P., ZHANG, Z. & ZONG, Y. 1986. Sea level changes along the coastal area of South China since the late Pleisocene. *In*: QIN, Y. & ZHAO, S. (eds) *Late Quatgernary Sea-Level Change*. China Ocean Press, Beijing, 142–154.

KÖGLER, F. C., ANDRESEN, H. & LINCKE, C. 1982. *RF Reedereigemeinschaft Forschungsschiffahrt GmbH*. Final Report SO 24

LAJ, C., KISSEL, C., MAZAUD, A., CHANNELL, J. E. T. & BEER, J. 2000. North Atlantic palaeointensity stack since 75 ka (NAPIS-75) and the duration of the Laschamp event. *Philosophical Transactions of the Royal Society of London, Series A*, **358**, 1009–1025.

LAMBECK, K. & CHAPPELL, J. 2001. Sea level change through the Last Glacial Cycle. *Science*, **292**, 679–686.

LANTZSCH, H., HANEBUTH, T. J. J. & BENDER, V. B. 2009. Holocene evolution of mud depocentres on a high-energy, low-accumulation shelf (NW Iberia). *Quaternary Research*, **72**, 325–336.

LE, J. & SHACKLETON, N. J. 1992. Carbonate dissolution fluctuations in the western equatorial Pacific during the late Quatemary. *Paleoceanography*, **7**, 21–42.

LI, F., DONG, T., JIANG, X., ZHUANG, J., YU, J. & DU, Q. 1990. Buried paleo-channel system and change of sea level on the continental shelf in Yinggehai BasinYinggehai-Song Hong Basin. *Oceanologia et Limnologia Sinica*, **21**, 356–363 [in Chinese].

LOBO, F. & RIDENTE, D. 2013. Stratigraphic architecture and spatio-temporal variability of high-frequency (Milankovitch) depositional cycles on modern continental margins: an overview. *Marine Geology*, **352**, 2015–247.

NAKADA, M. & LAMBECK, K. 1989. Late Pleistocene and Holocene sea level change in the Australian region and mantle rheology. *Geophysical Journal International*, **96**, 497–517.

NI, Y., ENDLER, R. *ET AL*. 2014. The 'butterfly delta' system of Qiongzhou Strait: morphology, seismic stratigraphy, sedimentation. *Marine Geology*, **355**, 361–368.

NIO, S. D. & YANG, C. S. 1991. Diagnostic attributes of clastic tidal deposits. *In*: SMITH, D. G., REINSON, B. A. & RAHMANI, R. A. (eds) *Clastic Tidal Sedimentology*. Canadian Society of Petroleum Geologists, Memoirs, **16**, 3–27.

PALAMENGHI, L., KEIL, H. & SPIESS, V. 2015. Sequencestratigraphic framework of a mixed turbidite–contourite depositional system along the NW slope of the South China Sea. *Geo-Marine Letters*, **35**, 1–21.

ROHLING, E. J., GRANT, K., BOLSHAW, M., ROBERTS, A. P., SIDDALL, M., HEMLEBEN, Ch. & KUCERA, M. 2009. Antarctic temperature and global sea level closely coupled over the past five glacial cycles. *Nature Geoscience*, **2**, 500–504.

SCHIMANSKI, A. & STATTEGGER, K. 2005. Deglacial and Holocene evolution of the Vietnam shelf: stratigraphy, sediments and sea-level change. *Marine Geology*, **214**, 365–387.

SHI, X., KOHN, B. *ET AL*. 2011. Cenozoic denudation of southern Hainan Island, South China Sea: constraints from low temperature thermochronology. *Tectonophysics*, **504**, 100–115.

SIDDALL, M., PRATT, L. J., HELFRICH, K. R. & GIOSAN, L. 2004. Testing the physical oceanographic implications of the suggested sudden Black Sea infill 8400 years ago. *Paleoceanography*, **19**, PA1024.

SKOOG, D. A. & WEST, D. M. 1969. *Fundamentals of Analytical Chemistry*, 2nd edn. Holt, Rinehart, Winston, New York.

STONER, J. S., CHANNELL, J. E. T. & HODELL, D. A. 2003. A ~580 kyr paleomagnetic record from the sub-Antarctic South Atlantic (Ocean Drilling Program Site 1089). *Journal of Geophysical Research*, **108**, 22–24.

SUN, Z., ZHONG, Z. *ET AL*. 2009. 3D analogue modeling of the South China Sea: a discussion on breakup pattern. *Journal of Asian Earth Sciences*, **34**, 544–556.

TANABE, S., HORI, K., SAITO, Y., HARUYAMA, S., VU, V. P. & KITAMURA, A. 2003. Song Hong (Red River) delta evolution related to millennium-scale Holocene sea-level changes. *Quaternary Science Reviews*, **22**, 2345–2361.

VAN HOANG, L., CLIFT, P. D., SCHWAB, A. M., HUUSE, M., NGUYEN, D. A. & SUN, Z. 2010. Large-scale erosional response of SE Asia to monsoon evolution reconstructed from sedimentary records of the Song Hong-Yinggehai and Qiongdongnan basins, South China Sea. *In*: CLIFT, P. D., TADA, R. & ZHENG, H. (eds) *Monsoon Evolution and Tectonic–Climate Linkage in Asia*. Geological Society, London, Special Publications, **342**, 219–244.

WAELBROECK, C., LABEYRIE, L. *ET AL*. 2002. Sea-level and deep water temperature changes derived from benthic formainifera isotopic records. *Quaternary Science Reviews*, **21**, 295–305.

WANG, P. 1999. Response of Western Pacific marginal seas to glacial cycles: paleoceanographic and sedimentological features. *Marine Geology*, **156**, 5–39.

WANG, P. & LI, Q. 2009. *The South China Sea*. Springer, Heidelberg.

WANG, X., HUTCHINSON, D. R., WU, S., YANG, S. & GUO, Y. 2011. Elevated gas hydrate saturation within silt and silty clay sediments in the Shenhu area, South China Sea. *Journal of Geophysical Research*, **116**, B0102.

WANG, Y. & ZHOU, L. 1990. The volcanic coast in the area of northwest Hainan Island. *Acta Geographica Sinica*, **45**, 321–330 [in Chinese].

XIE, X., MÜLLER, R. D., REN, J., JIANG, T. & ZHANG, C. 2008. Stratigraphic architecture and evolution of the continental slope system in offshore Hainan, northern South China Sea. *Marine Geology*, **247**, 129–144.

XU, J., WANG, P. X., HUANG, B. Q., LI, Q. Y. & JIAN, Z. M. 2005. Response of planktonic foraminifera to glacial cycles: mid-Pleistocene change in the southern South China Sea. *Marine Micropaleontology*, **54**, 89–105.

YAMAMOTO, Y., YAMAZAKI, T., KANAMATSU, T., IOKA, N. & MISHIMA, T. 2007. Relative paleointensity stack during the last 250 kyr in the northwest Pacific. *Journal of Geophysical Research*, **112**, B01104.

YANG, X. Q., FRIEDRICH, H., WU, N. Y., YANG, J. & SU, Z. H. 2009. Geomagnetic aleointensity dating of South China Sea sediments for the last 130 kyr. *Earth and Planetary Science Letters*, **284**, 258–266.

YAO, Y., HARFF, J., MEYER, M. & ZHAN, W. 2009. Reconstruction of paleocoastlines for the northwestern South China Sea since the last glacial maximum. *Science in China Series D, Earth Sciences*, **52**, 1127–1136.

ZHAO, Q. & WANG, P. 1999. Progress in Quaternary paleoceanography of the South China Sea: a review. *Quaternary Research*, **19**, 481–501 [in Chinese].

Late Quaternary tectonics, sea-level change and lithostratigraphy along the northern coast of the South China Sea

Y. ZONG[1]*, G. HUANG[2], X. Y. LI[1] & Y. Y. SUN[1]

[1]*Department of Earth Sciences, The University of Hong Kong, Hong Kong SAR, China*

[2]*Guangzhou Institute of Geography, Guangzhou 510070, China*

**Corresponding author (e-mail: yqzong@hku.hk)*

Abstract: This paper examines the Late Quaternary evolutionary history of the northern coast of the South China Sea by reviewing geological evidence from this coast. Results show that: (1) a lithostratigraphy with two marine sequences is observed in the deltaic basins of the Song Hong, Pearl and Han rivers, and these basins are all bounded by active faults; (2) whilst the upper marine sequence belongs to the present interglacial or the past 10.5 ka, the lower or older marine sequence was most likely deposited during the high sea level of the last interglacial period, *c.* 126–120 ka; (3) the burial depth of the older marine sequence is recorded at −15 m below modern sea level or deeper, implying localized subsidence of varying rates between these deltaic basins because the height of sea level during the last interglacial was close to that of the present; (4) this tectonic process is probably associated with the continuous long-term subsidence of the northern South China Sea continental shelf within the tectonic framework of southward continental extension of the Eurasian Plate during the Cenozoic; (5) fault activity has enhanced the localized subsidence of these deltaic basins, which led to marine inundation during interglacial high sea levels.

Coastal change is commonly considered to be a result of several geological processes, including sea-level change and fluvial discharge acting over various timescales and at different magnitudes in a given locality. However, for Late Quaternary evolution along the northern coast of the South China Sea, geological evidence suggests that tectonic subsidence may also have been an important factor. Previous studies in this region have focussed on addressing several key scientific matters, including the development of a sequence stratigraphic framework (e.g. Huang *et al.* 1982; Li *et al.* 1987; Yim 1994; Owen *et al.* 1998; Fyfe *et al.* 2000; Tanabe *et al.* 2006; Tran *et al.* 2007; Zong *et al.* 2009*a*) and the influences of natural (climate and sea level) and non-natural (human activity) drivers on coastal change (e.g. Fyfe *et al.* 1997, 1999; Owen 2005; Yim *et al.* 2008; Funabiki *et al.* 2012; Hu *et al.* 2013; Zong *et al.* 2013). These studies have provided information on the spatial distribution and characteristics of the Upper Quaternary sedimentary formations, which is useful to geo-engineering projects, including road and bridge construction, building-material extraction, port facilities and land reclamation. However, the effects of tectonic processes on the region's coastal change have not been thoroughly examined. As a result, future coastal change cannot be appropriately assessed if the long-term movement of the bedrock relative to sea-level change is unknown. Similarly, although the projection of future global mean sea-level rise has been refined by a series of reports from the Inter-Governmental Panel on Climate Change (e.g. IPCC 2014), the magnitude and pace of future sea-level change for the study area are still difficult to estimate if the rate of long-term tectonic movement is not quantified. In this paper we review the region's sedimentary evolution and investigate the history of tectonism, in particular, the question of how tectonic movement has interplayed with sea-level change during the Late Quaternary, as reflected by the coastal lithostratigraphy.

Geologically, the study area lies within the Cathaysia Block, bordered by the subduction boundary of the Philippine Sea Plate to the east and the Red River Fault with the Sundaland Block to the west (Fig. 1). The Cathaysia Block consists of the upland landscape of southern China and part of northern Vietnam, and extends southwards under the continental shelf, which is in contact with the oceanic crust of the South China Sea along the base of the continental slope (e.g. Sun *et al.* 2008). This section of the coast comprises three deltaic basins and many small estuaries. All these deltas and estuaries are separated by rocky headlands, and active normal faults are commonly found between these rocky headlands and deltaic/estuarine lowlands (e.g. Huang *et al.* 1987). The Upper Quaternary sedimentary sequences are well preserved in these deltaic/estuarine basins. The south China coast lies on the

From: Clift, P. D., Harff, J., Wu, J. & Qui, Y. (eds) 2016. *River-Dominated Shelf Sediments of East Asian Seas*. Geological Society, London, Special Publications, **429**, 123–136.
First published online September 3, 2015, updated May 26, 2016, http://doi.org/10.1144/SP429.1

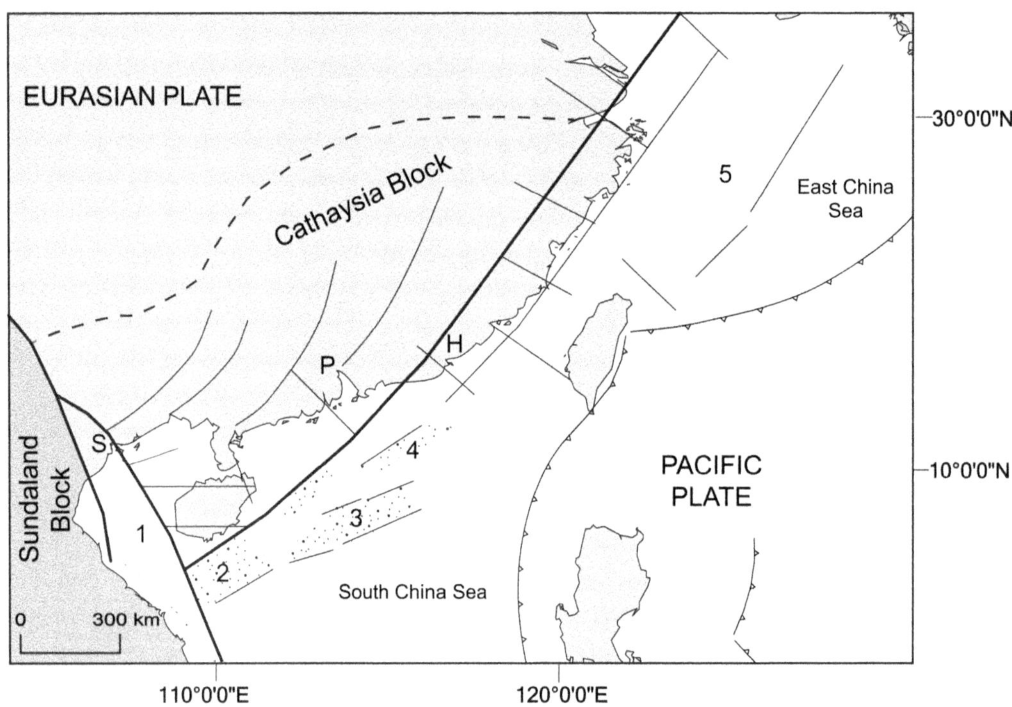

Fig. 1. A map showing generalized tectonic elements of the northern margin of the South China Sea with major faults and depocentres (after Nielsen *et al.* 1999; Pubellier & Chan 2006; Sun *et al.* 2008). Locations mentioned in the text are: (1) the Song Hong–Yinggehai Basin, (2) the Qiongdongnan Basin, (3) the Main Baiyun Sag, (4) Pearl River Mouth Basin and (5) the Yangtze Mouth Basin; and (S) the Song Hong River deltaic basin, (P) the Pearl River deltaic basin and (H) the Han River deltaic basin.

northern boundary of the tropics and is influenced by the Asian monsoon climate.

This paper reviews geological data and maps from a number of sources including the 1:5 000 000 scale structural geology of China and its marginal seas (Zhang 1983). Books and journal articles are consulted for details of fault activity and lithological data from the three deltaic basins, the Song Hong (Red River), Pearl and Han, including core records and the associated sedimentary analyses for the study area, with references for these sources given where appropriate.

The geological record

The Song Hong (Red River) deltaic basin lies within the Hanoi Trough, which is the most landward portion of the Song Hong–Yinggehai Basin, extending *c.* 1000 km southwards into the South China Sea (Fig. 1). The Hanoi Trough, and to some extent the northern and western boundaries of the present delta plain (Fig. 2a), are apparently confined by the active NW–SE-trending Song Lo and Song Chay Faults (Rangin *et al.* 1995; Nielsen *et al.* 1999). The sedimentary rocks under the delta plain consist of thick sequences of sandstone, siltstone and shale that date back to as early as the Late Eocene, while sedimentation has continued into the Quaternary in this basin (Clift *et al.* 2006, 2008). Resting on the Pliocene sediment, the Quaternary sediment shows five cycles of change, which have been correlated with river terraces located further inland along the Song Hong River (Tran *et al.* 1991). Sea-level change has been considered as the main factor driving the formation of these five sedimentary cycles within this deltaic basin (Tran *et al.* 2007). Details of the lithologies show that the lowermost two cycles comprise gravels and coarse sands deposited in a fluvial environment and are considered as pluvial and floodplain facies (C_1 and C_2, Fig. 2c) (Tran *et al.* 1991). The third cycle (C_3) displays characteristics of deltaic–lacustrine–swamp deposits, corresponding to the high sea level of the last interglacial or marine isotope stage (MIS) 5 (*c.* 129–74 ka) (Tran *et al.* 2007). The upper two cycles are both dated as being postglacial, including the transition from MIS 2

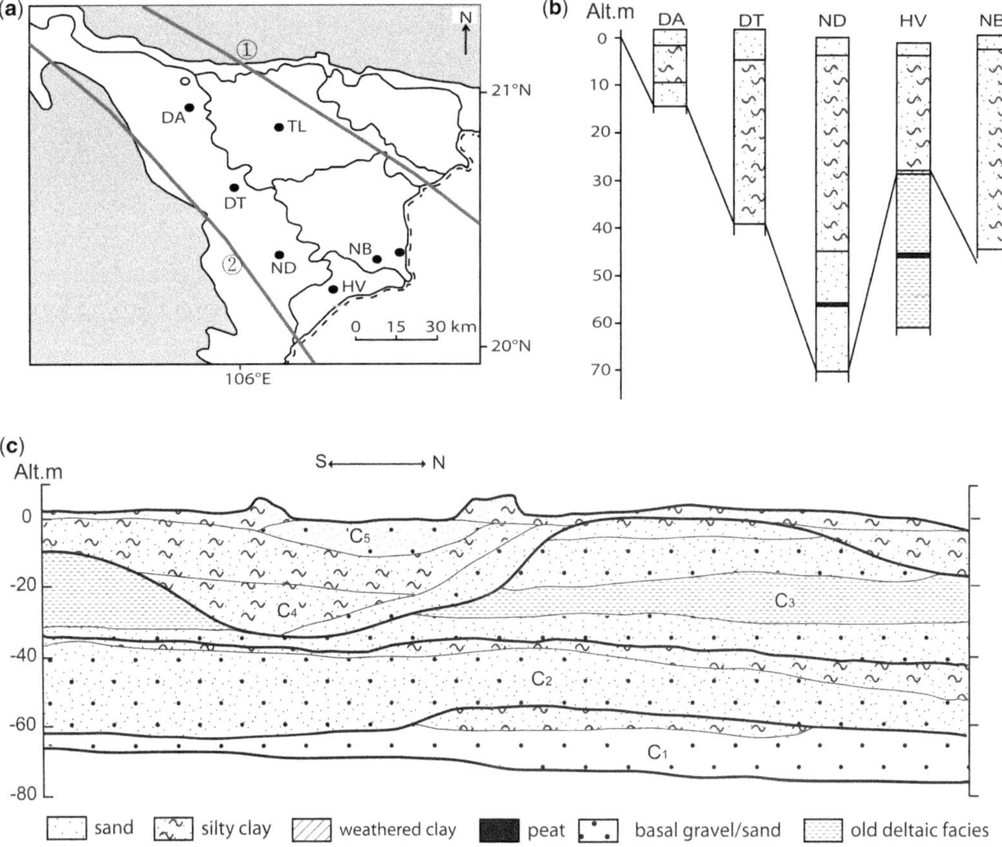

Fig. 2. (**a**) The Song Hong River deltaic basin, locations of key boreholes (black circles), the Song Lo Fault (line 1) and Song Chay Fault (line 2). (**b**) The lithology of key boreholes from Tanabe *et al.* (2003) and Hanebuth *et al.* (2006). (**c**) A south–north cross transect showing the Quaternary lithostratigraphy as modified after Tran *et al.* (1991).

(*c.* 23–12 ka) to MIS 1 (<12 ka), with the characteristics of deltaic transgressive facies (C_4) and deltaic regressive facies (C_5) (Tran *et al.* 1991). In summary, two deltaic/marine layers are found within this deltaic basin.

Detailed sedimentary characteristics of the lower deltaic sequence from a borehole (Core HV, Fig. 2b) drilled near the current deltaic shoreline show that this sequence is very thick (*c.* 32 m), and lies beneath 29 m of Holocene deltaic sediment. The lower sequence shows a deltaic origin, with shoreline-related characteristics, and is dated by anisotropy of magnetic susceptibility (AMS) ^{14}C methods to between >47.8 and 40.1 ka, which may be considered as MIS 3, or possibly older (Hanebuth *et al.* 2006). About 20 km NW from Core HV, another core (ND) revealed a Lower Holocene sandy sequence filling an incised channel down to *c.* 70 m below sea level (Fig. 2b, Tanabe *et al.* 2006). This incised channel is also reported by Tran *et al.* (1991) (Fig. 2c). Total thickness of the Quaternary sediment is *c.* 100–200 m, and fluvial sequences of Middle and Early Pleistocene ages are found beneath these two marine sequences (e.g. Mathers & Zalasiewicz 1999).

The present Pearl River delta plain is developed within a basin confined by the dominant NW–SE-trending normal faults, which intersect a set of older NE–SW-trending normal faults (Fig. 3a). These faults are found cutting through the underlying sedimentary bedrock, which consists of sandstone and mudstone of Late Eocene and Oligocene age (Fig. 3b, c), which in turn overlie Cretaceous and older volcanic and plutonic rocks of the Cathaysia Block. Amongst these faults, the Xijiang Fault, Shawan–Ningdingyang Fault and Luofushan–Dongjiang Fault are still active. Several earthquakes of surface-wave magnitude (M_s) >4

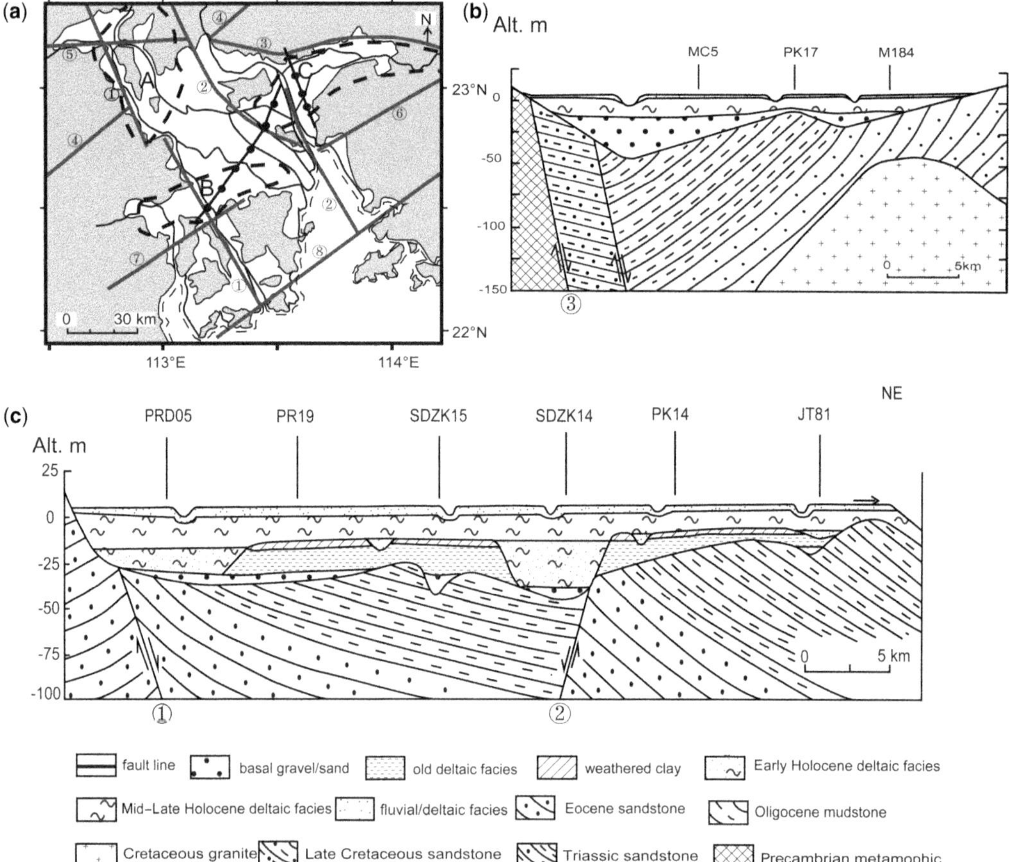

Fig. 3. (**a**) The Pearl River deltaic basin and geological sub-basins: A as Sanshui, B as Xinhui and C as Dongguan basins. Major faults: (1) Xijiang Fault; (2) Shawan–Ningdingyang Fault; (3) Luofushan–Dongjiang Fault; (4) Guangsong Fault; (5) Guangzhou–Sanshui Fault; (6) Dongguan Fault; (7) Hengli Fault; (8) Shenzhen Fault. (**b**) A simplified lithostratigraphy across the East River deltaic plain with borehole records from Zong *et al.* (2009*b*). (**c**) A simplified lithostratigraphy across the central part of the main deltaic plain with borehole records from Zong *et al.* (2012).

were generated on these structures during the twentieth century, with the biggest being M_S 5.5 recorded on the Xijiang Fault near Macau (e.g. Liu *et al.* 2002). Whilst the Luofushan–Dongjiang Fault controls the basement topography of the East River deltaic sub-basin and the orientation of the East River (Fig. 3a, b), the Shawan–Ningdingyang and Xijiang Faults confine the east and west borders of the main deltaic basin respectively, as well as the orientation of the West River and the present estuary (Fig. 3a, c).

Along the fringe of the deltaic basin, some outcrops of Jurassic–Cretaceous sandstone and granite are found. Structural geological data from the basin indicate a two-stage development model (e.g. Huang *et al.* 1982): (1) during the Cretaceous, fault activity caused localized subsidence at the same time as the emplacement of granitic intrusions; and (2) by the Late Paleocene and Early Eocene, three sub-basins had formed, i.e. the Sanshui basin in the NW, the Dongguan basin in the NE and the Xinhui basin in the SW (Fig. 3a). These three sub-basins received sediment during the Eocene and Oligocene, resulting in the accumulation of sandstones and mudstones. Immediately overlying the Eocene sandstone and Oligocene mudstone in the vicinity of the river mouth is an Upper Quaternary sequence (Fig. 3b, c), i.e. there is a sedimentary hiatus between the Oligocene and the Quaternary. In the main deltaic basin, two layers of marine sediment are found separated by a layer of weathered clay or fluvial sand. The lower marine layer consists

Table 1. *Radiocarbon dates of the older marine sequence reported from the study area*

Location	Core	Depth (m)	Material	No. of dates	Youngest (cal. BP)	Oldest (cal. BP)	References
Song Hong delta	HV	31.6–56.6	Plant fragments	8 (AMS)	40 060	>47 750	1
Pearl River delta	PK19	22.5	Plant fragments	1	32 900–34 300		2
	UV1	Below 10.5	Marine shell	1 (AMS)	40 900–41 200		2
Han River delta	HK25	27.8–81.2	Plant fragments	2	37 900	41 400	3
	HK26	20.4–101.6	Plant fragments	2	26 200	29 100	3
	E2	31.9–72.2	Plant fragments	1	29 300–30 100		3

References: (1) Hanebuth *et al.* 2006; (2) Zong *et al.* 2009*a*, *b*; (3) Li *et al.* 1987.

of silt and clay deposited in an estuarine environment and is dated to between 41.2 and 32.9 ka by radiocarbon methods (Table 1). This lower marine layer was incised by fluvial action during the last glacial period (*c.* 20 ka; e.g. Zong *et al.* 2012). The upper marine layer contains three sedimentary units: sandy silt with a mixture of freshwater and brackish water diatoms (tidal-channel facies of Early Holocene age) within the incised channels; silty clay with abundant brackish water diatoms (estuarine facies of Middle–Late Holocene age) within the broad deltaic basin; and silt and sands (deltaic floodplain facies of Late Holocene age) covering the surface of the entire deltaic plain. In the East River deltaic basin, only the Holocene marine layer is found overlying a thick layer of sand and gravel, which lies along the Luofushan–Dongjiang Fault (Fig. 3b). The maximum thickness of the Quaternary sediment within the basin is <70 m (e.g. Zong *et al.* 2009*a*), and this is found near the southern end of the Xijiang Fault.

Similarly, the Han River deltaic basin is confined by NE–SW- and NW–SE-trending normal faults (Fig. 4a). These faults, particularly the NW–SE-trending faults, are still active at present and have generated several earthquakes of Ms >6 in the past 200 years (e.g. Li *et al.* 1987). Whilst these faults have caused the deltaic basin to subside, localized uplift is also evident outside of the deltaic basin (e.g. Pedoja *et al.* 2008). The Quaternary sediment in the Han River basin is mainly underlain by Jurassic–Cretaceous granite (Fig. 4b). Although some outcrops of Jurassic sandstone, mudstone and volcanic conglomerate are found along the northern edge of the deltaic basin (Fig. 4c), Jurassic granite is the dominant bedrock-type of the region. The granite itself is cut by NW–SE normal faults (Li *et al.* 1987). Immediately overlying this bedrock is sediment of Late Quaternary age. The Quaternary lithostratigraphy is similar to that of the Pearl River delta, i.e. two marine layers separated by a layer of weathered clay or sandy sediment (Fig. 4b, c).

The lower marine layer overlies a thick sequence of fluvial sands and gravels and consists of sandy silt and clay with a rich assemblage of marine diatoms, the latter of which is considered as a deltaic facies (Zong 1987) and has been dated by conventional radiocarbon methods to between 41.4 and 26.2 ka (Table 1). The upper marine layer consists of three units: (1) sandy silt containing a mixture of freshwater and brackish water diatoms (tidal-channel deposit of Early Holocene age); (2) silt and clay, rich in brackish water diatoms (estuarine facies of Middle–Late Holocene age); and (3) sandy sediment dominated by freshwater diatoms (deltaic floodplain facies of Late Holocene age) (Li *et al.* 1987; Zong 1992). The maximum thickness of the Late Quaternary sediment in this basin is recorded as 168 m at a site close to one of the NW–SE-trending faults (Fig. 4b), and is much greater than the thicknesses recorded from the deltas of the Song Hong and Pearl rivers. However, like the Pearl River delta, the maximum thickness of the postglacial sediment is found at a site close to the basin-bounding fault (Zong 1992).

Discussion

Tectonic subsidence

As outlined above, a common feature of the three deltas is that their sedimentary histories are all associated with the region's tectonic activity which has resulted in large-scale, long-lived crustal subsidence within these deltaic basins. For instance, the Song Hong River has developed along the Red River shear zone, and its delta plain is formed within the Hanoi Trough at the NW end of the Song Hong–Yinggehai Basin, a depositional centre created by the activity of the Red River Fault system (e.g. Rangin *et al.* 1995; Z. Sun *et al.* 2003; Clift & Sun 2006; Zhu *et al.* 2009). Although the Red River Fault might have started its activity as early as the Late Cretaceous, the trough, or the basin, was generated by a phase of regional north–south extension,

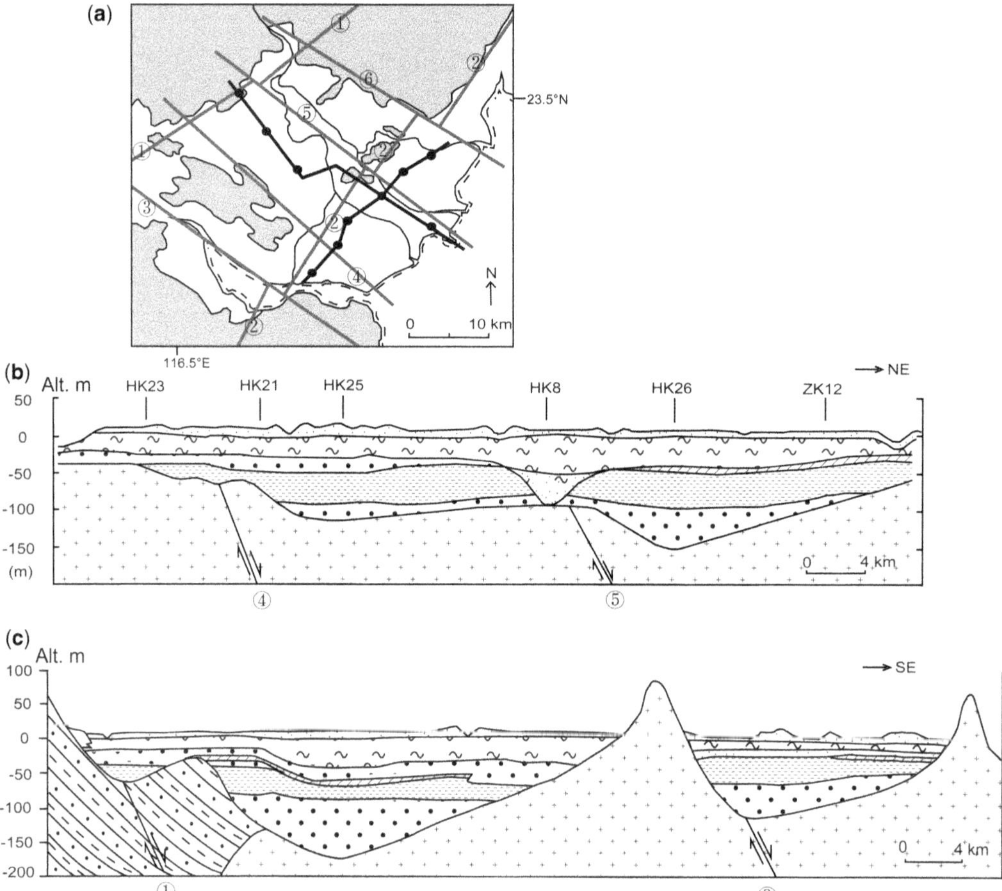

Fig. 4. (**a**) The Han River deltaic basin and major faults: (1) Puning–Chaoan Fault; (2) Santao–Raoping Fault; (3) Rongjiang Fault; (4) Yujiao–Xiapeng Fault; (5) Chenghai–Guxiang Fault; (6) Hanjiang Fault. (**b**) A simplified lithostratigraphy along a SW–NE cross transect (after Li *et al.* 1987). (**c**) A simplified lithostratigraphy along a SE–NW cross transect (after Li *et al.* 1987). Legends are as the same as in Figure 3.

initiated by collision of the Indian Plate with the Eurasian Plate after the Mid Eocene (e.g. Lee & Lawver 1995; Zhou *et al.* 1995), or the motion on the fault could have been driven by motion of Indochina linked to subduction to the south of Borneo and Indonesia (e.g. Peltzer & Tapponnier 1988; Morley 2002). Motion on the Red River Fault caused the Sundaland Block to move SE and subsequently rotate clockwise (Taylor & Hayes 1983; Nielsen *et al.* 1999). The left-lateral movements along the Red River Fault created the Song Hong basin, which subsequently preserved thick deposits during the Oligocene and Miocene, a result of several phases of rapid subsidence coupled with significant sediment delivery (Clift & Sun 2006; Hoang *et al.* 2010).

More importantly, the sedimentary facies of these rocks preserved within the Hanoi Trough change from dominantly fluvial–lacustrine types before the Late Oligocene to dominantly deltaic–estuarine types after the Late Oligocene (Nielsen *et al.* 1999). This requires a long history of tectonic subsidence of the Hanoi Trough, including the Song Hong deltaic basin. In other words, the preservation of Late Quaternary deltaic sequences in the Song Hong basin is inherently a result of long-term tectonic subsidence, coupled with sediment supply, which is possibly caused by a combination of three processes: the offshore continuation of the Red River Fault system and the resultant crustal thinning (Lei *et al.* 2011); the dextral movement of the fault zone that created large depositional centres (Zhu *et al.* 2009); and the very large volume of sediment eroded from the edges of the Tibetan Plateau delivered by the Red River to the basin (e.g. Hoang *et al.* 2010), resulting in largely continuous sedimentation in the deltaic basin throughout the Cenozoic.

The subsidence history of the Pearl River deltaic basin is slightly different. The Pearl River catchment extends westwards from the Cathaysia Block into the Yangtze Block. The Pearl River has been close to its modern configuration since the Miocene (e.g. Clift *et al.* 2006; Hoang *et al.* 2009), although it probably initiated during rifting of the South China Sea and has subsequently expanded its basin inland. However, before the initiation of the river catchment, the Cathaysia Block was affected by a number of tectonic and magmatic processes, the most important of which included folding/rifting along NE–SW-trending faults, granite intrusions during the Jurassic–Cretaceous, and fluvial sandstone/mudstone deposition during the Eocene and Oligocene (Chan *et al.* 2010). This sedimentation took place within folded basins such as the Palaeogene (Tertiary) Dongjiang, Sanshui and Xinhui sub-basins that make up the basement of the present Pearl River deltaic basin (Fig. 3) (Huang *et al.* 1982). These folding and rifting stresses along the regional NE–SW-trending faults and shear zones during the Early Cenozoic (e.g. Zhou *et al.* 1995; Sun *et al.* 2008; Zhao *et al.* 2011) also led to the creation of the super-deep Baiyun Sag sub-basin and other depositional centres on the northern continental shelf and slope of the South China Sea (Fig. 1).

Pre-Miocene folding/rifting activity coincided in part with the opening of the South China Sea, north–south tectonic extension of the southern edge of the Eurasian continent, and the formation of the Hanoi Trough and the Song Hong–Yinggehai Basin. Since the Miocene, the landscape of the Pearl River catchment has been slightly uplifted and influenced by long-term, relatively slow erosion (e.g. Yan *et al.* 2009), as demonstrated by the fact that fluvial sandstone/mudstone deposition ceased in many inland basins. As a result, sediment accumulation between the Miocene and Late Quaternary within the Pearl River deltaic basin (Fig. 3a) is absent because this is on the rift flank where thermal subsidence is negligible so that there was no post-rift accommodation space generated. The slight uplift of the continental landscape relative to the subsidence of the South China Sea, together with the development of a monsoon climate (e.g. Sun & Wang 2005; Clift *et al.* 2014) have led to an increased erosion of terrestrial sediment from the continental landscape (Clift *et al.* 2002), and sediment supply to the South China Sea filling the Baiyun Sag and other depositional centres (e.g. Sun *et al.* 2008). Since the Miocene, the NE–SW-trending faults became less active (Chan *et al.* 2010). Entering the Quaternary, the NW–SE-trending faults became active and resulted in localized subsidence. This neotectonic history is somewhat comparable with events offshore of the Pearl River deltaic basin and around the Dongsha Islands, where two uplift episodes are reported in the Late Miocene and Middle Pleistocene respectively (Lüdmann *et al.* 2001). The two sets of faults have acted together and created the fault-block basin.

The Han River deltaic basin is another fault-block basin as it is also bounded by two sets of normal faults (Fig. 4a). However, the Han River basin is slightly different from the Pearl River deltaic basin, because the Quaternary sediment in the Han River deltaic basin lies immediately on Jurassic–Cretaceous granite. The absence of Eocene sandstone and Oligocene mudstone from the deltaic area may have been a result of sediments having no accommodation space to occupy due to a lack of thermal subsidence in this area following break-up of the South China Sea, or possibly because they were eroded from the landscape since the Miocene. In spite of this contrast, the two marine lithostratigraphic units recognized in the Han River deltaic basin are comparable with those found in the deltas of the Song Hong and Pearl rivers.

In summary, the three deltas considered in this study are all inherently influenced by Cenozoic tectonic activity. The tectonic history can be divided into three major periods. First, during the Palaeogene, the collision of the Indian and Eurasian Plates and the opening of the South China Sea (e.g. Briais *et al.* 1993; Barckhausen *et al.* 2014) transformed the region's land and sea distribution, as well as some geological structures including folding and rifting along the NE–SW-trending faults and the generation of sags and depositional centres within the Cathaysia Block (e.g. Taylor & Hayes 1983; Nielsen *et al.* 1999; Sun *et al.* 2008; Chan *et al.* 2010). By the end of this period, the monsoon climate was established (Sun & Wang 2005; Clift *et al.* 2014), the rivers' catchments were initiated (e.g. Clift *et al.* 2006; Hoang *et al.* 2009), and the modern pattern of erosion and deposition was formulated.

Second, since the Middle Miocene, seafloor spreading ceased in the South China Sea, but the seafloor and continental margins continued to subside. The deepening of the South China Sea coincided with the slight uplift of the continental landscape of the Cathaysia Block – as suggested by the uplifted mountainous Eocene sandstone–Oligocene mudstone landscape in the upper reaches of the Pearl River – this being a consequence of the collision of the Indian, Pacific and Philippines Plates with the Eurasian Plate. The Asian monsoon, which may have been intensified by the uplift of the Tibetan Plateau, as well as other processes such as retreat of the Paratethys and uplift of the Himalaya, had caused erosion of the continental landscape, as well as transport and deposition of massive amounts of sediment on to the continental margins (e.g. Liu *et al.* 2003; Clift 2006). The absence of sedimentation between the Pliocene and Late Quaternary in the deltaic basins of the Pearl and Han rivers suggests two possibilities: firstly, no sedimentation took place during this period as sediments had simply bypassed the coastal regions, and secondly, there was sedimentation but it had been eroded. Either way, the coastal regions have been a transitional zone between the upland which has been under erosion and the continental margin which has continuously received sediment.

Finally, by the Middle Pleistocene, subsidence of the northern continental shelf had propagated progressively landwards and reached the present coast. Activity along the NW–SE-trending faults, together with the NE–SW-trending faults generated the fault-block basins. This tectonism continued into the Late Pleistocene, during which time global climatic variability was already greatly magnified (e.g. Linsley 1996) causing sea level to rise and fall over 100 m (e.g. Chappell 2002; Jouzel *et al.* 2002; Siddall *et al.* 2003). As all three deltaic basins had subsided below present sea level by the Late Pleistocene and then continued to subside, an accommodation space was opened and two periods of marine transgression and preservation took place.

The age of the older marine sequence

This older marine sequence is recorded in all three deltaic basins, and it is also widely reported across the globe. However, its age has long been unresolved. In addition to the three deltaic basins, such Late Quaternary marine sequences are also observed in many small estuaries along the south China coast (Chen 2014). The existence of these estuaries is related in part to the activities of normal faulting. In a small barrier coast on the SE corner of Hainan Island, for instance, a drill core has revealed two layers of marine sediment, the older one being 39 m thick (Wang 2014). Farther north towards the east coast of China, this Late Quaternary twin-marine sequence is also evident (e.g. Lin 1996). On both the north and south flanks of the Yangtze deltaic plain, the older marine sequence is rather thick, mostly >30 m, and it is overlain by a thin (<10 m) postglacial, estuarine sequence (e.g. Li *et al.* 2000). Along the Indochina peninsula, this older marine sequence was also widely reported as lying in coastal sites (e.g. Bosch 1988).

In the deltaic basins of the Pearl and Han rivers, dating results using conventional radiocarbon methods provided age estimates (Table 1) around MIS 3 (58–24 ka). These ages were considered as biased because they had all reached the dating limit of the method (Yim *et al.* 1990). With the application of an accelerated mass spectrometer, Hanebuth *et al.* (2006) produced age estimates also within MIS 3 for the older marine sequence in the Song Hong delta, although they hinted at a possible older age for this sequence. With the use of optically stimulated luminescence (OSL) dating, the age of this sequence in the Pearl River mouth region has been fixed at around MIS 5e (Yim *et al.* 2008). This older marine layer is also reported from Korea, where the sediment is dated to 138–110 ka, or around MIS 5e (Chang *et al.* 2014).

In addition to both the radiocarbon and OSL dating methods, a comparison between the current altitude of the older marine sequences from the three deltaic basins and the far-field sea-level curve (e.g. Siddall *et al.* 2003) offers an opportunity to elucidate the possibility of the older marine sequence being a product of high sea level during the last interglacial period, i.e. MIS 5. As illustrated in Figure 5, sea level in far-field regions during MIS 5e (*c.* 126–120 ka) was possibly 3–5 m higher than the present sea level. This sea-level highstand has been recorded in many far-field locations (e.g. Bard *et al.* 1990; Stirling *et al.* 1998; Chappell 2002; Dumas *et al.* 2006; Rohling *et al.* 2007). During MIS 5c and MIS 5a (*c.* 103–99 and 80–74 ka respectively), sea level varied around −40 m. During MIS 3, sea level in far-field regions is recorded to decrease to between −60 and −80 m (Bard *et al.* 1990; Yokoyama *et al.* 2001). Given the fact that these basins have been subject to continuous subsidence, the older marine sequence is unlikely to be a deposit of MIS 3. For instance, the current burial depth of the top of this marine sequence in the Song Hong delta is recorded at −15 to −30 m (Fig. 2c) (Tran *et al.* 1991), suggesting that the most probable age for this sequence is MIS 5e. Similarly, the top end of this sequence in the Pearl River deltaic basin is around −15 to −20 m (Zong *et al.* 2009*a*), and thus it is probably also a deposit of MIS 5e. In the Han River delta where the older marine sequence currently lies at −45 to −80 m (Fig. 4), with a history of strong subsidence (e.g. Li *et al.* 1987), it is again a deposit not likely to be younger than MIS 5.

Sedimentary facies data from this sequence provide further support for the above assessment. The older marine sequence in the Song Hong delta is marked by shoreline-related sedimentary facies (Hanebuth *et al.* 2006). Studies in the deltas of the Pearl and Han rivers confirmed that this sequence is an intertidal-to-subtidal deposit in those systems (Li *et al.* 1987; Zong *et al.* 2009*a*). In other words, during the deposition of the older marine

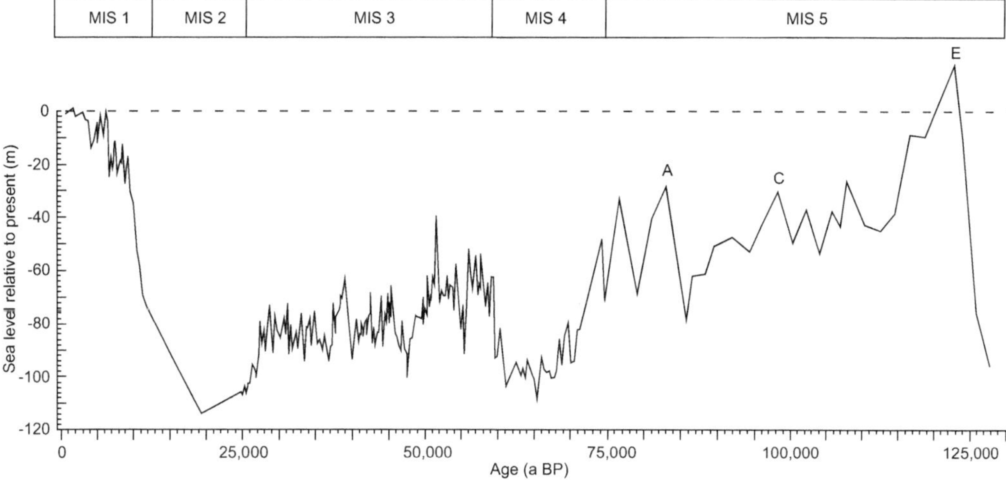

Fig. 5. The sea-level curves for the past 130 000 years from the Red Sea (after Siddall *et al.* 2003; Rohling *et al.* 2007), Barbados and Haiti (Bard *et al.* 1990; Dumas *et al.* 2006), Huon Peninsular (Yokoyama *et al.* 2001), and west coast of Australia (Stirling *et al.* 1998).

sequence in these three deltaic basins, the depositional surface was very close to the height of present sea level, given the fact that the MIS 5e sea level was a few metres higher than the present. Therefore, the difference in the burial depth of the older marine sequence relative to present sea level may provide an estimate of the amount of subsidence that has occurred since the sequence was laid down. The minimum subsidence since the end of MIS 5e (i.e. 120 ka) in the Han River delta (from 0 to -45 m at a rate of 37.5 cm ka^{-1}) is much larger than that seen in the deltas of the Song Hong and Pearl rivers (from 0 to -15 m at a rate of 12.5 cm ka^{-1}). These subsidence rates are much higher than the average long-term subsidence rate (between 1 and 2 cm ka^{-1}) of the Pearl River Mouth Basin (*c.* 100 km offshore from the current coast) for the past 10 Ma according to Clift & Lin (2001). This strong subsidence in the deltaic basins may thus imply a phase of increased fault activity along the coast during the Late Quaternary.

In summary, by about 130 ka, the three deltaic basins had subsided below present sea level. As sea level rose to the highstand during Stage E of MIS 5 (Fig. 5), an accommodation space was opened in these basins for marine transgression and sedimentation, resulting in the formation of the older marine sequence. After MIS 5e, sea level dropped below -40 m during the rest of MIS 5 and moved to much lower levels during MIS 4–2 between 74 and 20 ka (Fig. 5). These basins continued to subside since MIS 5e, but the subsidence rates were slower than the falling sea level. Thus there was no accommodation space for marine inundation and preservation of sediment during this period until the Holocene, *c.* 10.5 ka, since which time new accommodation space was created by the rising sea level, allowing a marine transgression into the basins in the Early and Middle Holocene.

The Late Quaternary lithostratigraphy and environmental history

Several authors have attempted to summarize the lithostratigraphy for the study area using detailed studies of sedimentary facies, microfossil analyses and lithostratigraphic surveys. For instance, Huang *et al.* (1982) proposed a scheme (Table 2) that summarizes the lithological data obtained from the Pearl River deltaic basin. This lithostratigraphic scheme shows six sedimentary formations, including the two marine sequences, and is largely comparable with schemes put forward by other authors. However, this scheme fails to recognize the Early Holocene fluvial–estuarine channel infill, which was recognized by later studies of the Han River delta (Li *et al.* 1987) and the Pearl River mouth region (Fyfe *et al.* 2000; Zong *et al.* 2009*a*). This Early Holocene unit was revealed by a series of seismic profiles showing rivers incised down to various depths, mostly -40 m to -50 m, within Hong Kong waters (e.g. Fyfe *et al.* 1997, 1999; Owen *et al.* 1998; Owen 2005).

In addition to these schemes, Yim (1994) proposed an interesting scheme, which included five marine sequences, but those sequences older than

Table 2. *A summary of the Late Quaternary lithostratigraphy for the study area*

Unit	Sediment	Li *et al.* (1987) for Han River delta	Huang *et al.* (1982) for Pearl River delta	Yim (1994) for Pearl River mouth region	Fyfe *et al.* (2000) for Pearl River mouth region	Mathers & Zalasiewicz (1999) for Song Hong delta	Estimated age (years BP)	MIS stages
M1b	Deltaic silt and clay (upper marine sequence)	Dong Li, Cheng Hai and Chao Zhou Formations	Deng Long Sha, Wan Qing Sha and Heng Lan Formations	First marine unit	Hang Hau Formation (Tseung Kwan O member & Kwo Chau Member)	Thai Binh and Haihung Formations	7000–0	1 (Mid–Late Holocene)
M1a	Fluvial–estuarine channel sandy silt	Lian Xia Formation			Hang Hau Formation (Pok Liu Member & Tung Lung Member)		10 500–7000	1 (Early Holocene)
T1	Terrestrial sand, silt and weathered clay	Tuo Pu Formation	San Jiao Formation	First terrestrial unit	Waglan Formation		73 000–10 500	4–2
M2	Shallow marine silt and clay (lower marine sequence)	Jia Li Formation	Xi Nan Formation	Second marine unit	Sham Wat Formation	Vinphuc Formation	125 000–73 000	5e–a
T2	Terrestrial silt, sand, gravel, cobbles and boulders	Nan She Formation	Shi Pai Formation	Second terrestrial unit	Chek Lap Kok & Tung Chung Formation	Hanoi & Lechi-Hanoi Formation	Prior to 125 000	6
Base	Bed rock	Bed rock	Bed rock	?	Bed rock	Early Quaternary		

MIS 5 are not readily present or reported from other parts of Hong Kong waters and elsewhere. However, along the east and north coasts of China, one more marine sequence was reported. In the Yangtze deltaic basin, this third marine layer was considered as deposited during the Middle Pleistocene (e.g. Chen *et al.* 1997).

Therefore, the region's lithostratigraphy (Table 2) reveals five stages of environmental history, which show two parallel processes, tectonic subsidence and sea-level change. Details are summarized as follows:

(1) *Prior to marine inundation (T2, before c. 126 ka).* The coastal area was under fluvial conditions. Fluvial sediment was transported by rivers through the current coastal area on to the continental shelf. A thick layer of terrestrial sediment is recorded in both the Song Hong deltaic basin and the Han River basin.
(2) *The older marine transgression (M2, 126–120 ka).* Tectonic subsidence had lowered the bedrock to a depth below sea level, providing accommodation space for marine sedimentation and its preservation. This took place during the last interglacial or MIS 5e, as sea level rose to a few metres above the present sea level (Fig. 5) and the sea flooded the deltaic basins. As a result, a marine sequence was deposited and preserved within the accommodation space created.
(3) *Terrestrial environment (T1, 120–10.5 ka).* By the start of the last glacial period (MIS 4), sea level dropped to below −40 m and eventually around −120 m during the last glacial maximum (MIS 2) (Yokoyama *et al.* 2000). During this period, tectonic subsidence continued in the deltaic basins. However, because the rate of subsidence was much slower than the rate of sea-level fall, there was no accommodation space within the deltaic basins. Exposed older marine sediment was either exposed to oxidation, leading to the formation of the weathered clay on top of the older sequence, or subject to fluvial incision. In some locations, rivers incised into the older marine sequence and created small channels. In other locations, rivers delivered fluvial (sandy) sediment overlying the older marine sequence.
(4) *The younger marine transgression (M1a, 10.5–8 ka).* Tectonic subsidence during the last glacial period had lowered the surface of the deltaic basins, which opened up accommodation space for marine inundation when postglacial sea level rose towards the present height. During this latest transgression, two phases of environmental change are recognized. Initially, around the beginning of the Holocene, *c.* 10.5 ka, sea level rose to about −40 m (e.g. Hanebuth *et al.* 2000; Bard *et al.* 2010), allowing the sea to inundate the incised channels. The precise timing for the onset of marine inundation is dependent on the depth of any given incised channel, with inundation occurring earlier in the lower part of incised channels. As sea level rose rapidly during the Early Holocene and as fluvial discharge was concentrated within incised channels, rapid sedimentation took place within these channels. Entering the period between 9 and 8 ka, with the exact timing dependent on the height of the deltaic–basin landscape surface, the sea flooded the basins, pushing river mouths to their most inland positions (e.g. Zong *et al.* 2009*b*).
(5) *Formation of deltaic plains (M1b, 8 ka to present).* By about 7 ka, sea level reached its present height and was stabilized. The accommodation space created within the deltaic basins was filled with sediment, forming the three deltaic plains during the past 7 ka (Zong 1992; Tanabe *et al.* 2006; Zong *et al.* 2009*b*).

We note that the Early Holocene was the time marked by the biggest recent change in the landscape of present coastal zones across the world (Smith *et al.* 2011). Before this time, the coastal seabed and the upper part of the continental shelves were exposed as sea level dropped to below −40 m for the majority of the last glacial period. The exposed landscape underwent several different processes for a long period of time, including the surface of former marine deposits being oxidized, the unconsolidated sediment sequences being incised by rivers which reached the outer part of continental shelves, and herbaceous vegetation colonizing parts of the exposed flat landscape (X. Sun *et al.* 2003, Z. Sun *et al.* 2003). After the last glacial maximum, sea level rose rapidly (Hanebuth *et al.* 2000), resulting in marine water flooding the continental shelves (Yim *et al.* 2006). As sea level continued to rise during the Early Holocene, marine transgression penetrated into the incised valleys (Li *et al.* 2000; Hori *et al.* 2001; Tamura *et al.* 2009; Zong *et al.* 2012). Some ocean passages were opened and this altered ocean circulation patterns (Smith *et al.* 2011), resulting in shifting rainfall patterns in some regions (Griffiths *et al.* 2009). Sea level then ceased to rise by the Middle Holocene in the low latitude regions. Coastal sedimentation filled the accommodation space created, forming deltas and estuarine wetlands (Woodroffe 2000; Hu *et al.* 2013).

Conclusions

The development of the deltaic landscape in the three delta plains considered here is a snap shot of long-term geological processes and is a result of the interplay between tectonic subsidence and climate-induced sea-level change, coupled with changing sediment supply. The twin-marine, Late Quaternary lithostratigraphy recognized in these three deltaic basins suggests that: (1) the deltaic basin landscape subsided to a depth below present sea level before or by the last interglacial, during which time a phase of marine transgression laid down the older marine sequence, so filling up the accommodation space created; (2) the deltaic basin landscape continued to subside and was exposed to oxidation and fluvial action during the last glacial as sea level dropped to −120 m; (3) when sea level rose to its present height during the Holocene, the deltaic accommodation space was opened again, allowing another phase of marine transgression which formed the present deltaic plains.

This research is partly supported by the Research Grant Council of Hong Kong (HKU707109P). The generous assistance from W.W.-S. Yim and L.S. Chan, the constructive comments from the two reviewers and the detailed advice from the editor, P. D. Clift, are gratefully acknowledged.

References

Barckhausen, U., Engels, M., Franke, D., Ladage, S. & Pubellier, M. 2014. Evolution of the South China Sea: revised ages for breakup and seafloor spreading. *Marine and Petroleum Geology*, **58**, 599–611.

Bard, E., Hamelin, B. & Fairbanks, R. G. 1990. U–Th ages obtained by mass spectrometry in corals from Barbados: sea level during the past 130 000 years. *Nature*, **346**, 456–458.

Bard, E., Hamelin, B. & Delanghe-Sabatier, D. 2010. Deglacial melwater pulse 1B and the Younger Dryas sea levels revisited with boreholes at Tahiti. *Science*, **327**, 1235–1237.

Bosch, J. H. A. 1988. *The Quaternary Deposits in the Coastal Plains of Peninsular (Malaysia)*. Geological Survey, Malaysia.

Briais, A., Patriat, P. & Tapponnier, P. 1993. Updated interpretation of magnetic anomalies and seafloor spreading stages in the South China Sea: implications for the Tertiary tectonics of Southeast Asia. *Journal of Geophysical Research: Solid Earth*, **98**, 6299–6328.

Chan, L. S., Shen, W. & Pubellier, M. 2010. Polyphase rifting of greater Pearl River delta region (south China): evidence for possible rapid changes in regional stress configuration. *Journal of Structural Geology*, **32**, 746–754.

Chang, T. S., Kim, J. C. & Yi, S. 2014. Discovery of Eemian marine deposits along the Baeksu tidal shore, southwest coast of Korea. *Quaternary International*, **349**, 409–418.

Chappell, J. 2002. Sea level changes forced ice breakouts in the Last Glacial cycle: new results from coral terraces. *Quaternary Science Reviews*, **21**, 1229–1240.

Chen, B. S. 2014. *Late Quaternary Stratigraphy and Palaeoenvironmental Reconstruction in Coastal Areas of East Guangdong Province (China)*. PhD thesis, Sun Yat-Sen University, Guangzhou, China.

Chen, Z., Chen, Z. & Zhang, W. 1997. Quaternary stratigraphy and trace-element indices of the Yangtze delta, eastern China, with special reference to marine transgressions. *Quaternary Research*, **47**, 181–191.

Clift, P. D. 2006. Controls on the erosion of Cenozoic Asia and the flux of clastic sediment to the ocean. *Earth and Planetary Science Letters*, **241**, 571–580.

Clift, P. D. & Lin, J. 2001. Preferential mantle lithospheric extension under the South China margin. *Marine and Petroleum Geology*, **18**, 929–945.

Clift, P. D. & Sun, Z. 2006. The sedimentary and tectonic evolution of the Yinggehai-Song Hong Basin and the southern Hainan margin, South China Sea: implications for Tibetan uplift and monsoon intensification. *Journal of Geophysical Research*, **11**, B06405.

Clift, P. D., Lee, J. I., Blusztajn, J. & Clark, M. K. 2002. Erosional response of South China to arc rifting and monsoonal strengthening recorded in the South China Sea. *Marine Geology*, **184**, 207–226.

Clift, P. D., Blusztajn, J. & Nguyen, D. A. 2006. Large-scale drainage capture and surface uplift in the eastern Tibet-SW China before 24 Ma inferred from sediments of the Hanoi Basin, Vietnam. *Geophysical Research Letters*, **33**, L19403.

Clift, P. D., Long, V. H. et al. 2008. Evolving East Asian river systems reconstructed by trace element and Pb and Nd isotope variations in modern and ancient Red River – Song Hong sediments. *Geochemistry, Geophysics, Geosystems*, **9**, Q04039.

Clift, P. D., Wan, S. & Blusztajn, J. 2014. Reconstructing chemical weathering, physical erosion and monsoon intensity since 25 Ma in the northern South China Sea: a review of competing proxies. *Earth Science Reviews*, **130**, 86–102.

Dumas, B., Hoang, C. T. & Raffy, J. 2006. Record of MIS 5 sea-level highstands based on U/Th dated coral terraces of Haiti. *Quaternary International*, **145–146**, 106–118.

Funabiki, A., Saito, Y., Phai, V., Nguyen, H. & Haruyama, S. 2012. Natural levees and human settlements in the Song Hong (Red River) delta, northern Vietnam. *The Holocene*, **22**, 637–648.

Fyfe, J. A., Selby, I. C., Shaw, R., James, J. W. C. & Evans, C. D. R. 1997. Quaternary sea-level change on the continental shelf of Hong Kong. *Journal of the Geological Society, London*, **154**, 1031–1038, http://doi.org/10.1144/gsjgs.154.6.1031

Fyfe, J. A., Selby, I. C., Plater, A. J. & Wright, M. R. 1999. Erosion and sedimentation associated with the last sea level rise offshore Hong Kong, South China Sea. *Quaternary International*, **55**, 93–100.

Fyfe, J. A., Shaw, R., Campbell, S. D. G., Lai, K. W. & Kirk, P. A. 2000. *The Quaternary Geology of Hong Kong*. Geotechnical Engineering Office, Civil

Engineering Department, The Government of the Hong Kong SAR.

Griffiths, M. L., Drysdale, R. N. *et al.* 2009. Increasing Australian – Indonesian monsoon rainfall linked to early Holocene sea-level rise. *Nature Geoscience*, **2**, 636–639.

Hanebuth, T. J. J., Stattegger, K. & Grootes, P. M. 2000. Rapid flooding of the Sunda Shelf: a late-glacial sea-level record. *Science*, **288**, 1033–1035.

Hanebuth, T. J. J., Saito, Y., Tanabe, S., Vu, Q. L. & Ngo, Q. T. 2006. Sea levels during late marine isotope stage 3 (or older?) reported from the Red River delta (northern Vietnam) and adjacent regions. *Quaternary International*, **145–146**, 119–134.

Hoang, L. V., Wu, F. Y., Vlift, P. D., Wysocka, A. & Swierczwska, A. 2009. Evaluating the evolution of the Red River system based on in-situ U–Pb dating and Hf isotope analysis of zircons. *Geochemistry, Geophysics, Geosystems*, **10**, Q11008.

Hoang, L. V., Clift, P. D., Schwab, A. M., Huuse, M., Nguyen, D. A. & Sun, Z. 2010. Large scale erosional response to SA Asian monsoon evolution reconstructed from sedimentary records of the Song Hong-Yinggehai and Qiongdongan Basins, South China Sea. *In*: Clift, P. D., Tada, R. & Zheng, H. (eds) *Monsoon Evolution and Tectonic-Climate Linkage in Asia*. Geological Society, London, Special Publications, **342**, 219–244.

Hori, K., Saito, Y., Zhao, Q., Cheng, X., Wang, P., Sato, Y. & Li, C. 2001. Sedimentary facies and Holocene progradation rates of the Changjiang (Yangtze) delta, China. *Geomorphology*, **41**, 233–248.

Hu, D., Clift, P. D. *et al.* 2013. Holocene evolution in weathering and erosion patterns in the Pearl River delta. *Geochemistry, Geophysics, Geosystems*, **14**, Q20166.

Huang, Z., Li, P., Zhang, Z., Li, K. & Qiao, P. 1982. *Zhujiang (Pearl) Delta*. General Scientific Press, Guangzhou [in Chinese].

Huang, Z., Li, P., Zhang, Z. & Zong, Y. 1987. Sea-level changes along the coastal area of South China since Late Pleistocene. *In*: Qin, Y. & Zhao, S. (eds) *Late Quaternary Sea-level Change*. China Ocean Press, Beijing, 142–154 [in Chinese].

IPCC (Inter-Governmental Panel on Climate Change) 2014. *Climate Change 2014: Impacts, Adaptation, and Vulnerability*. UNESCO, World Meteorological Organization, Geneva, Switzerland.

Jouzel, J., Hoffmann, G., Parrenin, F. & Wealbroeck, C. 2002. Atmospheric oxygen 18 and sea-level changes. *Quaternary Science Reviews*, **21**, 307–314.

Lee, T. Y. & Lawver, L. A. 1995. Cenozoic plate reconstruction of Southeast Asia. *Tectonophysics*, **251**, 85–138.

Lei, C., Ren, J., Clift, P. D., Wang, Z., Li, X. & Tong, C. 2011. The structure and formation of diapirs in the Yinggehai-Song Hong Basin, South China Sea. *Marine and Petroleum Geology*, **28**, 980–991.

Li, C., Chen, Q., Zhang, J., Yang, S. & Fan, D. 2000. Stratigraphy and paleoenvironmental changes in the Yangtze Delta during the Late Quaternary. *Journal of Asian Earth Sciences*, **18**, 453–469.

Li, P., Huang, Z., Zong, Y. & Zhang, Z. 1987. *Hanjiang Delta*. China Ocean Press, Beijing [in Chinese].

Lin, J. 1996. *Quaternary Environment of Eastern China*. China Seismological Press, Beijing.

Linsley, B. K. 1996. Oxygen isotope evidence of sea level and climatic variations in the Sulu Sea over the past 150 000 years. *Nature*, **380**, 234–237.

Liu, Z., Zhao, H., Fan, S. & Chen, S. 2002. *Geology of the South China Sea*. China Science Press, Beijing. [in Chinese].

Liu, Z., Trentesaux, A., Clemens, S. C., Colin, C., Wang, P., Huang, B. & Boulay, S. 2003. Clay mineral assemblages in the northern South China Sea: implications for East Asian monsoon evolution over the past 2 million years. *Marine Geology*, **201**, 133–146.

Lüdmann, T., Wong, H. K. & Wang, P. 2001. Plio-Quaternary sedimentation processes and neotectonics of the northern continental margin of the South China Sea. *Marine Geology*, **172**, 331–358.

Mathers, S. & Zalasiewicz, J. 1999. Holocene sedimentary architecture of the Red River delta, Vietnam. *Journal of Coastal Research*, **15**, 314–325.

Morley, C. K. 2002. A tectonic model for the Tertiary evolution of strike-slip faults and rift basins in SE Asia. *Tectonophysics*, **347**, 189–215.

Nielsen, L. H., Mathiesen, A., Bidstrup, T., Vejbæk, O. V., Dien, P. T. & Tiem, P. V. 1999. Modelling of hydrocarbon generation in the Cenozoic Song Hong basin, Vietnam: a highly prospective basin. *Journal of Asian Earth Sciences*, **17**, 269–294.

Owen, R. B. 2005. Modern fine-grained sedimentation – spatial variability and environmental controls on an inner pericontinental shelf, Hong Kong. *Marine Geology*, **214**, 1–26.

Owen, R. B., Neller, R. J., Shaw, R. & Cheung, P. C. T. 1998. Late Quaternary environmental changes in Hong Kong. *Palaeogeography, Palaeoclimatology, Palaeoecology*, **138**, 151–173.

Pedoja, K., Shen, J. W., Kershaw, S. & Tang, C. 2008. Coastal Quaternary morphologies on the northern coast of the South China Sea, China, and their implications for current tectonic models: a review and preliminary study. *Marine Geology*, **255**, 103–117.

Peltzer, G. & Tapponnier, P. 1988. Formation and evolution of strike-slip faults, rifts, and basins during the India-Asia Collision: an experimental approach. *Journal of Geophysical Research*, **93**, 15 085–15 117.

Pubellier, M. & Chan, L. S. 2006. *Morphotectonic Map of Cenozoic Structures of South China – North Vietnam Coastal Region*. Ecole Normale Supérieure, Paris, France, and The University of Hong Kong, Hong Kong.

Rangin, C., Klein, M., Roques, D., Le Pichon, X. & Trong, L. V. 1995. The Red River fault system in the Tonkin Gulf, Vietnam. *Tectonophysics*, **243**, 209–222.

Rohling, E. J., Grant, K., Hemleben, C. H., Siddall, M., Hoogakker, B. A. A., Bolshaw, M. & Kucera, M. 2007. High rates of sea-level rise during the last interglacial period. *Nature Geoscience*, **1**, 38–42.

Siddall, M., Rohling, E. J., Almogl-Labin, A., Hemleben, C. H., Melschner, D., Schmelzer, I. & Smeed, D. A. 2003. Sea-level fluctuations during the last glacial cycle. *Nature*, **423**, 853–858.

Smith, D. E., Harrison, S., Firth, C. R. & Jordan, J. T. 2011. The early Holocene sea level rise. *Quaternary Science Reviews*, **30**, 1846–1860.

STIRLING, C. H., ESAT, T. M., LAMBECK, K. & MCCULLOCH, M. T. 1998. Timing and duration of the Last Interglacial: evidence for a restricted interval of widespread coral reef growth. *Earth and Planetary Sciences Letters*, **160**, 745–762.

SUN, X. & WANG, P. 2005. How old is the Asian monsoon system? Palaeobotanical records from China. *Palaeogeography, Palaeoclimatology, Palaeoecology*, **222**, 181–222.

SUN, X., LUO, X., HUANG, F., TIAN, J. & WANG, P. 2003. Deep-sea pollen from the South China Sea: Pleistocene indicators of East Asian monsoon. *Marine Geology*, **201**, 97–118.

SUN, Z., ZHOU, D., ZHONG, Z., ZENG, Z. & WU, S. 2003. Experimental evidence for the dynamics of the formation of the Yinggehai basin, NW South China Sea. *Tectonophysics*, **372**, 41–58.

SUN, Z., ZHONG, Z. ET AL. 2008. Dynamics analysis of the Baiyun Sag in the Pearl River Mouth Basin, north of the South China Sea. *Acta Geologica Sinica*, **82**, 73–83.

TAMURA, T., SAITO, Y. ET AL. 2009. Initiation of the Mekong River delta at 8 ka: evidence from the sedimentary succession in the Cambodian lowland. *Quaternary Science Reviews*, **28**, 327–344.

TANABE, S., HORI, K., SAITO, Y., HARUYAMA, S., VU, V. P. & KITAMURA, A. 2003. Song Hong (Red River) delta evolution related to millennium-scale Holocene sea-level changes. *Quaternary Science Reviews*, **22**, 2345–2361.

TANABE, S., SAITO, Y., VU, Q. L., HANEBUTH, T. J. J., NGO, Q. L. & KITAMURA, A. 2006. Holocene evolution of the Song Hong (Red River) delta system, northern Vietnam. *Sedimentary Geology*, **187**, 29–61.

TAYLOR, B. & HAYES, D. E. 1983. Origin and history of the South China Sea basin. *AGU Geophysical Monograph Series, The Tectonic and Geologic Evolution of Southeast Asian Seas and Islands: Part 2*, **27**, 23–56.

TRAN, N., NGO, Q. T., DO, T. V. T., NGUYEN, D. M. & NGUYEN, V. V. 1991. Quaternary sedimentation of the principal deltas of Vietnam. *Journal of Southeast Asian Earth Sciences*, **6**, 103–110.

TRAN, N., NGUYEN, T. L., DIHN, X. T., PHAM, N. H. V., NGUYEN, H. S. & TRAN, T. T. N. 2007. Quaternary sedimentary cycles in relation to sea level change in Vietnam. *VNU Journal of Science, Earth Sciences*, **23**, 235–243.

WANG, M. Y. 2014. *Quaternary Palaeoenvironmental Records and MIS 5 SST Reconstruction from Sanya and Qinglangang of Hainan Island (China)*. PhD thesis, Sun Yat-Sen University, Guangzhou, China.

WOODROFFE, C. D. 2000. Deltaic and estuarine environments and their late Quaternary dynamics on the Sunda and Sahul shelves. *Journal of Asian Earth Sciences*, **18**, 393–413.

YAN, Y., CARTER, A., XIA, B., GE, L., BRICHAU, S. & HU, X. 2009. A fission-track and (U/Th)/He thermochronometric study of the northern margin of the South China Sea: an example of a complex passive marine. *Tectonophysics*, **474**, 584–594.

YIM, W. W.-S. 1994. Offshore Quaternary sediments and their engineering significance in Hong Kong. *Engineering Geology*, **37**, 31–50.

YIM, W. W.-S., IVANOVICH, M. & YU, K.-F. 1990. Young age bias of radiocarbon dates in pre-Holocene marine deposits of Hong Kong and implications for Pleistocene stratigraphy. *Geo-Marine Letters*, **10**, 165–172.

YIM, W. W.-S., HUANG, G., FONTUGNE, M. R., HALE, R. E., PATERNE, M., PIRAZZOLI, P. A. & RIDLEY THOMAS, W. N. 2006. Postglacial sea level changes in the northern South China Sea continental shelf: evidence for a post-8200 calendar year meltwater pulse. *Quaternary International*, **145–146**, 55–67.

YIM, W. W.-S., HILGERS, A., HUANG, G. & RADTKE, U. 2008. Stratigraphy and optically stimulated luminescence dating of subaeriallyexposed Quaternary deposits from two shallow bays in Hong Kong, China. *Quaternary International*, **183**, 23–39.

YOKOYAMA, Y., LAMBECK, K., DE DECKKER, P., JOHNSTON, P. & FIELD, L. K. 2000. Timing of the last glacial maximum from observed sea-level minima. *Nature*, **406**, 713–716.

YOKOYAMA, Y., ESAT, T. & LAMBECK, K. 2001. Coupled climate and sea-level changes deduced from Huon Peninsula coral terraces of the last ice age. *Earth and Planetary Science Letters*, **193**, 579–587.

ZHANG, W. Y. (ed.) 1983. *The 1:5 000 000 Structural Geology Map for China and the Marginal Seas*. Science Publishers, China.

ZHAO, Z., SUN, Z., XIE, H., YAN, C. & LI, Y. 2011. Cenozoic subsidence and lithospheric stretching deformation of the Baiyun deepwater area. *Chinese Journal of Geophysics*, **54**, 1161–1168.

ZHOU, D., RU, K. & CHEN, H. 1995. Kinematics of Cenozoic extension on the South China Sea continental margin and its implications for the tectonic evolution of the region. *Tectonophysics*, **251**, 161–177.

ZHU, M., GRAHAM, S. & MCHARGUE, T. 2009. The Red River Fault zone in the Yinggehai Basin, South China Sea. *Tectonophysics*, **476**, 397–417.

ZONG, Y. 1987. Depositional cycles of the Quaternary strata in the Han River Delta, south China. *Tropical Geography (China)*, **7**, 117–127.

ZONG, Y. 1992. Postglacial stratigraphy and sea-level changes in the Han River Delta, China. *Journal of Coastal Research*, **8**, 1–28.

ZONG, Y., YIM, W. W.-S., YU, F. & HUANG, G. 2009*a*. Late Quaternary environmental changes in the Pearl River mouth region, China. *Quaternary International*, **206**, 35–45.

ZONG, Y., HUANG, G., SWITZER, A. D., YU, F. & YIM, W. W.-S. 2009*b*. An evolutionary model for the Holocene formation of the Pearl River delta, China. *The Holocene*, **19**, 129–142.

ZONG, Y., HUANG, K. ET AL. 2012. The role of sea-level rise, monsoonal discharge and the palaeo-landspace in the early Holocene evolution of the Pearl River delta, southern China. *Quaternary Science Reviews*, **54**, 77–88.

ZONG, Y., ZHENG, Z., HUANG, K., SUN, Y., WANG, N., TANG, M. & HUANG, G. 2013. Changes in sea level, water salinity and wetland habitat linked to the late agricultural development in the Pearl River delta plain of China. *Quaternary Science Reviews*, **70**, 145–157.

Major sinks of the Changjiang (Yangtze River)-derived sediments in the East China Sea during the late Quaternary

SHOUYE YANG[1]*, LEI BI[1], CHAO LI[1], ZHONGBO WANG[2] & YANGUANG DOU[2]

[1]*State Key Laboratory of Marine Geology, Tongji University, Shanghai 200092, China*

[2]*The Key Laboratory of Marine Hydrocarbon Resources and Environmental Geology, Ministry of Land and Resources, Qingdao Institute of Marine Geology, Qingdao 266071, China*

**Corresponding author (e-mail: syyang@tongji.edu.cn)*

Abstract: The East China Sea (ECS) is a typical marginal sea located between the Eurasian continent and west Pacific Ocean. In this study, we review state-of-the-art research progress on the possible sinks of the Changjiang-derived sediments in the ECS during the late Quaternary. The major sinks of these sediments in the ECS are on the outer shelf and the Okinawa Trough during the last glacial maximum corresponding to a lowstand of sea level. During the deglacial marine transgression, the gently dipping shelf was rapidly inundated and strong tides prevented fine sediment from deposition on the open shelf, resulting in the development of a unique tidal sand ridge system. With sea level reaching the present situation and the modern marine environment being completed in the early Holocene, the Changjiang sediments mostly accumulated in the river's estuary to build a large delta, with only a fraction reaching the inner shelf and coastal embayments. The late-Quaternary changes in monsoon-climate-induced river flux, sea level and oceanic circulation primarily controlled the source-to-sink transport of the Changjiang sediments in the ECS, and further determined the stratigraphic framework and sedimentary facies on the shelf.

The collision between the Indian and Eurasian Plates and the opening of the west Pacific marginal seas in the Cenozoic led to events and processes that have become some of the key scientific issues for the Earth sciences and global change studies, in particular these include the Tibet uplift, monsoon evolution, continental weathering and sediment transport from land to ocean (Raymo & Ruddiman 1992). Flux and fate of terrestrial sediment in marginal seas have been the major scientific objectives of international research programmes, such as Land Ocean Interaction in the Coastal Zone (LOICZ) and MARGINS Program (Syvitski *et al.* 2003).

The East China Sea (ECS) is located in the East Asian continental margin, linking the largest continent, Eurasia, with the largest ocean, the Pacific. It is characterized by a broad continental shelf, huge terrigenous sediment input from surrounding rivers, striking land–sea interaction and palaeoenvironmental changes during the late Quaternary (Milliman & Farnsworth 2011; Li *et al.* 2014; Yang *et al.* 2014). Basically, two types of river systems and sediment source-to-sink processes dominate the sedimentation on the ECS shelf (Fig. 1). One is represented by the Changjiang (Yangtze River), one of the largest rivers in the world, and another by the small mountainous rivers on Taiwan. Both river systems deliver very large volumes of sediment directly into the ECS, despite their striking differences in drainage basin areas, weathering regimes, and sediment erosion and transport processes. The Changjiang has developed a complicated tributary system and one of the largest deltas in the world and thus the sediment transfer from the large catchment to the sea depends on many factors. In comparison, the strong land erosion and fast sediment transfer especially during extreme weather events characterize the small mountainous rivers in Taiwan (Kao & Milliman 2008; Liu *et al.* 2013). Over the geological past, another large river, the Huanghe (Yellow River), might have also emptied a large volume of sediment into the ECS (Yang *et al.* 2002, 2003; Liu *et al.* 2010; Li *et al.* 2014). In this sense, the ECS can be regarded as a typical river-dominated pericontinental sea, which makes it an ideal natural laboratory for the study of land–sea interaction during the late Quaternary.

The ECS links the west Pacific by the Okinawa Trough in the east which has a large section >1000 m deep and with a maximum depth of 2716 m at its southeastern extent near Taiwan. The Okinawa Trough is regarded as an incipient intracontinental basin formed behind the Ryukyu arc–trench system where the Philippine Sea Plate is subducting under the Eurasia Plate (Clift *et al.* 2003; Fig. 1). It has continuously received terrigenous sediments from the surrounding rivers and thus became a major depocenter of the ECS during the late Quaternary.

From: Clift, P. D., Harff, J., Wu, J. & Qui, Y. (eds) 2016. *River-Dominated Shelf Sediments of East Asian Seas*. Geological Society, London, Special Publications, **429**, 137–152.
First published online October 2, 2015, updated May 26, 2016, http://doi.org/10.1144/SP429.6

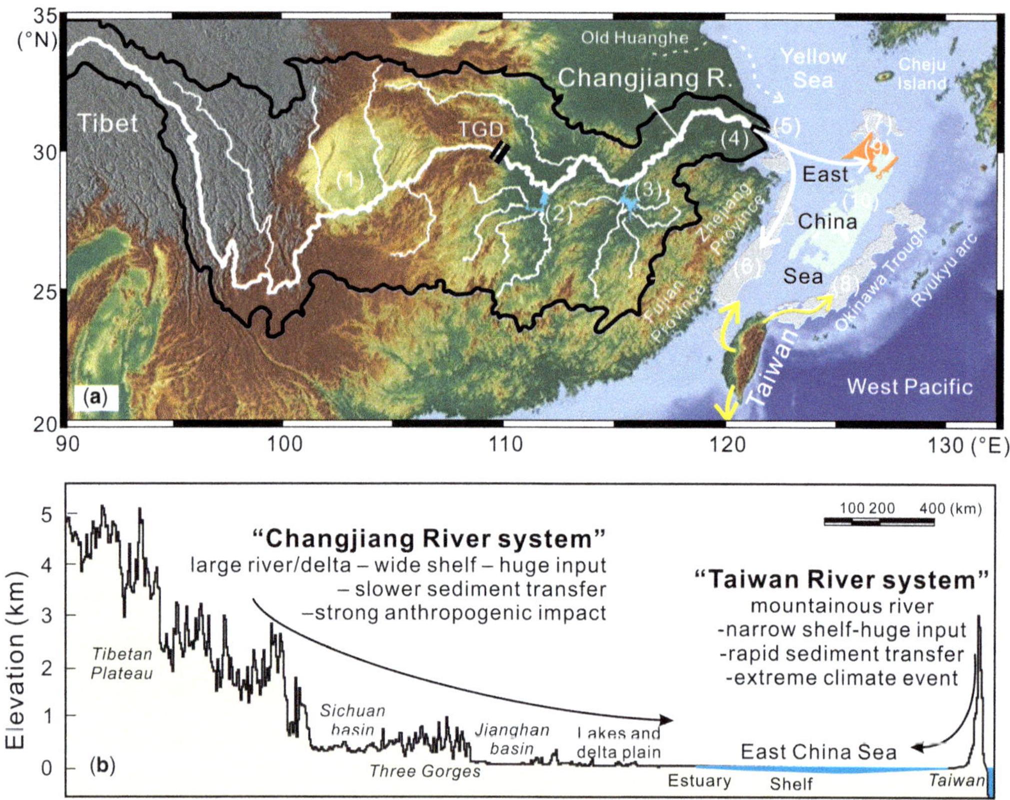

Fig. 1. A schematic model showing two types of river systems (Changjiang v. Taiwan mountainous rivers) and sediment transport processes in the ECS. (**a**) Possible sinks of the Changjiang sediments in the river basin and the ECS are indicated by numbers (1)– 8). Mud areas ((5)–(8)) in the ECS are indicated in light grey. (1) Sichuan Basin; (2) Dongting Lake and Jianghan Basin; (3) Poyang Lake; (4) delta plain; (5) the mud off the Changjiang Estuary; (6) the inner shelf mud; (7) the mud in the SW of Cheju Island (SWCIM); (8) mud deposition in the Okinawa Trough; (9) Hupijiao Rise; (10) Zhedong and Xihu depressions (modified after Li (2008)); TGD refers to Three Gorges Dam. (**b**) Cross section showing the contrasting topography and relief between the Changjiang basin, ECS shelf and Taiwan.

The oceanic current system in the ECS is dominated by the northward intrusion of the Kuroshio Current and its branch, the Taiwan Warm Current, and the southward Zhe-Min (Zhejiang and Fujian) Coastal Current (Fig. 2; Guan 1994; Lee & Chao 2003). In summer, the Changjiang Diluted Water plays an important role in the shelf circulation in the northern ECS (Hu & Li 1993). The complicated oceanic circulation and tidal currents, as well as sea-level variability, primarily determine the distribution and dispersal patterns of terrigenous sediments in the ECS, forming unique muddy and sandy depositional bodies on the shelf and the trough (Hu 1984; Qu & Hu 1993; Gao & Collins 2014; Li *et al.* 2014; Yang *et al.* 2014). Several muddy patches developed off the Changjiang estuary, on the SW shelf of Cheju Island and the inner shelf along the Zhejiang-Fujian coastal area, as well as on the west slope of the Okinawa Trough (Niino & Emery 1961; Yang & Milliman 1983; Hu 1984; Nittrouer *et al.* 1984; Beardsley *et al.* 1985; DeMaster *et al.* 1985; Qin *et al.* 1987; Milliman *et al.* 1989; Qu & Hu 1993; Guo *et al.* 2002; Lim *et al.* 2007; J. P. Liu *et al.* 2006, 2007; Xu *et al.* 2009, 2012; Dou *et al.* 2010*a*, *b*, 2012; Chen *et al.* 2011; Horng & Huh 2011; Gao & Collins 2014; Li *et al.* 2014; Yang *et al.* 2014). The sheet sand that blankets the ECS shelf has been considered to be made of relict deposits (Niino & Emery 1961; Emery 1968), primarily formed at the last glacial maximum (LGM) or deglacial period with a sea-level lowstand (Yang & Sun 1988; Yang 1989; Liu 1997; Berné *et al.* 2002; Z. X. Liu *et al.* 2007; Li *et al.* 2014; Wang *et al.* 2014; Yang *et al.* 2014). As a whole, the unique muddy and sandy sedimentary systems provide a typical example of the product of sediment accumulation on a broad shelf with large quantities of terrigenous sediment supply (Gao 2013).

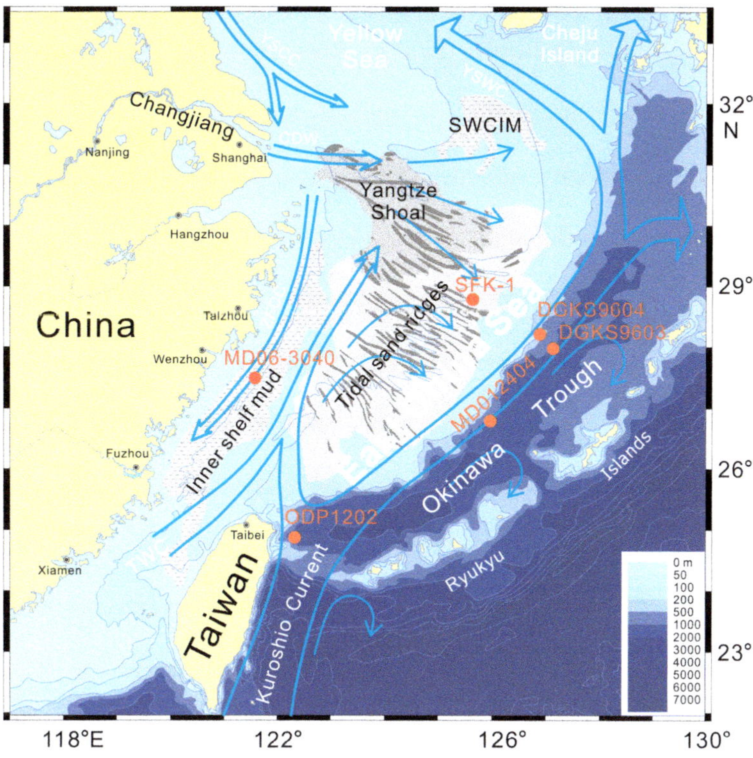

Fig. 2. A schematic map showing the ECS with its typical sedimentary systems. The positions of cores DGKS9604, DGKS9603, SFK-1, ODP1202, MD012404 and MD06-3040 are also indicated. ZFCC, Zhenjiang and Fujian Coastal Current; TWC, Taiwan Warm Current; YSCC, Yellow Sea Coastal Current; YSWC, Yellow Sea Warm Current; CDW, Changjiang Diluted Water. The dotted area showing the muddy sediment systems in the ECS is from Qin *et al.* (1987) and Li *et al.* (2005*b*). The shelf circulation was modified after Guan (1994) and Lee & Chao (2003).

Over the last decade, the provenances of these sediments and related depositional processes and palaeoenvironmental changes have been extensively investigated, but the sediment origins require further clarification, even though oceanographic, sedimentological, mineralogical and geochemical methods have been applied in the numerous studies mentioned above. In addition, appropriate interpretation of the late-Quaternary sedimentary record greatly depends on an improved understanding of the formation of these sedimentary systems and the reliable reconstruction of river–sea interaction at variable temporal and spatial scales (Gao 2013; Gao & Collins 2014; Li *et al.* 2014; Yang *et al.* 2014). In view of this, a fundamental scientific question for marine geological study of the river-dominated ECS is the transport and dispersal patterns of the fine-grained fluvial sediments from the surrounding Eurasian continent and Taiwan to the broad shelf.

In this study, we review state-of-the-art research progress on the possible sinks of the Changjiang-derived sediments in the ECS during the late Quaternary. In particular, the existing mineralogical and geochemical approaches for provenance discrimination of the fluvial sediments in the ECS were compiled with the major aim of providing more robust constraints on sediment source-to-sink transport.

The river setting and key questions on river–sea interaction during the late Quaternary

The Changjiang originates from the Tibetan Plateau and its catchment is primarily situated on the Yangtze Craton, with a basin area of *c.* 1.8×10^6 km^2, more than a fifth of the land area of China. Geologically, the Changjiang basin comprises complex rock types, including Palaeozoic carbonate rocks, Jurassic red sandstone, Mesozoic igneous rocks, Palaeozoic marine and Quaternary fluvio-lacustrine sedimentary rocks (Yang *et al.* 2004). Geographically, the Changjiang catchment covers variable topography and tectonic relief, with mountainous terrains dominant in the upper valley and lake basins and a delta plain in the middle and lower valley (Fig. 1), which may considerably influence the source-to-sink transport of river sediment.

The monsoon climate characterizes the catchment, resulting in uneven rainfall over the basin at seasonal scale, with a mean annual precipitation and temperature of 1100 mm and 15°C respectively (Yang *et al.* 2004). Based upon long-term hydrologic observation, the water and sediment discharges of the Changjiang average at 8964×10^8 km^3 a^{-1} and 390×10^6 t a^{-1} respectively, which makes the Changjiang the largest terrigenous source contributing to the ECS. Nevertheless, the sediment flux of the Changjiang has been rapidly decreasing since the impoundment of the Three Gorges Reservoir in 2003, averaging at only *c.* 155×10^6 t a^{-1} over the last ten years (2003–2013) (Changjiang Water Resources Committee 2013).

In the present day, a major part of the Changjiang sediment accumulates in the river's estuary to sustain a large delta, while some fine-grained sediment escapes from being trapped in the estuary and is transported southeastwards by the coastal current, finally developing into a unique mud belt on the inner shelf along the coasts of Zhejiang and Fujian Provinces (Yang & Milliman 1983; Nittrouer *et al.* 1984; DeMaster *et al.* 1985; Milliman *et al.* 1989; J .P. Liu *et al.* 2006, 2007; Xu *et al.* 2009, 2012; Gao 2013; Gao & Collins 2014; Li *et al.* 2014; Yang *et al.* 2014). Nevertheless, the influence of the Changjiang on the sedimentation of the ECS has not been resolved yet, and some fundamental questions to be answered include: (1) whether the palaeo-Changjiang once reached the outer shelf of the ECS by incising the broad shelf during the LGM; (2) if so, how to recognize the pathways and sediment distribution of the palaeo-Changjiang on the shelf; (3) how to differentiate the Changjiang sediment in the ECS from the terrigenous sediments from other rivers and aeolian input; (4) how to quantitatively estimate the sediment influx from the Changjiang to the shelf and to relate the river discharges to natural processes and anthropogenic activities in the catchment. Therefore, the gap in the estimation of the sediment budgets and dispersal pattern on the ECS shelf is mainly due to poor recognition of the river-flow pathways, and uncertainty on past sediment fluxes from the ancient Changjiang (Yang *et al.* 2003; J. P. Liu *et al.* 2007; Xu *et al.* 2012).

It is obvious that provenance identification of marine sediment is the prerequisite for the reliable reconstruction of the sediment source-to-sink transport from land to sea and of the palaeoenvironmental changes. Mineralogical and geochemical compositions of detrital sediments in fluvial and marine environments are basically determined by source rock type (petrographic composition in drainage basin), physical and chemical weathering, hydrodynamic sorting, depositional and post-depositional alterations, and human activities as well (Yang *et al.* 2003). As introduced above, although various approaches have been applied to the provenance study of the marine sediments in the ECS and Yellow Sea, the source-to-sink transport pathways of the Changjiang sediment in East China's marginal seas during the late Quaternary remain controversial. One major difficulty of the sediment provenance study lies in the poor characterization of the Changjiang-derived sediment that is discharged into the sea and the lack of reliable discrimination of the Changjiang sediment from the other river sediments in dynamic marine depositional environments.

For instance, a major challenge in reconstructing the source-to-sink transport of the Changjiang sediment in the ECS is the differentiation between the Changjiang- and Huanghe-derived sediments on the shelf and in the Okinawa Trough. As one of the largest rivers in the world, the Huanghe has high sediment load, reaching 10^9 t a^{-1} before the 1980s, and it once flowed into the southern Yellow Sea (during the period AD 1128–1855), which resulted in the formation of a large delta. Due to the shift of the course of the Huanghe northwards to North China, the abandoned Huanghe Delta in the southwestern Yellow Sea has been experiencing strong erosion, resulting in a large supply of the Huanghe-derived sediment into the northern ECS (Milliman *et al.* 1985*a*; Saito 1998; Yang *et al.* 2003; Liu *et al.* 2010; Gao & Collins 2014; Zhou *et al.* 2014).

In short, the river (Changjiang) and sea interaction on the wide continental shelf of the ECS was complicated by the changes in river flux, sea level and oceanic circulation during the last glacial period. Undoubtedly, more systematic seismic investigations combined with studies on high-resolution sediment cores are needed in order to better constrain the source-to-sink process of the fluvial sediments on the epicontinental shelf.

Discussion

Whether the palaeo-Changjiang entered the ECS during the LGM

As introduced above, there still exist different opinions on the localities and shifts of the palaeo-Changjiang river course and its mouth on the ECS shelf during the LGM. Two hypotheses have been proposed for the pathway of the palaeo-Changjiang during the last glaciation (Xiao *et al.* 2004). Some studies suggest that the palaeo-Changjiang entered a palaeolake north of Jiangsu Province and thus did not enter the sea during the LGM because the extremely arid climate caused low-water and sediment discharge at that time (Zhao 1984). Furthermore, the strong winter monsoon during the LGM might have introduced desertification of the open ECS shelf and thus prevented fluvial deposition (Zhao 1984).

A second possibility is that the palaeo-Changjiang incised the continental shelf, formed complicated incised river networks (Fig. 3), and delivered its sediment into the ECS shelf, Okinawa Trough or the Yellow Sea, during the lowest stand of sea level at the LGM (Qin *et al.* 1987; F. Li *et al.* 1993; C. X. Li *et al.* 2002; G. X. Li *et al.* 2005*a*, *b*, 2014; Li & Zhang 1995; Xia & Liu 2001; Warren & Bartek 2002; Wellner & Bartek 2003; K. Liu *et al.* 2009; Xu & Oda 2009; Dou *et al.* 2010*a*, *b*, 2012; Wang *et al.* 2013, 2014). The large sheet sand distributed on the mid to outer ECS shelf occupies an area >10 000 km^2, and is primarily composed of medium and fine sands with abundant fragments of microfossils. Early in 1968, Emery defined this sheet sand as 'relict sediment' formed under low sea level or the deglacial transgression (Niino & Emery 1961; Emery 1968). Radiocarbon dating results confirmed the deposition of the sheet sand during the last glaciation, yielding the ages of 20–15 ka (Qin *et al.* 1987). This sheet sand was also called the 'Yangtze Shoal' apparently suggesting a genetic relationship with the Changjiang (Yangtze River) sediment. It was regarded as as a palaeodeltaic or estuarine deposit while the palaeo-Changjiang river mouth extended to the mid–outer shelf with falling sea level (Qin *et al.* 1987; Liu 1997; Li *et al.* 2005*a*, *b*; Gao 2013; Wang *et al.* 2013). During the last deglacial sea-level rise, the pre-existing deposits were reworked by rotatory tidal currents, with the top layer of the old strata being transformed into tidal sand (Gao 2013).

Nevertheless, Zhao (1984) suggested that the Yangtze Shoal sand was not a deltaic deposit that had originated from the palaeo-Changjiang, but the weathering product of the Tertiary bedrock that was sub-aerially exposed on the open shelf. Xia & Liu (2001) further suggested that the tidal sheet sand on the mid–outer ECS shelf was formed by the postglacial transgression, and the palaeo-Changjiang probably flowed through the southern Yellow Sea and delivered its sediment to the northern Okinawa Trough during the LGM.

Therefore, direct evidence is still lacking that would enable this sheet sand to be related to the palaeo-Changjiang sediment. Furthermore, consensus on the pathway of the palaeo-Changjiang on the shelf during the last glacial period has never been reached. Li *et al.* (2001) and Yang *et al.* (2002) suggest that the Changjiang has always flowed into the ECS, without significant shift of river pathway since the LGM, whereas other researchers suggest that the Changjiang once debouched its sediments directly into the south Yellow Sea during the sea-level lowstand, forming a palaeo-Changjiang delta therein (Qin *et al.* 1987; Yang 1989; Zhu & An 1993). Thus far, most studies have suggested that the palaeo-Changjiang entered the ECS and left complicated river channels on the shelf (Chen & Yang 1991; Liu 1997; Warren & Bartek 2002; Wellner & Bartek 2003; Li *et al.* 2005*a*, *b*; Liu *et al.* 2009; Wang *et al.* 2013, 2014). A synthetic study by Li *et al.* (2005*a*) demonstrated that the incised palaeoriver channel system of the Changjiang during the last glaciation was along the Zhedong and Xihu depressions in the SE of the Hupijiao Rise (Fig. 1; Li 2008), but did not directly enter the Okinawa Trough, despite the surface-water freshening of the northern trough. Six palaeoriver-channel systems on the continental shelf were recognized as the major meandering tributaries of the palaeo-Changjiang (Fig. 3). Wellner & Bartek (2003) also revealed the incised valley system of the palaeo-Changjiang on the ECS shelf during the early LGM, and tidal sheet sands formed in the late LGM.

Recent studies by Wang *et al.* (2013, 2014) have revealed that the palaeo-Changjiang channels might have incised the ECS shelf during the LGM and directly supplied fluvial sediments to the outer shelf under the depositional environment of strong tidal reworking. This *c.* 3 m-thick tidally influenced palaeofluvial sediment can be clearly recognized on high-resolution seismic reflections and from the combined evidence from lithological character, microfossil assemblages, element geochemistry and chronology of borehole SFK-1 on the outer shelf (Fig. 4). This palaeofluvial sandy sediment, formed during the LGM, is thicker than that previously reported (*c.* 1 m) by Tang (1996), showing a fining-upward succession with an erosional surface at the bottom. The fine–medium sandy bedding with muddy flaser bedding and mud clasts all suggest typical river-channel deposition. Based on the high-resolution seismic profile, the LGM deposition is characterized by chaotic and irregular reflections, cutting across the underlying strata and locally pinching out on the outer shelf, which is interpreted as fluvial–estuarine facies of a lowstand systems tract (Wang *et al.* 2013, 2014). Interestingly, small fragments of marine microfossils often occur in this facies, showing an overall upward-increasing trend in its abundance. Most of them are heavily worn, and mixed with a small amount of larger and intact bodies of some benthic

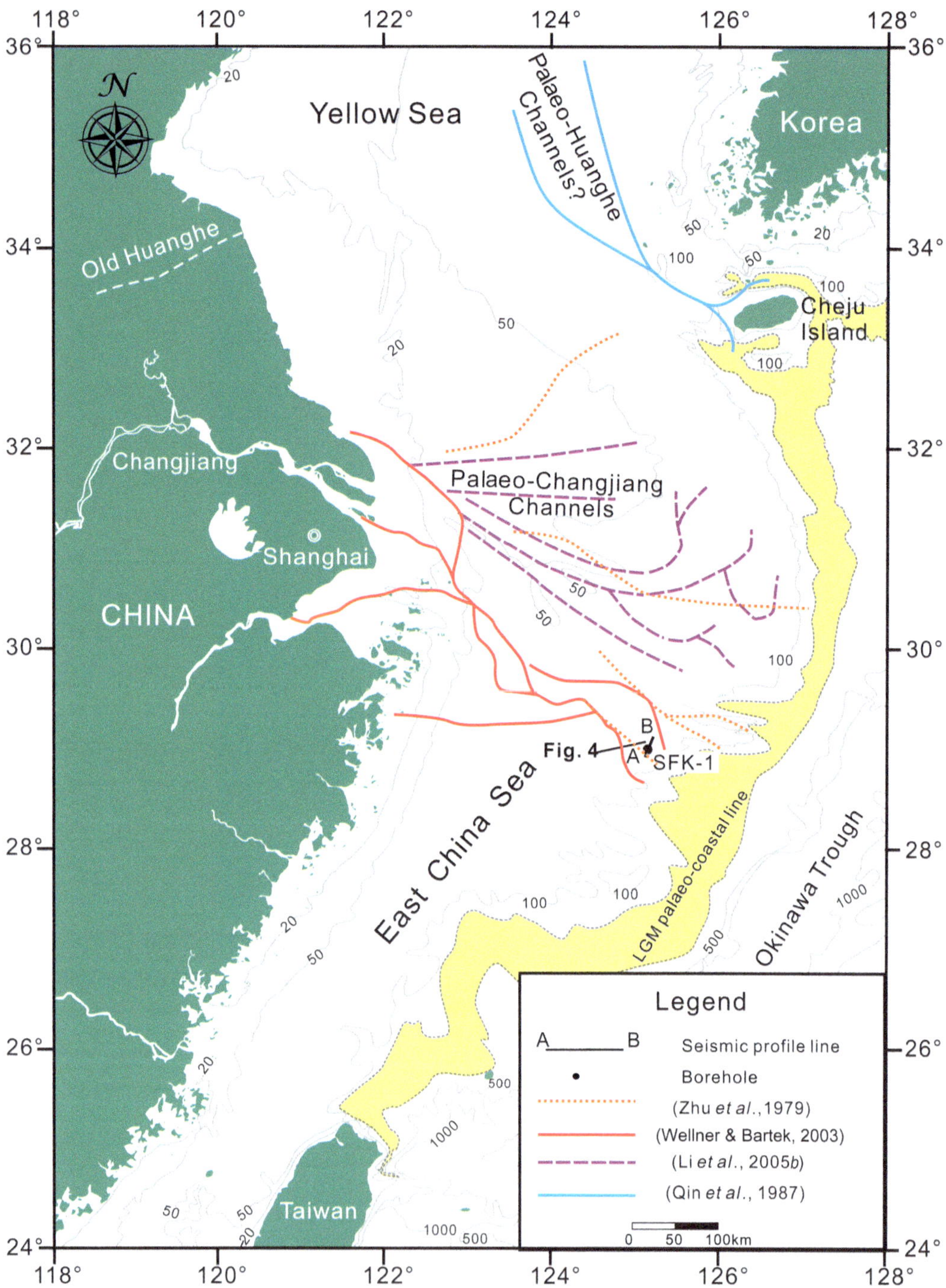

Fig. 3. Map showing the ECS with the possible pathways of the palaeo-Changjiang and palaeo-Huanghe channels (Zhu *et al.* 1979; Zhao 1984; Qin *et al.* 1987; Zhu & An 1993; Wellner & Bartek 2003; Li *et al.* 2005*b*) and the palaeocoastal line at the LGM (yellow) (Saito *et al.* 1998). The locations of seismic profile A–B and borehole SFK-1 are indicated.

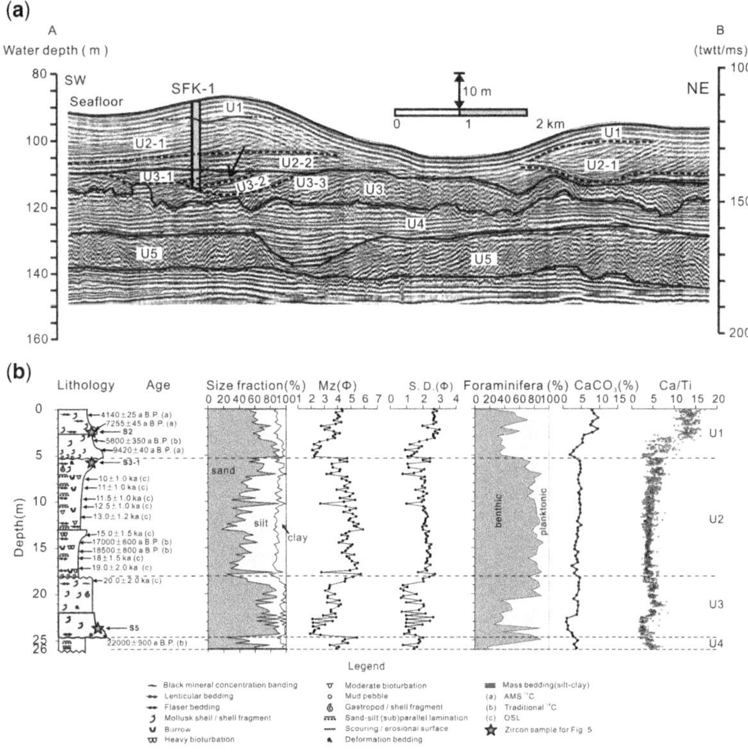

Fig. 4. Palaeochannel deposits accumulated in the outer shelf of the ECS during the LGM. (**a**) Depositional units (U1–U5) determined from a high-resolution seismic profile A–B across borehole SFK-1 on the ECS outer shelf (Wang *et al.* 2013) (positions in Fig. 3). The solid lines are unit boundaries, and the dotted lines are sub-unit boundaries. The yellow lines indicate the depositional units formed at the LGM. (**b**) Downcore variations of lithology, grain-size parameters, relative percentages of benthic and planktonic foraminifera, and Ca/Ti ratios in borehole SFK-1 (Wang 2014; Wang *et al.* 2014). Mz,mean grain size; S.D.,standard deviation; Ski, skewness; KG, kurtosis. U1, tidal sand ridges formed in the Holocene; U2, estuarine–tidal flat formed at *c.* 18–10 ka; U3, tidal-influenced palaeofluvial deposition formed at LGM; U4, nearshore deposition formed in the late marine isotope stage 3.

foraminifera. Based on the microfossil assemblage, and sedimentary structures and composition, this facies was interpreted to be tide-influenced fluvial facies; the fluvial discharge and process might have been weak compared to the present situation because of the cold and arid environment through the LGM (Saito *et al.* 1998; Wellner & Bartek 2003).

Geochronology of detrital zircon provides a more powerful and reliable tool for the provenance identification of sandy sediments than traditional heavy mineralogy, because the former is less influenced by hydrodynamic sorting in fluvial and marine environments. Three sandy samples taken from the early–middle Holocene and LGM deposits of borehole SFK-1 have very similar U–Pb age patterns of detrital zircons to the modern Changjiang sediment (Fig. 5). They all show distinct peak ages at *c.* 300–200 and 1000–800 Ma (Yang *et al.* 2012; Wang 2014), which are very different from the zircon ages of the Huanghe sediment with high proportions of old zircons (2500–2300 Ma) (Yang *et al.* 2009). Therefore, we suggest that the sandy palaeochannel sediments deposited on the ECS outer shelf during the LGM were probably derived from the palaeo-Changjiang.

The Okinawa Trough has been regarded as a major depocenter of the ECS and might have continuously received terrigenous sediment from the East Asian continent and Taiwan during the late Quaternary. Sedimentological and geochemical studies of several cores (e.g. DGKS 9603, 9604, MD012404, ODP Site 1202) suggest that during the last glaciation (*c.* 30–10 ka) the sediment that accumulated in the middle Okinawa Trough was predominantly sourced from the palaeo-Changjiang (Diekmann *et al.* 2008; Dou *et al.* 2010*a*, *b*, 2012; Chen *et al.* 2011), or partly from eastern Taiwan (Chen *et al.* 2011). All of these studies imply that even during the LGM with a severe monsoon

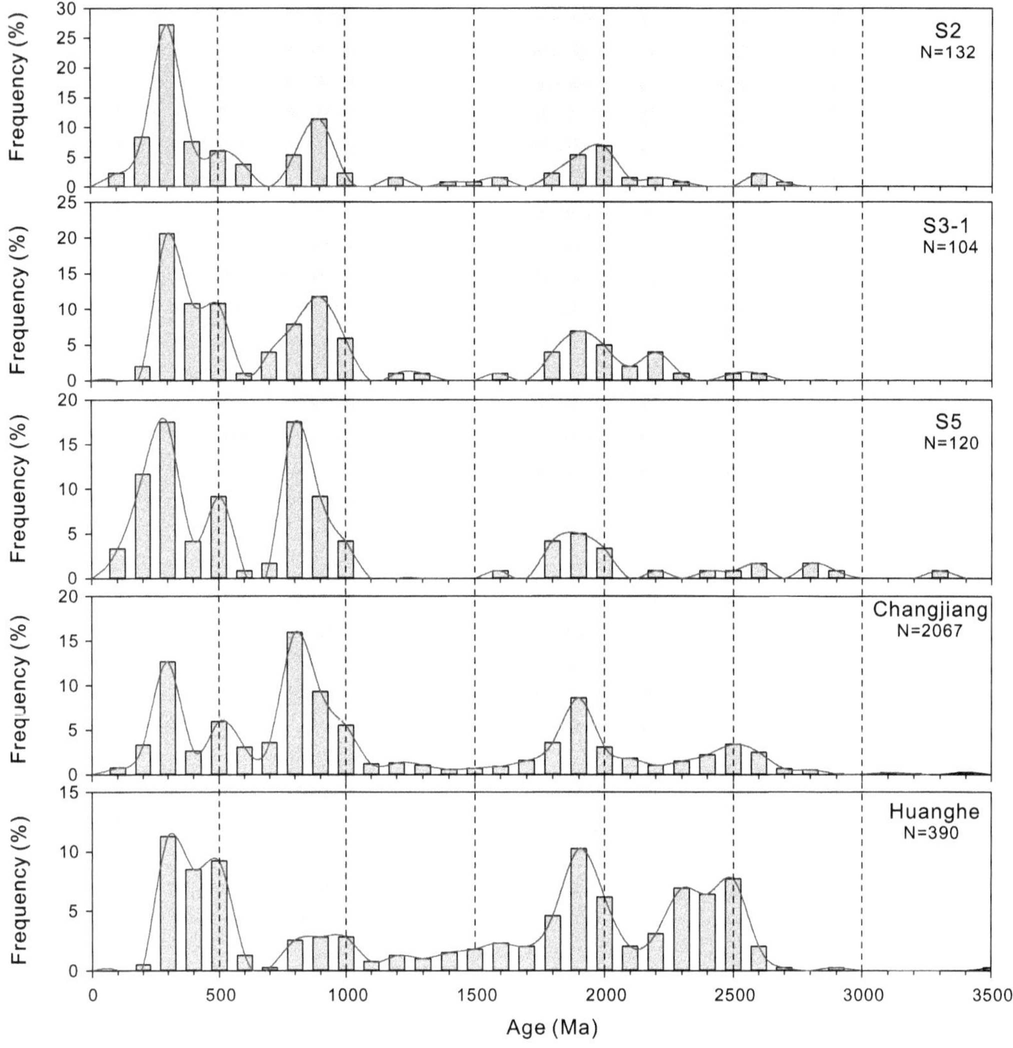

Fig. 5. Comparison of the U–Pb geochronology of detrital zircon grains between the selected sediments from Core SFK-1 (S2, S3-1 and S5) and the Changjiang and Huanghe river sediments. The samples of S2, S3-1 and S5 are respectively collected from the middle and early Holocene and LGM deposits. The locations for these three samples in the core SFK-1 are shown in Figure 4. Data sources: the zircon ages of the core, Changjiang and Huanghe sediments are respectively from Wang (2014), Yang *et al.* (2012) and Yang *et al.* (2009). N indicates the numbers of detrital zircon grains. Note the similarity of zircon geochronology between these core sediments and the Changjiang sediment, suggesting their genetic relationships.

climate, the palaeo-Changjiang river mouth might have extended to the present-day outer shelf, supplying terrigenous sediment both to the shelf itself and even the Okinawa Trough. Therefore, we infer that during the last glaciation with a sea-level lowstand, the outer shelf and the Okinawa Trough could have been the major sink of palaeo-Changjiang sediment into the ECS. In comparison, the inner and middle ECS shelf were exposed during the last glaciation and thus could not be the major depocenters of the palaeo-Changjiang river sediment because of insufficient accommodation space.

The shifts of the palaeo-Changjiang river course during the deglacial transgression

With the onset of last deglacial transgression, the gentle ECS shelf was rapidly flooded by rising seawater, which caused the retreat of the coastline as well as the palaeo-Changjiang river mouth. Correspondingly, the delivery and dispersal patterns of the Changjiang-derived sediment on the ECS shelf were also altered with changing sea level and oceanic circulation. Previous studies suggest that the tidal currents were indeed strong during the postglacial transgression and early periods of the Holocene when sea level was lower than present day (Uehara *et al.* 2002; Uehara & Saito 2003). The palaeo-Changjiang sediment accumulated on the mid–outer shelf during sea-level lowstands might have been reworked and winnowed by strong tidal currents, resulting in the formation of a tidal sand ridge system on the ECS shelf (Z. X. Liu *et al.* 2000, 2007; Berné *et al.* 2002; Yoo *et al.* 2002; Gao & Collins 2014; Li *et al.* 2014).

In the present day and under the present oceanic environment, the tidal sand ridges on the shelf are still in motion because of the strong tidal current (Z. X. Liu *et al.* 2007), unlike the so-called 'moribund' ridges described by Yang (1989). Nevertheless, the present-day sandy sediments from the modern Changjiang can hardly escape the estuary and inner shelf and replenish the sand ridges on the mid–outer shelf. Similar to the tidal sand ridges on the mid–outer ECS shelf, the Yangtze Shoal off the Changjiang river mouth is recognized as a tidal sheet sand (Liu 1997) with sandy sediments derived from the underlying pre-Holocene relict deposit, but with the tidal reworking here being much weaker than in the tidal sand ridges (Gao & Collins 2014).

In addition, during the deglacial sea-level rise, the fine-grained sediment winnowed from the open shelf by strong tide or wave currents might have been transported to the outer shelf and Okinawa Trough by shelf currents (Diekmann *et al.* 2008; Dou *et al.* 2010*a*, *b*, 2012), or to the inner shelf and coastal bays like the Hangzhou Bay mostly by tidal transport (Lin *et al.* 2005; Gao 2013; Zhang *et al.* 2014). Mineralogical and geochemical evidence of the core sediments confirm the accumulation of Changjiang-derived clay in the middle Okinawa Trough during the last deglacial period (Dou *et al.* 2010*a*, *b*, 2012; Chen *et al.* 2011).

The establishment of modern sediment source-to-sink transport of the Changjiang during the mid–late Holocene

With sea level rising to its present position in the early–mid Holocene at *c.* 7 ka, a normal marine environment characterized the ECS (Milliman *et al.* 1985*b*, 1989; Li *et al.* 2014), and modern sedimentary systems started to develop in the Changjiang river mouth and on the shelf. In response to the deceleration of global sea level in the early Holocene, the Changjiang Delta initiated by retaining more than half of the river sediment load in the large estuary, as with other large deltas (Stanley & Warne 1994; Hori *et al.* 2002; Li *et al.* 2002). In the present day, the Changjiang Delta consists of the sub-aerial and subaqueous parts with areas of *c.* 25×10^3 km^2 and 10×10^3 km^2 respectively, and the main body reaching a thickness of *c.* 30–50 m.

Over the last 6–7 kyr, the Changjiang fine-grained sediment that escaped the estuarine trapping was mostly transported southeastward and firstly accumulated in the Hangzhou Bay (Lin *et al.* 2005; Zhang *et al.* 2014), with the remainder being delivered further southwards and deposited on the inner shelf near the Zhejiang and Fujian coasts (Yang & Milliman 1983; Nittrouer *et al.* 1984; DeMaster *et al.* 1985), forming a unique *c.* 800 km-long mud wedge during the last 7 kyr (Figs 1 & 2). Meanwhile, only a small amount of the Changjiang-derived sediment was dispersed eastwards to the open shelf, and hardly reached the shelf beyond 124° E or over a distance of 250–300 km eastwards (Beardsley *et al.* 1985; Milliman *et al.* 1985*b*, 1989; Zhang 1999). This means that the mid–outer shelf was a 'starved shelf' in the sense of Gao & Collins (2014), only the relict sandy deposits occupying the present-day seafloor. It is noteworthy that the construction of the Three Gorges Dam in 2003 has exerted a significant impact on marine deposition in the Changjiang subaqueous delta, causing coarsening and erosion of the pro-delta mud in response to the decline of sediment flux from the Changjiang to the ECS (Yang *et al.* 2007, 2011; Luo *et al.* 2012).

During the postglacial period, the fate of the Changjiang-derived sediment (and Huanghe sediments too) to the ECS varied with changing sea level and estuarine–shelf circulation (Li *et al.* 2014). In consequence, the sediment retention index, defined as the ratio between sediment retention in the estuary and sediment that has escaped to the open sea, also changed with depositional environment and sediment-dynamic conditions (Gao 2013). The modern Changjiang Delta started to form at 7–6 ka and since then >50% of discharged sediment has been retained in its estuary (Li *et al.* 2002). Nevertheless, the sediment retention index in the Changjiang estuary has gradually decreased as the delta front has advanced into deeper waters (Gao & Collins 2014), and/or the Changjiang sediment discharge has declined considerably in recent years, and will in the future (Yang *et al.* 2007, 2011).

Physical oceanographic observation suggests that fine-grained sediment can be transported towards the mid–outer ECS shelf, or even across the entire shelf, to the continental slope (e.g. Okinawa Trough) by shelf currents (Hoshika *et al.* 2003;

Katayama & Watanabe 2003). Recent study further suggests that the bottom turbid layers or sediment gravity flows may be a dominant process for the cross-shelf transport of fine-grained sediment, especially during typhoon events (Li *et al.* 2012). During the early stages of the Changjiang Delta development, fluvial sediments flowing into the estuary were largely dispersed and the delta front slope did not develop, which suggest that sediment gravity flow might not have exist at that time (Gao & Collins 2014). However, the present-day Changjiang river subaqueous delta becomes an ideal area for observational and theoretical studies of sediment gravity-flow-induced transport (Wright & Friedrichs 2006).

Formation and sediment origins of muddy sedimentary systems in the ECS during the Holocene

As unique sedimentary systems on the ECS shelf, several mud zones formed in the Holocene have received considerable attention in terms of their depositional history, sediment provenances and palaeoenvironmental records (Figs 1 & 2). The source-to-sink transport of muddy sediments in the ECS is closely related to the current system driven by the interaction of the East Asian Monsoon and Kuroshio Current (Li *et al.* 2014). The mud patch off the Changjiang estuary, as the pro-delta deposition of the Changjiang subaqueous delta, had its sediment sources dominated by the Changjiang during the late Holocene. Whether and to what extent the palaeo-Huanghe sediments originating from the southwestern Yellow Sea contributed to the deposition of this mud patch are still open questions (Sun *et al.* 2000; Liu *et al.* 2010; Zhou *et al.* 2014).

Another mud patch in the SW of Cheju Island (SWCIM, Yang *et al.* 2003), at the northern margin of the ECS, has also been widely documented; this sediment was suggested to be predominantly derived from the palaeo-Huanghe delta in the southwestern Yellow Sea (Yang & Milliman 1983; DeMaster *et al.* 1985; Lee & Chough 1989; Milliman *et al.* 1989; Alexander *et al.* 1991; Saito 1998; Lim *et al.* 2007; Youn *et al.* 2007; Youn & Kim 2011). This mud patch was first found in 1960s and has an area >10 000 km^2. The SWCIM has been regarded as the distal end of the Huanghe dispersal system in the ECS (DeMaster *et al.* 1985; Alexander *et al.* 1991), and was formed in a counter-clockwise cyclonic eddy during the late Holocene (Hu 1984; Hu & Li 1993; Qu & Hu 1993). Interestingly, recent study by Kim *et al.* (2013) on magnetic properties of the surface sediments in and around this mud patch suggests mixing sources from the Changjiang palaeo-Huanghe Taiwanese rivers, or the NW Pacific and Korean rivers. Nevertheless, the boundaries between different sediment provenances are not always distinct.

The inner-shelf mud wedge of the ECS has been regarded as a good archive for palaeoenvironmental study, given its relatively continuous deposition and dominant sediment source from the Changjiang during the mid–late Holocene. The relatively continuous nature of the mud deposition with uniform sediment grain size and reliable constraints on radiocarbon dates imply that this elongated inner-shelf mud probably formed during the last 7–8 kyr (Xiao *et al.* 2006; Zheng *et al.* 2010). This suggests that the export of Changjiang sediment from its river mouth towards the open shelf probably occurred at an earlier stage, and not just within the last 2 kyr (Liu *et al.* 2006; Gao 2013; Gao & Collins 2014). The completion of the modern shelf-circulation system and favorable monsoon climate since the middle Holocene might account for the dispersal of the Changjiang-derived fine sediment to the open shelf, resulting in the formation of shelf mud.

Nevertheless, whether a part of the fine sediment in the inner-shelf mud belt was derived from the ‘older material’ previously accumulated on the mid–outer shelf during the sea-level lowstand (Gao & Collins 2014), or from the abandoned Huanghe Delta in the southern Yellow Sea, and local rivers in SE China and Taiwan remains to be clarified. Recently, Gao & Collins (2014) suggested that during the Holocene sea-level rise, intense tidal currents and tidally induced landward transport of fine-grained sediment might have played an important role in the formation of the shelf and coastal mud deposits. In addition, Shao (2012) investigated the sediment geochemistry of Core MD06-3040 (27° 43.37′ N, 121° 46.88′ E, 46 m seawater depth) retrieved from the inner-shelf mud area, and suggested that over the last 7 kyr, mud deposition was dominated by the Changjiang-borne sediment, while the sudden change in geochemical proxies (e.g. Al/Ti, Li/Sc, Co/Sc and Nb/Sc) at *c.* 1.5 cal. ka BP is related to the increasing sediment contribution from small local rivers in Zhejiang and Fujian Provinces (Fig. 6). In addition, deeper investigation is needed into the question of whether these abrupt changes in geochemical composition as well as magnetic property (hard isothermal remanent magnetization, HIRM) at *c.* 1.5 ka were caused by the evolution of the Asian summer monsoon, as suggested by Zheng *et al.* (2010), and/or by anthropogenic activities.

Although the inner-shelf mud of the ECS has been widely used for the palaeoenvironmental study of its relatively continuous deposition, resuspension under extreme conditions such as storm

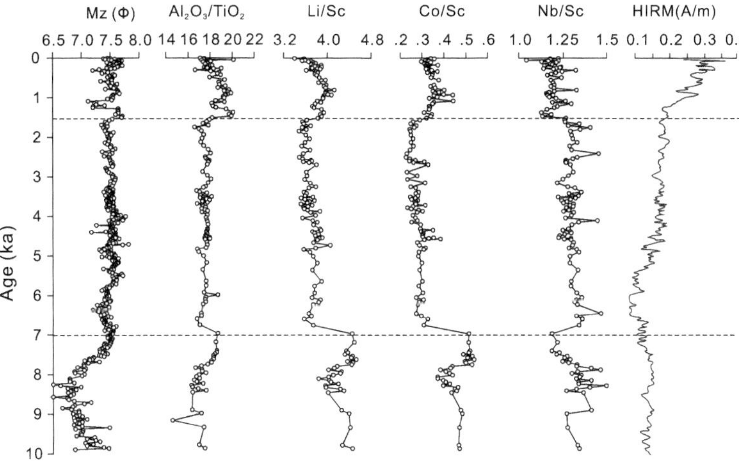

Fig. 6. Depth profiles of mean grain size (Mz), and ratios of major and trace elements in the sediments of Core MD06-3040 retrieved from the inner-shelf mud zone of the ECS (modified after Shao 2012). Data source of HIRM is from Zheng *et al.* (2010). The samples for geochemical measurement refer to the residual fraction after the leaching by 1N HCl. The core position is shown in Figure 2. The dashed lines indicate abrupt changes in geochemical compositions. Note the relatively uniform geochemical compositions between 7 and 1.5 ka suggesting the dominant sediment provenance is from the Changjiang. The significant shifts in geochemical proxies at 1.5–0 ka and at *c.* 10–7 ka suggest potential changes in sediment provenances. The change in HIRM is probably related to the variability of the Asian summer monsoon.

and typhoon events may result in a hiatus in the strata because the water depth (*c.* 20–60 m deep) of the mud belt lies within the range of wave action (Gao & Collins 2014). Therefore, the sampling resolution at an annual scale may be unrealistic, although results of both ^{210}Pb and ^{137}Cs dating of the muddy sediment indicate a general depositional rate of 10^0 cm a^{-1}. Under circumstances of

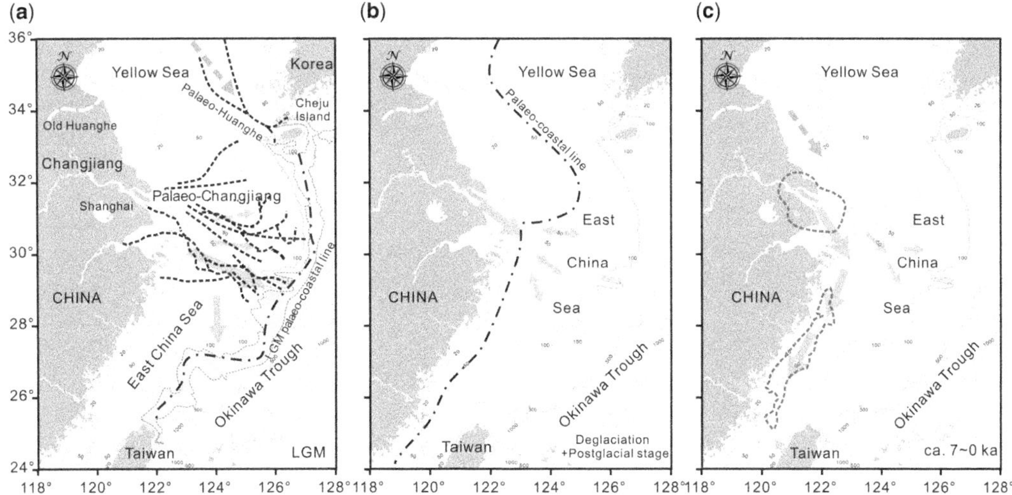

Fig. 7. A model showing the changes in major sinks and depocenters of the Changjiang sediment in the ECS since the LGM (modified after Dou *et al.* 2010*b*). The blue arrows indicate the possible transport pathways of the Changjiang sediment, while the yellow arrows indicate the possible dispersals of the modern and palaeo-Huanghe sediments. See the text for detailed explanation. In (**a**) dash lines indicate possible pathways of the palaeo-Changjiang and palaeo-Huanghe channels; (**b**) dot dash lines indicate palaeo-coastal line; (**c**) dash lines indicate Holocene depocentres of the Changjiang sediment in the ECS.

an ideal depositional environment, the resolution at decennial scale, or a 'high-resolution slice' in the sense of Gao & Collins (2014), can be expected for time-series palaeoenvironmental study in the mid-late Holocene.

Although modelling work suggests that the limit of delta growth has probably been reached already (Gao 2007), the distribution and dispersal patterns, as well as the budgets of the palaeo- and modern-Changjiang-borne sediment in the estuarine, coastal, shelf and continental slope regions of the ECS require further clarification. Apart from high-resolution core study with the development of new proxies for sediment provenance and palaeoenvironmental reconstruction, we urgently need more in-depth, *in situ* hydrodynamic observations in combination with geophysical investigations such as using a sub-bottom profiler for the better understanding of depositional processes and strata formation.

Concluding remarks

The ECS is characterized by a broad continental shelf, unique sedimentary systems, very large terrigenous sediment input and complex river–sea interactions during the late Quaternary. In terms of sediment source-to-sink studies, the ECS can thus be regarded as one of the best research sites for the investigation of sediment transfer from land to sea and for the study of land–sea interaction at various temporal and spatial scales.

In this contribution, we review state-of-the-art research progress in the sediment source-to-sink transport of the Changjiang in the ECS during the late Quaternary (Fig. 7). Various lines of evidence suggest that the palaeo-Changjiang incised the open shelf of the ECS and debouched its sediments to the shelf edge during the LGM with the lowest stand of sea level and a severe monsoon climate. With the onset of the last deglacial transgression, the major sinks and depocenters of the Changjiang sediment in the ECS changed in response to rising sea level. Due to the rapid sea-level rise and fast inundation of the gently dipping shelf, as well as strong tidal reworking during the late deglacial period to early Holocene, the Changjiang sediment was mostly dispersed or reworked and was hardly retained on the open shelf. The dispersed fluvial sediment was transported landwards (estuarine and coastal areas) and/or down slope (finally to the Okinawa Trough) by complicated marine processes, while the tidal sand ridges and sheets have dominated the mid–outer shelf up to the present day. Since the early–mid Holocene at *c.* 7 ka, with the gradual establishment of the modern marine environment and shelf circulation, the Changjiang sediment has been predominantly trapped in the estuary and has built a large delta, and the remainder of the sediment has been transported southeastwards by coastal currents and then deposited in the coastal bays and inner shelf. The overall geometry of major sedimentary systems in the ECS was reached in the late Holocene.

The establishment of Holocene shelf–coastal sedimentary systems associated with the Changjiang is strongly related to active sediment transport processes induced by tides and waves, shelf circulation and sediment gravity flow. For the river-dominated marginal sea, geochemical proxies and sedimentological and seismic profile observations may provide more robust constraints on the provenance and transport processes of these muddy and sandy sedimentary systems in the ECS at high-resolution spatial and temporal scales. In particular, the combination of *in situ* observation, experimental analysis and modelling work will significantly advance our understanding of the sediment source-to-sink process and river–sea interaction in this unique continental margin.

This work was supported by research funds awarded by the National Natural Science Foundation of China (Grants 41225020, 41376049), Continental Shelf Drilling Program (Grant No. GZH201100202), and China Geologic Survey (Grant No. GZH201100203).

References

ALEXANDER, C. R., DEMASTER, D. J. & NITTROUER, C. A. 1991. Sediment accumulation in a modern epicontinental-shelf setting: the Yellow Sea. *Marine Geology*, **98**, 51–72.

BEARDSLEY, R. C., LIMEBURNER, R., YU, H. & CANNON, G. A. 1985. Discharge of the Changjiang (Yangtze River) into the East China Sea. *Continental Shelf Research*, **4**, 57–76.

BERNÉ, S., VAGNER, P. *ET AL.* 2002. Pleistocene forced regressions and tidal sand ridges in the East China Sea. *Marine Geology*, **188**, 293–315.

CHANGJIANG WATER RESOURCE COMMITTEE 2013. *Changjiang Sediment Bulletin.* Changjiang Press, Wuhan, China.

CHEN, H. F., CHANG, Y. P. *ET AL.* 2011. Mineralogical and geochemical investigations of sediment-source region changes in the Okinawa Trough during the past 100 ka (IMAGES core MD012404). *Journal of Asian Earth Sciences*, **40**, 1238–1249.

CHEN, Z. Y. & YANG, W. D. 1991. Quaternary paleogeography and paleoenvironment of Changjiang River estuarine region. *Acta Geographica Sinica*, **46**, 436–448 [in Chinese with English abstract].

CLIFT, P. D., SCHOUTEN, H. & DRAUT, A. E. 2003. A general model of arc-continent collision and subduction polarity reversal from Taiwan and the Irish Caledonides. *In*: LARTER, R. D. & LEAT, P. T. (eds) *Intra-Oceanic Subduction Systems; Tectonic and*

Magmatic Processes. Geological Society, London, Special Publications, **219**, 81–98, http://doi.org/10.1144/GSL.SP.2003.219.01.04

DeMaster, D. J., McKee, B. A., Nittrouer, C. A., Qian, J. & Cheng, G. 1985. Rates of sediment accumulation and particle reworking based on radiochemical measurement from continental shelf deposit in the East China Sea. *Continental Shelf Research*, **4**, 143–158.

Diekmann, B., Hofmann, J., Henrich, R., Fütterer, D. K., Röhl, U. & Wei, K.-Y. 2008. Detrital sediment supply in the southern Okinawa Trough and its relation to sea-level and Kuroshio dynamics during the late Quaternary. *Marine Geology*, **255**, 83–95.

Dou, Y. G., Yang, S. Y., Liu, Z. X., Clift, P. D., Yu, H., Berné, S. & Shi, X. F. 2010*a*. Clay mineral evolution in the central Okinawa Trough since 28 ka: implications for sediment provenance and palaeoenvironmental change. *Palaeogeography, Palaeoclimatology, Palaeoecology*, **288**, 108–117.

Dou, Y. G., Yang, S. Y., Liu, Z. X., Clift, P. D., Shi, X. F., Yu, H. & Berne, S. 2010*b*. Provenance discrimination of siliciclastic sediments in the middle Okinawa Trough since 30 ka: constraints from rare earth element compositions. *Marine Geology*, **275**, 212–220.

Dou, Y. G., Yang, S. Y., Liu, Z. X., Li, J., Shi, X. F., Yu, H. & Berné, S. 2012. Sr-Nd isotopic constraints on terrigenous sediment provenances and Kuroshio Current variability in the Okinawa Trough during the late Quaternary. *Palaeogeography, Palaeoclimatology, Palaeoecology*, **356–366**, 38–47.

Emery, K. O. 1968. Relict sediments on continental shelves of the world. *AAPG Bulletin*, **52**, 445–464.

Gao, S. 2007. Modeling the growth limit of the Changjiang Delta. *Geomorphology*, **85**, 225–236.

Gao, S. 2013. Holocene shelf-coastal sedimentary systems associated with the Changjiang River: an overview. *Acta Oceanologica Sinica*, **32**, 4–12.

Gao, S. & Collins, M. B. 2014. Holocene sedimentary systems on continental shelves. *Marine Geology*, **352**, 268–294.

Guan, B. X. 1994. Pattern and structures of the current in Bohai, Huanghai and East China Seas. *In*: Zhou, D., Liang, Y. B. & Zeng, C. K. (eds) *Oceanology of China Seas*. Kluwer Academic Publication, Dordrecht, **1**, 3–16.

Guo, Z. G., Yang, Z. S., Zhang, D. Q., Fan, D. J. & Lei, K. 2002. Seasonal distribution of suspended matter in the northern East China Sea and barrier effect of current circulation on its transport. *Acta Oceanologica Sinica*, **24**, 71–80 [in Chinese with English abstract].

Hori, K., Saito, Y., Zhao, H. Q. & Wang, P. X. 2002. Architecture and evolution of the tide-dominated Changjiang (Yangtze) River delta, China. *Sedimentary Geology*, **146**, 249–264.

Horng, C. S. & Huh, C. A. 2011. Magnetic properties as tracers for source-to-sink dispersal of sediments: a case study in the Taiwan Strait. *Earth and Planetary Science Letters*, **309**, 141–152.

Hoshika, A., Tanimoto, T., Mishima, Y., Iseki, K. & Okamura, K. 2003. Variation of turbidity and particle transport in the bottom layer of the East China Sea. *Deep Sea Research Part II: Topical Studies in Oceanography*, **50**, 443–455.

Hu, D. X. 1984. Upwelling and sedimentation dynamics. *Chinese Journal of Oceanology and Limnology*, **2**, 12–19.

Hu, D. X. & Li, Y. X. 1993. Study of ocean circulation. *In*: Tseng, C. K., Zhou, H. O. & Li, B. C. (eds) *Marine Science Study and its Prospect in China*. Qingdao Publishing House, Qingdao, 513–516 [in Chinese].

Kao, S. J. & Milliman, J. D. 2008. Water and sediment discharge from small mountainous rivers, Taiwan: the roles of lithology, episodic events and human activities. *Journal of Geology*, **116**, 431–448.

Katayama, H. & Watanabe, Y. 2003. The Huanghe and Changjiang contribution to seasonal variability in terrigenous particulate load to the Okinawa Trough. *Deep-Sea Research (II): Topical Studies in Oceanography*, **50**, 475–485.

Kim, W. Y., Doh, S. J., Yu, Y. J. & Lee, Y. I. 2013. Magnetic evaluation of sediment provenance in the northern East China Sea using fuzzy c-means cluster analysis. *Marine Geology*, **337**, 9–19.

Lee, H. J. & Chao, S. Y. 2003. A climatological description of circulation in and around the East China Sea. *Deep-Sea Research, Part II: Topical Studies in Oceanography*, **50**, 1065–1084.

Lee, H. J. & Chough, S. K. 1989. Sediment distribution, dispersal and budget in the Yellow Sea. *Marine Geology*, **87**, 195–205.

Li, C. X. & Zhang, G. J. 1995. A sea-running Changjiang River during the last glaciation? *Acta Geographica Sinica*, **50**, 459–463 [in Chinese with English abstract].

Li, C. X., Zhang, J. Q., Fan, D. D. & Deng, B. 2001. Holocene regression and the tidal radial sand ridge system formation in the Jiangsu coastal zone, East China. *Marine Geology*, **173**, 97–120.

Li, C. X., Wang, P., Sun, H. P., Zhang, J. Q., Fan, D. D. & Deng, B. 2002. Late Quaternary incised-valley fill of the Yangtze delta (China): its stratigraphic framework and evolution. *Sedimentary Geology*, **152**, 133–158.

Li, F., Jiang, X. Y. & Song, H. N. 1993. Influence of the Huanghe and Changjiang rivers sediment loads on south Yellow Sea sedimentation since late Pleistocene. *Studia Marina Sinica*, **34**, 61–72 [in Chinese with English abstract].

Li, G. X., Liu, Y., Yang, Z. G., Yue, S. H. & Han, X. B. 2005*a*. Ancient Changjiang channel system in the East China Sea continental shelf during the last glaciation. *Science in China (Series D)*, **48**, 1972–1978.

Li, G. X., Yang, Z. G. & Liu, Y. 2005*b*. *Genetic Study of Submarine Sedimentary Environments in East China's Seas*. Science Press, Beijing [in Chinese with English abstract].

Li, G. X., Li, P., Liu, Y., Qiao, L. L., Ma, Y. Y., Xu, J. S. & Yang, Z. G. 2014. Sedimentary system response to the global sea level change in the East China Seas since the last glacial maximum. *Earth-Science Reviews*, **139**, 390–405.

Li, J. B. 2008. *Regional Geology of the East China Sea*. Ocean Press, Beijing.

Li, Y. H., Wang, A. J., Qiao, L., Fang, J. Y. & Chen, J. 2012. The impact of typhoon Morakot on the modern sedimentary environment of the mud deposition center off the Zhejiang–Fujian coast, China. *Continental Shelf Research*, **37**, 92–100.

Lim, D. I., Choi, J. Y., Jung, H. S., Rho, K. C. & Ahn, K. S. 2007. Recent sediment accumulation and origin of shelf mud deposits in the Yellow and East China Seas. *Progress in Oceanography*, **73**, 145–159.

Lin, C. M., Zhuo, H. C. & Gao, S. 2005. Sedimentary facies and evolution in the Qiantang River incised valley, eastern China. *Marine Geology*, **219**, 235–259.

Liu, J., Saito, Y., Kong, X., Wang, H., Xiang, L. H., Wen, C. & Nakashima, R. 2010. Sedimentary record of environmental evolution off the Yangtze River estuary, East China Sea, during the last ~13 000 years, with special reference to the influence of the Yellow River on the Yangtze River delta during the last 600 years. *Quaternary Science Reviews*, **29**, 2424–2438.

Liu, J. P., Li, A. C., Xu, K. H., Velozzi, D. M., Yang, Z. S., Milliman, J. D. & DeMaster, D. J. 2006. Sedimentary features of the Yangtze River-derived along-shelf clinoform deposit in the East China Sea. *Continental Shelf Research*, **26**, 2141–2156.

Liu, J. P., Xu, K. H., Li, A. C., Milliman, J. D., Velozzi, D. M., Xiao, S. B. & Yang, Z. S. 2007. Flux and fate of Yangtze River sediment delivered to the East China Sea. *Geomorphology*, **85**, 208–224.

Liu, J. T., Kao, S. J., Huh, C. A. & Hung, C. C. 2013. Gravity flows associated with flood events and carbon burial: Taiwan as instructional source area. *Annual Review of Marine Science*, **5**, 12.1–12.22.

Liu, K., Zhuang, Z. Y., Liu, D. Y., Ye, Y. C. & Hu, G. Y. 2009. Study of the buried ancient channels in the continental shelf area out of the mouth of the Changjiang River in China. *Acta Oceanologica Sinica*, **31**, 80–88 [in Chinese with English abstract].

Liu, Z. X. 1997. Yangtze Shoal – a modern tidal sand sheet in the northwestern part of the East China Sea. *Marine Geology*, **137**, 321–330.

Liu, Z. X., Berné, S., Saito, Y., Lericolais, G. & Marsset, T. 2000. Quaternary seismic stratigraphy and palaeoenvironments on the continental shelf of the East China Sea. *Journal of Asian Earth Science*, **18**, 441–452.

Liu, Z. X., Berné, S. et al. 2007. Internal architecture and mobility of tidal sand ridges in the East China Sea. *Continental Shelf Research*, **27**, 1820–1834.

Luo, X. X., Yang, S. L. & Zhang, J. 2012. The impact of the Three Gorges Dam on the downstream distribution and texture of sediments along the middle and lower Yangtze River (Changjiang) and its estuary, and subsequent sediment dispersal in the East China Sea. *Geomorphology*, **179**, 126–140.

Milliman, J. D. & Farnsworth, K. L. 2011. *River Discharge to the Coastal Ocean: A Global Synthesis*. Cambridge University Press, Cambridge.

Milliman, J. D., Beardsley, R. C., Yang, Z. S., Limeburner, R. 1985*a*. Modern Huanghe-derived muds on the outer shelf of the East China Sea: identification and potential transport mechanisms. *Continental Shelf Research*, **4**, 175–188.

Milliman, J. D., Shen, H. T., Yang, Z. S. & Meade, R. H. 1985*b*. Transport and deposition of river sediment in the Changjiang estuary and adjacent continental shelf. *Continental Shelf Research*, **4**, 37–45.

Milliman, J. D., Qin, Y. S. & Park, Y. A. 1989. Sediments and sedimentary processes in the Yellow and East China seas. *In*: Taira, A. & Masuda, F. (ed.) *Sedimentary Facies in the Active Plate Margin*. Terry Scientific Publishing Company, Tokyo, 233–249.

Niino, H. & Emery, K. O. 1961. Sediments of shallow portions of the East China Sea and South China Sea. *Geological Society of America Bulletin*, **72**, 731–762.

Nittrouer, C. A., DeMaster, D. J. & McKee, B. A. 1984. Fine-scale stratigraphy in proximal and distal deposits of sediment dispersal systems in the East China Sea. *Marine Geology*, **61**, 13–24.

Qin, Y. S., Zhao, Y. Y., Chen, L. R. & Zhao, S. L. 1987. *Geology of the East China Sea*. Science Press, Beijing [in Chinese].

Qu, T. D. & Hu, D. X. 1993. Upwelling and Sedimentation dynamics II. A simple model. *Chinese Journal of Oceanology and Limnology*, **11**, 289–295.

Raymo, M. E. & Ruddiman, W. F. 1992. Tectonic forcing of late Cenozoic climate. *Nature*, **359**, 117–122.

Saito, Y. 1998. Sedimentary environment and budget in the East China Sea. *Bulletin on Coastal Oceanography*, **36**, 43–58 [in Japanese].

Saito, Y., Katayama, H. et al. 1998. Transgressive and highstand systems tracts and post-glacial transgression, the East China Sea. *Marine Geology*, **122**, 217–232.

Shao, J. Q. 2012. *Sedimentological Records of the Changjiang Sediment and its Paleoenvironment Response in the East China Sea During the Holocene*. MSc thesis, Tongji University, Shanghai, China [in Chinese with English abstract].

Stanley, D. J. & Warne, A. G. 1994. Worldwide initiation of Holocene marine deltas by deceleration of sea-level rise. *Science*, **265**, 228–231.

Sun, X. G., Fang, M. & Huang, W. 2000. Spatial and temporal variations in suspended particulate matter transport on the Yellow Sea and the East China Sea shelf. *Oceanologia Limnologia Sinica*, **31**, 581–587 [in Chinese with English abstract].

Syvitski, J. P. M., Peckham, S. D., Hilberman, R. D. & Mulder, T. 2003. Predicting the terrestrial flux of sediment to the global ocean: a planetary perspective. *Sedimentary Geology*, **162**, 5–24.

Tang,, B. G. 1996. Quaternary stratigraphy on the shelf of the East China Sea. *In*: Yang, Z. G. & Liu, H. M. (eds) *Quaternary Stratigraphy in China and its International Correlation*. Geology Press, Beijing, 56–75 [in Chinese].

Uehara, K. & Saito, Y. 2003. Late Quaternary evolution of the Yellow/East China Sea tidal regime and its impacts on sediment dispersal and seafloor morphology. *Sedimentary Geology*, **162**, 25–38.

Uehara, K., Saito, Y. & Hori, K. 2002. Palaeotidal regime in the Changjiang (Yangtze) Estuary, the East China Sea, and the Yellow Sea at 6 and 10 ka estimated from a numerical model. *Marine Geology*, **183**, 179–192.

Wang, Z. B. 2014. *The Late Quaternary Stratigraphy on the Outer Shelf of the East China Sea and its Provenance-paleoenvironmental Study*. PhD thesis, Tongji University, Shanghai, China [in Chinese with English abstract].

Wang, Z. B., Yang, S. Y., Zhang, Z. X., Lan, X. H., Gu, Z. F. & Zhang, X. H. 2013. Palaeo-fluvial sedimentation on the outer shelf of the East China

Sea during the last glacial maximum. *Chinese Journal of Oceanology and Limnology*, **31**, 886–894.

WANG, Z. B., YANG, S. Y. *ET AL.* 2014. Late Quaternary stratigraphic evolution on the outer shelf of the East China Sea. *Continental Shelf Research*, **90**, 5–16, http://doi.org/10.1016/j.csr.2014.04.015

WARREN, J. D. & BARTEK, L. R. 2002. The sequence stratigraphy of the East China Sea, Where are the incised valleys? *In*: ARMENTROUT, J. & ROSEN, N. (eds) *Sequence Stratigraphic Models for Exploration and Production*. Gulf Coast SEPM Conference Proceedings, Houston, TX, 729–738.

WELLNER, R. W. & BARTEK, L. R. 2003. The effect of sea level, climate, and shelf physiography on the development of incised-valley complexes: a modern example from the East China Sea. *Journal of Sedimentary Research*, **73**, 926–940.

WRIGHT, L. D. & FRIEDRICHS, C. T. 2006. Gravity-driven sediment transport on continental shelves: a status report. *Continental Shelf Research*, **26**, 2092–2107.

XIA, D. X. & LIU, Z. X. 2001. Tracing the Changjiang River's route entering the sea during the last ice age maximum. *Acta Oceanologica Sinica*, **23**, 87–94 [in Chinese with English abstract].

XIAO, S. B., LI, A. C., JIANG, F. Q., LI, T. G., WAN, S. M. & HUANG, P. 2004. The history of the Yangtze River entering sea since the last glacial maximum: a review and look forward. *Journal of Coastal Research*, **20**, 599–604.

XIAO, S. B., LI, A. C. *ET AL.* 2006. Coherence between solar activity and the East Asian winter monsoon variability in the past 8000 years from Yangtze River-derived mud in the East China Sea. *Palaeogeography, Palaeoclimatology, Palaeoecology*, **237**, 293–304.

XU, K. H., MILLIMAN, J. D., LI, A. C., LIU, J. P., KAO, S. J. & WAN, S. M. 2009. Yangtze- and Taiwan-derived sediments on the inner shelf of East China Sea. *Continental Shelf Research*, **29**, 2240–2256.

XU, K. H., LI, A. C. *ET AL.* 2012. Provenance, structure, and formation of the mud wedge along inner continental shelf of the East China Sea: a synthesis of the Yangtze dispersal system. *Marine Geology*, **291–294**, 176–191.

XU, X. D. & ODA, M. 2009. Surface-water evolution of the eastern East China Sea during the last 36 000 years. *Marine Geology*, **156**, 285–304.

YANG, C. S. 1989. Active, moribund and buried tidal sand ridges in the East China Sea and the Southern Yellow Sea. *Marine Geology*, **88**, 97–116.

YANG, C. S. & SUN, J. S. 1988. Tidal sand ridges on the East China Sea shelf. *In*: DE BOER, P. L., VAN GELDER, A. & NIO, S. D. (eds) *Tide-Influenced Sedimentary Environments and Facies*. Reidel Publishing Company, Dordrecht, 23–38.

YANG, J., GAO, S. *ET AL.* 2009. Episodic crustal growth of North China as revealed by U-Pb age and Hf isotopes of detrital zircons from modern rivers. *Geochimica et Cosmochimica Acta*, **73**, 2660–2673.

YANG, S. L., ZHANG, J. & XU, X. J. 2007. Influence of the Three Gorges Dam on downstream delivery of sediment and its environmental implications, Yangtze River. *Geophysical Research Letters*, **34**, L10401, http://doi.org/10.1029/2007GL029472

YANG, S. L., MILLIMAN, J. D., LI, O. & XU, K. H. 2011. 50,000 dams later: erosion of the Yangtze River and its delta. *Global and Planetary Change*, **75**, 14–20.

YANG, S. Y., LI, C. X., JUNG, H. S. & LEE, H. J. 2002. Discrimination of elemental compositions between the Changjiang and Huanghe sediments and identification of sediment source in northern Jiangsu coast plain, China. *Marine Geology*, **186**, 229–241.

YANG, S. Y., JUNG, H. S., LIM, D. I. & LI, C. X. 2003. A review on the provenance discrimination of the Yellow Sea sediments. *Earth-Science Reviews*, **63**, 93–120.

YANG, S. Y., JUNG, H. S. & LI, C. X. 2004. Two unique weathering regimes in the Changjiang and Huanghe drainage basins: geochemical evidence from river sediments. *Sedimentary Geology*, **164**, 19–34.

YANG, S. Y., ZHANG, F. & WANG, Z. B. 2012. Grain size distribution and age population of detrital zircons from the Changjiang (Yangtze) River system, China. *Chemical Geology*, **296–297**, 26–38.

YANG, S. Y., WANG, Z. B., DOU, Y. G. & SHI, X. F. 2014. A review of sedimentation since the Last Glacial Maximum on the continental shelf of eastern China. *In*: CHIVAS, A. R. & CHIOCCI, F. L. (eds) *Continental Shelves During Last Glacioeustatic Cycle: Shelves of the World*. Geological Society, London, Memoirs, **41**, 293–303, http://doi.org/10.1144/M41.21.

YANG, Z. S. & MILLIMAN, J. D. 1983. Fine-grained sediments of Changjiang and Huanghe River and sediment sources of the East China Sea. *In*: *Sedimentation on the Continental Shelf: with Special Reference to the East China Sea 2*. China Ocean Press, Qingdao, 436–446.

YOO, D. G., LEE, C. W., KIM, S. P., JIN, J. H., KIM, J. K. & HAN, H. C. 2002. Late Quaternary transgressive and highstand systems tracts in the northern East China Sea mid-shelf. *Marine Geology*, **187**, 313–328.

YOUN, J. S. & KIM, T. J. 2011. Geochemical composition and provenance of muddy shelf deposits in the East China Sea. *Quaternary International*, **230**, 3–12.

YOUN, J. S., YANG, S. Y. & PARK, Y. A. 2007. Clay minerals and geochemistry of the bottom sediments in the northwestern East China Sea. *Chinese Journal of Oceanology and Limnology*, **25**, 235–246.

ZHANG, J. 1999. Heavy metal compositions of suspended sediments in the Changjiang estuary: significance of riverine transport to the ocean. *Continental Shelf Research*, **19**, 1521–1543.

ZHANG, X., LIN, C. M., DALRYMPLE, R. W., GAO, S. & LI, Y. L. 2014. Facies architecture and depositional model of a macrotidal incised-valley succession (Qiantang River estuary, eastern China), and differences from other macrotidal systems. *Geological Society of America Bulletin*, **126**, 499–522.

ZHAO, S. L. 1984. Quaternary geological questions of the Yangtze delta. *Marine Sciences*, **5**, 15–20 [in Chinese].

ZHENG, Y., KISSEL, C., ZHENG, H. B., LAJ, C. & WANG, K. 2010. Sedimentation on the inner shelf of the East China Sea: magnetic properties, diagenesis and paleoclimate implications. *Marine Geology*, **268**, 34–42.

ZHOU, L. Y., LIU, J., SAITO, Y., ZHANG, Z. X., CHU, H. X. & HU, G. 2014. Coastal erosion as a major sediment supplier to continental shelves: example from the abandoned Old Huanghe (Yellow River) delta. *Continental Shelf Research*, **82**, 43–59.

ZHU, D. K. & AN, Z. S. 1993. Formation and evolution of radial sand ridges in Jiangsu offshore zones. *In*: *Collection of Geography papers to mark the 80th birthday of Professor M. E. Ren.* Nanjing University Press, China, 142–147 [in Chinese].

ZHU, Y. Q., LI, C. Y., ZENG, C. K. & LI, B. G. 1979. Lowest sea level about the continental shelf of the East China Sea during late Pleistocene. *Chinese Science Bulletin*, **7**, 317–320 [in Chinese].

Trapping and escaping processes of Yangtze River-derived sediments to the East China Sea

JIAXUE WU*, JIE REN, HUAN LIU, CHUNHUA QIU,
YONGSHENG CUI & QIANJIANG ZHANG

The Centre for Coastal Ocean Science and Technology, School of Marine Sciences, Sun Yat-sen University, Guangzhou 510275, China

**Corresponding author (e-mail: wujiaxue@mail.sysu.edu.cn)*

Abstract: Contour-parallel sediment dispersal from the Yangtze Estuary into the East China Sea develops a large-scale mud belt on the inner shelf. The sediment dynamics of long-distance dispersal is, however, still an open question. This was investigated by field observations in the 2013 wet season. To clarify the physics of the large-scale mud belt, we examined: (a) shelf circulation currents and their interaction with the Yangtze River; (b) small-/meso-scale processes including bottom boundary-layer flows, stratification and mixing, upwelling, and fronts; and (c) river-borne sediment gravity and contour currents. Field observations demonstrated that estuarine turbidity maxima can trap benthic concentrated suspensions in the near-bed layer and move these downslope of the subaqueous delta, forming sediment gravity currents supported by tidal currents. Compared with near-bed sediment transports, the buoyant coastal current cannot be a controlling factor in the mud belt formation. A constant along-shelf flux of near-bed sediment transport is responsible for the long-distance dispersal of the large-scale mud belt on the East China Sea inner shelf. The upwelling events provide more turbulent energy to sediment suspension under unstably stratified boundary flow. Our recognition of a contour-parallel 'sediment channel' has deep implications for understanding this inner-shelf mud belt and ancient mud deposits.

The mud deposits on the continental shelf, no matter whether they are modern or ancient, can be viewed as environmental archives that can be used to reconstruct sedimentary environments, palaeoclimate changes and even human activities (e.g. Dojiri *et al.* 2003; Sommerfield & Lee 2003; Bianchi & Allison 2009). These depositional systems may be controlled by different combinations of sediment supply and redistribution processes, induced by waves, currents, tides and sediment gravity currents (Nittrouer & Wright 1994; Wright & Nittrouer 1995; Fagherazzi & Overeem 2007). Furthermore, these agents may be highly variable over a wide range of temporal and spatial scales, dependent on accommodation conditions (slope, width and depth), climate and sea-level changes (Fagherazzi & Overeem 2007). Walsh & Nittrouer (2009) summarized five distinct types of river dispersal systems: estuarine accumulation dominated; marine dispersal dominated; proximal accumulation dominated; canyon captured; and subaqueous delta clinoform. The fate of sediment dispersed from the river to the coastal ocean involves at least four processes: supply via plumes; initial deposition; resuspension and transport by marine processes; and long-term net accumulation (Wright & Nittrouer 1995). Among them, gravity-driven sediment transport across continental shelves within negatively buoyant (hyperpycnal) layers is likely to be a primary mechanism that supports turbid hyperpycnal layers on shelves (Mulder *et al.* 2003; Puig *et al.* 2004; Wright & Friedrichs 2006). This does not require sediment to be supplied by hyperpycnal river effluents, but that near-bed sediments can be suspended by wave- or current-induced velocity shear (Traykovski *et al.* 2000; Wright & Friedrichs 2006; Ayranci *et al.* 2012).

Numerous studies of locally confined depocentres on the continental shelf documented occurrences of well-developed along-shelf mud clinoforms originating from estuaries: for example, the Amazon, Changjiang (Yangtze), Huanghe (Yellow), Po, Ebro, Eel and Columbia (Wright *et al.* 1988, 1990; Kineke & Sternberg 1995; Wright & Nittrouer 1995; Nittrouer *et al.* 1996; Driscoll & Karner 1999; Allison *et al.* 2000; Cattaneo *et al.* 2003; Liu *et al.* 2006, 2007; Gao & Collins 2014). River-derived sediments can be transported great distances from the river mouth, forming a large-scale mud belt along the continental shelf. The mechanism for long-distance along-shelf delivery of river-borne sediments, however, is subject to considerable scientific debate in the source-to-sink dynamics community. Sediment gravity currents appear to be a reasonable mechanism for understanding the formation of the

From: Clift, P. D., Harff, J., Wu, J. & Qui, Y. (eds) 2016. *River-Dominated Shelf Sediments of East Asian Seas*. Geological Society, London, Special Publications, **429**, 153–169.
First published online August 27, 2015, updated May 26, 2016, http://doi.org/10.1144/SP429.7

mid-shelf mud, downslope of the continental shelf. Meanwhile, wave- or current-supported sediment gravity currents seem not to provide a valid mechanism for the along-contour transport and deposition of river-borne sediments without significant slope disturbance.

In general, the principle estuary-derived mud depocentres include the mid-estuary shoal (usually termed a river-mouth bar), the deltaic clinoform, and the contour-parallel mud belt on the inner and/or mid-shelf. On the wide East China Sea shelf, however, there exists no mid-shelf mud belt offshore the modern Yangtze River delta, but only a large-scale mud belt on the inner shelf. It is probable that most of the mass of the mud belt was originally brought downslope by gravity events and then redistributed by contour currents. The purpose of this study is to test this hypothesis from the Yangtze River Estuary to its adjoining coastal ocean, especially from the perspective of small-scale estuarine and marine processes, and sediment transport.

Regional background

A unique system of shelf circulation occurs in the East China Seas (Fig. 1). The East China Sea circulation currents move dominantly along the contours of the subaqueous topography, while cross-shelf flows are relatively weaker (see Yuan & Hsueh 2010). This circulation system is highly subject to the strength of East Asian monsoon, the riverine input, tidal mixing and external boundary forcings from the open ocean (Su 2001). Among these forcings, the interaction between boundary forcings and the subaqueous topography is a major factor in forming the unique circulation system in this area (Su 2001; Ma *et al.* 2010). The most important shelf current, which is involved in the Yangtze-borne sediment dispersal on the East China Shelf, is the buoyant coastal current, usually termed the Zhejiang–Fujian Coastal Current (ZFCC). The ZFCC can reach an along-shelf velocity of 50 cm s^{-1} in the winter (Wu *et al.* 2013). The northwards-flowing Taiwan Warm Current can reach the offshore waters of the Yangtze Estuary (Zhu *et al.* 2004). Tides are highly energetic in the Yangtze Estuary and the adjoining shelf waters, with an M2 astronomical constituent tide dominating (Su & Yuan 2005).

The Yangtze River is a major source of terrigenous material input into the East China Sea. Based on hydrological data from 1950 to 2010 at the Datong station in the lower Yangtze River, the largest, the smallest and the multi-year averaged river discharges from the estuary to the coastal ocean are 92 600 $m^3 s^{-1}$ (in 1954), 4620 $m^3 s^{-1}$ (in 1979) and 29 500 $m^3 s^{-1}$, respectively. The annual runoff is 9.24×10^{10} m^3, of which 71.7% occurs in the wet season (from May to October) and 28.3% in the dry season (from November to April). The yearly averaged sediment load delivered to the coastal ocean amounts to 4.86×10^8 *t*, with 87.2% in the wet season. The Yangtze-borne buoyancy input into the East China Sea forms a distinct river plume. This river plume extends to the NE (even as far as the Cheju island: Fig. 1) during the late spring and summer seasons, and it propagates to the south during the autumn and winter seasons, joining the buoyant coastal current (Zhu *et al.* 2004; H. Wu *et al.* 2011). The expansion of the Yangtze River plume causes seasonal stratification induced by temperature and salinity in the East China Sea. Intense stratification occurs over the whole of the shelf waters in the wet season, but this disappears in the dry season with vertically uniform structures (Tang *et al.* 2000). Moreover, stratification exhibits a notable yearly variability (Bai *et al.* 2013). Density stratification and turbulent mixing in the shelf waters usually plays an important role in vertical transfers of momentum, heat and mass. However, how these small-scale processes adjust or control the dispersal and trapping of the Yangtze river-derived sediments on the East China Shelf is still unknown.

The Zhejiang–Fujian mud belt (see ZFMB in Fig. 1), one of the longest mud belt in the world (*c.* 800 km in length: Liu *et al.* 2007), is a large-scale mud depocentre derived from the Yangtze River Estuary. This mud depocentre has been influenced by the ZFCC since 12.3 ka, and the Yangtze-derived materials have been transported southwards since that time (Xu *et al.* 2009). The sedimentary environment of the depocentre since 7.3 ka has been dominated by the ZFCC, and a homogeneous mud deposit began to develop (Xu *et al.* 2009). The mud belt was traditionally supposed to have been induced by marine dynamical structures over the same scale as the depositional system, such as the southwards-flowing coastal current driven by the winter monsoon (Liu *et al.* 2007; Xu *et al.* 2012), the along-shelf ocean front separating the river-affected buoyant coastal current and the intruding high-salinity water masses of the Taiwan Warm Current (Huh & Su 1999; Liu *et al.* 2007). In addition, Hu (1984) debated the role of upwelling currents in controlling shelf sedimentation in the East China Sea. He found that the spatial distribution of the eddy-induced upwelling is in a good agreement with that of shelf mud deposits (see Fig. 1). These large-/meso-scale marine dynamical structures are not the only processes responsible for sediment transport and sequestration in the mud belt. Sediment transport processes from the estuary to the coastal ocean must be directly investigated so that we can understand the underlying physics of

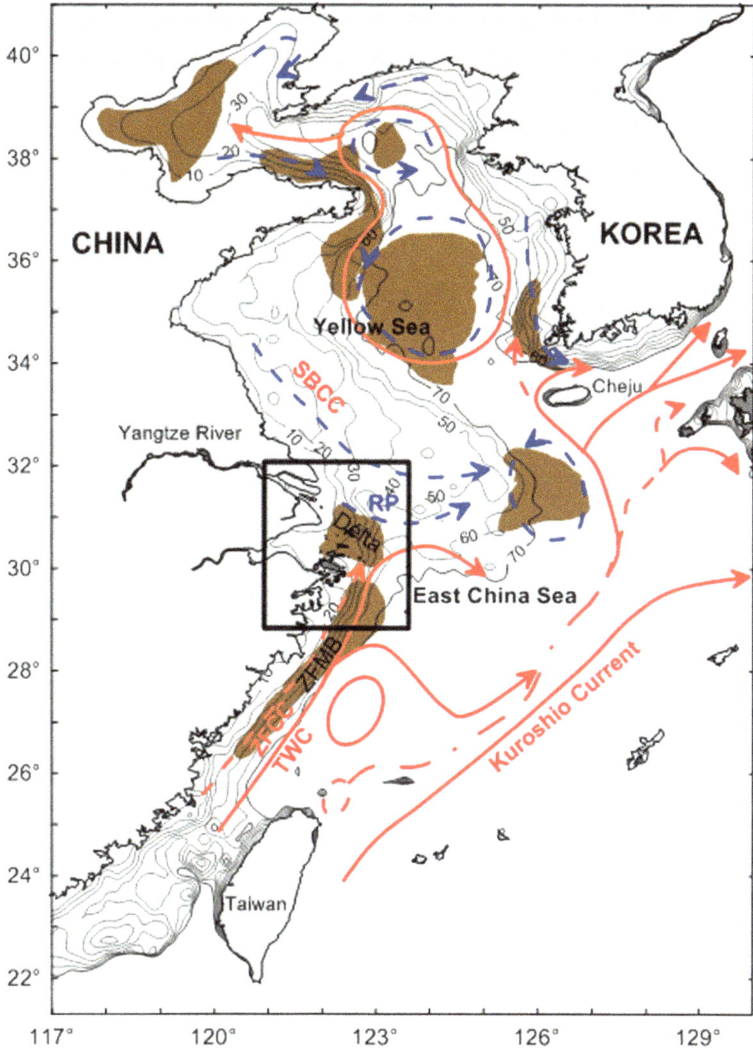

Fig. 1. Mud depocentres (shaded areas) and shelf circulation (solid and dashed arrow lines) in the summer season in the East China Sea. For the East China Sea shelf circulation system (after Su 2001), SBCC is the northern Jiangsu Coastal Current, RP represents the river plume, TWC represents the Taiwan Warm Current, ZFCC is the Zhejiang–Fujian Coastal Current. For the mud depocentres (after Gao & Collins 2014), ZFMB indicates the Zhejiang–Fujiang Mud Belt. The solid square indicates the study area. The depth contour is in metres.

the long-distance along-shelf mud dispersal and trapping on the inner shelf.

Field observations and data collection

Field observations

To understand the along-estuary and along-shelf dispersal and sequestration of the Yangtze-derived sediments in the coastal ocean, we conducted field hydrological observations from 3 to 9 July 2013 in the Yangtze Estuary and the Zhoushan waters (Fig. 2). The yearly averaged river discharge in 2013 was 24 984 $m^3\ s^{-1}$ at the Datong hydrological station, 624 km upstream of the river mouth, and the monthly averaged river discharge in July 2013 was 41 341 $m^3\ s^{-1}$. Large floods due to heavy rainfall occurred in July 2013 (Fig. 3). Field observations included three transects (marked as A, B and C) of moving shipboard surveys, three couples (marked as L, M and N) of moored benthic tripods and shipboard profiling over tidal cycles.

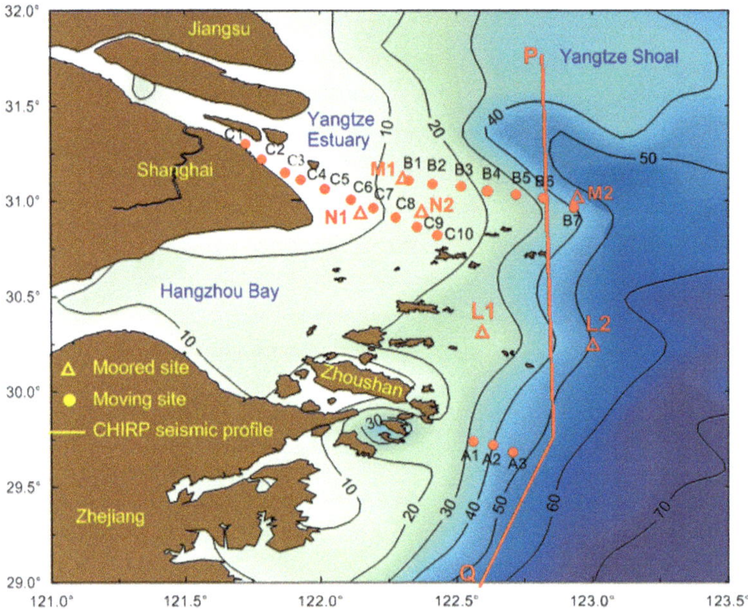

Fig. 2. Field hydrological observations from 3 to 9 July 2013 in the Yangtze Estuary and the Zhoushan waters. The line P–Q indicates the sub-bottom track of CHIRP seismic profiles. Three transects of the moving profiles are indicated by A, B and C. Three couples of moored surveys are marked by L, M and N. The depth contour is in metres.

Moving shipboard surveys were performed using one acoustical Doppler current profiler (ADCP) for measuring layered mean currents, one CTD (conductivity–temperature–depth) sensor for measuring water temperature, salinity and depth, one optical backscatter sensor (OBS) for measuring water turbidity, one microstructural profiler MSS-90L for measuring density stratification and turbulence mixing parameters, and one BioSonics digital echosounder with a transducer frequency of 430 kHz for identifying density interfaces in the stratified waters. The cell size of the echosounder was set at 0.5 m and the sampling rate was 5 Hz.

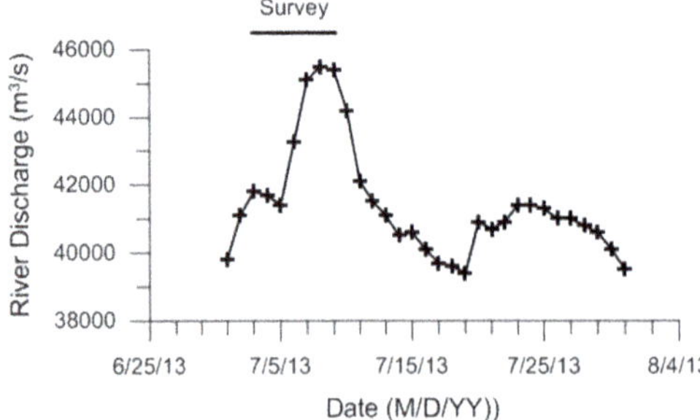

Fig. 3. River discharge at the Datong hydrological station in July 2013. The field survey period is indicated.

Moored shipboard profiling was achieved using one ADCP for measuring layered mean currents, one CTD for measuring water temperature, salinity and depth, and one OBS for measuring water turbidity. Concurrent water sampling was carried out at three depths ($0.2H$, $0.6H$ and $0.8H$, where H is water depth) in the water column and was performed hourly to calibrate the OBS turbidity. Two instrumental tripods were lowered to the seafloor to observe bottom boundary-layer flows and sediment transport at each of the two moored sites. One high-resolution acoustical Doppler profiler (HR-ADP) with a central frequency of 1.0 MHz was fastened at the top of the tripod at an elevation of 1.5 m above the seafloor to measure the mean currents, with a bin size of 0.5 m. One acoustical Doppler velometer (ADV) with a central frequency of 6.0 MHz was located at an elevation of 0.45 m. Two OBS sensors were located at elevations of 0.4 and 1.0 or 1.3 m above the bed, respectively.

Microwave OI-SST dataset

To check the distinct water masses in the estuarine, coastal and open oceans under all-weather conditions, we used the products of daily and 9 km spatial resolution microwave Optimal Interpolation Sea-Surface Temperature (OI-SST), which were derived from the Global Ocean Data Assimilation Experiment (GODAE) High-Resolution SST Pilot Project. The OI-SST products were produced by the microwave and infrared sensors, including the Tropical Rainfall Measuring Mission (TRMM) microwave Imager (TMI), the Advanced Microwave Scanning Radiometer for Earth (AMSR-E), AMSR2, WindSat, Terra Moderate Resolution Imaging Spectroradiometer (MODIS) and Aqua MODIS. Product errors have been evaluated to control the data quality. Details are provided on the website http://www.remss.com/measurements/sea-surface-temperature/oisst-description. Based on these calibrated OI-SST products, we obtained sea-surface temperature (SST) maps of the East China Sea during the field surveys. Because the error in the SST data analysis in our study area is approximately 1°C, we conclude that this does not strongly influence the spatial distribution of SST.

CHIRP seismic transect

One CHIRP seismic transect (line P–Q in Fig. 2) surveyed in 2003 was collected from the Jiangsu coast through the Yangtze delta and Zhoushan waters southwards to the Zhejiang coast. The Holocene mud belt in the topmost layer can be clearly identified from the reflection interfaces of the sub-bottom strata on the inner shelf.

Methods

The bottom stress can be roughly estimated from the quadratic drag law using the bottom current equation:

$$\tau_b = \rho C_d U_b^2 \tag{1}$$

where ρ is water density, U_b is the near-bed flow velocity and C_d is the drag coefficient ($C_d = 3.1 \times 10^{-3}$ on the mud seafloor). The pre-processing of the ADV dataset is required before the estimate of turbulence parameters (see Liu *et al.* 2009; J. Wu *et al.* 2011). An independent estimate of the bottom stress was conducted by the ADV-based eddy-correlation technique:

$$\tau_z = -\rho(\overline{u'w'} + \overline{v'w'}) \tag{2}$$

where the turbulent components (u', v', w') are the spanwise, streamwise and vertical directions, respectively. The Reynolds stress, τ_z, at the ADV height represents the bottom stress, τ_z, relying on the 'law of the wall'.

For the microstructure profiler MSS-90L, pre-processing of noise is also needed before the viscous dissipation can be estimated, including despiking, band-pass filter and acceleration-coherence. The viscous dissipation of isotropic turbulence kinetic energy per unit mass was estimated from the velocity shear spectrum:

$$\varepsilon = 7.5\nu\overline{\left(\frac{\partial u'}{\partial z}\right)^2} \tag{3}$$

where $\overline{(\partial u'/\partial z)^2}$ is the variance in the vertical gradient of turbulent fluctuations in the horizontal direction and ν is the kinematic viscosity. The Kolmogorov length scale was used to determine the viscous cutoff wavenumber:

$$k_s = (\varepsilon\nu^{-3})^{1/4} \tag{4}$$

To integrate the turbulent shear spectrum, the high wavenumber would be close to k_s (Oakey 1982), and the low wavenumber ranges from 1 to 2 cpm (Osborn 1980). The effect of stratification can be parameterized by the dimensionless ratio $\varepsilon/(\nu N^2)$, often referred to as buoyancy Reynolds number, where N is the buoyancy frequency: $N^2 = (-g/\rho)\,d\rho/dz$. When $\varepsilon/(\nu N^2) > 200$, the turbulence can be considered isotropic (Yamazaki & Osborn 1990). Dissipation estimated using equation (3) was in a good agreement with those estimated from the anisotropic formula for $\varepsilon/(\nu N^2) > 20$, but their discrepancy increases with decreasing values of $\varepsilon/(\nu N^2)$. In our study waters, the average values of ε and N within the halocline are 7.5×10^{-8} W kg^{-1} and 5.8×10^{-2} s^{-1}, respectively. Therefore, the average values of $\varepsilon/(\nu N^2)$ amounts to 1.5. This may lead to a notable error when we estimate dissipation within the halocline. The gradient Richardson number is defined as $Ri_g = N^2/S^2$, where the mean shear is $S = dU/dz$, in which U is the mean velocity.

Results and discussion

Large-scale mud belt on the East China Sea inner shelf

The CHIRP seismic track of sub-bottom strata profiles extends from the Jiangsu coast, via the subaqueous Yangtze delta and Zhoushan waters, southwards to the Zhejiang coast (Fig. 4a). This track is roughly along the subaqueous depth contour between 40 and 50 m on the inner shelf (Fig. 2). The mud belt, which has a substantial thickness, can be clearly identified from the subaqueous Yangtze delta, southwards to the Zhejiang coast. However, no mud deposit appears off the Jiangsu coast, although unidirectional oblique bedding is observed and directed to the south (Fig. 4b), which indicates that the sediment transport originated from the Yangtze Estuary to the north. This suggests that the Yangtze River-derived

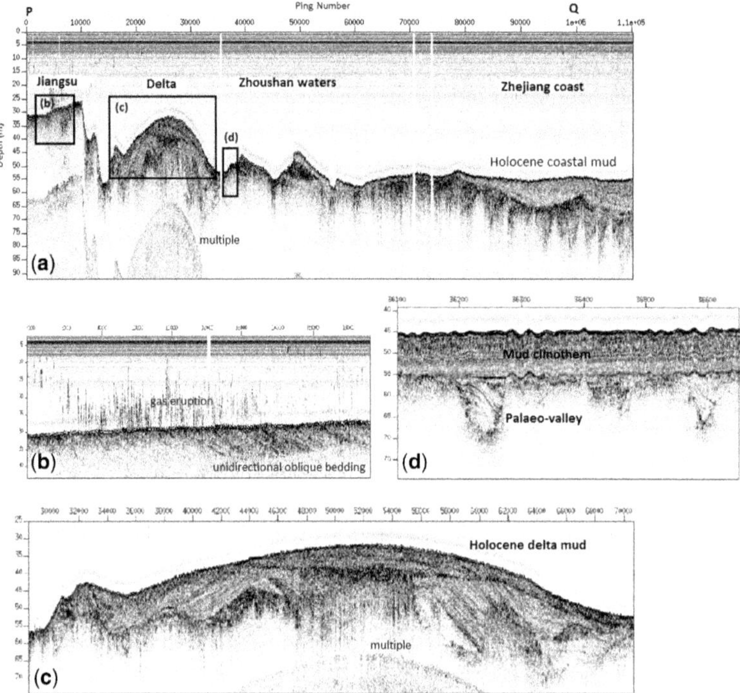

Fig. 4. Shallow strata structures of the mud belt on the inner shelf of the East China Sea. (**a**) The whole CHIRP seismic profile, and its details (**b**) off the Jiangsu coast, (**c**) in the Yangtze delta and (**d**) of the mud clinothem and underlying palaeo-valley infill in the Yangtze subaqueous delta.

sediments are discharged dominantly to the south, with little sediment transport to the north. The mud belt is seen to originate from the Yangtze Estuary, and it forms the topmost layer of Holocene strata on the East China inner shelf. Moreover, the thickness of the mud belt strongly reflects the palaeo-topography during the last deglaciation (Fig. 4c). The Yangtze palaeo-valleys can be identified below the bottom of the mud belt (Fig. 4d). The mud belt appears in Zhoushan waters where sand ridges also occur. Further to the south, the mud infill occurs over the relatively flat palaeo-topography offshore the Zhejiang coast, where the mud belt is thicker than in the Yangtze subaqueous delta or the Zhoushan mud belt. The mechanism that disperses mud far from its estuarine source needs to be identified, so that we can understand the physics of long-distance sediment transport on the sloping inner shelf.

Sediment trapping and escaping from the estuary to the coastal ocean

Turbidity maxima in the Yangtze Estuary. Field moving surveys along the Yangtze River Estuary (Fig. 5) indicate that estuarine turbidity maxima (ETM) occur markedly on the seawards slope of the river-mouth bar. The ETM lies immediately on the landwards side of the salinity front induced by saline intrusion. Meanwhile, a localized patch of high dissipation appears in the ETM. This indicates that both density stratification (turbulence suppression) and turbulent mixing are stronger in the ETM. Accordingly, flocculation settling and resuspension are both trapping mechanisms for fine-grained estuarine sediments (see also J. Wu *et al.* 2012). A secondary high-concentration sediment cloud appears on the landwards slope of the river mouth, where a high dissipation zone develops under weak salinity stratification. We infer that sediment resuspension dominates this sediment cloud. This process cannot localize the high-concentration suspension to the near-bed layer as the ETM, so sediment tends to escape rather than be trapped within the river mouth. The river-mouth bar is under erosion on the landwards slope but undergoes siltation on its seawards slope. The net effect is to cause the river-mouth bar to move seawards.

Sediment gravity current over the subaqueous delta. Field surveys over the Yangtze subaqueous delta (Fig. 6) indicate that sediment gravity currents

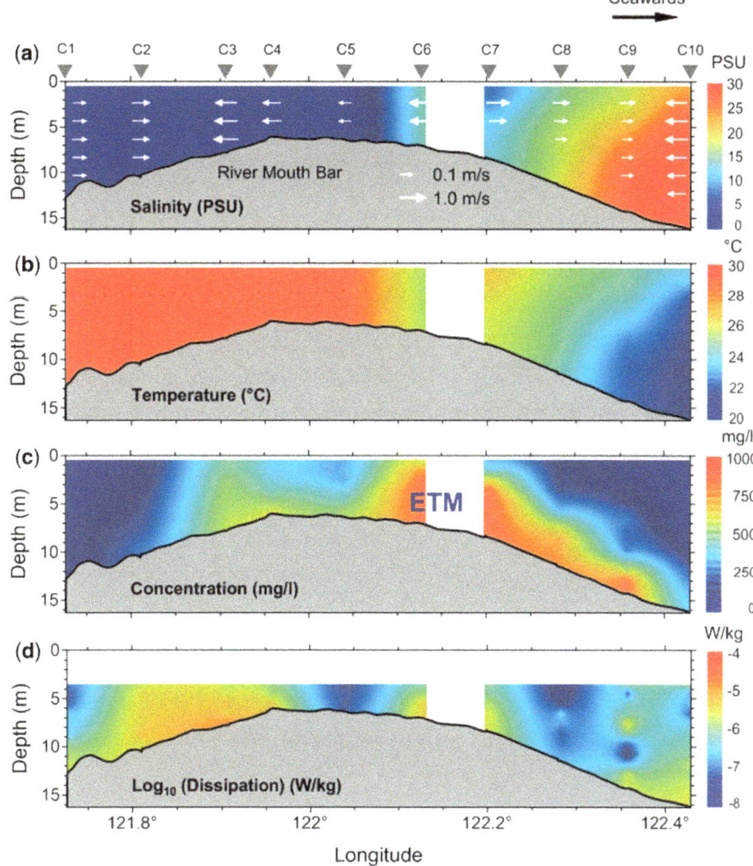

Fig. 5. Vertical transects of salinity in practical salinity units (PSU), temperature in °C, suspended sediment concentration in mg l^{-1} and viscous dissipation in W kg^{-1} along the Yangtze River Estuary in the 2013 wet season. Transect C1–C6 was conducted on 6 July, and the transect C10–C7 was performed on 7 July.

were particularly active over the delta-front seafloor during the 2013 wet season. These flows can move downslope to a water depth of at least 30 m. Evidence of a sediment gravity current is that anomalous events of warm, low-salinity water masses appear in the near-bed layer (Ayranci *et al.* 2012). A peak of suspended sediment concentration at the tidal reversal (*c.* 19:30) from ebb to flood at site M1 corresponds to the warm, low-salinity water mass, when the bottom stress is very small (see later). Estimates of bottom stresses induced by tidal currents at sites N1 and N2, and M1 and M2 (see later), show that sediment resuspension predominates over the upper slope of the delta front, while sediment deposition mainly appears over the lower slope, if the erosion limit of the silty seabed is 0.2 Pa. This suggests that the bottom stresses can support sediment gravity current over the upper slope, but they cannot maintain it over the lower slope. The near-bed sediment concentration of the gravity current on the Yangtze River delta is much lower than the value of >30 kg m^{-3} found in mountainous estuaries (e.g. Mulder & Syvitski 1995; Traykovski *et al.* 2000), but the gravity-driven sediment flow, supported by tidal currents, indeed exists over the subaqueous delta during the large floods.

Along-shelf currents and river-borne sediment dispersal

Complex water masses off the Yangtze Estuary. Complex water masses appear off the Yangtze Estuary (Fig. 7), where low-temperature water masses from the northern source meet high-temperature water masses from southern waters. Meanwhile, the control of the subaqueous topography on the

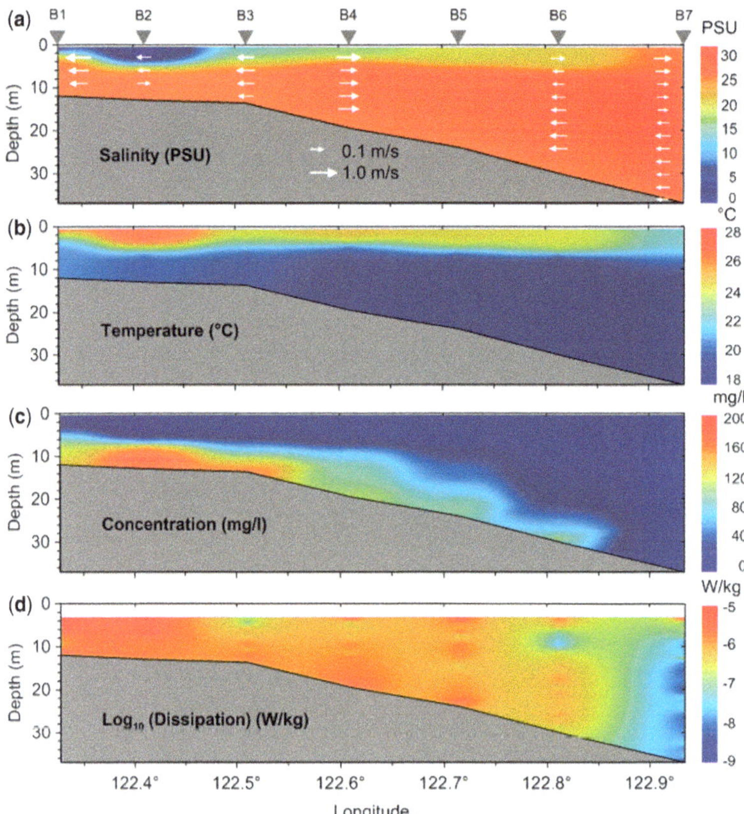

Fig. 6. Vertical transects of salinity in practical salinity units (PSU), temperature in °C, suspended sediment concentration in mg l^{-1} and viscous dissipation in W kg^{-1} over the subaqueous delta front in the 2013 wet season.

distribution of water masses can be clearly discerned. Cold water masses appear on the Yangtze Shoal, while warm water masses can protrude into the front of the Yangtze delta through the deep channel between the Yangtze Shoal and the Yangtze River delta. An upwelling event was visible from 6 to 9 July 2013. It appeared as a tiny spot on 5 July in the Zhoushan waters, extended quickly into a semi-circle shape from 6 to 8 July and connected to cold water masses on the Yangtze Shoal.

A diagram of temperature and salinity at six moored sites shows that estuarine water masses are quite different from coastal water masses (Fig. 8). Estuarine waters are of high temperature and low salinity, whereas coastal waters are of low temperature and high salinity. These distinct water characteristics produce a clear interface between water masses, probably forming density fronts on the inner shelf (Huh & Su 1999; Liu *et al.* 2007). In addition, two peaks in temperature on the L2 curve exhibits the controlling effect of upwelling events on water masses.

Mean flows and residual currents. The mean current flow off the Yangtze Estuary and in the adjoining coastal waters is rotatory, and is dominantly controlled by rotatory tidal currents. Residual currents can be obtained from an elimination of periodic tidal currents from the instantaneous mean currents. Residual currents over the whole column indicate that a SE flow predominates in the river delta and coastal waters during the wet season (Fig. 9). This makes a great contribution to the along-shelf coastal current on the inner shelf of the East China Sea.

Interface microstructures and sediment trapping. Echograms can represent the interface between the upper diluted waters and the lower saline waters (Fig. 10). The salinity interface between river plume and saline coastal waters by salinity intrusion can be clearly identified (Fig. 10a). Meanwhile, the salinity interface between the buoyant coastal current and more saline shelf waters is also visible (Fig. 10b). A large-scale oscillation occurs on the

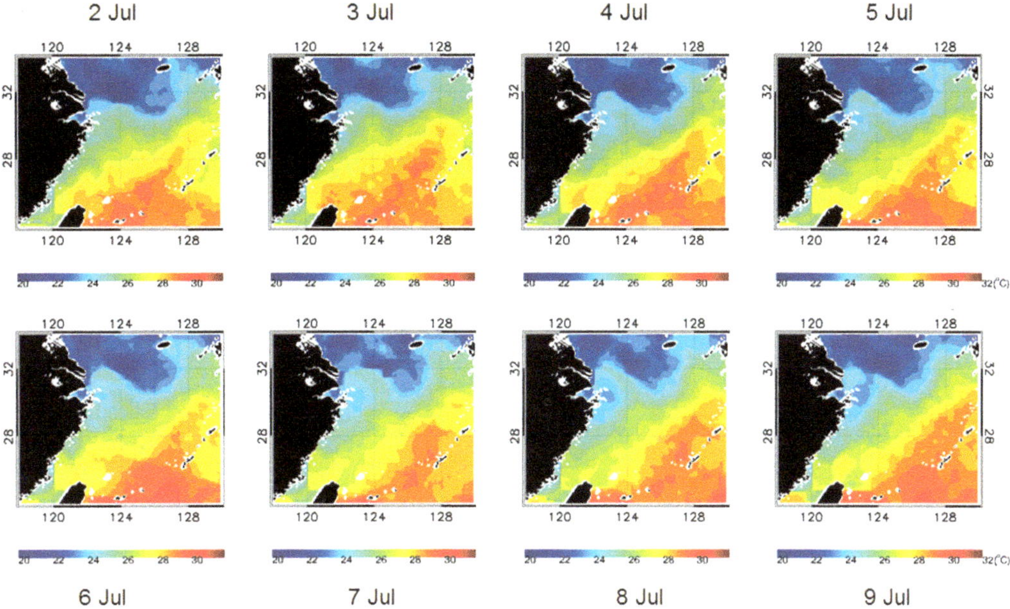

Fig. 7. Sea-surface temperature on 2–9 July 2013 in the East China Sea. An upwelling event occurred in the Zhoushan waters, causing an ever-increasing patch of low sea-surface temperature.

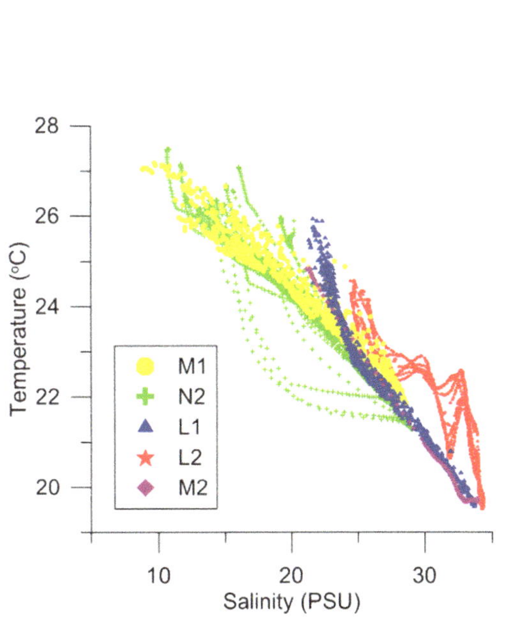

Fig. 8. Diagram of temperature and salinity in the study region. Sites M1 and N2 indicate estuarine stations, while sites L1, L2 and M2 represent coastal stations; for more details see Figure 2.

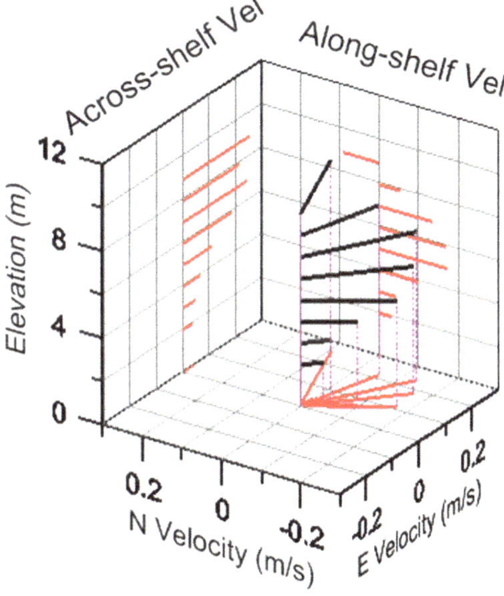

Fig. 9. Along-shelf and cross-shelf residual currents over the water column at site N2 in the Yangtze River Estuary. A positive value of the velocity component represents the northern (eastern) velocity in the along-shelf (cross-shelf) direction. A rotary current is visible from the mapping of current velocities over the water column on the bed.

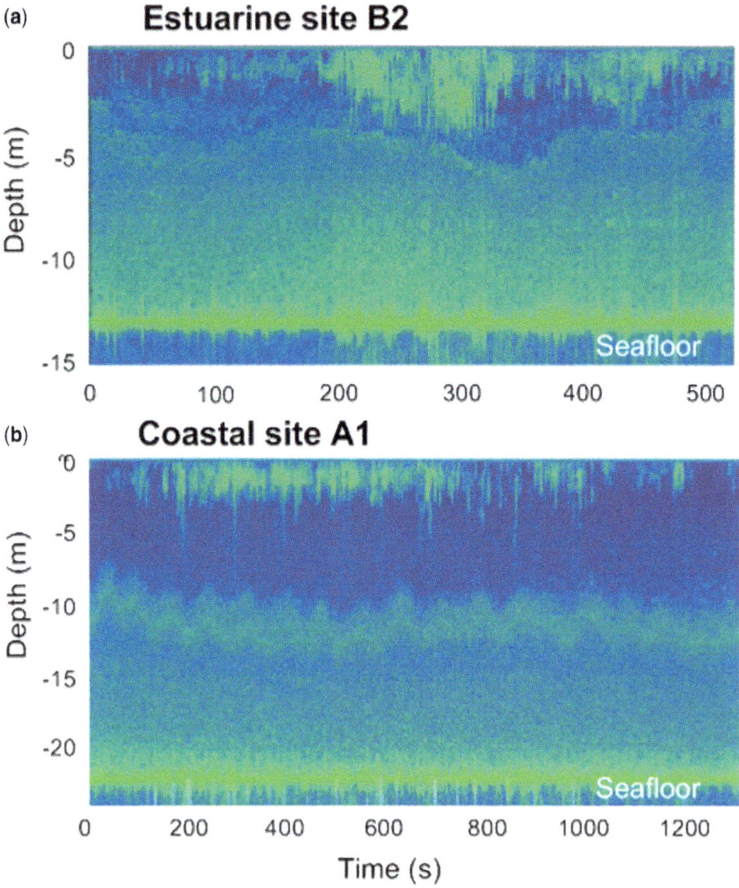

Fig. 10. Echograms obtained by a BioSonics digital echosounder with a transducer frequency of 430 kHz at (**a**) estuarine site B2 and (**b**) coastal site A1.

estuarine interface, while a relatively flat interface appears on the coastal interface. It is noteworthy that high-frequency fluctuations on the density interface and the subsurface patches with a high backscatter intensity are caused by the oscillation of the research vessel induced by surface waves.

Core profiles of stratification and mixing show the vertical distributions of temperature, salinity, stratification, mean shear, Richardson number and viscous dissipation (Fig. 11). At the estuarine sites, intense stratification and high shear occur along the interface, but viscous dissipation is vertically uniform over the water column (Fig. 11a, b). At the coastal sites, however, density stratification and shear are relatively weak on the interface, but the turbulence depression by density stratification is greater (Fig. 11c, d). Stratification and mixing under different conditions can be understood from the relationship between the gradient Richardson number and the viscous dissipation (Fig. 12). Stable stratification represents a rapidly decreasing tendency of dissipation with the Richardson number. Unstable stratification indicates a highly scattered distribution of dissipation, although a relatively flat tendency of dissipation appears with the Richardson number. Temperature-induced stratification due to upwelling events on the sloping inner shelf is unstable. This will adjust turbulence mixing and sediment diffusion in the coastal ocean.

The halocline of the river plume can trap suspended sediments, which are dispersed to the coastal ocean (Ren & Wu 2014). Under present conditions, the sediment transport along the halocline is confined to the river plume over the subaqueous delta, but it does not extend to those interfaces in the buoyant coastal current (Fig. 13). This indicates that the buoyant coastal current cannot be a major factor controlling river-derived sediment dispersal to the coastal ocean. Under this condition, benthic currents and sediment transport should play a

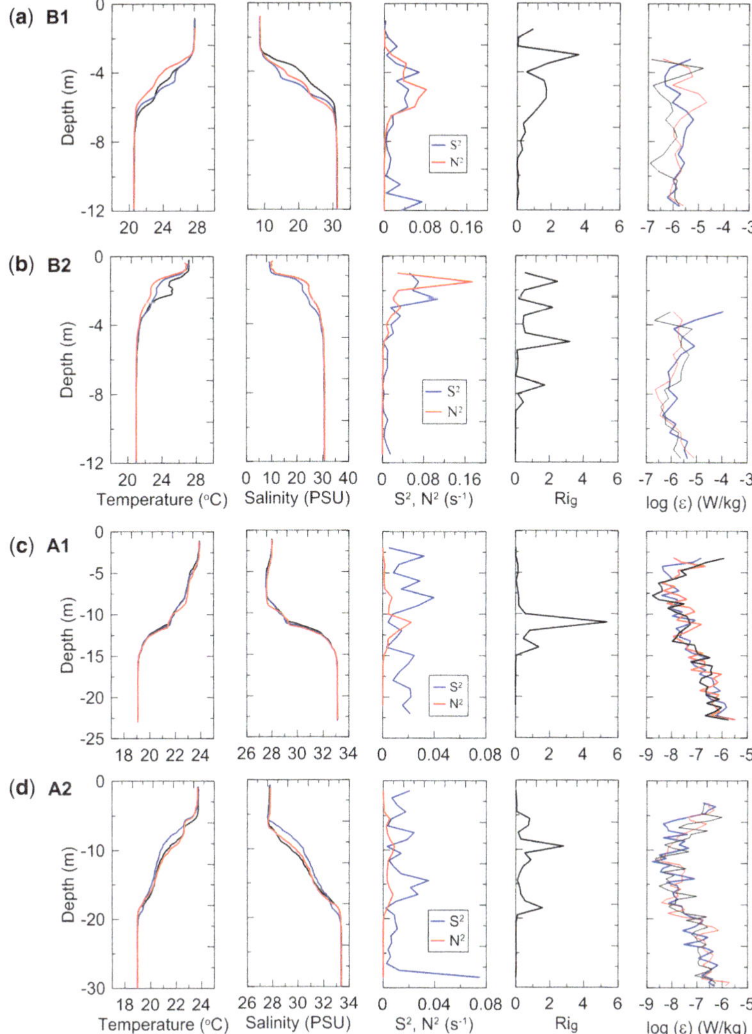

Fig. 11. Core profiles of stratification and mixing at (**a**) & (**b**) estuarine sites and (**c**) & (**d**) coastal sites. The variables of mean shear (S^2), buoyancy frequency (N^2), gradient Richardson number (Ri_g) and dissipation (ε) are defined in the 'Methods' section of this paper.

more important role in along-shelf sediment dispersal (see also Geyer *et al.* 2004).

Bottom boundary-layer flows and sediment transport. Tidal oscillations of temperature, salinity, suspended sediment concentration and bottom stress are evident at the estuarine sites (Fig. 14). This indicates that the tidal forcing is dominant in controlling sediment transport in the Yangtze Estuary. At the front of the subaqueous delta, tidal oscillation at the inshore site is also evident, but the variations in temperature, salinity and suspended sediment concentration at the offshore site appear stationary (Fig. 15). The sediment concentration in the near-bed layer is approximately steady over half the tidal cycle, although the bottom stress is only greater than the critical value of 0.2 Pa for mud bed resuspension at the time of tidal reversal. This stationary characteristic of suspension can clearly be identified from two levels of OBS measurements in the Zhoushan mud belt (Fig. 16), where the bottom stress is less than the critical value of 0.2 Pa. Two couples of moored tripod surveys on the delta front and in Zhoushan waters indicate that the near-bed sediment suspension in the mud belt mostly results in deposition, and has a constant flux of along-shelf

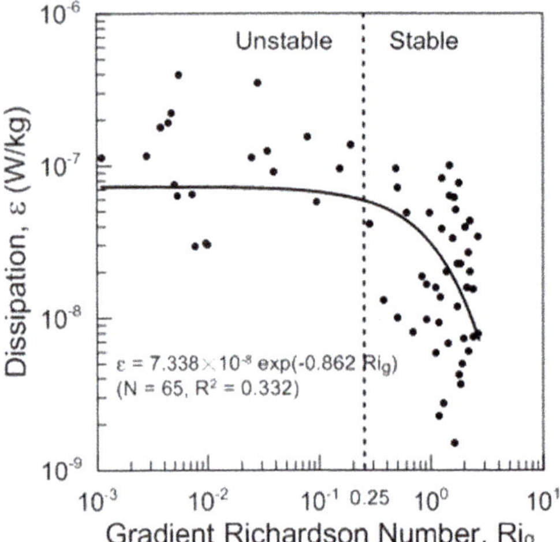

Fig. 12. Relationship between the gradient Richardson number (Ri_g) and viscous dissipation (ε) at three moored coastal sites (A1, A2 and A3) on 3 July 2013.

dispersal, termed a contour-parallel 'sediment channel'. This probably provides a more reasonable physical explanation of why the mud belt extends so far along the East China Sea inner shelf.

Furthermore, CTD data from offshore site L2, located in the mud belt, shows that more saline waters at 1.0 m above the seafloor occurred than at 0.4 m above the sea floor (Fig. 16). This indicates that an upwelling event occurred over the sloping inner shelf, and that vertical density stratification was unstable, resulting in more buoyancy flux. Such an upwelling can provide more turbulent kinetic energy to the near-bed sediment suspension under unstably stratified boundary flows.

Conclusions

The river delta and along-shelf clinoform are two modern shelf depocentres for Yangtze-derived fine-grained sediments, thus forming a large-scale mud belt on the East China Sea inner shelf. The mechanism for long-distance along-shelf delivery of river-borne sediments, however, is subject to considerable scientific debate in the source-to-sink dynamics community. Sediment dynamics of long-distance along-shelf dispersal is still an open question, although some marine physical processes, such as drift coastal current, ocean fronts, upwelling currents and the cross-shelf sediment gravity current, have been applied in order to understand the sediment-transport processes of the mud belt.

This study systematically investigated sediment dispersal from the Yangtze Estuary to the adjoining mud belt controlled by both large-/meso-scale coastal currents and small-scale marine processes. The mechanisms causing long-distance dispersal

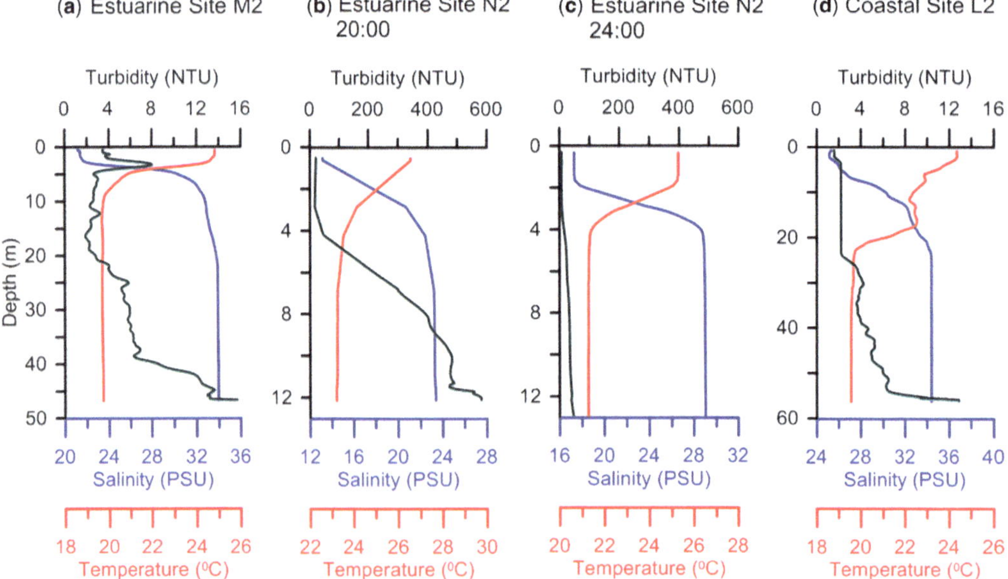

Fig. 13. Core profiles of salinity, temperature and turbidity at estuarine sites M2 and N2, and at coastal site L2. A turbidity peak occurs at the estuarine site M2, but does not appear at the other sites. This means that sediment trapping can be induced by the haloclines of the river plume, rather than by the buoyant coastal current.

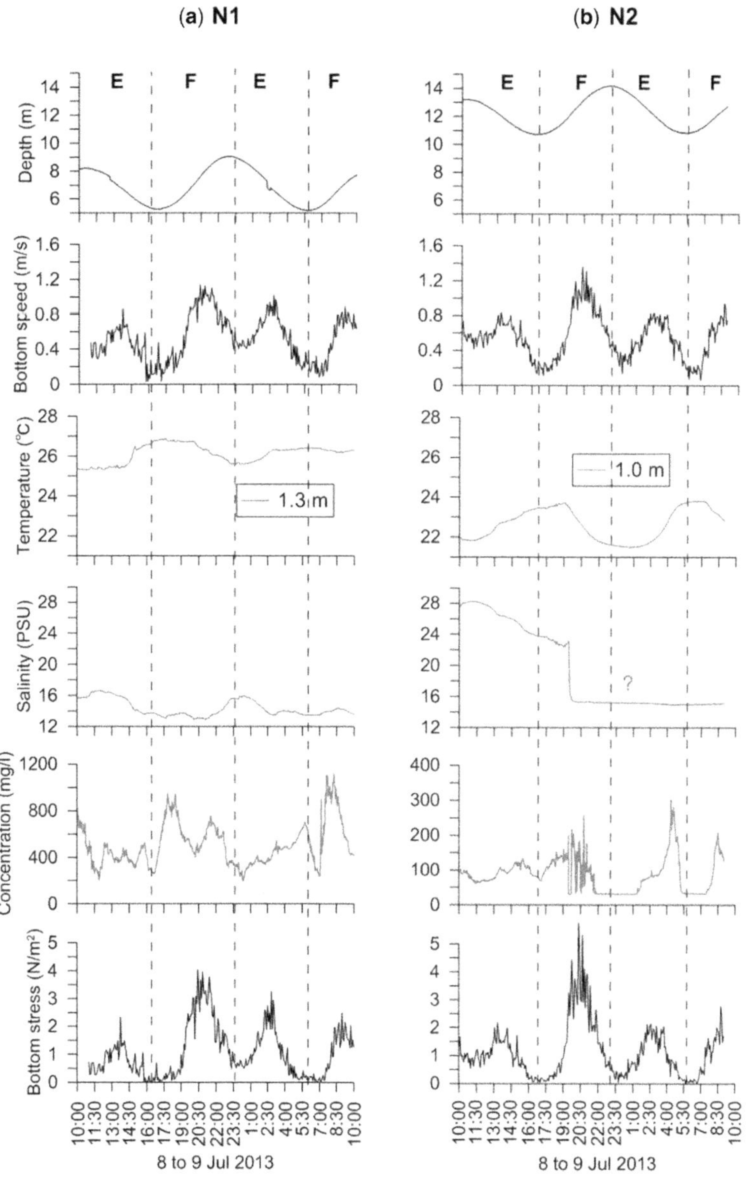

Fig. 14. Time series of water depth, flow speed, temperature, salinity, suspended sediment concentration and bottom stress at the top of the subaqueous delta. Solid lines in the bottom stress are based on the quadratic drag law, with a drag coefficient of $C_d = 3.1 \times 10^{-3}$. The given 1.3 and 1.0 m indicate the elevation above the seabed.

of river-derived sediments to the East China Sea inner shelf were investigated using field observations in the Yangtze Estuary and the adjoining mud belt. Field observations demonstrated that fine-grained suspended sediments from the Yangtze Estuary was originally brought downslope by sediment gravity currents over the subaqueous delta and then redistributed by benthic contour currents on the inner shelf. The near-bed sediment suspension in the mud belt mostly results in deposition, and sediment is dispersed along-shelf with a constant flux, forming the contour-parallel 'sediment channel'. Upwelling events provide more turbulent energy for sediment suspension under unstable stratified boundary flows. These findings provide a reasonable physical explanation for how the mud belt has developed so far along the inner shelf.

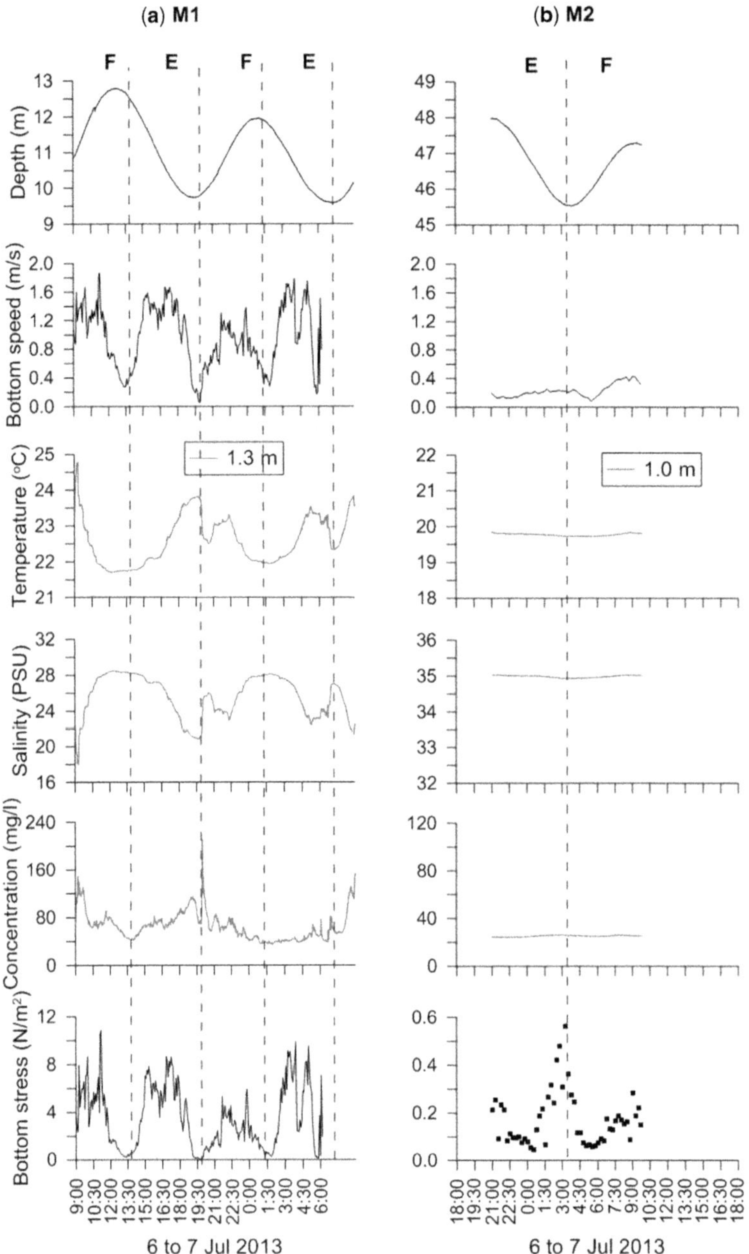

Fig. 15. Time series of water depth, flow speed, temperature, salinity, suspended sediment concentration and bottom stress at the front of the subaqueous delta. Solid lines in the bottom stress are based on the quadratic drag law, with a drag coefficient of $C_d = 3.1 \times 10^{-3}$, while the scattered points are derived from the ADV-based eddy-correlation technique. The given 1.3 and 1.0 m indicate the elevation above the seabed.

Fine-grained suspended sediment dispersal by the Yangtze River plume is confined to the subaqueous delta as long as it cannot spread to the buoyant coastal current. As a result, neither the river plume nor the buoyant coastal current can be the major carrier of river-derived sediments to the coastal ocean over long distances.

Two important mechanisms of transport for the Yangtze River-derived sediments over the East China Sea inner shelf in the wet season were

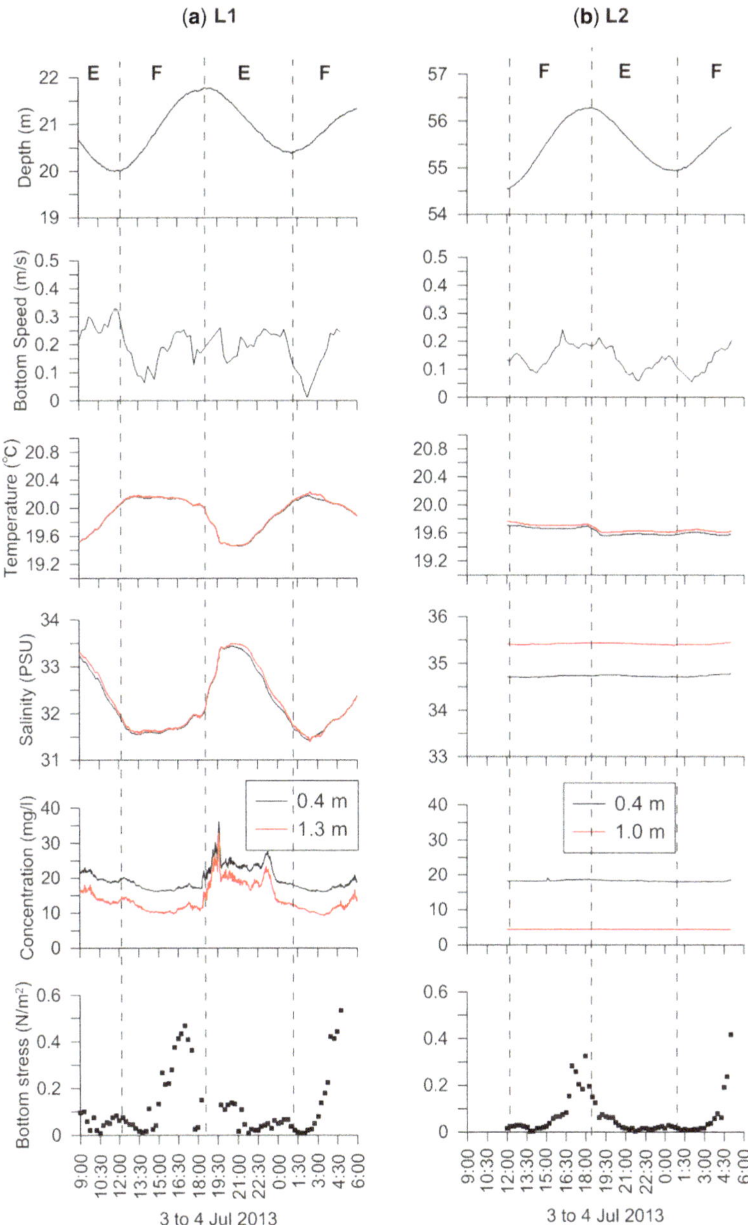

Fig. 16. Time series of water depth, flow speed, temperature, salinity, suspended sediment concentration and bottom stress of the Zhaoushan waters in the mud belt. The scattered points in the bottom stress are derived from the ADV-based eddy-correlation technique. The given 1.3, 1.0 and 0.4 m indicate the elevation above the seabed.

identified: (i) sediment gravity currents over the subaqueous delta; and (ii) a 'sediment channel' with the constant flux of along-shelf benthic sediment dispersal in the mud belt. Together these provide an interpretation that is quite different from the traditional view that shelf sedimentation is controlled by the buoyant coastal current or the widespread evidence of sediment gravity flows from the estuary downslope to the mid-shelf seafloor. However, whether the stationary contour-parallel dispersal of the river-borne sediments occurs in the dry season or whether it also affects other mud belts still needs to be investigated. Crucially, the physics of this contour-parallel 'sediment channel' on the sloping continental shelf needs to be clarified, so that we can better understand the depositional environments and formation mechanism of ancient mud belts.

This work was financially supported by the National Basic Research Programme of China (grant number 2013 CB956502) and the National Science Foundation of China (grant numbers 41276079 and 41176067). The valuable comments and suggestions on the original manuscript by Peter D. Clift and another anonymous referee are greatly appreciated. Guofu Pan is greatly thanked for providing the CHIRP seismic data. Echosounder surveys were performed by Yu Zhang and Zongxi Zhao.

References

ALLISON, M. A., LEE, M. T., OGSTON, A. S. & ALLER, R. C. 2000. Origin of Amazon mudbanks along the northeastern coast of South America. *Marine Geology*, **163**, 241–256.

AYRANCI, K., LINTERN, D. G., HILL, P. R. & DASHTGARD, S. E. 2012. Tide-supported gravity flows on the upper delta front, Fraser River delta, Canada. *Marine Geology*, **326–328**, 166–170.

BAI, Y., PAN, D., CAI, W-J., HE, X., WANG, D., TAO, B. & ZHU, Q. 2013. Remote sensing of salinity from satellite-derived CDOM in the Changjiang River dominated East China Sea. *Journal of Geophysical Research: Oceans*, **118**, 227–243.

BIANCHI, T. S. & ALLISON, M. A. 2009. Large-River Delta-Front Estuaries as Natural 'Recorders' of global environmental change. *Proceedings of the National Academy of Sciences of the United States of America*, **106**, 8085–8092.

CATTANEO, A., CORREGGIARI, A., LANGONE, L. & TRINCARDI, F. 2003. The late-Holocene Gargano subaqueous delta, Adriatic shelf: sediment pathways and supply fluctuations. *Marine Geology*, **193**, 61–91.

DOJIRI, M., YAMAGUCHI, M., WEISBERG, S. B. & LEE, H. J. 2003. Changing anthropogenic influence on the Santa Monica Bay watershed. *Marine Environmental Research*, **56**, 1–14.

DRISCOLL, N. W. & KARNER, G. D. 1999. Three-dimensional quantitative modeling of clinoform development. *Marine Geology*, **154**, 383–398.

FAGHERAZZI, S. & OVEREEM, I. 2007. Models of deltaic and inner continental shelf landform evolution. *Annual Review of Earth and Planetary Sciences*, **35**, 685–715.

GAO, S. & COLLINS, M. B. 2014. Holocene sedimentary systems on continental shelves. *Marine Geology*, **352**, 268–294.

GEYER, W. R., HILL, P. S. & KINEKE, G. C. 2004. The transport, transformation and dispersal of sediment by buoyant coastal flows. *Continental Shelf Research*, **24**, 927–949.

HU, D.-X. 1984. Upwelling and sedimentation dynamics. I. The role of upwelling in sedimentation in the Huanghai sea and East China Sea – a description of general features. *Chinese Journal of Oceanology and Limnology*, **2**, 12–19.

HUH, C.-A. & SU, C.-C. 1999. Sedimentation dynamics in the East China Sea elucidated from ^{210}Pb, ^{137}Cs and ^{239,240}Pu. *Marine Geology*, **160**, 183–196.

KINEKE, G. C. & STERNBERG, R. W. 1995. Distribution of fluid muds on the Amazon continental shelf. *Marine Geology*, **125**, 193–233.

LIU, H., WU, C., XU, W. & WU, J. 2009. Contrasts between estuarine and river systems in near-bed turbulent flows in the Zhujiang (Pearl River) Estuary, China. *Estuarine, Coastal and Shelf Science*, **83**, 591–601.

LIU, J. P., LI, A. C., XU, K. H., VELOZZI, D. M., YANG, Z. S., MILLIMAN, J. D. & DEMASTER, D. J. 2006. Sedimentary features of the Yangtze River-derived along-shelf clinoform deposit in the East China Sea. *Continental Shelf Research*, **26**, 2141–2156.

LIU, J. P., XU, K. H., LI, A. C., MILLIMAN, J. D., VELOZZI, D. M., XIAO, S. B. & YANG, Z. S. 2007. Flux and fate of Yangtze River sediment delivered to the East China Sea. *Geomorphology*, **85**, 208–224.

MA, C., WU, D., LIN, X., YANG, J. & JU, X. 2010. An open-ocean forcing in the East China and Yellow seas. *Journal of Geophysical Research*, **115**, C12056, http://doi.org/10.1029/2010JC006179

MULDER, T. & SYVITSKI, J. P. M. 1995. Turbidity currents generated at river mouths during exceptional discharges to the world oceans. *Journal of Geology*, **103**, 285–299.

MULDER, T., SYVITSKI, J. P. M., MIGEON, S., FAUGERES, J.-C. & SAVOYE, B. 2003. Marine hyperpycnal flows: initiation, behavior and related deposits. A review. *Marine and Petroleum Geology*, **20**, 861–882.

NITTROUER, C. A. & WRIGHT, L. D. 1994. Transport of particles across continental shelves. *Reviews of Geophysics*, **32**, 85–113.

NITTROUER, C. A., KUEHL, S. A. ET AL. 1996. The geological record preserved by Amazon shelf sedimentation. *Continental Shelf Research*, **16**, 817–841.

OAKEY, N. S. 1982. Determination of the rate of dissipation of turbulent energy from simultaneous temperature and velocity shear microstructure measurements. *Journal of Physical Oceanography*, **12**, 256–271.

OSBORN, T. R. 1980. Estimates of the local rate of vertical diffusion from dissipation measurements. *Journal of Physical Oceanography*, **10**, 83–89.

PUIG, P., OGSTON, A. S., MULLENBACH, B. L., NITTROUER, C. A., PARSONS, J. D. & STERNBERG, R. W. 2004. Storm-induced sediment gravity flows at the head of the Eel submarine canyon, northern California margin. *Journal of Geophysical Research*, **109**, C03019, http://doi.org/10.1029/2003JC001918

REN, J. & WU, J. 2014. Sediment trapping by haloclines of a river plume in the Pearl River Estuary. *Continental Shelf Research*, **82**, 1–8.

SOMMERFIELD, C. K. & LEE, H. J. 2003. Magnitude and variability of Holocene sediment accumulation in Santa Monica Bay, California. *Marine Environmental Research*, **56**, 151–176.

SU, J. 2001. A review of circulation dynamics of the coastal oceans near China. *Acta Oceanologica Sinica*, **23**, 1–16 [in Chinese with English abstract].

SU, J. & YUAN, Y. 2005. *Oceanography of China Seas*. China Ocean Press, Beijing.

TANG, Y., ZOU, E., LIE, H.-J. & LIE, J.-H. 2000. Some features of circulation in the southern Huanghai Sea. *Acta Oceanologica Sinica*, **22**, 1–16 [in Chinese with English abstract].

TRAYKOVSKI, P., GEYER, W. R., IRISH, J. D. & LYNCH, J. F. 2000. The role of wave-induced density-driven fluid mud flows for cross-shelf transport on the Eel River continental shelf. *Continental Shelf Research*, **20**, 2113–2140.

WALSH, J. P. & NITTROUER, C. A. 2009. Understanding fine-grained river-sediment dispersal on continental margins. *Marine Geology*, **263**, 34–45.

WRIGHT, L. D. & FRIEDRICHS, C. T. 2006. Gravity-driven sediment transport on continental shelves: a status report. *Continental Shelf Research*, **26**, 2092–2107.

WRIGHT, L. D. & NITTROUER, C. A. 1995. Dispersal of river sediments in coastal seas: six contrasting cases. *Estuaries*, **18**, 494–508.

WRIGHT, L. D., WISEMAN, W. J. ET AL. 1988. Marine dispersal and deposition of Yellow River silts by gravity-driven underflows. *Nature*, **332**, 629–632.

WRIGHT, L. D., WISEMAN, W. J., JR, YANG, Z. S., BORNHOLD, B. D., KELLER, G. H., PRIOR, D. B. & SUHAYDA, J. N. 1990. Processes of marine dispersal and deposition of suspended silts off the modern mouth of the Huanghe (Yellow River). *Continental Shelf Research*, **10**, 1–40.

WU, H., ZHU, J., SHEN, J. & WANG, H. 2011. Tidal modulation on the Changjiang River plume in summer. *Journal of Geophysical Research*, **116**, C08017, http://doi.org/10.1029/2011JC007209

WU, H., DENG, B. ET AL. 2013. De-tiding measurement on transport of the Changjiang-derived buoyant coastal current. *Journal of Physical Oceanography*, **43**, 2388–2399.

WU, J., LIU, H., REN, J. & DENG, J. 2011. Cyclonic spirals in tidally accelerating bottom boundary layers in the Zhujiang (Pearl River) Estuary. *Journal of Physical Oceanography*, **41**, 1209–1226.

WU, J., LIU, J. T. & WANG, X. 2012. Sediment trapping of turbidity maxima in the Changjiang Estuary. *Marine Geology*, **303–306**, 14–25.

XU, F., LI, A., XIAO, S., WAN, S., LIU, J. & ZHANG, Y. 2009. Paleoenvironmental evolution in the inner shelf of the East China Sea since the last deglaciation. *Acta Sedimentological Sinica*, **27**, 118–127 [in Chinese with English abstract].

XU, K., LI, A. ET AL. 2012. Provenance, structure, and formation of the mud wedge along inner continental shelf of the East China Sea: a synthesis of the Yangtze dispersal system. *Marine Geology*, **291–294**, 176–191.

YAMAZAKI, H. & OSBORN, T. 1990. Dissipation estimates for stratified turbulence. *Journal of Geophysical Research*, **95**, 9739–9744.

YUAN, D. & HSUEH, Y. 2010. Dynamics of the cross-shelf circulation in the Yellow and East China Seas in winter. *Deep-Sea Research II: Topical Studies in Oceanography*, **57**, 1745–1761.

ZHU, J., CHEN, C., DING, P., LI, C. & LIN, H. 2004. Does the Taiwan warm current exist in winter? *Geophysical Research Letters*, **31**, L12302, http://doi.org/10.1029/2004GL019997

Lingdingyang Bay, Pearl River Estuary (China): geomorphological evolution and hydrodynamics

PEIHONG JIA[1,2,3], ZHEN XIA[4]*, YONG YIN[1,2,3] & QIAO XUE[4]

[1]*Key Laboratory of Coast and Island Development (Nanjing University), Ministry of Education, Nanjing, 200023, China*

[2]*Collaborative Innovation Centre for the South China Sea Studies, Nanjing University, Nanjing, 200023, China*

[3]*School of Geographical and Oceanographic Sciences, Nanjing University, Nanjing, 200023, China*

[4]*Guangzhou Marine Geological Survey, Guangzhou, GD 510760, China*

**Corresponding author (e-mail: xia-zhen@163.com)*

Abstract: The Pearl River is the largest river in southern China; it has eight outlets flowing into the South China Sea, four of which (Humen, Jiaomen, Hongqili and Hengmen) empty via Lingdingyang Bay. Tides, discharge strength and the presence of islands control sediments within these four outlets. Sediments are deposited, and form three shoals and two channels. The shoreline and its seabed topography have changed dramatically with increasing human impact.

In order to understand the geomorphological evolution of Lingdingyang Bay and its hydrodynamics, we integrated data obtained from two marine environment investigations conducted by the Guangzhou Marine Geological Survey (GMGS) in 2003 and in 2012, together with remote sensing multi-temporal images and historical materials, and analysed them using a geographical information system (GIS).

The results obtained show that the outlets are becoming narrower; the surface water area of Lingdingyang Bay is diminishing. Meanwhile, the sediment flux is reducing and becoming finer grained because the currents are being blocked by shoals and islands. Some parts of the bay are becoming deeper (27–37 m in the Chuanbi Channel). Correspondingly, the hydrodynamics in the bay are becoming weaker and sediments easier to deposit. Water exchange between the inner and outer sea is decreasing under the influence of unrestricted anthropogenic activities.

Lingdingyang Bay should be managed and protected continuously based on a scientific framework: otherwise, it will become narrower and, perhaps, even evolving eventually into a river, which may result in serious futures disasters through flooding and tides.

The Pearl River, the largest river in southern China, empties into Lingdingyang Bay through four of eight outlets and, ultimately, into the South China Sea. The Pearl River Estuary is the second largest estuary in China. Correspondingly, Lingdingyang Bay is the largest estuary of the Pearl River system, with an area of approximately 1041 km^2, and extends about 5.2 km from the northern Humen outlet to the inner Lingding Island (Fig. 1). As the only sea route for the ports of Guangzhou and Huangpu, it is deeper in the east than in the west, is funnel-shaped and opens to the south. Four outlets (Humen, Jiaomen, Hongqili and Hengmen) are located close together in the north and west of the bay. The geo-environment of Lingdingyang Bay is very sensitive and fragile owing to the interaction between land, ocean and the anthroposphere (Chen 1995; Glibert *et al.* 2001).

Interestingly, the shoreline of Lingdingyang Bay has been changing since prehistoric time. In general, the world's coastal areas are exposed to rapid urban growth, the pressure of increasing population, the expansion of major industries (particularly tourism) and the extensive exploitation of marine resources. As a result, the geomorphology and land use of the coastal areas are changing dramatically in response to natural processes, as well as to human activities. As such, a similar situation occurs in Lingdingyang Bay as in many Chinese developed cities around it (e.g. Hong Kong, Shenzhen, Dongguan, Guangzhou, Zhongshan, Zhuhai and Macao): dramatic economic expansion brings the largest rapidly growing populations, land reclamation, sand mining and numerous construction projects of factories, pipe lines, bridges and harbours. Such developments exert more and more pressures on the coastal

From: Clift, P. D., Harff, J., Wu, J. & Qui, Y. (eds) 2016. *River-Dominated Shelf Sediments of East Asian Seas*. Geological Society, London, Special Publications, **429**, 171–184.
First published online May 31, 2016, http://doi.org/10.1144/SP429.14

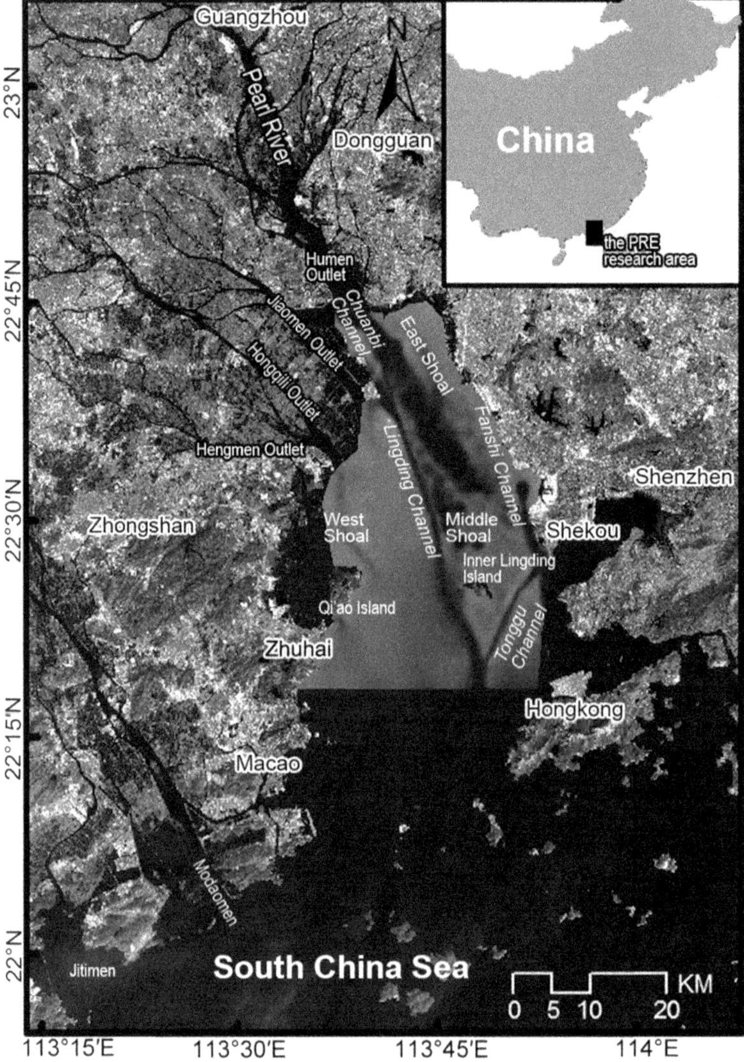

Fig. 1. Research area of Lingdingyang Bay located in the Pearl River Estuary, China.

environment. With anthropogenic impact increasing, the surrounding land cover, shoreline and topography are changing dramatically.

At present, the geomorphological characteristics of Lingdingyang Bay comprises three shores (West Shore, Middle Shore and East Shore) and two channels (Lingding Channel and Fanshi Channel) (Fig. 1) that have formed in response to interactions between tides and discharges from the Pearl River (Xia 2005). In general, shorelines and land use are the key elements in the study of geomorphological change; as such, these have become a central component for managing natural resources and monitoring environmental changes (Muttitanon & Tripathi 2005; Hanebuth *et al.* 2011). Large-scale land reclamation has been carried out in Inner Lingdingyang Bay for more than 20 years, so that the water area of the inner bay has now been reduced. Meanwhile, the water level has risen, so that floodwaters cannot drain smoothly as before (Xia *et al.* 2013).

All the above characteristics pose considerable hazards to sea routes, shores, shoals, tidal flats and the mangrove ecosystem. For example, in the summer of 1997, the floods in the upper reaches of the Pearl River resulted in damage estimated at one billion yuan (*c.* $150 million) (Statistical Bureau of Guangdong Province 1998). Some researchers have proposed that Lingdingyang is changing into a river as a result of sedimentation in the bay (Ying 1995; Wen *et al.* 2003; Zong *et al.* 2009; Li & Qiao 2012; Zhang *et al.* 2012). In this state, because of deposition, it would not be possible to navigate the Pearl River without radical measures being undertaken. The length of time required for the bay to sediment up has been calculated at between 150 and 490 years (Wen *et al.* 2003). At the same time, remote sensing data have been used to analyse shoreline changes: these offer a historical and recent perspective on landscape dynamics (Gu *et al.* 1990; Frihy *et al.* 1998; Weng 2002; Peterson *et al.* 2004). The changing coastline of Lingdingyang Bay was monitored using Thematic Mapper (TM) imagery (Chen *et al.* 2005).

The present study focuses on the coastal geomorphological change and dynamics in Lingdingyang Bay, using technologies developed for satellite remote sensing, field surveying and a geographical information system (GIS), to carry out a detailed characterization of the morphologies and their evolution through tracing historical shorelines and topographical changes with time. Such approaches provide the appropriate management strategies to governments and coastal administration departments, at all levels, with a scientific framework on which sound administration policies and measures could be based in order to harmonize human activities and natural processes.

Data and methods

Location

The study area, Lingdingyang Bay and its surroundings are located in the northern part of the South China Sea; it is a part of the southern coastline of China, ranging from 22° 12′–22° 45′ N to 113° 30′–114° 00′ E (Fig. 1).

Data description

This study integrates all available data in order to study the geomorphological evolution. The data consist mainly of bathymetric data, remote sensing images and historical topographical maps from old British Admiralty charts or documents (Table 1). An accurate mathematical theory is fundamental to establishing a geospatial database. All the data are overlain in order to facilitate spatial analysis and simulation. Digital elevation maps (DEMs) are the most important data sources in the database: 1:10 000 and 1:50 000 scale maps were used to identify the water systems and to build the DEM. The Guangzhou Marine Geological Survey (GMGS) carried out an integrated mapping project of marine geo-environment and geo-hazards surveys in Lingdingyang Bay in 2003–04. Another mapping project was undertaken in 2012. Consequently, there are two sets of bathymetric data for Lingdingyang Bay, dating from 2003 and 2012. The satellite remote sensing images were based on Landsat TM and ETM (Enhanced Thematic Mapper) images, with ranges collected in July 1986, March 1996, November 2005 and October 2009. Data on population and economics were standardized, according to that published in Statistical Bureau of Guangdong Province (1988, 1998). The data of the historical shorelines were obtained from British Admiralty charts.

Methodology

The techniques applied in this study include advanced marine environment surveying and charting technologies, computer-aided remote sensing image interpretation, GIS spatial analysis and data expression, GPS and remote sensing integrated techniques, and other related mathematical statistical methods.

Surveying and investigative methods

The surveying and investigative methods used in this study include sub-bottom profiling, single-channel seismic profiling and use of a side-scan sonar system. The combination of the different seismic sources and measured parameters permits detailed

Table 1. *Data sources and their characteristics*

Type	Sensor*	Data (time)	Accuracy (m)	Acquisition source
Survey data	–	2003–05	10	GMGS mapping project (1:100 000)
Survey data	–	2012	5	GMGS mapping project (1:50 000)
Landsat	MSS	1978	79	Aerial photographs
Landsat	TM	1988	30	Aerial photographs
Landsat	TM	30 July 1986	30	Aerial photographs
Landsat	TM	13 June 1996	30	Aerial photographs
Landsat	ETM	2003	15	Aerial photographs

*MSS, Multi-Spectral Scanner; TM, Thematic Mapper; ETM Enhanced Thematic Mapper.

information to be obtained about the seabed and the underlying strata in the area under investigation. By using 'state-of-the-art' geophysical equipment, a total 575.8 km of survey data was collected. Acoustical instruments, sampling and drilling approaches were carried out regularly, and at the same time (Xia 2005).

The single-channel seismic system used was the Delph Seismic System produced by Triton Elics (USA), operated at 800 J, with a 1000 ms trigger interval, a 250 ms scan width for the printer, a 250 ms record length for data and a 2 s record length, with SEG-D format. The side-scan sonar MS992 system, produced by Simrad Mesotech Systems Ltd, Norway, was used at a 120 kHz operating frequency, with a 100 m scan range for each side, a resolution of 8 bits and a 22.37 kHz sampling frequency. During the survey, the differential global positioning system (DGPS), manufactured by NavCom Technology, Inc. (USA), SF2050 DGPS receiver was operated with Hypack software: this can reach an accuracy of better than 15 cm in the horizontal and 30 cm in the vertical (Fig. 2).

As a result of the surveying, the bathymetry of Lingdingyang Bay showed the following characteristics: the eastern part of the bay is deeper than the west. Many shoals, troughs or channels and islands are co-located on the undulating seabed (Xia 2005). There are four major outlets, which act as entrances for the Pearl River into the South China Sea via the bay. The Tonggu Channel was dredged in 2007, separating the middle shoal into two shoals. Hence, four shoals and three channels now exist in Lingdingyang Bay.

Remote sensing analysis

The interpretation and information extraction of the computer-aided multi-temporal remote sensing images were carried out using ERDAS IMAGINE8.7 software, developed by ERDAS Company (USA). The technical procedure follows a certain sequence (Fig. 3): first, the remote sensing images from different times are pre-processed using multispectral compositions, in order to achieve true colour images. Then, the water borders on the images are enhanced, to improve their interpretation and identification. Secondly, the enhanced images are corrected geometrically and rectified in terms of spatial location. Subsequently, the image data, of different phases or different resolutions, are then merged together. Thirdly, the merged images are improved further in order to enhance the texture features of the shoreline components to be identified (using histogram adjustment, unsharp mask sharpen, colour balance, hue/saturation adjustment and contrast enhancement). Lastly, by comparing the pixels between images, one by one, shoreline changes between two different times can be detected. In the process of shoreline boundary detection, the spatial rectification between images of two different phases should be accurate to within 3 m.

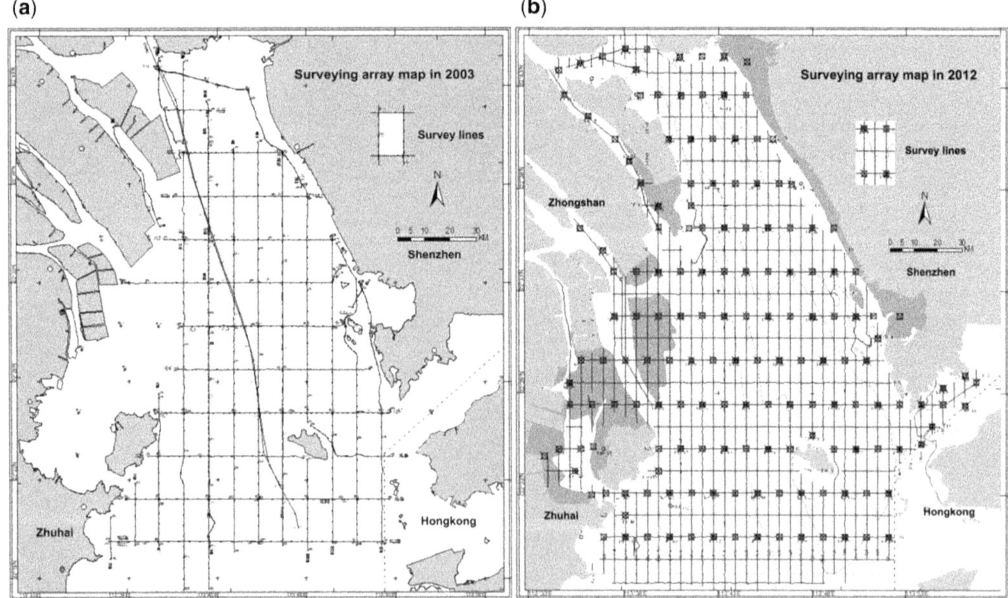

Fig. 2. Grid spacing of bathymetry surveying in (**a**) 2003 and (**b**) 2012.

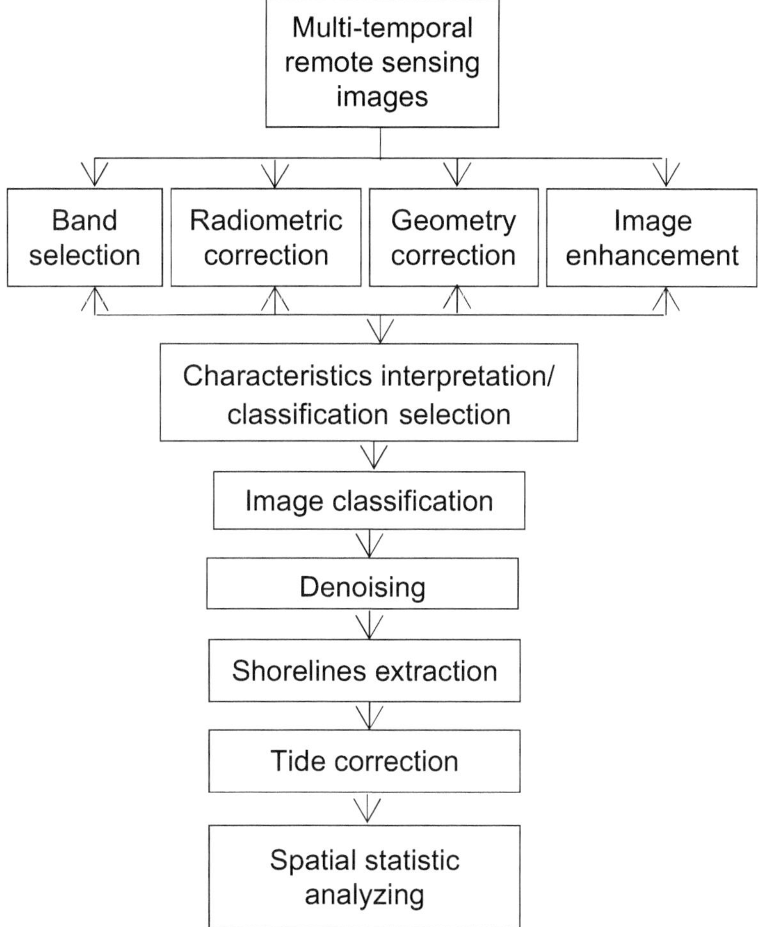

Fig. 3. Technical flow chart of multi-temporal remote sensing images pre-processing.

Radiometric correction is employed in the research to guarantee the accuracy of the spatial rectification. Stable distinctive features in the images are also used as control points to manipulate image-to-image rectification, with an error of less than 0.2 pixels (3 m).

Integrated analysis with GIS

Field surveying is a realistic approach for collecting real-time, detailed data, but it is expensive and time-consuming. Remote sensing data can be obtained relatively easily, and this is a useful and creative technology that, combined with GIS, can be utilized to study coastal geomorphology with access to the necessary surveying data. Under these circumstances, field surveying over long times and wide area can be reduced to save both money and time.

The objective in selecting data, as mentioned earlier, is to consider fully the need to integrate conveniently the data in the GIS environment, especially in terms of quality and operation. Such an approach also allows full advantage of GIS to be taken in data storage, collection and analysis, in order to effectively monitor littoral exploitation and clarify the decline of the waterways. As a result, the techniques applied in this study include computer-aided remote sensing image interpretation, GIS spatial analysis, GPS techniques (or the integrated application of 3S technology) and related mathematical statistics methods.

Images were matched to the bathymetric map, then overlain by the research boundary, in order to fuse all types of images and to analyse coastline flux. Subsequently, a man–machine interactive interpretation approach is performed in ARCGIS9.3, professional GIS software, developed by ESRI Company (USA), to delineate the various boundaries of the seashore components. The graphic interpretation results are compared further with field and GPS data in order to validate, check and correct the classification and identification performed by computer, integrated with satellite remote sensing, GIS and DGPS. The vector shorelines at different times are then generated. Finally, after being compiled and topologically related, the vector data layers at various times are overlain, so that the changing areas of the shorelines, between every two different periods, can be calculated. The resultant sketch of the interpretation and analysis of shorelines shift during the 24 years from 1986 to 2012 are shown in Figure 4. Furthermore, several historical shorelines are also drawn, converted from historical documents (Fig. 5).

Meanwhile, the spectral characteristics of the shorelines (such as rocky, sand, mud and artificial coastline) are distinct from one to another, so that they are easily interpreted and extracted from the images. In general, it is very difficult to obtain the correct elevation of the shoreline whilst the exact sea level is unknown. Likewise, multi-temporal images were gained at different times, with different tides and sea levels. As such, all the shoreline data extracted from the images should be corrected and adjusted with the tidal data at the same time, in order to establish accurate elevation. Generally, mean sea level is used to correct the tidal data and the shoreline elevation. Thus, the shifts of the shorelines over different periods can be detected effectively and calculated. Even land-use conversion information can be established, with regard to water area loss and artificial land gain in the bay.

However, the DEM of the topography of Lingdingyang Bay in 2003 and in 2012 were built separately using GIS technology. Reconstruction of where and how the topography and bathymetry changed is made possible by comparing the DEM of these two phases (Figs 6 & 7). The surface analyst tool of ARCGIS 10.1 was applied to analyse and calculate the bathymetric change (Fig. 8).

Results

In general, the evolution of any geomorphology can be very complicated; it is a multi-factor process that includes bathymetry and shoreline shifts, as well as geology and geophysics. The present study focuses on two major factors, and provides accurate quantities from temporal–spatial analysis by GIS and remote sensing.

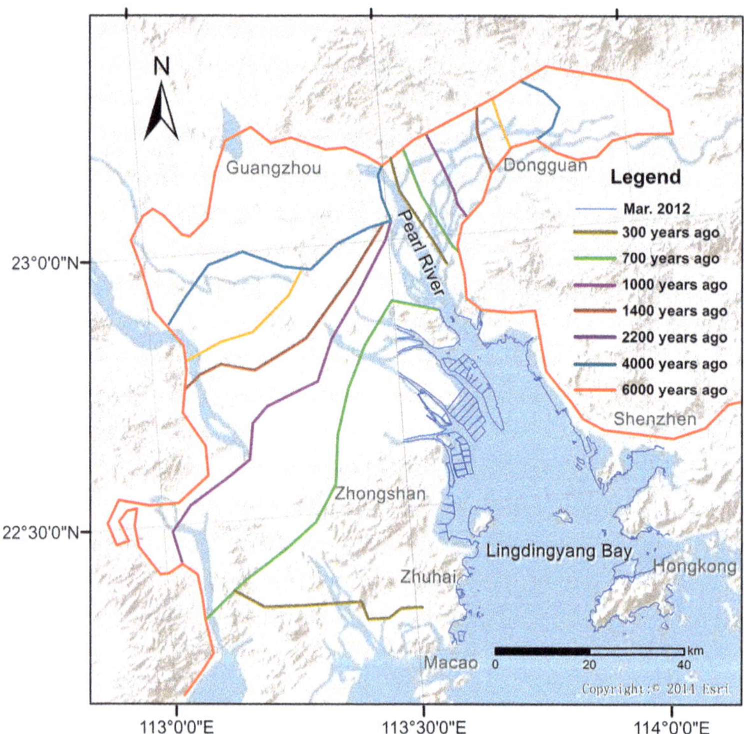

Fig. 4. Shoreline shift of the Pearl River Delta over the past 6 kyr (6000 years) (revised from Wang *et al.* 2005).

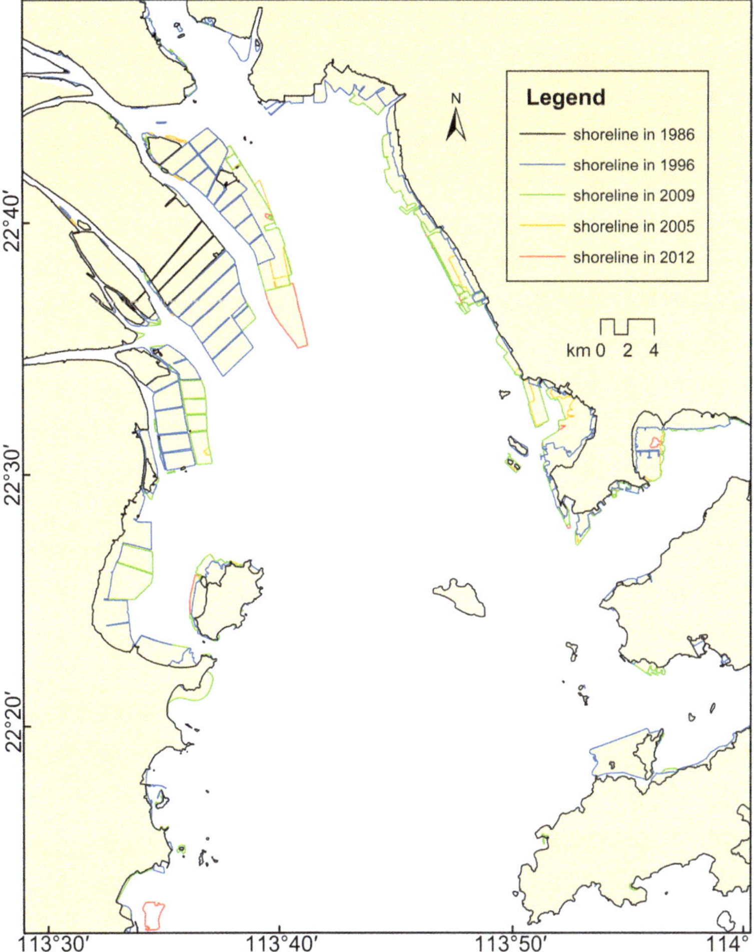

Fig. 5. Shoreline shift within Lingdingyang Bay over these 20 years.

Bathymetric variation

Bathymetric variation is a very important factor in analysing the geomorphological evolution of the region. As the bathymetric DEM shows, the troughs or channels, with an average depth of more than 10 m, are much deeper than the shoals (0–8 m) (Fig. 6). According to the DEM obtained in 2012, the Middle Fanshi Shoal is shown to be concavo-convex, with an average depth of 4 m. The East Shoal is less than 10 m in depth, whilst the West Shoal is flat with an average depth of 5 m (Fig. 7).

Comparing these two phases of DEM for Lingdingyang Bay, the west channel (the Lingding Channel), which is the main sea route, is shown to become shallower and narrower under the influence of sedimentary materials transported from the West Shoal. The channel was 10–15 m deep in 2003 and 4–10 m deep in 2012. The channel is also moving to the east (Fig. 8). The West Shoal is less than 6 m deep, with an average depth of 4 m, and at the same time is extending to the east. Depositional areas are shown as a yellow to red colour in Figure 8 (0–15.1 m depth). Areas of erosion are shown in dark blue in Figure 8 (−19.2 to 0 m depth).

The West Shoal is expanding towards the SE, with sediments accumulating in a seasonal fashion. The shoal even dries out in the autumn. The mid-shoal, the Fanshi Shoal, is flat. Deposition occurs readily when the tidal current is decelerated by the inner part of Lingding Island. The east shoal is changing and extending slowly, especially along the east bank, because of an absence of deposition, and so its topography shows less variability. At the same time, the scale of sand excavation is expanding considerably, and there are numerous and

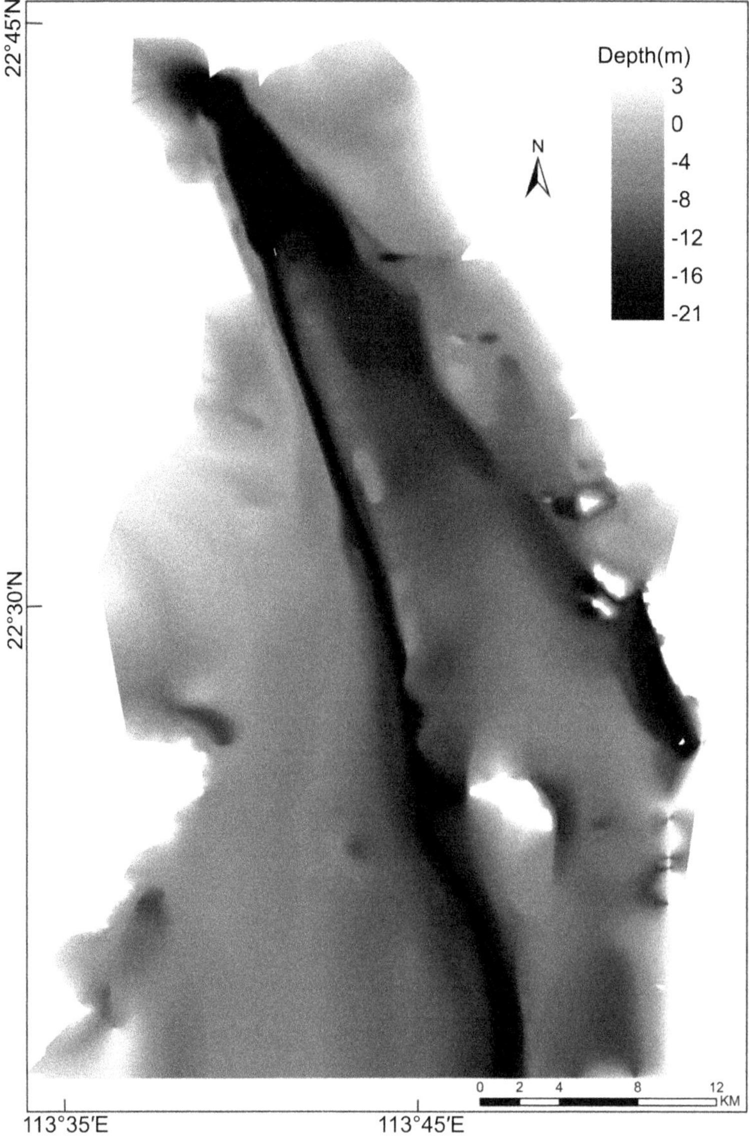

Fig. 6. The DEM of bathymetry in Lingdingyang Bay in 2003.

distinctive excavation holes scattered on the seabed of the West Shore of Lingdingyang Bay. The excavated area increased by about 57 042 m^2 from 2003 to 2012. The east channel, the Fanshi Channel, is about 5 km wide. Its northern part is about 730 m wider than its southern part, and was more than 24 m deep at its deepest point in the north in 2003. Since then, it has been generally stable due to strong tidal current action and a reduction in sediment supply. The Fanshi Channel was 7–9 m deep in 2003 and 8–20 m in 2012. However, parts of this channel are reducing in width and also deepening, with many depressions occurring because of irregular sand quarrying and the spread of aquaculture.

The Chuanbi Channel is the largest channel in Lingdingyang Bay, and represents a confluence between the Lingding Channel and the Fanshi Channel. The deepest point in the Chuanbi Channel changed from 27 m in 2003 to 36.9 m in 2012, in response to tidal dynamics.

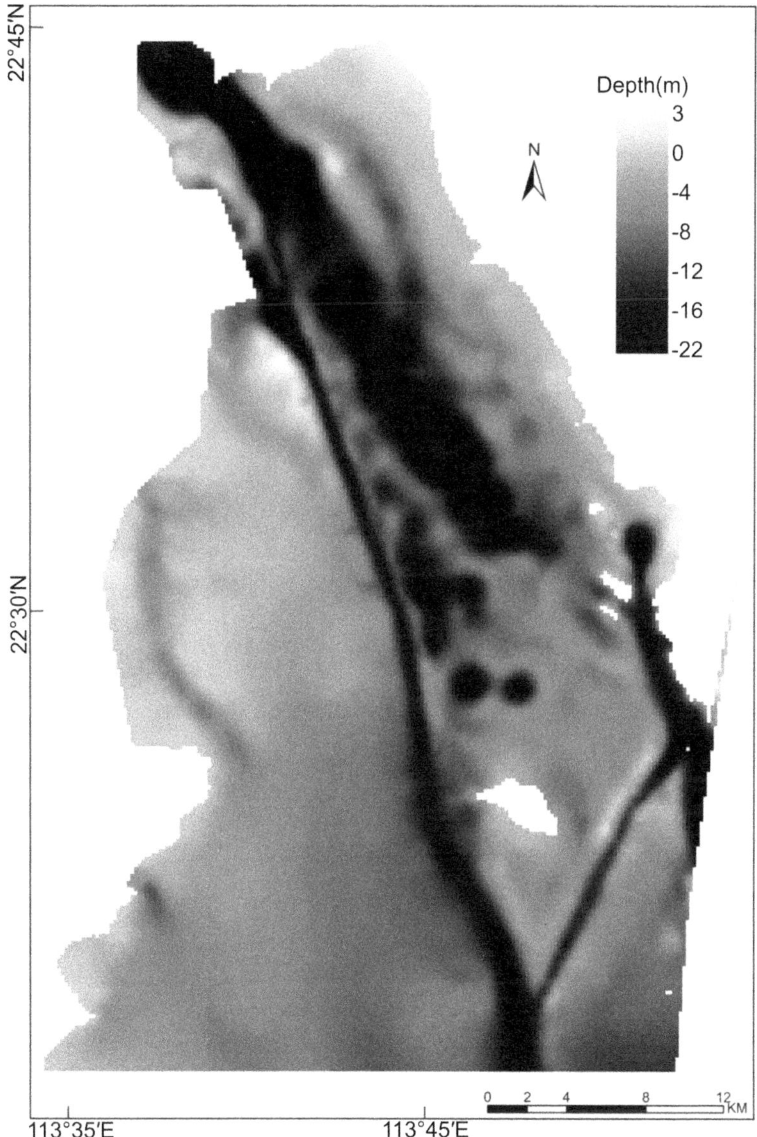

Fig. 7. The DEM of bathymetry in Lingdingyang Bay in 2012.

Geomorphological variations

Changes in shorelines are another element that can be analysed in order to reconstruct the geomorphological evolution. According to the statistical data for 1974 and 1984, obtained from the Guangdong Statistical Bureau, the area of shoals above the −2 m isobath in Inner Lingdingyang Bay has increased annually by 337 ha. Of this, 283 ha was ascribed to the West Shoal, the accretion rate of which has amounted to 1.3% annually. Furthermore, in response to the prevailing hydrodynamic conditions, the sediment discharge and the seabed characteristics vary from one outlet to another (Xia *et al.* 2013). During the tenth and the fifteenth centuries, Lingdingyang Bay was shaped as a large estuary and was affected primarily by the formation of delta plains from the west and north rivers, including the areas along Wugui Hill, Huangyang Hill and Niukuling Hill. Thereafter, the submarine plain gradually became exposed due to sediment discharges from the Humen, Jiaomen, Hongqili and

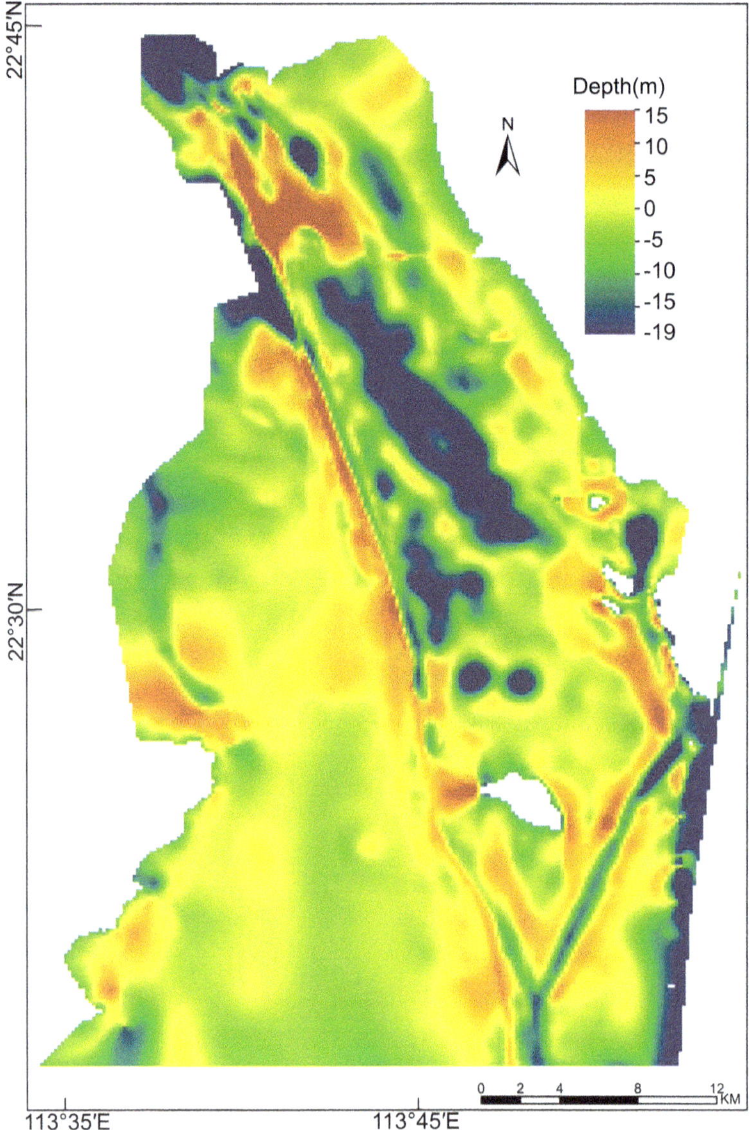

Fig. 8. The bathymetric change in the Lingdingyang Bay from 2003 to 2012.

Hengmen outlets. The exposed plain has been extending seawards constantly, leading to the shallowing of Lingdingyang Bay (Li *et al.* 1991).

By detailed spatial analysis of the changes in the shorelines of Lingdingyang Bay using two groups of data (one of which comprises six phases of historical data from 6 kyr ago to the present time, and another which has five phases, from 1986 to 2012), we can see that Lingdingyang Bay was part of the estuary around 5–6 kyr ago (Fig. 4). The shoreline in the NW was located about: 64–127 km from its present position 6 kyr ago; 60–100 km distant from its present location around 4 kyr ago; 44–86 km away from its present position 2.26 kyr ago; 34–78 km from its present location 1.46 kyr ago; 32–70 km away from its present position 16 kyr ago; 16–53 km away from its present location 700 years ago; and a distance of 9–31 km from its present position around 300 years ago. The overall tendency in the reduction of the Lingdingyang Bay is shoreline migration from the west to the SW, towards the South China Sea. In response to the abundant sand infilling from the four outlets, the land area has increased by 6261 km^2 during the past 6 kyr. Within the last 50 years, the western shoreline of the bay has moved forward continuously, in response to land reclamation. Xia *et al.* (2007) has calculated, on the basis of TM images, that 146 km^2 of land has been reclaimed on the West Shore and 42.6 km^2 on the East Shore. Over the past 30 years, the western shoreline has moved 1330 m seawards, and the land area has had equal annual increases of 11 km^2. The shoreline has moved seawards by 10 km in the Hongqili Outlet area, 13 km seawards in the Jiaomen Outlet area and 10 km seawards in Henmen Outlet area. The east shoreline of the bay remained relatively stable from 1966 to 1996.

This research has established that, between 1986 and 2012, about 250 km^2 of land was reclaimed on the West Shore and about 55 km^2 on the East

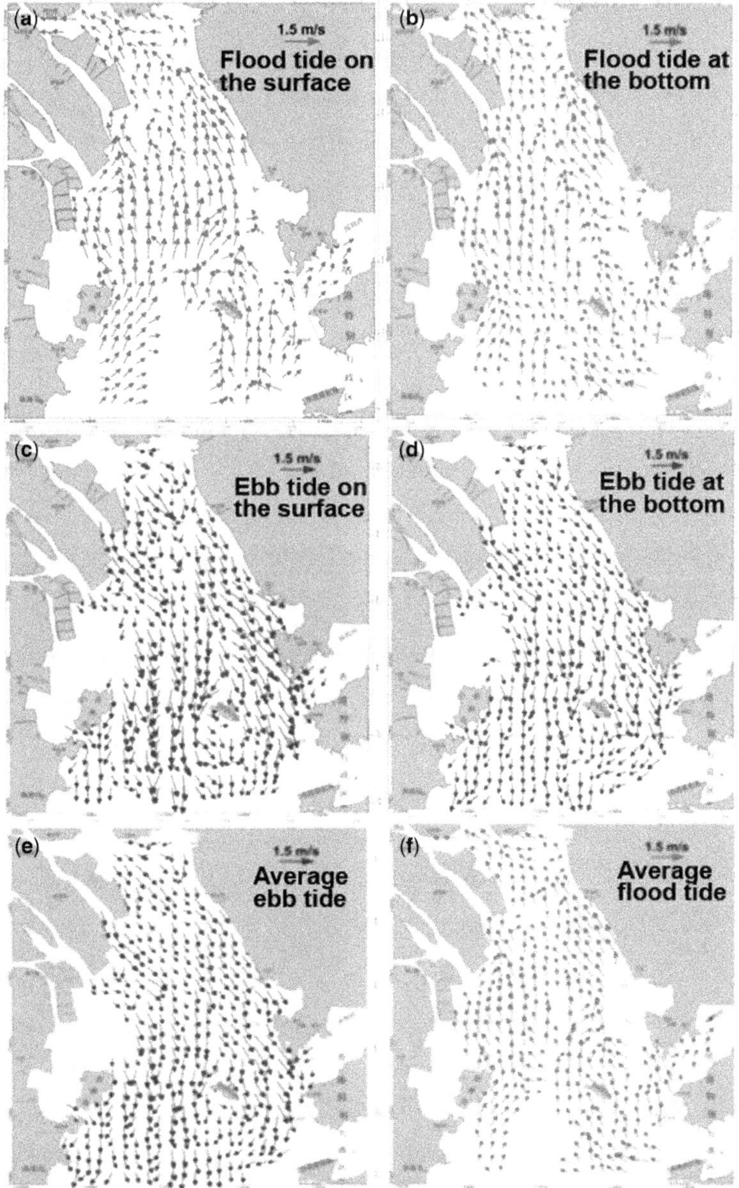

Fig. 9. Tidal currents in Lingdingyang Bay.

Shore. The shoreline has moved seawards by some 6.6 km in the Hongqili Outlet area, 17.0 km in the Jiaomen Outlet area and 9.4 km in the Henmen Outlet area. This amounts to about 320 m of seawards movement annually, and was similar to the west between 1986 and 2012. The comparative reclaimed land on the East Shore was 1.8 km during this time. Thus, large-scale land reclamation has been carried out in Lingdingyang Bay for more than 20 years, causing a reduction in the area of the bay. Large areas of reclaimed land can be found in the Chiwan and Shekou districts of Shenzhen. Furthermore, many embankments and ports have been established along the coast. As this rate, Lingdingyang will be largely infilled and will become a river within about 100 years.

Hydrodynamics

On the basis of the above descriptions, it can be seen that the geomorphology of Lingdingyang Bay has changed dramatically from that of 6 kyr ago. Various factors have contributed to these changes: for example, tidal currents, Pearl River discharges, rising local sea level, land reclamation, sand dredging and aquaculture. All of these can be divided into two processes: natural and human impacts.

The hydrographic conditions in Lingdingyang Bay are complicated, and are influenced by the bathymetry, tidal currents and discharges from the rivers. The tidal currents are the main forcing mechanism in the bay: they ranged between 0.4 and 178 cm s^{-1} in 2003–04, predominantly in a NW–SE direction (Fig. 9). The two troughs in the bay were created by the regular periodic movement of tidal currents. The tide is irregular and of a semi-diurnal type. In summer, the surface current speed is 21–190 cm s^{-1} on a spring tide and 6–52 cm s^{-1} on a neap tide (Fig. 9a). Bottom current speeds are 29–107 cm s^{-1} on a spring tide and 8–29 cm s^{-1} on a neap tide (Fig. 9b). In winter, surface currents are 14–157 cm s^{-1} on spring tides and 6–67 cm s^{-1} on neap tides, while bottom currents are 38–118 cm s^{-1} on a spring tide and 16–51 cm s^{-1} on a neap tide. The corresponding flood tide velocity is 0.3–84 cm s^{-1}, while the ebb tide reaches 18–92 cm s^{-1}. The survey data carried out by GMGS, on 7 June 2012 using ADCP (Acoustic Doppler Current Profilers), indicates that the largest flood tide current reaches 1.49 m s^{-1} at the surface and 1.93 m s^{-1} for an ebb tide. In contrast, a flood tide is 1.88 m s^{-1} at the bottom and 1.18 m s^{-1} for ebb tides (Fig. 9a–d).

The tide in Lingdingyang Bay is irregular semi-diurnal, with mixed currents from outlets and the open sea. The current increases from the north to the south of the bay due to the combined effects of the trumpet-shaped estuary and the strong flow, which worsens sediment sorting. Coarse sediments are deposited in the north and fine sediments are deposited in the south. As a result, the sorting improves towards the mouth. The tidal currents in the Humen Outlet are much stronger than in the western outlets (the Jiaomen, Hongqili and Hengmen outlets), which in turn influences the Chuanbe Channel.

All of the above research demonstrates that anthropogenic impacts are important in controlling the evolution of Lingdingyang Bay, not only because of land reclamation, but also due to dredging and other constructions (Xia *et al.* 2007). On the one hand, land reclamation has reduced the width of the bay mouth, but also constrained the outlets, and decreased the current strength. On the other hand, the dredging of sand for reclamation leaves behind silt re-deposition and disrupts the balance of water flux in the bay.

With the outlets extending seawards because of the constant deposition and land reclamation, the western slopes of the West Channel have tended to collapse easily to form many sandbanks, bars, new erosion troughs and gullies. The environment of the area is influenced by continuously increasing usage of the waterway and the utilized capacity of the port and aquaculture, as well as related problems. The Pearl River runoff extends in a tongue shape into Lingdingyang Bay, where it intensifies the flow during the flood season, depositing sediment near the outlets as the flow decelerates.

For example, during the summer of 1997, floods in the upper reaches of the Pearl River (resulting from unplanned coastal land reclamation, watercourse siltation and diminished drainage capacity) caused damage estimated at 1 billion yuan (*c.* $150 million).

Conclusions

In this study, time-series and high-resolution remote sensing techniques were combined with GIS to investigate the geomorphological evolution of Lingdingyang Bay by mapping and seafloor characterization. Such approaches have shown that Lingdingyang Bay is characterized by a heterogeneous geomorphology where, over a relatively short time and spatial scale, diverse coastal seascapes have been present and have originated from the interaction of different processes.

We conclude that, with the dramatic economic expansion around Inner Lingdingyang, the bay of the Pearl River Estuary, regional socioeconomic activity and other anthropogenic activities have become influential geomorphological agents and drivers of vigorous land-use change. Before 1970, natural sediment depositional processes were important elements affecting shoreline changes. However, in the past 50 years, anthropogenic activities such as land reclamation, sand excavation and sand dredging (for navigation purposes), as well as engineering constructions, have had a critical influence on the configuration of the coastal geomorphology. Consequently, the terrain and the bathymetric and hydrographic conditions have all changed. Such changes are clearly reflected in modifications to coastal geomorphology, as shorelines shifted and underwater bathymetry has been modified.

In response to an expansion of thc ports and industrial sites around the bay, the major land-use pattern in the Inner Lingdingyang area has been converted from early agriculture and aquaculture exploitation to more profitable high-technology industries. At the same time, unplanned land reclamation has caused intensified watercourse siltation, leading to the extension of rivers where sand transport is arrested. In response to the shallowing of the riverbed, the drainage capacity of the river is diminished and this results in a greater risk of flooding because the waters cannot be released freely to the open sea. As the bay is reducing in area, the water exchange in the bay is lessened and the hydrodynamics are weakening. Human-induced shoreline changes are more significant along the West Bank than on the East Bank. The land is expanding from the west, to the south and SW, at a historical average rate of $8\ km^2\ a^{-1}$, with an average of $8.93\ km^2\ a^{-1}$ over the past 50 years. As this rate, without human intervention, Lingdingyang Bay is likely to become completely infilled and the flow restricted to rivers within the next 100 years. With the Western Shoal extending and the Lingding Channel moving to the east, the Fanshi Channel in the east is about to become the major channel. Parts of Fanshi Channel have already become deeper over the past 30 years.

This study was supported by the ecological survey project (No. 200311000006) funded by the China Geographical Survey, the Ministry of Land and Resources PRC, and was carried out by the Guangzhou Marine Geographical Bureau. All the survey data used in the research were provided by the project. The authors gratefully acknowledge all the support.

References

Chen, S.S., Chen, L.F., Liu, Q.H., Li, X. & Tan, Q.Y. 2005. Remote sensing and GIS-based integrated analysis of coastal changes and their environmental impacts in Lingding Bay, Pearl River Estuary, South China. *Ocean & Coastal Management*, **48**, 65–83.

Chen, Y.T. 1995. Sedimentation divisions of Pearl River mouth. *Acta Scientiarum Naturalium Universitatis Sunyatseni*, **34**, 109–114.

Frihy, E., Dewidar, K.M., Nasr, S.M. & Raey, M.M.El. 1998. Change detection of the northeastern Nile Delta of Egypt: shoreline changes, spit evolution, margin changes of Manzala lagoon and its islands. *International Journal of Remote Sensing*, **19**, 1901–1912.

Glibert, P.M., Magnien, R. *et al.* 2001. Harmful algal blooms in the Chesapeake and coastal bays of Maryland, USA: comparison of 1997, 1998, and 1999 events. *Estuaries and Coasts*, **24**, 875–883.

Gu, Q.S., Rao, X.Y. & Li, X. 1990. *Remote Sensing in Lingdingyang Estuary*. Science Press, Beijing.

Hanebuth, T.J.J., Voris, H.K., Yokoyama, Y., Saito, Y. & Okuno, J.i. 2011. Formation and fate of sedimentary depocentres on Southeast Asia's Sunda Shelf over the past sea-level cycle and biogeographicimplications. *Earth-Science Reviews*, **104**, 92–110.

Li, P.R. & Qiao, P.N. 2012. Calculation of shrinking and vanishing of Lingdingyang Sea at the Pearl River Estuary. *Tropical Geography*, **32**, 260–262.

Li, P.R., Qiao, P.N. & Zheng, H.H. 1991. *Environment Evolution of Zhujiang Delta in the Past 10 000 Years*. China Ocean Press, Beijing.

Muttitanon, W. & Tripathi, N.K. 2005. Land use/land cover changes in the coastal zone of Ban Don Bay, Thailand using Landsat 5 TM data. *International Journal of Remote Sensing*, **26**, 2311–2323.

Peterson, D.L., Egbert, S.L., Price, K.P. & Martinko, E.A. 2004. Identifying historical and recent landcover changes in Kansas using post-classification change detection techniques. *Transactions of the Kansas Academy of Science*, **107**, 105–118.

Statistical Bureau of Guangdong Province 1988. *Guangdong (China) Yearbook of Statistics. October 1988*. China Statistics Press, Beijing.

Statistical Bureau of Guangdong Province 1998. *Guangdong (China) Yearbook of Statistics. December 1998*. China Statistics Press, Beijing.

Wang, Z.Y., Cheng, D.S. & Liu, C. 2005. Impacts of human activities on typical delta processes – the Yangtze and Pearl River deltas. *Journal of Sediment Research*, **12**, 76–79.

Wen, P., Liu, P. & Lei, Y.P. 2003. An analysis of the tendency of erosion and sedimentation on Foreshore and Channels of Lingding estuary in past fifty years. *ACTA Scientiarum Naturalium Universitatis Sunyatseni*, **42**, 240–243.

Weng, Q.H. 2002. Land rise change analysis in the Zhujiang Delta of China using satellite remote sensing, GIS and stochastic modeling. *Journal of Environmental Management*, **64**, 273–284.

Xia, Z. 2005. Characters of underwater topography and geomorphology in inner Lingdingyang firth of the Pearl River (Zhujiang River) estuary. *Marine Geology & Quaternary Geology*, **25**, 19–24.

Xia, Z., Jia, P.H., Lei, Y. & Chen, Y.Z. 2007. Dynamics of coastal land use patterns analyses of Inner Lingdingyang Bay in the Zhujiang River Estuary. *Chinese Geographical Science*, **17**, 222–228.

Xia, Z., Jia, P.H., Ma, S.Z., Liang, K., Shi, Y.H. & Joanna, J.W. 2013. Sedimentation in the Lingdingyang Bay, Pearl River Estuary, southern China. *Journal of Coastal Research*, **66**, 12–24.

Ying, Z.F. 1995. A study on notable features of sedimentary dynamics in Lingdingyang estuary of Zhujiang River. *Tropic Oceanology*, **14**, 76–82.

Zhang, J., Pu, L.J., Shan, Y.J., Zhang, R.S., Xu, Y., Zhu, M. & Peng, B.Z. 2012. Progress of the research on land exploitation and its use and eco-environmental effects of coastal zone. *Resources and Environment in the Yangtze Basin*, **21**, 36–43.

Zong, Y., Huang, G., Switzer, A., Yu, F. & Yim, W.W.S. 2009. An evolutionary model for the Holocene formation of the Pearl River delta, China. *The Holocene*, **19**, 129–141.

Stratigraphic variations in the Diaokou lobe area of the Yellow River delta, China: implications for an evolutionary model of a delta lobe

YONGHONG WANG[1]*, XIUJIN LIU[1,2], GUANGXUE LI[1] & WEIGUO ZHANG[3]

[1]*Key Laboratory of Submarine Science and Prospecting Techniques, Ministry of Education, Ocean University of China, Qingdao 266100, China*

[2]*Qinhuangdao Mineral Resources and Hydrogeological Brigade, Hebei Geological Prospecting Bureau, Qinhuangdao 066001, China*

[3]*State Key Laboratory of Estuarine and Coastal Research, East China Normal University, Shanghai 200062, China*

**Corresponding author (e-mail: yonghongw@ouc.edu.cn)*

Abstract: Characterization of the evolution of delta lobes has theoretical significance for the formation of entire deltas. The Diaokou lobe of the Yellow River delta provides a typical example. The Yellow River transported a great deal of sediment to the Diaokou area during the period of 1964–76 after the river changed its course to the Diaokou River and formed a new delta lobe – the Diaokou lobe. The Diaokou lobe reflects the evolution of the modern Yellow River delta and contains a record that represents the complete modern Yellow River delta depositional system. Using grain-size characteristics, magnetism, and the accelerator mass spectrometry (AMS) ^{14}C dating of cores ZK10-3 and ZK30 in the northern part of the Diaokou lobe, combined with collected data from cores ZK227, ZK1 and ZK228, which are located further south on the Diaokou lobe, we analysed the Holocene and recent stratigraphy of the Diaokou lobe area and its evolution. The Holocene stratigraphy of the Diaokou lobe and the nearby area contains upwards-succession, shallow-marine, river and lake, salt-marsh, and delta facies. This area received deltaic deposits beginning in 1855 and experienced prodelta, delta-front (lateral) and delta-plain deposition. When the Diaokou lobe began to form, this area experienced four stages of deposition after 1964: (1) dispersed-flow deposition; (2) single-channel deposition; (3) diversion deposition; and (4) abandonment and erosion. Compared with the evolution of a lobe of the Mississippi delta, the Yellow River delta contains thick, laterally extensive deposits as a result of the higher sediment load.

Characterization of the evolution of delta lobes has theoretical significance for the formation of entire deltas. A lobe formed by a single stream is an important element of a delta, and study of the stratigraphic evolution of a single stream can help us to understand the development of river-dominated deltas. The Yellow River delta in China is a typical river-dominated delta that has developed in conditions of weak tides and sand deposition. The river shifts its course frequently and sediments accumulate in the estuary (Chen *et al.* 2005; Wang 2012), thereby forming a large and thick delta that has consisted of several superlobes since the middle Holocene (Qiao *et al.* 2011). The overall Yellow River delta, like the Mississippi River delta (Roberts 1997), is composed of juxtaposed and imbricated deltaic lobes that were formed around mouths. Older deltaic deposits of this type are favourable targets for petroleum exploration; thus, the record of stratigraphic evolution of this type of delta during the Holocene has an important theoretical and practical significance. Eleven stratigraphic units, including six marine layers and five terrestrial layers, have been deposited near the estuary of the Yellow River since the middle Pleistocene (Zhang *et al.* 2004). The stratigraphic evolution of the nearby area during the Quaternary includes a repeating succession of shallow-marine and prodelta deposits (Liu *et al.* 2006). Analysis of shallow sediments on the NE side of the Yellow River delta revealed that the sedimentary facies consist primarily of riverbed, crevasse-splay, floodplain and tidal-flat facies (Wang & Ye 1990), and the primary types of vertical sequences include: (1) estuarine sand sheets and delta lateral area deposits; and (2) estuarine sand sheets, delta lateral area deposits and sand sheets (Fan 2009). Although each of the Yellow River delta lobes developed at different times, each lobe developed following a pattern of growth, seawards extension, lateral expansion and retrogradation (Cheng *et al.* 1987).

The sediments of the Diaokou lobe were deposited in the Shenxian Channel during the period of 1934–64 and in the Diaokou Channel during

From: Clift, P. D., Harff, J., Wu, J. & Qui, Y. (eds) 2016. *River-Dominated Shelf Sediments of East Asian Seas*. Geological Society, London, Special Publications, **429**, 185–195.
First published online October 2, 2015, updated May 26, 2016, http://doi.org/10.1144/SP429.8

1964–76 (Fig. 1). After the Yellow River shifted artificially to the Qingshui Channel in 1976, the sediment supply was nearly cut off and the coast rapidly eroded landwards (Gao 2011). This paper discusses the recent and Holocene stratigraphic evolution of the Diaokou lobe area and its development based on grain-size characteristics, and on the magnetism of two cores (ZK10-3 and ZK30) obtained from the northern part of the Diaokou lobe and three cores (ZK227, ZK1 and ZK228) collected further south on the Diaokou lobe.

Geographical setting

Geographical setting of the Yellow River Estuary

The Yellow River originates on the eastern Qinghai–Tibet Plateau and flows eastwards into the Bohai Sea, and has a total length of 5464 km. Geologically, the basin above Lanzhou is situated on the uplifting Qinghai–Tibet Plateau, where the channels are mostly valley shaped and composed of bedrock, while the middle reaches meander on the loess regions and the lower reaches on the submerging North China Plain. The study area is located in the Bohai Basin area. The Bohai Basin is a cratonic fault-bounded basin of Mesozoic–Cenozoic age. Faulted depressions and deformations of strata are controlled by block-faulting activities. Two fault systems, which strike NE–ENE and NW–WNW, control the distribution of the uplifts and depressions in the basin. The NE-striking fault is the Tancheng–Lujiang Fault and the study area is in the middle part of the Tancheng–Lujiang Fault System (Qin *et al.* 1990).

The Bohai Basin has been strongly affected by Quaternary sea-level changes because the maximum water depth reaches only 30 m in the area and the average depth is only 18 m. There were four relative sea-level rises during the Late Pleistocene as identified onshore by Qin *et al.* (1990), and they were Cangzhou (108–70 ka BP: which had two pulses), Bohai (65–53 ka BP) and Xianxian (39–22 ka BP) (Qin *et al.* 1990; Marsset *et al.* 1996). Then the Bohai Sea was subaerially exposed during the last glacial maximum (LGM), when the sea level was 120–130 m lower than the present level (Yang & Lin 1993; Hori *et al.* 2001). The palaeo-shoreline reached the present coastal position at 8 ka BP and the Holocene maximum transgression at 6–7 ka BP in the Bohai Sea area. After the sea level was highest during the middle Holocene, it oscillated slightly and then fell to the present level (Zhao *et al.* 1979).

The modern Yellow River Estuary has formed since 1855 and is located on the west coast of the

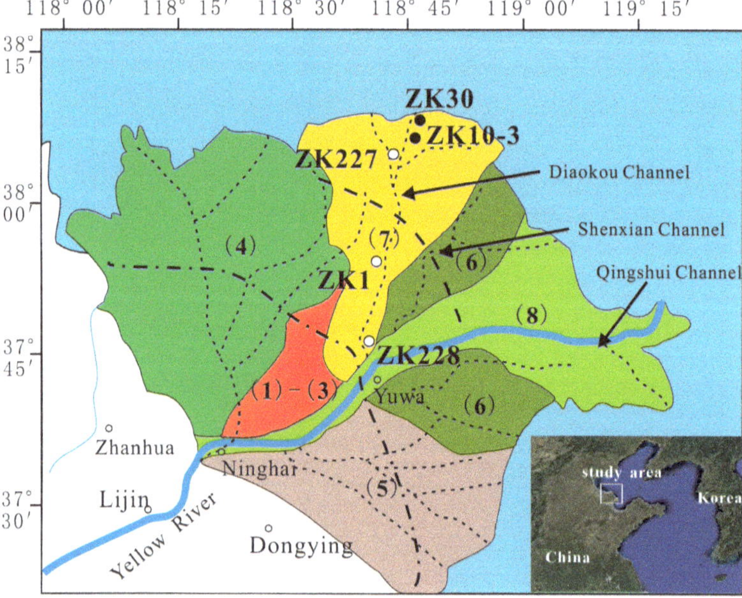

Fig. 1. Study area and locations of the cores. Lobe (7) indicates the Diaokou Lobe that formed during 1964–76. Black dots show the core positions, and white circle show the collected cores. -------, indicates the abandoned channel; – – –, indicates the 1937 coastline; – · – · –, indicates the 1855 coastline. (1) 1855–89; (2) 1889–97; (3) 1897–1904; (4) 1904–29; (5) 1929–34; (6) 1934–53 (north), 1934–64 (south); (7) 1964–76; (8) 1976–present.

Bohai Sea (Fig. 1). It is composed of juxtaposed and imbricated deltaic lobes, which formed around the mouth of the river at different periods (Shi *et al.* 2003) as a result of the vertical aggradations of the channels, and the increasing channel instability and avulsion (Gelder *et al.* 1994). In contrast to most other large estuaries, the modern Yellow River Estuary has only one active distributary channel at any given time (Wang *et al.* 2006). Although the watercourse in the Yellow River Estuary changed frequently before 1950, it only had three major switches after 1950 (the Shenxian Channel, the Qingshui Channel and the Diaokou Channel), all of which were the result of artificial diversion. The river mouth changed to the Diaokou Channel in 1964. Since 1976, the Yellow River has flowed through the other channels (the Shenxian Channel and the Qingshui Channel) into the Bohai Sea.

Water and sediment discharges

Since the late Holocene, the Yellow River has exhibited heavy sediment load and relatively low water discharge. It has annually delivered more than 1000 Mt of sediment to the Yellow or Bohai seas (Milliman *et al.* 1987; Saito *et al.* 2001; Wang *et al.* 2007). Based on the sediment flux data from the Lijin Hydrological Station, which is the station that is closest to the sea (about 100 km upstream from the mouth of the river), the amount of sediment brought by the Yellow River to the sea from 1952 to 2012 was 7.0×10^8 t a^{-1}, which is nearly twice the amount brought by the Yangtze River. However, the water discharge at Lijin was 298×10^8 m^3 a^{-1}, which was only 3.4, 17 and 16% of the values for the Yangtze River (BCRS 2000–12), the Mississippi River and Mekong River, respectively. The water discharge and sediment load from the river to the sea have high seasonal variability; more than 60% of the annual water and sediment discharge to the sea occurs during the summer monsoon season (July–October) (Wang *et al.* 2007). The annual average sediment discharge was 12×10^8 t a^{-1} from 1950 to 1970 (Wang *et al.* 2007) and 4.9×10^8 t a^{-1} from 1970 to 2012. The corresponding water discharges were 500×10^8 and 220×10^8 m^3 a^{-1}. The runoff changes from 1976 to 1986 were influenced mainly by natural processes, and human disturbances began to grow in significance after that time (Ding & Pan 2007).

Tidal, wind and particle sizes of bed sediments

Within Bohai Bay, surface-current velocities are about 2–3 knots (kn) in the central part of the sea, but reach more than 5 kn in Yellow River Estuary. In addition, currents related to the extension of the Kuroshio Warm Current induce an anticlockwise circulation in the Bohai Sea. However, a clockwise circulation may develop seasonally in the NE part of the Bohai Sea. The maximum surface velocity related to this circulation does not exceed 0.4 kn (Marsset *et al.* 1996). The Yellow River Estuary is a weak tidal estuary, with a mean tidal range near the river mouth of less than 0.8 m (Wang *et al.* 2007). The waves are primarily governed by local winds in the river-mouth area, and the mean wave height is less than 0.3 m. The wind is usually from the SSE in summer, which generates a northwards alongshore current, and to the NW–NE in winter, which produces a southwards alongshore current.

Sampling and analytical methods

Core data

Two cores, designated ZK10-3 (38° 03′ 06.3″ N, 118° 46′ 24.8″ E; surface elevation 1.337 m; length 10.3 m) and ZK30 (38° 08′ 20.0″ N, 118° 48′ 29.7″ E; surface elevation 0.279 m; length 30.4 m), were drilled from the Diaokou lobe in April 2010. The cores were extruded from their PVC core tubes and were sliced into 2 cm-thickz segments (above 2 m) and 5 cm segments (deep portion) using a stainless steel knife in the laboratory. A total of 857 samples were obtained (Fig. 1). Additional cores, designated ZK227 (Cheng *et al.* 1987; Xue *et al.* 2009), ZK1 (Gong 2010) and ZK228 (Li *et al.* 1998; Xue *et al.* 2008), were collected further south on the Diaokou lobe from published references (Table 1).

Particle size and magnetic analysis

The particle sizes were measured using a laser particle-size analyser (The Malvern Mastersizer 2000) at the Qingdao Institute of Marine Geology. The measurement range spanned 0.02–2000 μm, the grain-size resolution was 0.01Φ and the relative error was less than 3%. Each sample weighed approximately 0.5 g. The samples were placed in a 10% H_2O_2 solution for 24 h to remove organic matter before the analysis. The magnetic characteristics were measured at the East China Normal University, State Key Laboratory of Estuarine and Coastal Research. The samples were dried at 40°C and then gently ground in an agate mortar. Approximately 7 g of sample material was placed into the sample compartment of the apparatus, and the low- (0.47 kHz) and high-frequency (4.7 kHz) susceptibilities (χ_{lf} and χ_{hf}, respectively) were measured using a Bartington MS2 dual-frequency susceptibility meter. The frequency-dependent susceptibility (χ_{fd}%) was calculated based on the relationship $\chi_{lf} = 100(\chi_{lf} - \chi_{hf})/\chi_{lf}$. The anhysteretic

Table 1. *Information on cores in this study*

No.	Core name	Positions		Surface elevation (m)	Core length (m)	Material	Calibrated age (cal ka BP)	Sources
		Longitude	Latitude					
1	ZK10-3	118° 46′ 24.8″	38° 03′ 06.3″	1.337	10.3	*		This paper
2	ZK30	118° 48′ 29.7″	38° 08′ 20.0″	0.279	30.4	Peat	8800 ± 100	This paper
3	ZK1	118° 40′ 50.7″	37° 53′ 49.6″	4.931	24.6	*		Gong (2010)
4	ZK227	118° 42′ 24.0″	38° 05′ 06.0″	2.432	28.3	Foraminifera	8800 ± 100	Cheng *et al.* (1987); Xue *et al.* (2009)
5	ZK228	118° 39′ 54.0″	37° 46′ 48.0″	5.081	27.2	Shell	8835 ± 100	Li *et al.* (1998); Xue *et al.* (2008)

*Using the changes in colour, lithology and historical records to determine the ages and sedimentary stages.

remnant magnetization (ARM) was acquired in a 0.04 mT DC field superimposed on a peak AF demagnetization field of 100 mT using a Molspin pulse magnetizer (Dtech 2000) and was expressed as ARM (χ_{ARM}) susceptibility, which was measured using a Molspin Minispin fluxgate magnetometer (MMPM 10). Both forward-field (1 T) and back-field (−20 and −300 mT) isothermal remnant magnetizations (IRMs) were induced. The low-field IRM was obtained at −20 mT and was defined as the 'Soft', where Soft = (SIRM − $IRM_{-20\,mT}$)/2. The saturation isothermal remnant magnetization (SIRM) of each sample was induced in a steady 1.0 T field. Reverse fields were then applied to evaluate the S_{-300} ratio parameters.

AMS ^{14}C and ^{210}Pb testing

We selected two samples of thick black layers (depths 22.0 and 23.6 m) from core ZK30 that contain a large amount of organic matter for accelerator mass spectrometry (AMS) ^{14}C testing at the Qingdao Institute of Marine Geology. Dating methods and results in cores ZK10-3, ZK227, ZK1 and ZK228 were summarized in Table 1. Another 37 samples were selected for ^{210}Pb testing at the East China Normal University, State Key Laboratory of Estuarine and Coastal Research.

Results and analysis

Based on the lithology, granularity and magnetic characteristics, we divided core ZK10-3 into four sedimentary facies, designated A–D (Fig. 2), and divided ZK30 into 10 sedimentary facies, designated A–J (Fig. 3).

The analysis of core ZK10-3

Core ZK10-3 can be divided into sections above (A and B) and below (C and D) the depth of 6.3 m. In the lower section, layer C (6.30–7.10 m) consists primarily of tawny clayey silt and minor yellow silty sand, and is interpreted as sheet-flood sediments. Layer D (7.10–10.3 m) consists primarily of rust-coloured clayey silt and is interpreted as deltaic lateral area deposits. The upper parts are relatively fine grained: that is, primarily silt and clay, and the median particle size ranges from 6.5Φ to 7.2Φ (average 6.9Φ). The average proportions of sand, silt and clay are 1.8, 69 and 29.3%, respectively. The χ, χ_{ARM}, χ_{fd}%, χ_{ARM}/χ and χ_{ARM}/SIRM values of layer C are slightly elevated, and the average S_{-300} is 93%, which indicates that the ferrimagnetic mineral fraction is greater than in the overlying layers, that the amount of SP grains is greater and that the amount of PSD/MD grains is much less. The values of χ, χ_{ARM}, χ_{fd}%, χ_{ARM}/χ and χ_{ARM}/SIRM in layer D are the highest of those throughout the core, and the SIRM/χ value is the least, which indicates that the amount of ferrimagnetic minerals is the greatest of that found throughout the core and that there are more SP grains.

Layer A (0–2.76 m) consists of light-yellow silty sand with parallel bedding, and is interpreted as riverbed and foreshore sediments. Layer B (2.76–6.3 m) consists primarily of light-yellow silty sand and thin layers of tawny clayey silt with parallel bedding, and is interpreted as riverbed sediments. These two layers consist primarily of sand and silty sand, and the median particle size ranges from 4Φ to 6Φ (average 4.9Φ). The average proportions of sand, silt and clay are 22.7, 68 and 9%, respectively. The values of χ and χ_{ARM} are low, and their averages of 35.3×10^{-8} and $125.8 \times 10^{-8}\ m^3\ kg^{-1}$ are the lowest measured throughout the core. Because the χ_{ARM} is sensitive to stable single-domain ferrimagnetic minerals, the magnetic mineral fractions in these two sections are low (Maher 1988). Although the SIRM varies abruptly, the average value is stable. The ferrimagnetic mineral concentration varies little throughout the core. The S_{-300} values exceed 90%, indicating that there are few incomplete anti-ferromagnetic minerals. The χ_{fd}% values indicate the concentration of SP grains in the samples. Usually, when

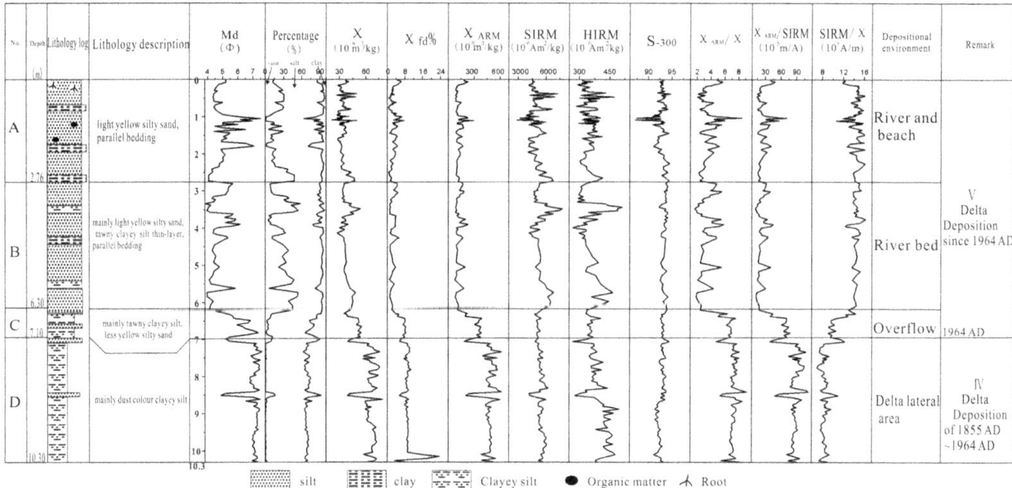

Fig. 2. Summary diagram of core ZK10-3.

the $\chi_{fd}\%$ is approximately 5%, the samples contain more SP grains (Zhang *et al.* 1995). In contrast, the χ_{ARM}/χ and χ_{ARM}/SIRM ratios indicate the concentration of MD/PSD grains in the samples. When the χ_{ARM}/χ values are <10 and the χ_{ARM}/SIRM values are $<60 \times 10^{-5}$ m A^{-1}, there tend to be more MD/PSD grains (Wang *et al.* 2004; Shi *et al.* 2007). In summary, the values of $\chi_{fd}\%$, χ_{ARM}/χ and χ_{ARM}/SIRM in these two layers are all low; thus, this section contains more MD/PSD grains.

The analysis of core ZK30

Core ZK30 can be divided into 10 layers. Layer J (29.70–30.40 m) consists of tawny, compacted soil sediments containing small shell fragments and is interpreted as a shallow-marine deposit. The grain size abruptly decreases at a depth of 29.7 m, and the average median particle size is 6.5Φ. The average proportions of sand, silt and clay are 9.7, 66 and 24%, respectively. The magnetic parameter characteristics are lower than, but similar to, those

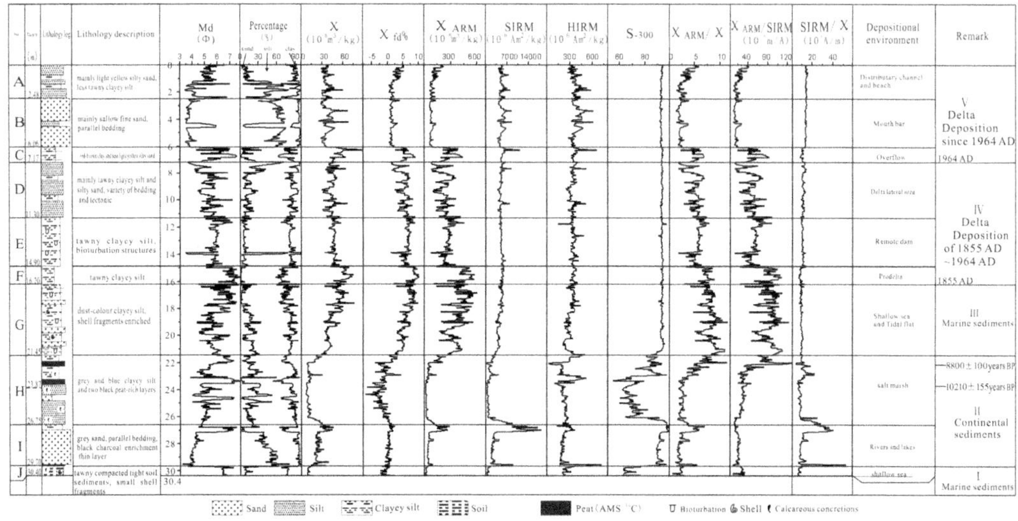

Fig. 3. Summary diagram of core ZK30.

of layer F, except that the $\chi_{fd}\%$, and certain magnetic parameters are the lowest observed in the core. The anti-ferromagnetic mineral fraction is the greatest in the core, the magnetic mineral particles are primarily PSD/MD grains and there are nearly no SP grains.

Layer I (26.52–29.70 m) consists of grey sand, displays parallel bedding, contains thin layers rich in black charcoal and is interpreted as lacustrine sediments. The upper portion of the layer is fine grained, but the lower portion is coarse grained. The average median particle size is 4.4Φ, and the average proportions of sand, silt and clay are 44.1, 50 and 8.5%, respectively. The χ, χ_{ARM} and SIRM values exceed those of layer J, but are still generally low. Therefore, there is little magnetic mineral material. The S_{-300} is only 91.5%, which indicates that there is little anti-ferromagnetic mineral material. The average $\chi_{fd}\%$ is only 0.3%, and thus there are nearly no SP grains.

Layer H (21.45–26.52 m) consists of grey and blue clayey silt, and contains two black peat-rich layers at depths of 22 and 23.5 m. The median particle size ranges from 3.8Φ to 7.5Φ (average 5.5Φ). The average proportions of sand, silt and clay are 17.9, 65.6 and 15.2%. Most of the magnetic parameter values vary little, except for $\chi_{fd}\%$ and S_{-300}, and the parameter values are low. These findings indicate that there are few magnetic minerals. The value of S_{-300} is only 77% and varies in a complex fashion, which indicates that there is little anti-ferromagnetic material. The average $\chi_{fd}\%$ is only −1.1%, and there are almost no SP grains.

Layer G (16.2–21.45 m) consists of rust-coloured clayey silt with abundant shell fragments, and is interpreted as shallow-marine to tidal-flat deposits. The grain-size characteristics are similar to those of layer C, and the median particle size ranges from 4.3Φ to 7.5Φ (average 5.9Φ). The magnetic parameter variations are the most complex: χ, χ_{ARM}, and SIRM decrease with depth, which indicates that the magnetic mineral fraction decreases with depth. The average S_{-300} is 93% and the SIRM/χ ratio is only 9.7, which indicates that there are fewer anti-ferromagnetic minerals and more ferrimagnetic minerals. The average $\chi_{fd}\%$ is 6% and decreases with depth, which indicates that there are few SP grains and that this content decreases with depth.

Layer F (14.90–16.20 m) consists primarily of tawny clayey silt and is interpreted as prodelta sediments. This layer is the finest part of the core, and the average median particle size is 7Φ. The sand content is extremely low: only 0.8%. The amount of clay represents an abrupt increase v. that of the overlying layers, and the average clay content is 31%. The magnetic parameter values are relatively high, and most of these values are the highest of those found in the core (except for the values of SIRM/χ), which indicates that there are few incomplete anti-ferromagnetic minerals, more ferrimagnetic minerals, more SP grains and fewer PSD/MD grains.

Layer E (11.30–14.90 m) consists primarily of tawny clayey silt, displays bioturbation structures and is interpreted as distal bar deposits. Layer D (7.17–11.30 m) consists primarily of tawny clayey silt and silty sand displays a variety of bedding and structures, and is interpreted as deltaic lateral area sediments. Layer C (6.06–7.17 m) consists primarily of red-brown clay and minor light-yellow silty sand, and is interpreted as sheet-flood sediments. The grain sizes of these three layers change abruptly, and the median particle size ranges from 5Φ to 6.5Φ (average 5.6Φ). The average proportions of sand, silt and clay are 12.1, 71.2 and 16.7%, respectively. The χ, χ_{ARM} and S_{-300} values are much greater than those of the upper layers, which indicates that there are more ferrimagnetic minerals here than in the upper layers. The average of $\chi_{fd}\%$ is approximately 5.8%, which indicates that there is a greater amount of SP grains. The χ_{ARM}/χ and χ_{ARM}/SIRM values are higher than those of the upper layers, which indicate that there are fewer PSD/MD grains than in the upper layers.

Layer B (2.48–6.06 m) consists primarily of pale fine sand, displays parallel bedding and is interpreted as mouth-bar deposits. The median particle size ranges from 3.4Φ to 6.7Φ (average 4.0Φ), and the average proportions of sand, silt and clay are 56, 38.1 and 5.6%, respectively. The sand content is the highest of the core. Layer A (0.0–2.48 m) consists primarily of light-yellow silty sand and minor tawny clayey silt, and is interpreted as distributary-channel and marginal-bank sediments. The median particle size ranges from 3.3Φ to 6.7Φ (average 4.8Φ). The average proportions of sand, silt and clay are 27.5, 62.1 and 10.9%, respectively. The χ, χ_{ARM} and SIRM values of layers A and B are low, which indicates that there are small amounts of magnetic minerals. The S_{-300} values exceed 90%, which indicates that there are few incomplete anti-ferromagnetic minerals and more ferrimagnetic minerals. The average $\chi_{fd}\%$ value is 2%; thus, there are few SP grains.

Stratigraphy of the Diaokou lobe

After comparing the findings from cores ZK227, ZK1 and ZK228 in the southern Diaokou lobe, and those of cores ZK10-3 and ZK30, we divided the stratigraphy of the Diaokou lobe into five layers (Figs 2 & 3): layer I, which consists of shallow-marine deposits; layer II, which consists of lacustrine deposits; layer III, which consists of shallow tidal-flat deposits; layer IV, which consists of deltaic deposits from 1855–1964; and layer V, which consists of post-1964 deltaic deposits

Dating of Holocene and modern sediments

AMS ^{14}C dating was performed on samples (depths 22 and 23.6 m) containing abundant black organic matter. The analysis yielded the calibrated dates of 8800 ± 100 and 10 210 ± 155 years BP. We chose 37 samples from core ZK10-3 and the upper layer of core ZK30 for analysis of ^{210}Pb. Unfortunately, no surplus ^{210}Pb was present, and we were thus unable to obtain accurate ages using this method.

Discussion

Modern sediment ages of the Diaokou lobe

The hydrodynamic conditions of the study area are very complex, which resulted in very little surplus ^{210}Pb in the samples. Thus, we were unable to determine age dates using the ^{210}Pb and ^{137}Cs methods. However, we were able to use the abrupt changes in colour and lithology to determine the ages (Xue 1994). Abrupt changes in the core can indicate changes in the position of the river mouth. There are clear historical records of distributary channel shifts (Xue 1994; Xue *et al.* 2009), particularly after 1934, and all of these are useful for determining the ages. Therefore, this method is better and less expensive than the ^{210}Pb method. Xue (1994) analysed the locations of distributary channels based on sudden changes in sediment cores and thereby determined the ages of the lobes of the Yellow River delta. There is a sudden change at a depth of 16.3 m for the core ZK30. The colour changes from greyish black to tawny, the clay content increases from 20 to 30% and the sand content decreases from 12.2 to 0.8%. Simultaneously, the χ, $\chi_{fd}\%$ and χ_{ARM} values change abruptly, and shell fragments and bioturbation are present below 16.2 m, which indicates that these sediments were deposited in a shallow-marine environment. Based on the extensive deposition of eroded fine loess particles in the prodelta, we can identify this depth as corresponding to the year 1855. The Yellow River shifted in 1964, thereby changing its pattern of flow into the sea, and fine-grained sediments were deposited at the river mouth. There is an abrupt change at a depth of 7.17 m in core ZK30: the clay content increases from 14.4 to 21.7%, the median grain size decreases from 5.2Φ to 6.0Φ and the lithological description changes from one of tawny silt to one of red-brown clayey silt. The χ, $\chi_{fd}\%$ and χ_{ARM} values also change abruptly. These indications are evidence of deposition of sediments of the Diaokou River. Based on the grain-size data of Figures 2 and 3, the deposits are coarse below the 1855 horizon and fine above.

Stratigraphic evolution of the Diaokou lobe since 1855

The Yellow River has shifted seven times since 1855 and has formed eight lobes during that time (Cheng 1991; Xue 1994; Huang & Fan 2004; Li *et al.* 2004; Wang *et al.* 2010) (Fig. 1). Each lobe was an active locus of deposition for a short time (i.e. years to decades). During the 12 years of the Diaokou lobe's development, 83 × 10^9 t of sediment was deposited, of which 34 × 10^9 t was deposited above sea level (Huang 1991). The data from the five cores indicate that each period of development was different (Table 2).

Stratigraphic evolution of the Diaokou lobe area

The local record exhibits evidence of two transgressions and two regressions in the study area. The sea level was high during the approximate period of 23–35 ka, at which time layer I was deposited (Figs 4 & 5). This layer primarily formed during the XianXian transgression (Zhang *et al.* 2004). The sediments of layer I are fine grained and contain abundant magnetic minerals. The last glacial maximum subsequently occurred, and the sea level fell by more than 100 m. The area was above sea level, and rivers and lakes developed, as indicated by layer II. AMS ^{14}C analysis was performed on samples (at depths of 22 and 23.6 m) that contained large amounts of black organic matter. The analysis yielded age dates of 8800 ± 100 and 10 210 ± 155 years BP (i.e. Holocene ages). The sediments are coarse near the bottom and fine near the top, the magnetic mineral fraction is low, and the magnetic mineral particles are coarse. The bottom part contains a greater ferrimagnetic fraction, and the upper part contains a greater incomplete anti-ferromagnetic mineral fraction. Approximately 8500 kyr ago, the climate warmed and the sea level rose. The Huanghua transgression (from 10 kyr ago to the present) occurred: this event was accompanied by 5.2 m of deposition (Cheng *et al.* 2010). This deposit developed in a tidal-flat environment and consisted of fine-grained sediment with abundant ferrimagnetic minerals. This deposit was blanketed by the modern Yellow River delta sediments of layers IV and V. Layer IV was deposited during the period of 1855–1964: this sediment is fine-grained and represents an abrupt change from the underlying material. The magnetic minerals are primarily ferrimagnetic minerals. The layer V was developed since 1964: the sediment grain size is

Table 2. *Evolution of the Diaokou lobe since 1855*

Time/year	River	Estuary and deposit characteristics
1855	Yellow River flow into the Bohai Sea	Large amounts of eroded loess were deposited on the shallow-marine sediments and a prodelta developed. The thickness of the prodelta deposit is approximately 1.3 m in core ZK30
1934		The Shenxian Channel lobe began to develop
1953	Artificial diversion	The estuary was at the SE end of the study area, and a distal bar and delta lateral areas developed; their combined thickness is approximately 7.7 m in core ZK30 and 3 m in core ZK10-3
1964	Artificial diversion	The estuary belonged to the Diaokou River, and the Diaokou lobe began to develop
1964–67	Overflow siltation	There was no fixed single river at the beginning of the diversion, and the river flowed into the Bohai Sea. The associated sediment thickness is approximately 0.8 m in core ZK30 and 1.1 m in core ZK10-3
1967–72	Fixed single-river deposition	The Yellow River gradually formed a single river and flowed into the sea through the central watercourse. The hydrodynamic conditions were relatively strong, and the grain-size characteristics changed abruptly from fine to coarse, and parallel bedding developed. The thickness of these sediments is approximately 2.5–2.8 m in the two cores
1972–76	Branching, shifting and associated deposition	The Yellow River shifted to the right watercourse of the lobe in 1972 and maintained this position for 2 years. The river shifted to the lobe's left watercourse in 1974 and maintained this position until 1976. Sediment deposition was constant and accumulated up to sea level, forming a delta plain. The thickness of these sediments is approximately 3.5 m in both cores
1976–	Abandoned retrogradation	The Yellow River shifted to the Qingshui Channel in 1976, and the sediment supply to the lobe was cut off. The lobe stopped growing, and experienced strong retrogradation under wave and tidal action

coarse and the magnetic minerals are mainly ferrimagnetic minerals.

The Yellow River delta and the Mississippi River delta are both composed of juxtaposed and imbricated deltaic lobes. The modern Yellow River delta is river-controlled, and the evolution of the river played an important role in the formation, development and sedimentary composition of the delta. The Yellow River delta is a shallow-water delta, and it has a vast shunt system and a large sediment load (which is approximately 42 times that of the Mississippi), which has resulted in its development into a fan-shaped delta. In contrast, the Mississippi River flows into the deep sea (thereby forming a deep-water delta), has well-developed distributary channels, carries a small sediment load and has developed a pedate structure (Ren 1989). The developments of the two deltas are similar: they are both composed of superimposed lobes. Eight lobes of the Yellow River have developed since 1855, and the average period of activity of each lobe was less than 20 years. For example, the Diaokou lobe is the youngest lobe of the Yellow River delta, and its period of activity has been only approximately 12 years. The Mississippi delta has been active for more than 7500 years, during which it has developed seven large lobes. The period of activity of these lobes has historically been approximately 1000–1500 years (Coleman & Prior 1980; Ye 1982). The changes in the Yellow River delta were dramatic because of its wide, shallow river channel, large sediment load and high rates of sediment deposition. In contrast, the Mississippi River is narrow and deep, and has a smaller sediment load and rate of sedimentation.

The sedimentary structures of the Yellow River delta in (upwards) vertical succession represent the following depositional environments: continental shelf, prodelta, distal bar, lateral delta, overflow, river and mouth bar, distributary channel, and beach (Penland *et al.* 1988). The corresponding succession of the Mississippi River delta represents the following depositional environments: continental shelf, collapse block, prodelta, distal bar, river and mouth bar, overbank splay, interdistributary bay, splay, bay, and marsh (Coleman & Wright 1975). The sediments are fine grained near the bottom and coarse grained near the top. The sedimentation models of the two deltas are very similar: they both begin with prodelta deposition, and end with river abandonment and erosion. However, there was no fixed channel during the initial stage of the Yellow

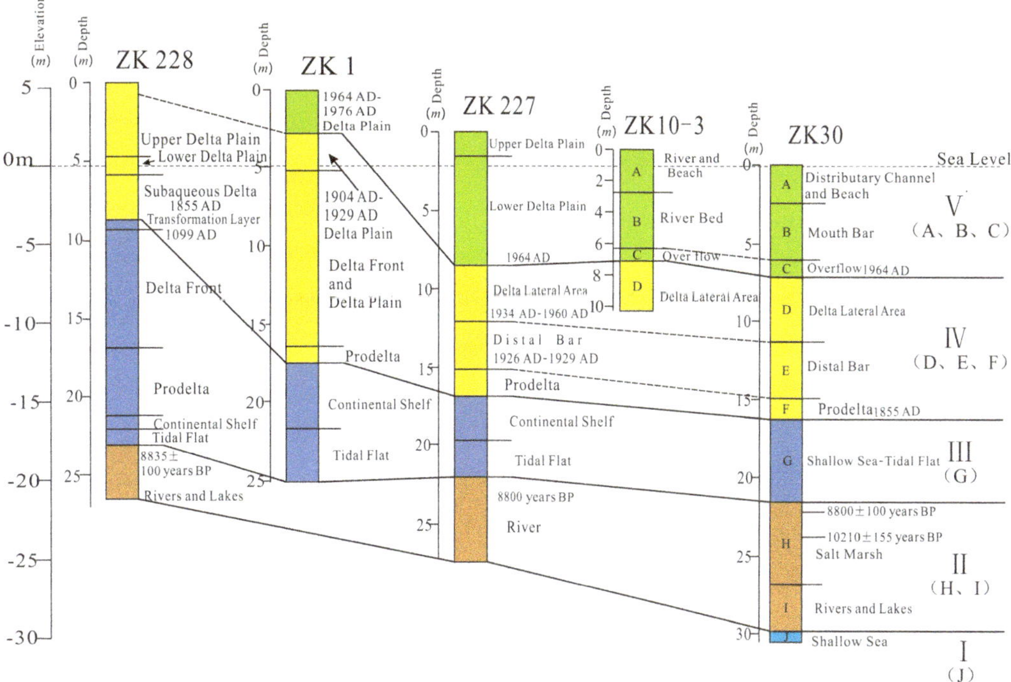

Fig. 4. Correlation chart of shallow sediments in the Diaokou lobe.

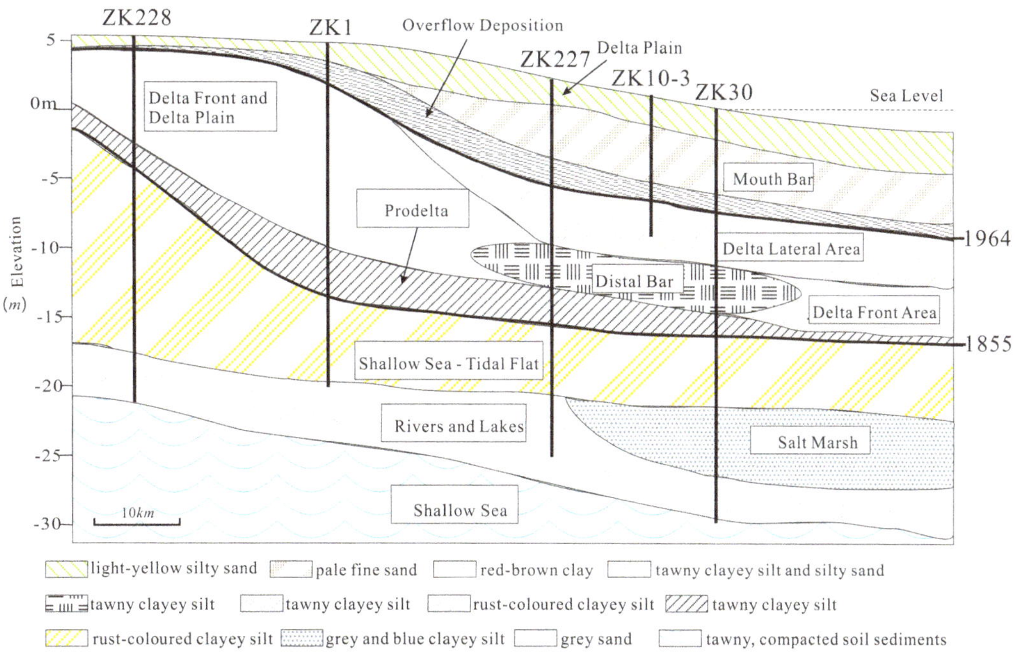

Fig. 5. Profile of the Diaokou lobe.

River delta development, and overflows went into the sea: thus, there were overflow deposits in the two cores. There are no overflow deposits in the Mississippi River delta sediments, which is possibly a result of the low sediment load (the Yellow River sediment concentration is 25.2 kg m^{-3}, whereas that of the Mississippi River is only 0.6 kg m^{-3}: Ye 1982). The Yellow River delta deposits contain a well-developed mouth bar that gradually extended seawards and covered the prodelta. This mouth-bar deposit coarsens upwards, which is the most important characteristic of the Yellow River delta (Cheng *et al.* 1986).

In summary, the evolutionary model of the Yellow River delta is that of a high-sediment-load river delta. Its development is controlled by the substantial amount of sediment, shallow estuary, weak wave and tidal action, and frequent channel shifts. These processes controlled the lobate development of the Yellow River delta.

Summary

- The silt fraction exceeds 50%, and certain sections contain more than 70% silt. The sand and clay fractions are less and vary greatly. The sediments are fine grained near the bottom and coarse grained near the top. The magnetic fraction consists primarily of ferrimagnetic minerals. Only a small fraction of the sediment samples consist of anti-ferromagnetic minerals – namely, in certain portions of core ZK30 – and the magnetic mineral particles are primarily PSD/MD grains.
- Based on analysis of the lithology and contrasting magnetic and grain-size parameters, we infer that the delta experienced shallow-marine and fluvial deposition, as well as lake, marsh, shallow-marine and delta deposition during the Holocene.
- Based on analysis of the lithology, magnetic parameters and grain-size parameters of cores ZK10-3 and ZK30, and collected data of the three southern cores ZK227, ZK1 and ZK228, we conclude that the delta deposits represent the following environments since 1855: prodelta, distal bar, lateral delta, overflow, river and mouth bar, distributary channel, and beach.
- The Yellow River delta contains a complex set of depositional environments. Based on the delta sediments and their sedimentary structures, we divided the development of the Diaokou lobe into four stages: overflow siltation; fixed single-channel deposition; branching shifting deposition; and channel abandonment and retrogradation. Compared to the evolution of a delta lobe of the Mississippi delta, the Yellow River delta contains thick, laterally extensive deposits as a result of the Yellow River's higher sediment load.

This study was funded by the National Natural Science Foundation of China (grant numbers 41176039, 41376054 and 41410304022) and the Ocean Public Welfare Scientific Research Project, State Oceanic Administration of China (grant number 201405037). The authors appreciate the assistance of Dr Liu Yong, Dr Xu Jisang and Dr Gao Wei from the Ocean University of China, and Dr Dong Chenyin from East China Normal University during the field and laboratory work.

References

BCRS 2000–12. *Yellow River Sediment Bulletin*. Bulletin of Chinese River Sediment, Ministry of Water Resources Conservancy, Beijing, http://www.yellowriver.gov.cn/nishagonggao/ [in Chinese].

Chen, S. L., Zhang, G. A. & Chen, X. Y. 2005. Coastal erosion feature and mechanism at Feiyantan in the Yellow River Delta. *Marine Geology & Quaternary Geology*, **25**, 9–14 [in Chinese with English abstract].

Cheng, G. D. 1991. *Modern Sedimentation and its Model of the Huanghe Delta*. Geological Publishing House, Beijing [in Chinese with English abstract].

Cheng, G. D., Ren, Y. C. & Li, S. Q. 1986. Channel evolution and sedimentary sequence of modern Huanghe River Delta. *Marine Geology & Quaternary Geology*, **6**, 1–12 [in Chinese with English abstract].

Cheng, G. D., Xue, C. T. & Zhou, Y. Q. 1987. Late-late Pleistocene and Holocene stratigraphy in Huanghe River Delta area. *Marine Geology & Quaternary Geology*, **7**, (Suppl.), 63–73 [in Chinese with English abstract].

Cheng, H. Y., Jiang, S. H. & Li, A. L. 2010. Analysis of the shallow stratigraphic division and sea-level variation in Yellow River Delta since the last glacial maximum. *Marine Geology Letters*, **26**, 31–36 [in Chinese with English abstract].

Coleman, M. J. & Prior, B. D. 1980. *Deltaic Sand Bodies*. American Association of Petroleum Geologists, Course Notes, **15**.

Coleman, M. J. & Wright, L. D. 1975. *Modern River Deltas: Variability of Process and Sand Bodies*. Houston Geological Society, Houston, TX, 99–149.

Ding, Y. F. & Pan, S. M. 2007. Evolutionary characteristics of runoff into the sea of the Huanghe River and their causes in recent 50 years. *Quaternary Sciences*, **27**, 709–717 [in Chinese with English abstract].

Fan, D. H. 2009. Stratigraphic characteristics and sedimentary models of near-surface formations in the estuary area of the Yellow River Delta. *Oil & Gas Geology*, **30**, 282–286.

Gao, W. 2011. Stratigraphy sequence of Diaokou lobe in the modern Yellow River Delta. PhD dissertation, Shandong Qingdao: Ocean University of China [in Chinese with English abstract].

Gelder, A. V., Berg, J. H. V. D., Cheng, G. & Xue, C. 1994. Overbank and channel fill deposits of the modern Yellow River Delta. *Sedimentary Geology*, **90**, 293–305.

GONG, S. J. 2010. *The sedimentary record and environmental analysis of ZK1, ZK3, ZK5 drilling cores in latter-day Yellow River Delta wetland*. PhD dissertation, China University of Geosciences (Beijing), Beijing [in Chinese with English abstract].

LI, A. L., LI, G. X. & CAO, L. H. 2004. The coast erosion and evolution of the abandoned lobe of the Yellow River Delta. *Acta Geographica Sinica*, **59**, 731–737 [in Chinese with English abstract].

LI, G. X., ZHUANG, Z. Y. & HAN, D. L. 1998. Stratigraphic sequences and characteristics of geological environment since the late period of last glacial age along southern shore of Bohai Sea. *Journal of the Ocean University of Qingdao*, **28**, 161–166 [in Chinese with English abstract].

LIU, S. F., ZHUANG, Z. Y., LV, H. Q. & FAN, D. H. 2006. The strata and environmental evolution in the late quaternary in the Chengdao area and modern Yellow River Delta coast. *Transactions of Oceanology and Limnology*, **2006**, 32–37.

HORI, K., SAITO, Y., ZHAO, Q., CHENG, X., WANG, P., SATO, Y. & LI, C. 2001. Sedimentary facies of the tide-dominated paleo-Changjiang (Yangtze) estuary during the last transgression. *Marine Geology*, **177**, 331–351.

HUANG, H. J. & FAN, H. 2004. Monitoring changes of nearshore zones in the Huanghe (Yellow River) Delta since 1976. *Oceanologia et Limnologia Sinica*, **35**, 306–314.

HUANG, S. G. 1991. The characteristics and evolution of scouring and silting in the Diaokou channel of the Huanghe River. *Coastal Engineering*, **10**, 10–24.

MAHER, B. A. 1988. Magnetic properties of some synthetic sub-micron magnetic. *Geophysical Journal*, **94**, 83–96.

MARSSET, T., XIA, D., BERNE, S., LIU, Z., BOURILLET, J. F. & WANG, K. 1996. Stratigraphy and sedimentary environments during the Late Quaternary, in the Eastern Bohai Sea (North China Platform). *Marine Geology*, **135**, 97–114.

MILLIMAN, J. D., QIN, Y. S., REN, M. E. & SAITO, Y. 1987. Man's influence on the erosion and transport of sediment by Asian rivers: the Yellow River (Huanghe) example. *Journal of Geology*, **95**, 751–762.

PENLAND, S., BOYD, R. & SUTER, R. J. 1988. Transgressive depositional systems of the Mississippi Delta plain: a model for barrier shoreline and shelf sand development. *Journal of Sedimentary Petrology*, **58**, 932–949.

QIAO, S. Q., SHI, X. F. *ET AL*. 2011. Sedimentary records of natural and artificial Huanghe (Yellow River) channel shifts during the Holocene in the southern Bohai Sea. *Continental Shelf Research*, **31**, 1336–1134.

QIN, Y., ZHAO, Y., CHEN, L., ZHAO, S. 1990. Geology of the Bohai Sea. Institute of Oceanology, Academia Sinica, Guangzhou [in Chinese].

REN, M. E. 1989. Man's impact on the coastal zone of the Mississippi River Delta. *Acta Geographica Sinica*, **44**, 221–229 [in Chinese with English abstract].

ROBERTS, H. H. 1997. Dynamic changes of the Holocene Mississippi River delta plain: the delta cycle. *Journal of Coastal Research*, **13**, 605–627.

SAITO, Y., YANG, Z. S. & HORI, K. 2001. The Huanghe (Yellow River) and Changjiang (Yangtze River) deltas: a review on their characteristics, evolution and sediment discharge during the Holocene. *Geomorphology*, **41**, 219–231.

SHI, C., ZHANG, D. D. & YOU, L. 2003. Sediment budget of the Yellow River Delta, China: the importance of dry bulk density and implications to understanding of sediment dispersal. *Marine Geology*, **199**, 13–25.

SHI, L. Q., LI, J. F. & ZHANG, W. G. 2007. Magnetic properties of core HF from Feiyantan tidal flat, the Huanghe River Delta and its environmental significance. *Journal of Marine Sciences*, **25**, 13–23.

WANG, A. H. & YE, Z. Z. 1990. Framework, developing processes and sedimentary model of the modern Huanghe River Delta. *Marine Geology & Quaternary Geology*, **10**, 1–12 [in Chinese with English abstract].

WANG, H. J., YANG, Z. S., SAITO, Y., LIU, J. P., SUN, X. X. & WANG, Y. 2007. Stepwise decreases of the Huanghe (Yellow River) sediment load (1950–2005): impacts of climate change and human activities. *Global and Planetary Change*, **57**, 331–354.

WANG, H. J., YUAN, X. J. & WANG, Y. 2010. Evolution of the abandoned Shenxiangou–Diaokou Delta lobe: processes and mechanism. *Journal of Sediment Research*, **4**, 51–60.

WANG, S. J., HASSAN, A. M. & XIE, X. P. 2006. Relationship between suspended sediment load, channel geometry and land area increment in the Yellow River Delta. *Catena*, **65**, 302–314.

WANG, Y. H. 2012. *Coastal Dynamic Geomorphology*. Science Press, Beijing [in Chinese].

WANG, Y. H., SHEN, H. T. & ZHANG, W. G. 2004. A preliminary comparison of magnetic properties of sediments from the Changjiang and the Huanghe estuaries. *Acta Sedimentologica Sinica*, **22**, 658–663 [in Chinese with English abstract].

XUE, C. T. 1994. Division and recognition of modern Yellow River Delta lobes. *Geographical Research*, **13**, 59–66 [in Chinese with English abstract].

XUE, C. T., LI, S. Q. & ZHOU, Y. Q. 2008. Sedimentary record of Yellow River Delta super lobe in 11 ~ 1099. *Acta Sedimentologica Sinica*, **26**, 804–812 [in Chinese with English abstract].

XUE, C. T., YE, S. Y. & GAO, M. S. 2009. Determination of depositional age in the Huanghe Delta in China. *Acta Oceanologica Sinica*, **31**, 117–124 [in Chinese with English abstract].

YANG, Z. G. & LIN, H. M. 1993. *Quaternary Process of China and Internal Analogy*. Geological Publishing House, Beijing [in Chinese].

YE, Q. C. 1982. The geomorphological structure of the Yellow River Delta and its evolution model. *Acta Geographica Sinica*, **37**, 349–363 [in Chinese with English abstract].

ZHANG, J., LIU, H. S. & TONG, S. Y. 2004. High-resolution seismic stratigraphic research of modern marine deposit in the Yellow River estuary area. *Marine Geology Letters*, **20**, 1–5 [in Chinese with English abstract].

ZHANG, W. G., YU, L. Z. & XU, Y. 1995. Brief reviews on environmental magnetism. *Progress in Geophysics*, **10**, 95–105.

ZHAO, X. T., GENG, X. S. & ZHANG, J. W. 1979. Sea level changes of the eastern China during the past 20,000 years. *Acta Oceanologica Sinica*, **1**, 269–281 [in Chinese with English abstract].

Palaeoproductivity linked to monsoon variability in the northern slope of the South China Sea from the last 290 kyr: evidence of benthic foraminifera from Core SH7B

YANG ZHOU[1,2]*, FANG CHEN[1,2], CONG WU[1,2], SHAOHUA YU[1,2] & CHANG ZHUANG[1,2]

[1]*Guangzhou Marine Geological Survey, Guangzhou 510760, China*

[2]*Key Laboratory of Marine Mineral Resources, Ministry of Land and Resources, Guangzhou 510760, China*

**Corresponding author (e-mail: zhouyang@hydz.cn)*

Abstract: A benthic foraminiferal proxy record of 290 kyr, acquired from Core SH7B in the northern continental slope of the South China Sea, was studied to identify the bottom-water environment changes since marine isotope stage (MIS) 8. The changes, including oxygenation and types of organic matter flux to the seafloor, reflect the palaeoproductivity fluctuations linked to monsoon variability. Four assemblages, characterizing different palaeoenvironmental changes, have been recognized by factor and cluster analysis with 32 foraminiferal species in 93 samples (>150 μm size fraction). Assemblage *Pyrgo* spp.–*Hoeglundina elegans* mainly dominates during interglacial periods of MIS 7, MIS 5 and MIS 3, suggesting well-oxygenated bottom environments with low sea surface productivity linked to a weak East Asian Winter Monsoon (EAWM). Assemblage *Uvigerina* spp.–*Globocassidulina subglobulosa* is composed of a constantly high percentage of shallow infaunal species, occurring in intervals of MIS 4, MIS 5, MIS 7 and MIS 8, which indicates that an enhanced EAWM led to a low seasonality of palaeoproductivity, a constant high flux of fresh and labile marine-derived organic particles to the seafloor and a low oxygen bottom environment. Assemblage *Melonis barleeanus*–*Clavulina* spp. is characterized by species that depend on seasonal supplies of more altered refractory organic matter and is mainly distributed in interglacial periods (late MIS 5, MIS 3 and MIS 1), suggesting a high seasonality of palaeoproductivity associated with a seasonal intensification of the EAWM. With low benthic foraminiferal diversity and abundance, assemblage *Globobulimina affinis*–*Chilostomella mediterranensis* was identified during the intermediate MIS 8 and early MIS 2. Both substantial input of terrigenous nutrients from an increased river run-off and an increased primary productivity correlated with an enhanced EAWM have led to abundant nutrient supplies and severe bottom-water oxygen depletion. As suggested by the composition of benthic foraminiferal in Core SH7B, changes in bottom-water environments of the northern South China Sea over the last 290 kyr were driven by the fluctuating palaeoproductivity linked to the high variability of the EAWM.

As the largest marginal sea in the subtropical part of the world ocean, the surface-water circulation, hydrography and primary production of the South China Sea are strongly influenced by the East Asian Monsoon system, which is characterized by the seasonal reversal of monsoonal winds (Shaw *et al.* 1996; Jian *et al.* 2001). The present-day monsoonal setting shows that the local winter monsoon exceeds the summer monsoon in strength and in duration on an annual basis (Wang & Li 2009), and that the primary production in the South China Sea also exhibits a clear spatial distribution and seasonal variation. Winter is the most productive season due to increased upwelling/mixing in response to the strengthened winter monsoon in the northern South China Sea (Liu *et al.* 2002; Chen *et al.* 2007). In summer, however, the overall productivity in the area is relatively low, and is intermittently influenced by increased rainfall and river discharge. Abundant evidence supports the idea that the monsoon climate not only determines the seasonal variations in modern productivity, but also the glacial–interglacial contrasts in productivity. Detailed studies have been performed to understand the links between palaeoproductivity and the East Asian Monsoon system based on different proxies (Jian & Wang 1997; Jian *et al.* 1999, 2000, 2001; Kuhnt *et al.* 1999; Wang *et al.* 2005; Huang *et al.* 2007; Jian *et al.* 2009). Most of the previous studies in the South China Sea have focused on the reconstruction of changes in the primary production as driven by changes in the intensity of the East Asian Monsoon over glacial–interglacial cycles. Different proxies indicate that the winter monsoon and high surface primary productivity are intensified in glacial periods but reduced in interglacial periods (Chen *et al.* 2003; Jian *et al.* 2009; Löwemark *et al.* 2009). However, the fluctuating

From: Clift, P. D., Harff, J., Wu, J. & Qui, Y. (eds) 2016. *River-Dominated Shelf Sediments of East Asian Seas*. Geological Society, London, Special Publications, **429**, 197–210.
First published online November 6, 2015, updated May 26, 2016, http://doi.org/10.1144/SP429.10

palaeoproductivity connected with various bottom-water environments has not been fully investigated, and the glacial–interglacial changes in the intensity and seasonality of the palaeoproductivity associated with monsoonal variability on the northern continental slope of South China Sea also need more proof to support the existing evidence (Jian *et al.* 2000; Su *et al.* 2013).

Because benthic foraminifera are widely distributed in seafloor sediments and are sensitive to environmental changes in bottom water, their assemblages can record these changes. Benthic foraminiferal assemblages have commonly been used as an indicator to reconstruct bottom-water environments and palaeoproductivity in different ocean basins (Jorissen 1999; Kaiho 1999; Murray 2006; Gupta *et al.* 2008). Furthermore, significant relationships have been found between benthic foraminiferal assemblages and seasonality of ocean productivity (Loubere & Fariduddin 1999; Sun *et al.* 2006). The composition and quality of the organic material has great impact on the composition and maintenance of species composition of the assemblages. Some ‘phytodetritus species’ that prefer labile organic matter, such as *Epistominella exigua* and *Alabaminella weddellensisare*, have been considered as potential indicators for reconstructing the seasonality of surface productivity (Loubere & Fariduddin 1999; Sun *et al.* 2006). Different benthic foraminiferal species respond to specific food particles, providing us with the opportunity to gain insight into the characteristics of past organic particle supplies, to estimate the main factors influencing the sea surface productivity and, consequently, to reconstruct the evolution of the monsoon (Caralp 1984, 1989; Goldstein & Corliss 1994; Fariduddin & Loubere 1997; Gupta & Thomas 2003; Nomaki *et al.* 2005; Jorissen *et al.* 2007).

Core SH7B is located on the northern continental slope of South China Sea, at a water depth of 1108 m (Fig. 1). Based on the studied results of Li & Qu (2006), the bottom water at Core SH7B presently corresponds to the intermediate water of the South China Sea and is located in the lower part of the modern oxygen minimum zone (OMZ). The range of the OMZ in the northern South China Sea is approximately from 700 to 1200 m. At a relatively low oxygen content, the benthic foraminiferal community structure in Core SH7B is sensitive to the input of nutrients and can record changes in palaeoproductivity linked to monsoonal climate variability. Thus, the benthic foraminiferal composition of Core SH7B could be used as a good indicator for exploring the relationship between the seasonality of palaeoproductivity and the dynamics of the East Asian Monsoon in the northern South China Sea. In this study, we performed a quantitative analysis of the benthic foraminiferal faunal composition in Core SH7B and compared these results with other proxies to identify the relationship between the faunal compositions and the ambient bottom-water environments. In addition, the quality and origin of organic matter can be used to estimate the intensity and seasonality of palaeoproductivity connected with the variability of the East Asian Monsoon.

Material and methods

Benthic foraminiferal assemblages were analysed from the samples selected from Core SH7B over the interval 0–23 m. Except for two distinct laminated silt layers at 2.95–4.80 and 20.00–22.10 m, the lithology of Core SH7B generally consists of silty clay. No visible turbidity layers could be detected. The core was sampled approximately every 10 cm for benthic foraminifera, resulting in a total of 93 samples. Each sample was soaked in water for 24 h prior to washing over a 63 μm size mesh. The fraction >63 μm was then oven-dried at 60°C. In the present study, only the fraction >150 μm was examined without any partitioning or splitting to pick approximately 200 individuals from each sample, or as many as present. Individuals of each species were identified and counted, and their percentages were calculated. The total abundance, simple diversity and ratio of infaunal to epifaunal species in each sample was calculated for environmental interpretations.

In addition to the faunal census, CABFAC factor analysis and cluster analysis were performed using the PAST (Palaeontology Statistic, version 2.0) software package (Hammer *et al.* 2001) to extract the dominant benthic foraminiferal assemblages present in Core SH7B. To reduce the variable dimensions, enhance the eigenvalues and strengthen the effect of statistical analysis, benthic foraminiferal species with percentages ≥5% in at least in five samples were considered. In addition, we grouped different species with similar ecological habitats in the same genus to increase the effect of the statistical analysis. Finally, 32 species in the 93 samples were used to perform factor and cluster analysis.

Oxygen and carbon stable isotope analyses were performed on the planktonic foraminifer species *Globigerinoides ruber* that was picked from the 250–300 μm size fraction. Analysis was performed using the MAT 352 mass spectrometer at the Laboratory for Radiometric Dating and Stable Isotope Research at Tongji University (Shanghai, China). Only fresh and unpolluted shells were used. The analytical precision of oxygen and carbon isotope analyses is ±0.08‰ (VPDB) and ±0.06‰ (VPDB), respectively.

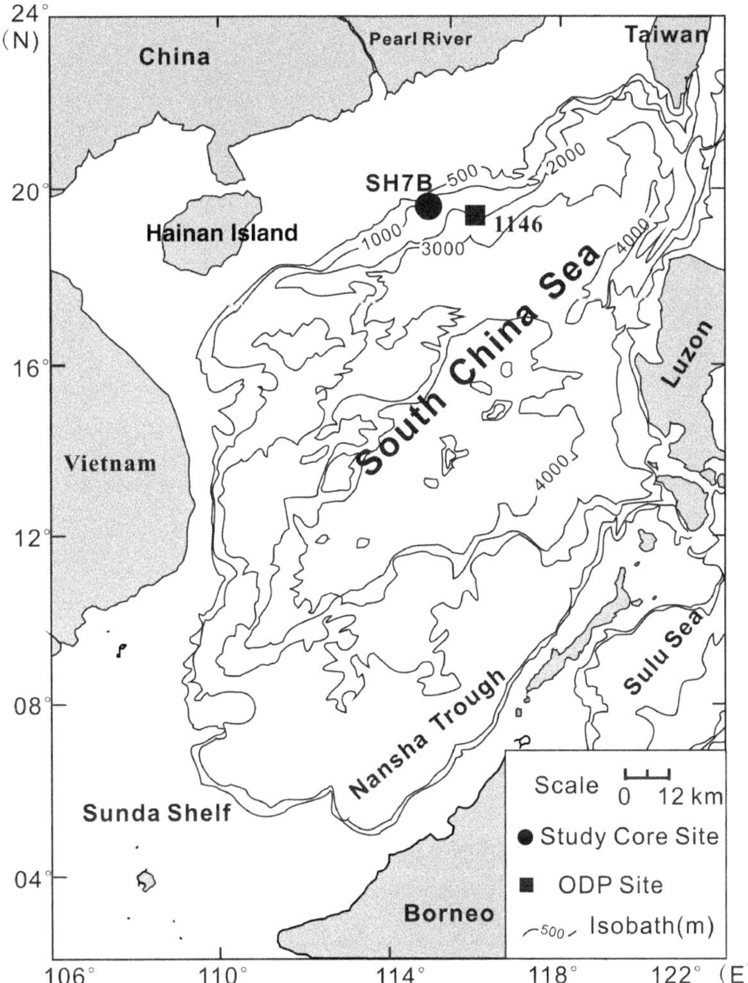

Fig. 1. Location of Core SH7B (after Wang *et al.* 2012).

Core stratigraphy

The chronology of Core SH7B was established by combining the stable isotope curve (*Globigerinoides ruber*, white, 250–300 μm), the carbonate content curve, four AMS (accelerator mass spectrometry)-^{14}C dates and one ^{10}Be date, as well as one biostratigraphy event of the last occurrence (LO) of planktonic foraminifera *Globigerinoides ruber* (pink), and correlating these with isotope events from ODP Site 1146 (Clemens & Prell 2003) located near Core SH7B (Fig. 2).

The results show consistently heavier $\delta^{18}O$ but lighter $\delta^{13}C$ values in glacial periods, and the opposite trend in interglacial intervals (Fig. 2). The variations in carbonate content ($CaCO_3\%$) run roughly parallel to the $\delta^{18}O$ curve. The carbonate curve show 'Atlantic-type' cycle, suggesting that the $CaCO_3\%$ was diluted by the discharged terrigenous materials in glacial periods (Liu *et al.* 2009). In general, lighter $\delta^{18}O$ stages are correlated with high $CaCO_3\%$ assigned to interglacial intervals, whereas heavier $\delta^{18}O$ stages correspond to low $CaCO_3\%$ belonging to glacial periods. Comparing the $\delta^{18}O$ curves of Core SH7B with ODP Site 1146 (Clemens & Prell 2003), the age at the bottom of Core SH7B was extrapolated to marine isotope stage (MIS) 8, covering the last 290 kyr.

Benthic foraminiferal abundance and diversity

In total, 58 taxa were identified, with simple diversity varying from a minimum of 7 species to a maximum of 58 species. Benthic foraminiferal

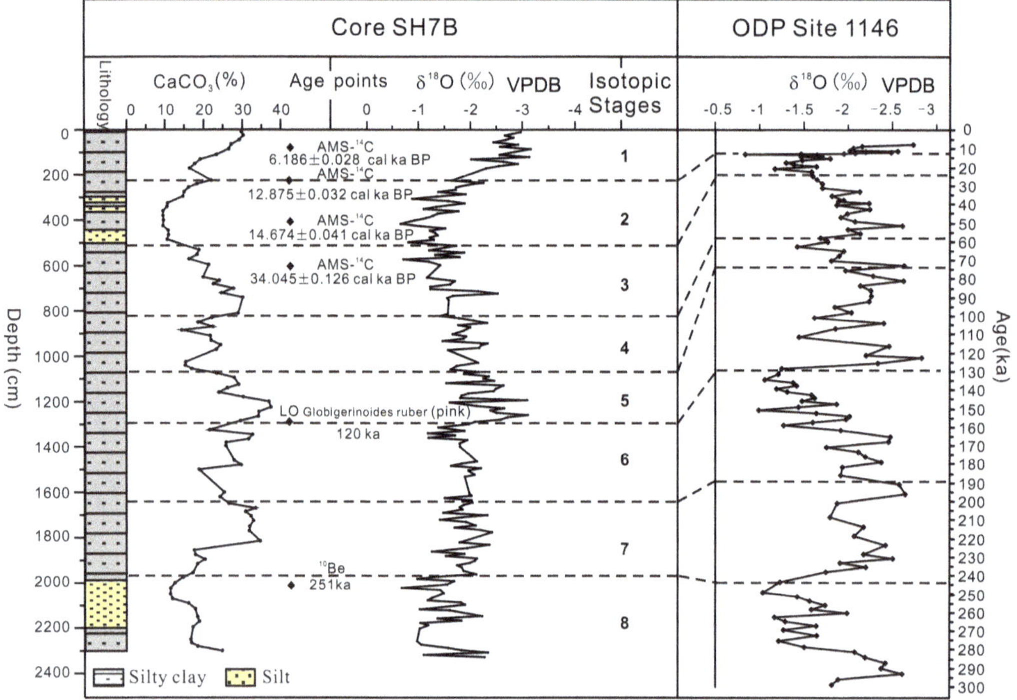

Fig. 2. Variations in $\delta^{18}O$ and $CaCO_3$% of Core SH7B in contrast with the $\delta^{18}O$ curves of ODP Site 1146 in the northern South China Sea. The diamonds indicate the age points used to establish the stratigraphic timescale. LO *G. ruber* (pink) indicates the biostratigraphic event of the last occurrence of planktic foraminifer *Globigerinoides ruber* (pink).

abundance varies between 13 and 200 individuals per 50 g of dried sediment, and exhibits similar trends with simple diversity (Fig. 3), indicating relatively small fluctuations before MIS 5. Overall, infaunal species dominate the assemblages throughout Core SH7B, with percentages of >50%, and their variations being opposite to that of the epifaunal species (Fig. 3). Except for the period of MIS 6, the ratios of infaunal to epifaunal species also show clear glacial–interglacial cycles, with values in excess of 1.8 during glacial periods.

Benthic foraminiferal assemblages and their ecological significance

Benthic foraminiferal assemblages were mainly determined based on the analysis results of factor analysis. According to the plot of eigenvalues v. the number of factors, the variance with eigenvalues >5% was defined as indicating significant loading, and four dominant factors were considered that account for 68.8% of the total variance (Factor 1: 47.26%; Factor 2: 9.97%; Factor 3: 6.08%; Factor 4: 5.53%).

Corresponding to these four factors, four assemblages were defined. For each sample, the assemblage (or corresponding factor) was determined based on the maximum contributing factor with an absolute factor score of >0.5 in these four significant factors (Fig. 4). The corresponding relationship between the factors and assemblages was defined as the following relationship: Factor 1 to Assemblage 2, Factor 2 to Assemblage 4, Factor 3 to Assemblage 1 and Factor 4 to Assemblage 3 (Fig. 4; Table 1). The assemblage names were determined after the acronyms of the two dominant species with maximum factor loadings in the corresponding significant factor (Table 1; Fig. 4). Finally, the downcore variations of the benthic foraminiferal assemblages in Core SH7B are plotted in Figure 4.

To evaluate the reliability of the benthic foraminiferal assemblage extracted from factor analysis, cluster analysis was performed on the percentages of the 32 species in the 93 samples using Ward's Minimum Variance method. Finally, four major clusters were identified on the basis of the similarity v. the number of clusters (Fig. 5), representing assemblages 1–4 (Fig. 5: column A). Comparing the benthic foraminiferal assemblages extracted

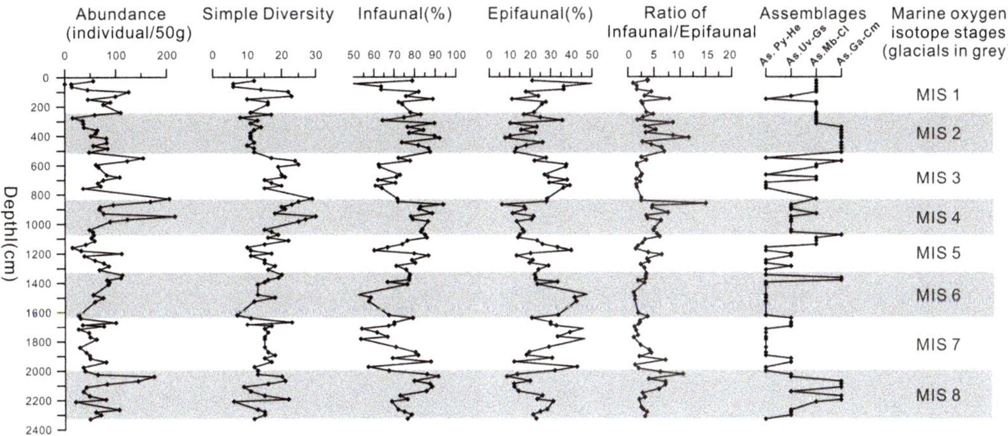

Fig. 3. Downcore variations of benthic foraminiferal abundance (individuals per 50 g of dried sediment), simple diversity, percentage of infaunal (%), epifaunal (%) species with their ratios, and benthic foraminiferal assemblages of the studied Core SH7B in the northern South China Sea.

from cluster analysis (Fig. 5: column A) with that from factor analysis (Fig. 5: column B), only 12 samples are inconsistent with different assemblages (Fig. 5: column C). Comparing results indicate that 87% of the samples display the same results, suggesting that the benthic foraminiferal assemblages (Table 1; Figs 3, 4 & 5: column A) extracted from factor analysis are reasonable and may be used as indicators for interpreting environmental changes.

Based on the ecological preferences of the most dominant species (Table 2; Fig. 6) in each significant factor that has absolute values of factor loading of >1, the environment of each assemblage was inferred (Table 1).

The dominant species *Pyrgo depressa* (Fig. 6: 1), *Pyrgo* sp. (Fig. 6: 2 & 3), *Hoeglundina elegans* (Fig. 6: 4 & 5), *Nodosaria communis* (Fig. 6: 26), *Nodosaria* sp. (Fig. 6: 25) and *Lenticulina* sp.

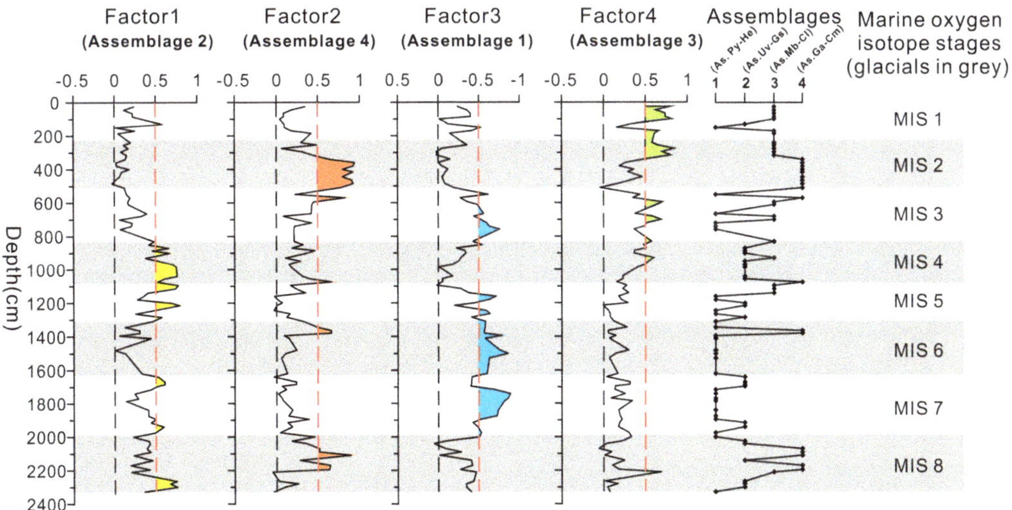

Fig. 4. Downcore distribution of benthic foraminiferal assemblages extracted from the four significant factors based on the results of factor analysis in Core SH7B. Assemblage was defined by the maximum contributing factor score of each sample with an absolute value of >0.5, and the corresponding relationship between the factors and assemblages was designated as Factor 1 to Assemblage 2, Factor 2 to Assemblage 4, Factor 3 to Assemblage 1 and Factor 4 to Assemblage 3.

Table 1. *Benthic foraminiferal assemblages with their respective factor loadings and interpreted environments based on modern or recent ecological preferences of different species*

Assemblages	Factor loadings	Environments
As. Py–He (Assemblage 1)	Factor 3	Epifaunal microhabitat
Pyrgo spp.	(−3.88)	High oxygenation
Hoeglundina elegans	(−2.56)	Low–intermediate organic flux (overall oligotrophic)
Nodosaria spp.	(−1.39)	
Lenticulina spp.	(−1.39)	No seasonality
As. Uv–Gs (Assemblage 2)	Factor 1	Shallow infaunal microhabitat
Uvigerina spp.	(3.01)	Low oxygenation
Globocassidulina subglobulosa	(2.48)	Intermediate–high flux of organic matter
Cassidulina carinata	(2.27)	
Bulimina spp.	(1.63)	Low seasonality
Bolivina spp.	(1.22)	
As. Mb–Cl (Assemblage 3)	Factor 4	Opportunistic species, epifaunal and infaunal microhabitat
Melonis barleeanus	(4.52)	
Clavulina spp.	(1.89)	Low–high organic flux, but variable flux of organic matter
Pyrgo spp.	(1.65)	
Cibicides wuellerstorfi	(1.41)	Variable oxygenation
Globobulimina globra	(1.17)	High seasonality
As. Ga–Cm (Assemblage 4)	Factor 2	Deep infaunal microhabitat
Globobulimina affinis	(4.12)	Low oxygen level
Chilostomella mediterranensis	(3.23)	Sustained flux of organic matter year-round
Bolivina spp.	(1.81)	No seasonality

(Fig. 6: 6 & 7) characterize assemblage *Pyrgo* spp.–*Hoeglundina elegans* (As. Py–He), with high negative loadings on Factor 3 (Table 2). The miliolid species *Pyrgo* spp., together with the epifaunal species, are generally found in the topmost layer of the sediment, and these species have been regarded as being sensitive to oxygen markers in the reconstruction of well-oxygenated bottom environments in the northern Arabian Sea (den Dulk *et al.* 2000). Thus, As. Py–He is suggestive of low–intermediate organic flux and well-oxygenated environments with a low seasonality of productivity.

The assemblage *Uvigerina* spp.–*Globocassidulina subglobulosa* (As. Uv–Cs) is dominated by the infaunal species *Uvigerina hispido-costata* (Fig. 6: 16), *U. peregrina* (Fig. 6: 21), *Globocassidulina subglobulosa* (Fig. 6: 11 & 12), *Bulimina costata* (Fig. 6: 17), *B. aculeate* (Fig. 6: 18), *Bolivinita spathulata* (Fig. 6: 22) and *B. quadrilatera* (Fig. 6: 23 & 24), and has high positive loadings on Factor 1. Some reports indicated that species such as *Uvigerina* spp. and *G. subglobulosa* in this assemblage have advantages over other species in conditions of high inputs of fresh, labile phytoplankton-derived organic matter (Fontanier *et al.* 2002; Suhr *et al.* 2003; Jorissen *et al.* 2007). In the Gulf of Biscay, *Cassidulina carinata* (Fig. 6: 13) shows a higher absolute and relative density after the spring and autumn phytoplankton blooms, probably responding to the freshly deposited organic matter from a reproductive event (Fontanier *et al.* 2003). These dominant species commonly occur in areas with high continuous fluxes of organic matter to the seafloor, and are often associated with reduced bottom-water oxygen concentrations, characterizing an environment with a high–intermediate sustained food supply and a low seasonality (Caralp 1989; Sen Gupta & Machain-Castillo 1993; Schmiedl *et al.* 1997). Therefore, As. Uv–Cs indicates that a sustained flux of marine organic matter (e.g. phytoplankton algae blooms) to the seafloor is caused by an enhanced sea surface primary productivity.

The assemblage *Melonis barleeanus*–*Clavulin* spp. (As. Mb–Cl) is dominated by *Melonis barleeanus* (Fig. 6: 10), *Clavulina* spp., *Pyrgo depressa* (Fig. 6: 1), *Pyrgo* sp. (Fig. 6: 2 & 3) and *Cibicides wuellerstorfi* (Fig. 6: 8 & 9). These species have high positive loadings on Factor 4. Previous studies have shown that the dominant species of this assemblage, *M. barleeanum*, is adapted to organic matter in a more altered form (Caralp 1984, 1989). More specifically, the study by Fontanier *et al.* (2002) suggested a relationship of this species with the zone of nitrifying bacteria that degrades more or less refractory organic matter. A 2-year experiment by Alve (2010) also suggested that *M. barleeanus* co-occurs with other ‘opportunistic species’ or 'phytodetritus taxa' (e.g. *Epistominella exigua* and *Alabaminella weddellensis*) probably as a feeding strategy that is not dependent on a regular supply of fresh organic material but, instead, utilizes refractory organic material and degradation products. As an agglutinated species, *Clavulina* spp., in general, appears to rely less on labile phytodetritus. The epifaunal species *C. wuellerstorfi* occupies elevated habitats and feeds on small organisms and detritus suspended in the water column (Gooday 1993; Rasmussen *et al.* 2002). The presence of these species in As. Mb–Cl suggests that pulsed inputs of labile or refractory organic matter to the seafloor are linked with highly seasonal productivity and a fluctuating oxygenated environment.

Assemblage *Globobulimina affinis–Chilostomella mediterranensis* (As. Ga–Cm) is defined by species having high positive loadings on Factor 2 (e.g. *Globobulimina affinis* (Fig. 6: 19), *Chilostomella mediterranensis* (Fig. 6: 14 & 15), *B. spathulata* (Fig. 6: 22) and *B. quadrilatera* (Fig. 6: 23 & 24)). Most of these species normally adopt a deep infaunal microhabitat, can survive in low oxygen conditions and feed on degraded rather than fresh organic material at the dysoxic–anoxic boundary (Fontanier *et al.* 2002). Nomaki *et al.* (2005) noted that *G. affinis* can consume sedimentary organic material when fresh food (organic material) is not available, although they prefer fresh phytodetritus. Thus, As. Ga–Cm indicates increasing preservation of organic matter at the seafloor in a dysoxic environment and is found abundantly in low-oxygen sediments, accompanied by increasing organic flux to the seafloor with no seasonality (Mackensen & Douglas 1989; Jorissen *et al.* 1998; Fontanier *et al.* 2002).

Discussion

The distribution of benthic foraminifera in sea surface sediment is controlled by environmental factors such as food supply (i.e. the organic carbon flux reaching the seafloor from surface-water productivity) and dissolved oxygen in bottom-water and/or pore-water (Sen Gupta & Machain-Castillo 1993; Fontanier *et al.* 2002). In regions where oxygenation is not the main determining factor, benthic foraminifera are commonly controlled by the food supply. Epifaunal species are often enriched in food-limited environments (oligotrophic) with high oxygen levels (Corliss & Emerson 1990; Jorissen *et al.* 1992). Shallow-infaunal species dominate eutrophic regions with no food limitation but with an increasing potential for an oxygen-limited environment. Deep-infaunal taxa are found where there is an abundance of food associated with lower pore-water oxygen concentrations (Corliss & Chen 1988). In contrast, only some deep-infaunal taxa can live in dysoxic environments within a distinct oxygen-limited environment. Different bottom-water environments control benthic foraminiferal compositions including abundance, diversity and assemblages in samples (Abu-Zied *et al.* 2008), which can be used as proxies for reconstructing the bottom-water environments and palaeoproductivity.

Corresponding to the lower part of the modern oxygen minimum zone (OMZ) in the northern South China Sea, the recently dissolved oxygen content of the bottom water in Core SH7B is approximately 2.0–2.8 ml l^{-1} (Chen *et al.* 2006; Li & Qu 2006), which belongs to a low oxygenation level for benthic foraminiferal microhabitats, based on the classification of Jorissen *et al.* (2007). Because infaunal species are dominant in low oxygen environments, the percentage of infaunal species in Core SH7B sediments is >50% (Fig. 3), indicating that the low oxygen conditions of the bottom water (corresponding to the intermediate water of the South China Sea) have had a weak influence on the faunal composition since 290 ka.

The decomposition and preservation of organic matter at the seafloor was mainly influenced by the oxygen levels in the water. With a background of low oxygen bottom-water environments, benthic foraminiferal composition is sensitive to the nutrient supply and the ventilation of the intermediate water (including intensity and extent). Different quantities and qualities of organic carbon flux to the seafloor could cause different rates of oxygen consumption and could influence the oxygen in bottom-water environments. Thus, the benthic foraminiferal composition in Core SH7B is sensitive to changes in organic matter supply between different sources of organic particles (e.g. terrigenous nutrients or marine-derived organic material), which is in turn linked to the intensity and seasonal behaviour of the East Asian Monsoon. These relationships make it reasonable to use these data to understand changes in the sea surface productivity and variability of the East Asian Monsoon.

Productivity in the study area has a clear seasonal variation related primarily to the nutrient supply, which is driven by the seasonal alternation of monsoonal winds, riverine input or run-off, sea-level changes and other regional oceanographic settings. The East Asian Winter Monsoon (EAWM) profoundly contributes to the sea surface productivity in the study area (Chen *et al.* 2003; Löwemark *et al.* 2009; Jian *et al.* 2009). Using the present-day pattern as an analogue and assuming that the same mechanisms operated during geological glacial–interglacial cycles, we reconstruct the evolution of palaeoproductivity and the related East Asian Monsoon dynamics. Different benthic foraminiferal assemblages in association with other proxies in

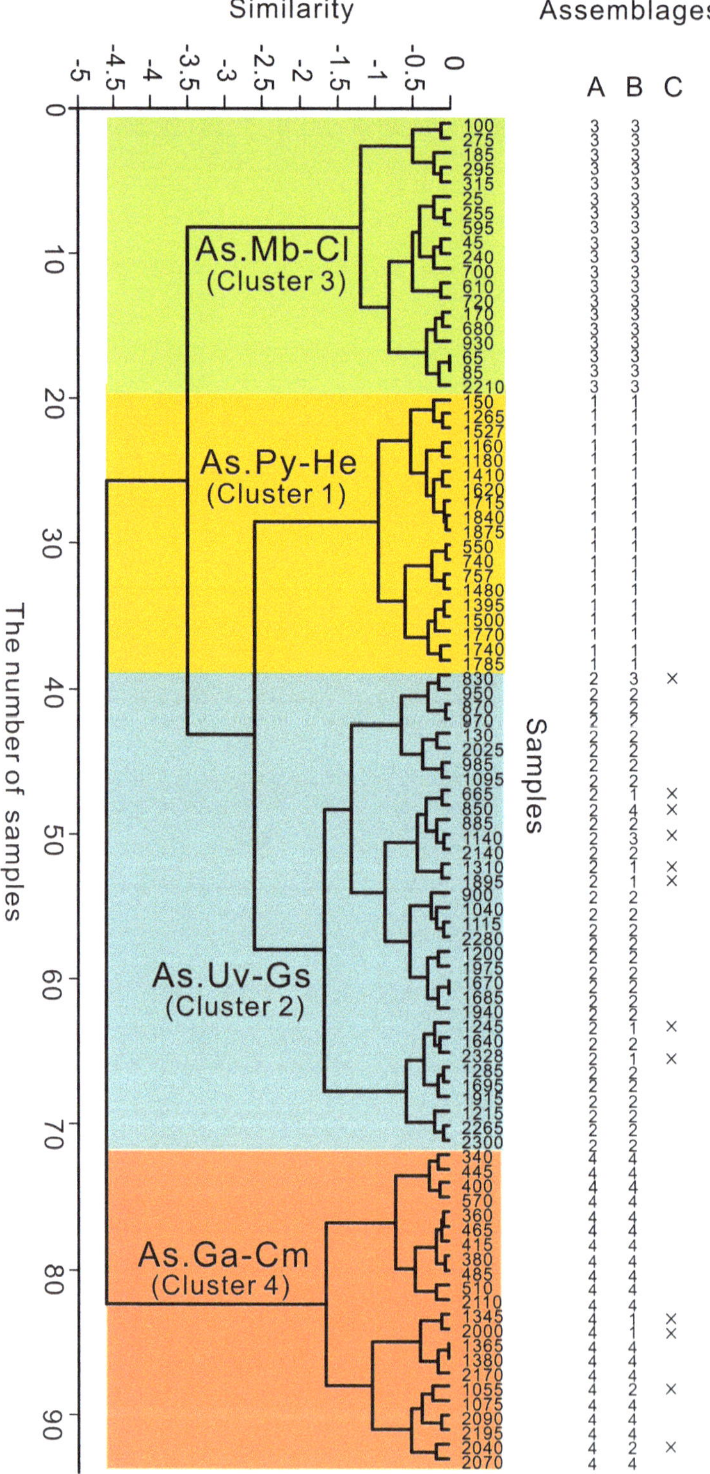
Similarity
Assemblages
A B C
As.Mb-Cl
(Cluster 3)
As.Py-He
(Cluster 1)
As.Uv-Gs
(Cluster 2)
As.Ga-Cm
(Cluster 4)
Samples
The number of samples

Table 2. *Ecological preferences and interpretation of some of the most dominant benthic foraminiferal species composing different assemblages in Core SH7B*

Taxa	Ecological preferences	References
Pyrgo spp.	Miliolids shell, epifaunal, well oxygenated, oligotrophic	Jorissen (1999), Kuhnt *et al.* (1999)
Hoeglundina elegans	Shallow infaunal or epifaunal, suboxic, carbonate well-preserved water	Jorissen *et al.* (2007)
Globocassidulina subglobulosa	Epifaunal and shallow infaunal, oxic, phytodetritus feeder, high level of food supply, strong bottom current, suboxic food supply	Gooday (1994), Fontanier *et al.* (2003), Suhr *et al.* (2003)
Uvigerina spp.	Shallow infaunal, suboxic, high level of food supply, low seasonality	Altenbach & Sarnthein (1989), Rathburn & Corliss (1994), Ohkushi *et al.* (2000), Fontanier *et al.* (2003)
Cassidulina carinata	Shallow infaunal, suboxic, high level of freshly deposited organic matter	Morigi *et al.* (2001), Fontanier *et al.* (2003)
Bulinina spp.	Infaunal, high level of food supply, different oxygen levels, dysoxic, opportunistic species	Van der Zwaan & Jorissen (1991), Sen Gupta & Machain-Castillo (1993)
Cibicides wuellerstorfi	Epifaunal, oxic, low level of food supply, suspension feeder, high bottom current, high seasonal food supply under oligotrophic	Lutze & Thiel (1989), Gupta (1997), Kaiho (1999), Loubere & Fariduddin (1999)
Melonis barleeanus	Intermediate infaunal, suboxic–oxic, high level of food supply, may prefer abundant but refractory organic matter	Caralp (1984, 1989), Morigi *et al.* (2001), Fontanier *et al.* (2003), Abu-Zied *et al.* (2008)
Globobulimina affinis	Deep infaunal, dysoxic	Fontanier *et al.* (2002, 2003), Jorissen *et al.* (2007)
Chilostomella mediterranensis	Deep infaunal, dysoxic	Fontanier *et al.* (2005), Jorissen *et al.* (2007)
Bolivina spp.	Infaunal, low oxygen and high level of sustained food supply	Thomas & Gooday (1996), Gupta (1997), Gupta & Satapathy (2000), Gooday (2003), Jorissen *et al.* (2007)

Core SH7B have revealed various bottom-water environments, and different quantities and qualities of organic matter associated with the fluctuating palaeoproductivity.

Assemblage Py–He, with relatively low abundance and diversity, dominated the interglacial periods of MIS 7, middle MIS 6, and some intervals of MIS 5, MIS 3 and MIS 1 (Fig. 3), suggesting a food-limited, rather oligotrophic environment with high bottom-water oxygen concentration. The oligotrophic environments support few benthic foraminifera, with relatively low abundance and diversity during these stages. The low sea surface palaeoproductivity with low seasonality decreased the quantity and quality of available food at the seafloor and probably resulted in a reduction of the foraminiferal biomass. These characteristics are similar to the modern sea surface productivity during summer, suggesting that a weaker EAWM and intensified water stratification block the delivery of nutrient transfer from deep to surface waters and depress ocean productivity. Thus, the distribution

Fig. 5. Dendrogram based on cluster analysis of the 93 samples from Core SH7B using Ward's Minimum Variance method. The three right-hand columns show benthic foraminiferal assemblages extracted from cluster analysis (column A), factor analysis (column B) and the contrasted results (column C) between the above two methods. The numbers '1–4' in columns A and B correspond to the designated benthic foraminiferal assemblages, which were named after the acronyms of the first two dominant species in each cluster and factor (1: As. Py–He; 2:As. Uv–Gs; 3: As. Mb–Cl; 4: As. Ga–Cm). The 'x' in column C indicates the samples with different benthic foraminiferal assemblages extracted from the factor analysis and cluster analysis.

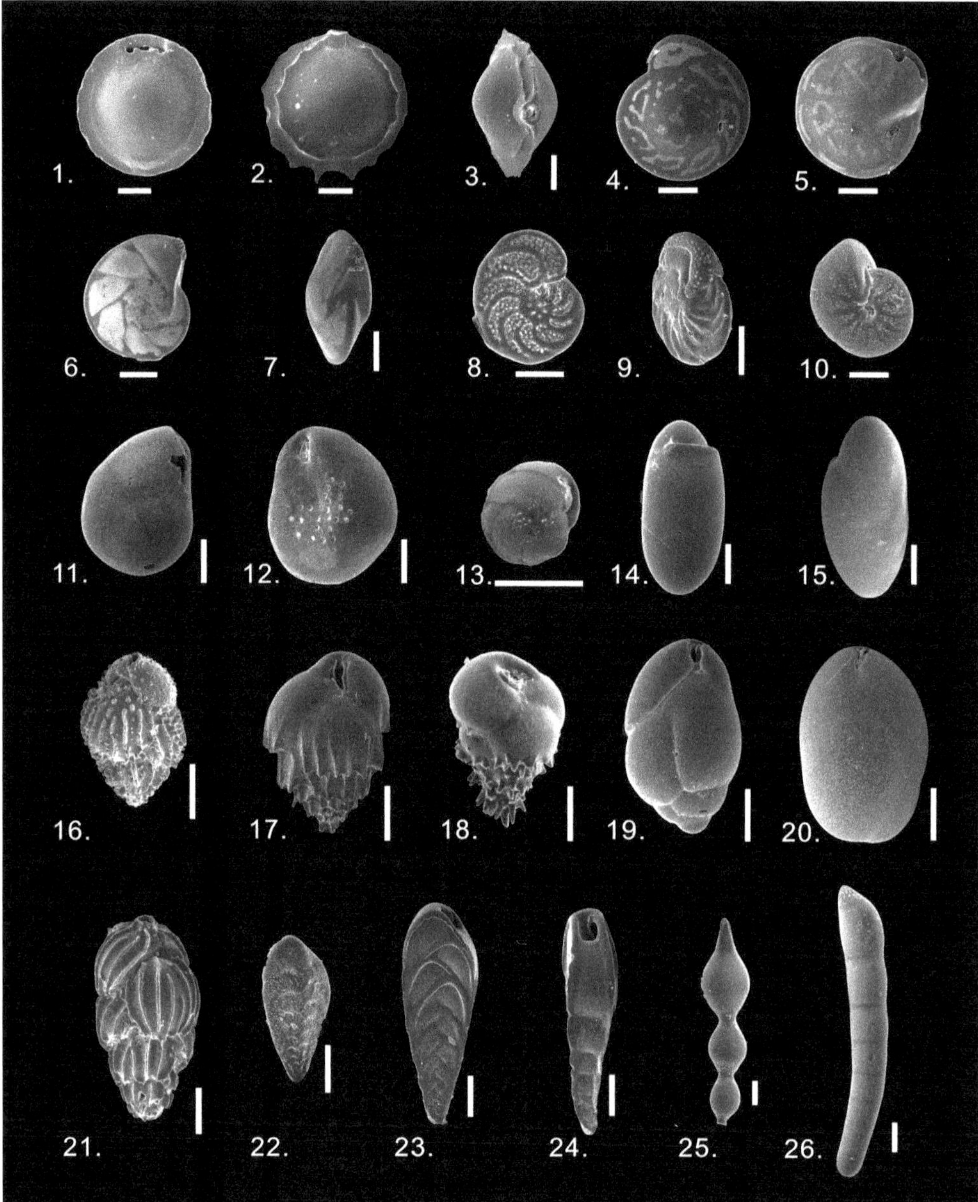

Fig. 6. Scanning electron microscope photographs of dominant foraminiferal species. (**1**) *Pyrgo depressa* (d'Orbigny), side view. (**2**) & (**3**) *Pyrgo* sp.: (2) is the side view and (3) is the apertural view. (**4**) & (**5**) *Hoeglundina elegans* (d'Orbigny): (4) is the spiral view and (5) is the umbilical and apertural view. (**6**) & (**7**) *Lenticulina* sp.: (6) is the side view and (7) is the peripheral view. (**8**) & (**9**) *Cibicides wuellerstorfi* (Schwager): (8) is the spiral view and (9) is the umbilical and peripheral view. (**10**) *Melonis barleeanum* (Williamson), umbilical and peripheral view. (**11**) & (**12**) *Globocassidulina subglobosa* (Brady): (11) is the side view and (12) is the side view showing the aperture. (**13**) *Cassidulina carinata* Silvestri, side view showing the aperture. (**14**) & (**15**) *Chilostomella mediterranensis* Cushman & Todd, peripheral views. (**16**) *Uvigerina hispido-costata* (Cushman & Todd), side view. (**17**) *Bulimina costata* d'Orbigny, side view. (**18**) *Bulimina aculeate* d'Orbigny, side view. (**19**) *Globobulimina affinis* (d'Orbigny), side view. (**20**) *Globobulimina glabra* Cushman & Parker, side view showing the aperture. (**21**) *Uvigerina peregrina* Cushman, side view. (**22**) *Bolivina spathulata* (Williamson), side view. (**23**) & (**24**) *Bolivinita quadrilatera* (Schwager): (23) side view and (24) peripheral view. (**25**) *Nodosaria* sp., side view. (**26**) *Nodosaria communis* d'Orbigny, side view. Each scale bar represents 200 μm.

of assemblage Py–He is suggestive of a weaker EAWM leading to low sea surface productivity and seasonality.

Assemblage Uv–Gs implies a continuous supply of marine-derived organic particles with a low seasonality during the cold glacial periods of MIS 4, MIS 8, and some cold sub-intervals in the interglacial periods of MIS 5 and MIS 7 (Fig. 3; Table 1). The constantly increasing marine-derived organic material resulted from the phytoplankton responding to the intensified sea surface primary productivity. We infer that an enhanced northeastern EAWM prevailed over the northern South China Sea during these periods, leading to a strong mixing of seawater, reinforcing nutrient supplies and high productivity. Overall, an intensified EAWM was identified during these periods.

Assemblage Mb–Cl indicates a relatively low but fluctuating environment, with a high seasonality of palaeoproductivity during interglacial stages MIS 1, MIS 3 and some intervals of MIS 5 (Fig. 3; Table 1). We infer that this situation could be the result of either a change in the amount and/or quality of organic matter delivered to the seafloor or in the frequency of phytoplankton blooms. The occurrence of 'opportunist' species can be viewed as an adaptation to a seasonally fluctuating food supply in a food-limited environment. These species adapted to a fluctuating nutrient supply, and were able to grow and reproduce rapidly when a suitable food source (refractory organic material or degradation products) was present. The origin of this food source remains unclear but may have been related to an abrupt increase in organic matter from the reworking of labile organisms caused by bottom currents, which may be associated with a seasonal intensification of the EAWM. Assemblage Mb–Cl mainly occurs during the interglacial stages after late MIS 5, especially in the later MIS 2 and early MIS 1 (Fig. 3), indicating a variable monsoonal system and a seasonal intensification of the EAWM, and resulting in a high seasonality of palaeoproductivity.

Assemblage Ga–Cm was dominant in the cold glacial stages during early MIS 2, and in some intervals of MIS 4, MIS 6 and MIS 8 (Fig. 3), indicating a high palaeoproductivity without seasonality and a constantly severe oxygen-depleted environment. Benthic foraminiferal composition in this assemblage is similar to that found in records from the northern Arabian Sea (den Dulk *et al.* 2000), which indicates that increased organic matters flux to the seafloor may have increased the oxygen consumption through the decomposition of organisms and led to the expansion of the OMZ. If high input of organic matter was only provided by the marine organic matter derived from the enhanced EAWM, it would not be sufficient to create severely oxygen-depleted environments where only low-oxygen-tolerant deep-infaunal species could survive. We suggest that nutrients originating from terrigenous input also play a key role in promoting the total organic flux to the seafloor during glacial periods.

A possible candidate could be the terrigenous input from the continent and exposed shelf via adjacent rivers, especially the Pearl River. Core SH7B is not located too far away from the Pearl River to be directly fed by riverine sedimentary discharge. The exposed continental shelf may lead the study area to receive more input of terrigenous nutrient supplies during the glacial periods of MIS 2 and MIS 8 (Huang *et al.* 2011). The flux of terrigenous dissolved organic matter from the continent can also be exported to the seafloor and increase oxygen consumption. Constantly increasing organic matter flux to the seafloor from different sources would reduce oxygen content in the bottom water and result in the expansion of the OMZ. We infer that the OMZ expanded to a level close to or below 1108 m (the modern depth of the core) at least during MIS 2 and in some intervals of MIS 8 due to the intensification of the EAWM and the drop in sea level.

Conclusions

Factor and cluster analyses performed on 32 benthic foraminiferal species in 93 samples (>150 μm size fraction), as well as benthic foraminiferal ecological preferences in combination with other proxies, help to explain various environments of bottom-water oxygenation, palaeoproductivity and different types of organic particle fluxes to the seafloor that correspond to different East Asian Monsoon-related signals in the northern South China Sea since 290 ka (from MIS 8 to MIS 1).

Benthic foraminiferal Assemblage Py–He revealed an overall well-oxygenated benthic environment during the interglacial periods of MIS 7, middle MIS 6, and some intervals of MIS 5, MIS 3 and MIS 1, reflecting a low seasonality of palaeoproductivity associated with a weak EAWM.

Assemblage Uv–Gs indicates that an enhanced EAWM led to a high and continuous palaeoproductivity with low seasonality, which constantly provided high quantities of marine-derived organic material to the seafloor with low oxygenated environments during glacial periods MIS 4, MIS 8, and some cold sub-intervals in the interglacial periods of MIS 5 and MIS 7.

Assemblage Mb–Cl, frequently accompanied by 'opportunistic' species, dominated during the late MIS 5, MIS 3 and MIS 1, reflecting a high seasonality of palaeoproductivity-linked fluctuating and seasonally enhanced EAWM.

With lower diversity and abundance, deep infaunal species were dominant in As. Ga–Cm during MIS 2 and MIS 8, indicating a higher glacial productivity with weak or no seasonality, which is attributed to an intensified EAWM and expansion of the OMZ. Meanwhile, the base of the OMZ has strengthened to at least a depth of 1108 m during early MIS 2 and middle MIS 8.

This research used samples and dates provided by Guangzhou Marine Geological Survey, Ministry of Land and Resources, China. We sincerely thank Xinrong Chen for laboratory assistance with isotope tests of oxygen and carbon. We are indebted to Professor Jan Harff and Zhuo Zheng for their critical reviewing of the manuscript. This study was supported by the National Natural Science Foundation of China (grants: 41506065 & 41372012) and Chinese–Polish co-operation project 'Sedimentary Environment and Climate Evolution since the Late Pleistocene in the Beibu-Gulf and its Adjacent Area'.

References

ABU-ZIED, R. H., ROHLING, E. J., JORISSEN, F. J., FONTANIER, C., CASFORD, J. S. L. & COOKE, S. 2008. Benthic foraminiferal response to changes in bottom-water oxygenation and organic carbon flux in the eastern Mediterranean during LGM to Recent times. *Marine Micropaleontology*, **67**, 46–68.

ALTENBACH, A. V. & SARNTHEIN, M. 1989. Productivity record in benthic foraminifera. *In*: BERGER, W. H., SMETACEK, V. S. & WEFER, G. (eds) *Productivity of the Ocean: Present and Past*. Wiley, New York, 255–270.

ALVE, E. 2010. Benthic foraminifera responses to absence of fresh phytodetritus: a two-year experiment. *Marine Micropaleontology*, **76**, 67–75.

CARALP, M. H. 1984. Quaternary Calcareous Benthic Foraminifers, Leg 80. *In*: DE GRACIANSKY, P. C., POAG, C. W. ET AL. (eds) *Initial Reports of the Deep Sea Drilling Project, Volume* **80**. United States Government Printing Office, Washington, DC, 725–755, http://www.deepseadrilling.org/80/volume/dsdp80 pt2_26.pdf

CARALP, M. H. 1989. Abundance of *Bulimina exilis* and *Melonis barleeanum*: relationship to the quality of marine organic matter. *Geo-Marine Letters*, **9**, 37–43.

CHEN, A., HOU, W. P., GAMO, T. & WANG, S. L. 2006. Carbonate-related parameters of subsurface waters in the West Philippine, South China and Sulu Seas. *Marine Chemistry*, **99**, 151–161.

CHEN, M. T., SHIAU, L. J., YU, P. S., CHIU, T. C., CHEN, Y. G. & WEI, K. Y. 2003. 500 0000-year records of carbonate, organic carbon, and foraminiferal sea-surface temperature from the southeastern South China Sea (near Palawan Island). *Paleogeography, Paleoclimatology, Paleoecology*, **197**, 113–131.

CHEN, Y. L., CHEN, H. Y. & CHUNG, C. W. 2007. Seasonal variability of coccolithophore abundance and assemblage in the northern South China Sea. *Deep-Sea Resarch Part II: Topical Studies in Oceanography*, **54**, 1617–1633.

CLEMENS, S. C. & PRELL, W. L. 2003. Date report: oxygen and carbon isotopes from Site 1146, northern South China Sea. *In*: PRELL, W. L., WANG, P., BLUM, P., REA, D. K. & CLEMENS, S. C. (eds) *Proceedings of the Ocean Drilling Program, Scientific Results, Volume* **184**. Ocean Drilling Program, College Station, TX, 1–8, http://www-odp.tamu.edu/publications/184_SR/214/214.htm

CORLISS, B. H. & CHEN, C. 1988. Morphotype patterns of Norwegian Sea deep-sea benthic foraminifera and ecological implications. *Geology*, **16**, 716–719.

CORLISS, B. H., & EMERSON, S. 1990. Distribution of rose bengal stained deep-sea benthic foraminifera from the Nova Scotia continental margin and Gulf of Maine. *Deep-Sea Research Part A. Oceanographic Research Papers*, **37**, 381–400.

DEN DULK, M., REICHART, G. J., VAN HEYST, S., ZACHARIASSE, W. J. & VAN DER ZWAAN, G. J. 2000. Benthic foraminifera as proxies of organic matter flux and bottom-water oxygenation? A case history from the northern Arabian Sea. *Paleogeography, Paleoclimatology, Paleoecology*, **161**, 337–359.

FARIDUDDIN, M. & LOUBERE, P. 1997. The surface ocean productivity response of deeper water benthic foraminifera in the Atlantic Ocean. *Marine Micropaleontology*, **32**, 289–310.

FONTANIER, C., JORISSEN, F. J., LICARI, L., ALEXANDRE, A., ANSCHUTZ, P. & CARBONEL, P. 2002. Live benthic foraminiferal faunas from the Bay of Biscay: faunal density, composition, and microhabitats. *Deep-Sea Research Part I: Oceanographic Research Papers*, **49**, 751–785.

FONTANIER, C., JORISSEN, F. J., CHAILLOU, G., DAVID, C., ANSCHUTZ, P. & LAFON, V. 2003. Seasonal and interannual variability of benthic foraminiferal faunas at 550 m depth in the Bay of Biscay. *Deep-Sea Research Part I: Oceanographic Research Papers*, **50**, 457–494.

FONTANIER, C., JORISSEN, F. J., CHAILLOU, G., ANSCHUTZ, G., GRÉMARE, A. & GRIVEAUD, C. 2005. Live foraminiferal faunas from a 2800 m deep lower canyon station from the Bay of Biscay: faunal response to focusing of refractory organic matter. *Deep-Sea Research Part I: Oceanographic Research Papers*, **52**, 1189–1227.

GOLDSTEIN, S. T. & CORLISS, B. H. 1994. Deposit feeding in selected deep-sea and shallow-water benthic foraminifera. *Deep-Sea Research Part I: Oceanographic Research Papers*, **41**, 229–241.

GOODAY, A. J. 1993. Deep-sea benthic foraminiferal species which exploit phytodetritus: characteristic features and controls on distribution. *Marine Micropaleontology*, **22**, 187–205.

GOODAY, A. J. 1994. The biology of deep-sea Foraminifera: a review of some advances and their applications in paleoceanography. *Palaios*, **9**, 14–31.

GOODAY, A. J. 2003. Benthic foraminifera (Protista) as tools in deep-water paleoceanography: environmental influences on faunal characteristics. *Advances in Marine Biology*, **46**, 1–90.

GUPTA, A. K. 1997. Paleoceanographic and paleoclimatic history of the Somali Basin during the Pliocene–Pleistocene: multivariate analyses of benthic foraminifera from DSDP site 241 (Leg 25). *Journal of Foraminiferal Research*, **27**, 196–208.

GUPTA, A. K. & SATAPATHY, S. K. 2000. Latest Miocene–Pleistocene abyssal benthic foraminifera from west-central Indian Ocean DSDP Site 236: paleoceanographic and paleoclimatic inferences. *Journal of the Paleontological Society of India*, **45**, 33–48.

GUPTA, A. K. & THOMAS, E. 2003. Initiation of Northern Hemisphere glaciation and strengthening of the northeast Indian monsoon: Ocean Drilling Program Site 758, eastern equatorial Indian Ocean. *Geology*, **31**, 47–50.

GUPTA, A. K., DAS, M., CHEMENS, S. C. & MUKHERJEE, B. 2008. Benthic foraminiferal faunal and isotopic changes as recorded in Holocene sediments of the northwest Indian Ocean. *Paleoceanography*, **23**, PA2214, http://doi.org/10.1029/2007PA001546

HAMMER, Ø., HARPER, D. A. T. & RYAN, P. D. 2001. Past: paleontological statistics software package for education and data analysis. *Paleontologia Electronica*, **4**, 1–9.

HUANG, B., JIAN, Z. & WANG, P. 2007. Benthic foraminiferal fauna turnover at 2.1 Ma in the northern South China Sea. *Chinese Science Bulletin*, **52**, 839–843.

HUANG, J., LI, A. & WAN, S. 2011. Sensitive grain-size records of Holocene East Asian summer monsoon in sediments of northern South China Sea slope. *Quaternary Research*, **75**, 734–744.

JIAN, Z. & WANG, L. 1997. Late Quaternary benthic foraminifera and deep-water paleoceanography in the South China Sea. *Marine Micropaleontology*, **32**, 127–154.

JIAN, Z., WANG, L., KIENAST, M., SARNTHEIN, M., KUHNT, W., LIN, H. & WANG, P. 1999. Benthic foraminiferal paleoceanography of the South China Sea over the last 40 000 years. *Marine Geology*, **156**, 159–186.

JIAN, Z., WANG, P. ET AL. 2000. Foraminiferal responses to major Pleistocene paleoceanographic changes in the southern South China Sea. *Paleoceanography*, **15**, 229–243.

JIAN, Z., HUANG, B., KUHNT, W. & LIN, H. L. 2001. Late Quaternary upwelling intensity and East Asian Monsoon forcing in the South China Sea. *Quaternary Research*, **55**, 363–370.

JIAN, Z., TIAN, J. & SUN, X. 2009. Chapter 3: upper water structure and paleo-monsoon. *In*: WANG, P. & LI, Q. (eds) *The South China Sea: Paleoceanography and Sedimentology*. Developments in Paleoenvironmental Research, **13**, Springer Science, Dordrecht, 25–73.

JORISSEN, F. J. 1999. Benthic foraminiferal microhabitats below the sediment-water interface. *In*: SEN GUPTA, B. K. (ed.) *Modern Foraminifera*. Kluwer Academic, Dordrecht, 161–179.

JORISSEN, F. J., BARMAWIDJAJA, D. M., PUSKARIC, S. & VAN DER ZWAAN, G. J. 1992. Vertical distribution of benthic foraminifera in the northern Adriatic Sea: the relation with the organic flux. *Marine Micropaleontology*, **19**, 131–146.

JORISSEN, F. J., WITTLING, I., PEYPOUQUET, J. P., RABOULLE, C. & RELEXANS, J. C. 1998. Live benthic foraminiferal faunas off Cape Blance, NW-Africa: community structure and microhabitats. *Deep-Sea Research I*, **45**, 2157–2188.

JORISSEN, F. J., FONTANIER, C. & THOMAS, E. 2007. Paleocenographical proxies based on deep-sea benthic foraminiferal assemblage characteristics. *In*: HILLAIRE-MARCEL, C. & DE VERNAL, A. (eds) *Proxies in Late Cenozoic Paleoceanography: Patr 2: Biological Tracers and Biomarkers*. Elsevier, Amsterdam, 263–326.

KAIHO, K. 1999. Effect of organic carbon flux and dissolved oxygen on the benthic foraminifral oxygen index (BFOI). *Marine Micropaleontology*, **37**, 67–76.

KUHNT, W., HESS, S. & JIAN, Z. 1999. Quantitative composition of benthic foraminiferal assemblages as a proxy indicator for organic carbon flux rates in the South China Sea. *Marine Geology*, **156**, 123–157.

LI, L. & QU, T. 2006. Thermohaline circulation in the deep South China Sea basin inferred from oxygen distributions. *Journal of Geophysical Research: Oceans*, **111**, C05017, http://doi.org/10.1029/2005JC003164

LIU, K. K., CHAO, S. Y., SHAW, P. T., GONG, G. C., CHEN, C. C. & TANG, T. Y. 2002. Monsoon-forced chlorophy II distribution and primary production in the South China Sea: observations and a numerical study. *Deep Sea Research Part I: Oceanographic Research Papers*, **49**, 1387–1412.

LIU, Z., HUANG, W. ET AL. 2009. Chapter 4: sedimentology. *In*: WANG, P. & LI, Q. (eds) *The South China Sea: Paleoceanography and Sedimentology*. Developments in Paleoenvironmental Research, **13**, Springer Science, Dordrecht, 171–295.

LÖWEMARK, L., STEINKE, S. ET AL. 2009. New evidence for a glacioeustatic influence on deep water circulation, bottom water ventilation and primary productivity in the South China Sea. *Dynamics of Atmospheres and Oceans*, **47**, 138–153, http://doi.org/10.1016/j.dynatmoce.2008.08.004

LOUBERE, P. & FARIDUDDIN, M. 1999. Quantitative estimation of global patterns of surface ocean biological productivity and its seasonal variation on timescales from centuries to millennia. *Global Biochemical Cycles*, **13**, 115–133.

LUTZE, G. F. & THIEL, H. 1989. Epibenthic foraminifera from elevated microhabitats: *cibicides wuellerstorfi* and *Planulina ariminensis*. *Journal of Foraminiferal Research*, **19**, 153–158.

MACKENSEN, A. & DOUGLAS, R. G. 1989. Down-core distribution of live and dead deep-water benthic foraminifera in box cores from the Weddell Sea and the California continental borderland. *Deep-Sea Research Part A. Oceanographic Research Papers*, **36**, 879–900.

MORIGI, C., JORISSEN, F. J., GERVAIS, A., GUICHARD, S. & BORSETTI, A. M. 2001. Benthic foraminiferal faunas in surface sediments off NW Africa: relationship with organic flux to the ocean floor. *Journal of Foraminiferal Research*, **31**, 350–368.

MURRAY, J. W. 2006. *Ecology and Applications of Benthic Foraminifera*. Cambridge University Press, Cambridge.

NOMAKI, H., HEINZ, P., HEMLEBEN, C. & KITAZATO, H. 2005. Behavior and response of deep-sea benthic foraminifera to freshly supplied organic matter: a laboratory feeding experiment in microcosm environments. *Journal of Foraminiferal Research*, **35**, 103–113.

OHKUSHI, K., THOMAS, E. & KAWAHATA, H. 2000. Abyssal benthic foraminifera from the northwestern Pacific (Shatsky Rise) during the last 298 kyr. *Marine Micropaleontology*, **38**, 119–147.

RASMUSSEN, T. L., BÄCKSTRÖM, D. *ET AL.* 2002. The Faroe–Shetland gateway: late Quaternary water mass exchange between the Nordic seas and the northeastern Atlantic. *Marine Geology*, **188**, 165–192.

RATHBURN, A. E. & CORLISS, B. H. 1994. The ecology of living (stained) deep-sea benthic foraminifera from the Sulu Sea. *Paleoceanography*, **9**, 87–150.

SCHMIEDL, G., MACKENSEN, A. & MÜLLER, P. J. 1997. Recent benthic foraminifera from the eastern South Atlantic Ocean: dependence on food supply and water masses. *Marine Micropaleontology*, **32**, 249–287.

SEN GUPTA, B. K. & MACHAIN-CASTILLO, M. L. 1993. Benthic foraminifera in oxygen-poor habitats. *Marine Micropaleontology*, **20**, 183–201.

SHAW, P. T., CHAO, S. Y., LIU, K. K., PAI, S. C. & LIU, C. T. 1996. Winter upwelling off Luzon in the northeastern South China Sea. *Journal of Geophysical Research*, **101**, 16,435–16,448.

SU, X., LIU, C., BEAUFORT, L., TIAN, J. & HUANG, E. 2013. Late Quaternary coccolith records in the South China Sea and East Asian monsoon dynamics. *Global and Planetary Changes*, **111**, 88–96.

SUHR, S. B., POND, D. W., GOODAY, A. J. & SMITH, C. R. 2003. Selective feeding by benthic foraminifera on phytodetritus on the western Antarctic Peninsula shelf: evidence from fatty acid biomarker analysis. *Marine Ecology Progress Series*, **262**, 153–162.

SUN, X., CORLISS, B. H., BROWN, C. W. & SHOWER, W. J. 2006. The effect of primary productivity and seasonality on the distribution of deep-sea benthic foraminifera in the North Atlantic. *Deep-Sea Research Part I: Oceanographic Research Papers*, **53**, 28–47.

THOMAS, E. & GOODAY, A. J. 1996. Cenozoic deep-sea benthic foraminifera: tracers for changes in oceanic productivity? *Geology*, **24**, 355–358.

VAN DER ZWAAN, G. J. & JORISSEN, F. J. 1991. Biofacial patterns in river-induced shelf anoxia. *In*: TYSON, R. V. & PEARSON, T. H. (eds), *Modern and Ancient Continental Shelf Anoxia*. Geological Society, London, Special Publications, **58**, 65–82, http://doi.org/10.1144/GSL.SP.1991.058.01.05

WANG, P. & LI, Q. 2009. Chapter 2: oceanographical and geological background. *In*: WANG, P. & LI, Q. (eds) *The South China Sea: Paleoceanography and Sedimentology*. Developments in Paleoenvironmental Research, **13**, Springer Science, Dordrecht, 297–394.

WANG, P., CLEMENS, S. *ET AL.* 2005. Evolution and variability of the Asian monsoon system: state of the art and outstanding issues. *Quaternary Science Reviews*, **24**, 595–629.

WANG, X., LEE, M., WU, S. & YANG, S. 2012. Identification of gas hydrate dissociation from wireline-log data in the Shenhu area, South China Sea. *Geophysics*, **77**, 125–134.

Significance of the *Paralia sulcata* fossil record in palaeoenvironmental reconstructions of the SE Asia marginal seas over the Last Glacial Cycle

JINPENG ZHANG[1,2*], MICHAŁ TOMCZAK[2,3], CHAO LI[4], ANDRZEJ WITKOWSKI[2], YAN QIU[1,3], HONGJUN CHEN[1,3] & HONGFANG GAO[1,3]

[1]*Guangzhou Marine Geological Survey, China Geological Survey, Guangzhou 510760, China*

[2]*University of Szczecin, Institute of Marine and Coastal Sciences, Szczecin 70-383, Poland*

[3]*Key Laboratory of Marine Mineral Resources, MLR, Guangzhou 510760, China*

[4]*Xiamen University, College of Ocean and Earth Sciences, Xiamen 361005, China*

**Corresponding author (e-mail: jinpenggmgs@sina.com)*

Abstract: *Paralia sulcata* (Ehr.) Cleve, a common diatom microfossil species in shelf and adjacent deep-sea sediments, is becoming potentially useful in palaeoenvironmental reconstructions due to its coarsely silicified exoskeleton being well preserved in sediments. This paper presents the distributions of *P. sulcata* in sediment cores from the SE Asian marginal seas taken from formerly published and original materials. *Paralia sulcata* achieved variable abundance in the glacial–interglacial sediment analysed, showing spatial changes in the core profiles of the South China Sea, with lower abundances during the last glacial stage and higher values in the post-glacial. Conversely, in the East China Sea, it attains lower abundances during the Holocene, while higher ones in last glacial stage. Over a short timescale, variations in the abundance of this species responded quickly to the climatic and oceanographic fluctuations. We illustrate a 'seesaw' effect that its high abundances revealed in the East China Sea at 15–11 cal ka BP, whereas this occurred during period 11–8 cal ka BP in the South China Sea. The rapid increase in abundance of this species in the South China Sea is interpreted as an indication of a plausible link with the opening of the Taiwan Strait and the general coastal water circulation pattern during deglaciation.

The Chinese marginal seas of SE Asia are located in an area of transition between Eurasia and the Pacific. The South China Sea, the East China Sea and the Yellow Sea, as the major parts of the East Asia marginal seas, receive huge amounts of terrigenous material discharged by numerous large rivers (e.g. the Yellow, Yangtze, Pearl and Red rivers). These basins contain significant amounts of marine authigenic matter deposited within an annual oceanic productivity cycle. This is predominantly organic detritus resulting from phytoplankton blooms (e.g. Stramska 2009), and, to a certain extent, from lateral transport within the marine littoral and neritic zones (e.g. Guo *et al.* 2010). Finally, the marginal seas of SE Asia are subject to complicated interactions between the land and ocean (Wang 1999), mainly due to reversing monsoon circulation patterns. The diatom assemblages in the sediments have been formed from phytoplankton taxa depositing from the open-ocean water column, from instability of benthos/periphyton in disturbed shallow-water environments or from redeposition by riverine discharge. The species composition of the microfossil assemblages and their origin provide palaeoenvironmental information after being deposited in the sedimentary record. This information is one of the important aspects of marginal-sea sedimentology used to reconstruct past environmental conditions (e.g. Guo *et al.* 2010).

The palaeooceanographic conditions of marginal seas of SE Asia, which are largely controlled by climatic factors and revealed by sedimentary sequences and their cyclic nature, have been greatly affected by regional and global sea-level fluctuation (e.g. Wang 1999; Wang & Li 2009; Yao *et al.* 2009; Katsuki *et al.* 2010).

Paralia sulcata (Ehr.) Cleve, 1873 (=*Melosira sulcata* (Ehr.) Kützing, 1844) is a marine species with relatively small cylindric cells with thickly silicified frustules (Round *et al.* 1990; Zong 1997; McQuoid & Nordberg 2003). *Paralia sulcata* is one of the dominant microfossil species within diatom assemblages of the continental slope that is further transported from the shelf down the slope to the deep sea by surface currents and deep-water circulation (Jiang 1987; Lan *et al.* 1995; Hasle & Syvertsen 1997; Shiga & Koizumi 2000; Witon *et al.* 2006; Ryu *et al.* 2008). This species occurs in offshore benthic habitats, but has been also reported from phytoplankton (meroplanktonic) hauls (Hasle &

From: CLIFT, P. D., HARFF, J., WU, J. & QUI, Y. (eds) 2016. *River-Dominated Shelf Sediments of East Asian Seas*. Geological Society, London, Special Publications, **429**, 211–221.
First published online December 9, 2015, updated May 26, 2016, http://doi.org/10.1144/SP429.11

Syvertsen 1997). It is widely distributed in global offshore and upwelling water columns (Blasco *et al.* 1980; Abrantes 1988; Karpuz & Schrader 1990; Lange *et al.* 1998), adapting to temperatures exceeding 3°C (Zong 1997), low illumination and fairly low salinity (brackish water) (Blasco *et al.* 1980; Jin *et al.* 1982), although its abundance drops when water salinity decreases below 3 psu and depth exceeds 100 m (Zong 1997). The most suitable environment for its growth is observed in the mid- and low-latitude, subtropical and temperate offshore sea areas (Abrantes 1988; Abrantes *et al.* 2007). The optimum temperature conditions of *P. sulcata* in Chinese seas range between 11 and 20°C (Tanimura *et al.* 2002). This species is observed in highest abundances in surface sediments within water depths ranging from 50 to 100 m (Jiang 1987; Wang *et al.* 1990). Owing to relatively narrow range of ecological preferences and good preservation status in marine sediments, *P. sulcata* seems to be a potentially important proxy indicator of environmental changes in marine ecosystems. However, the reconstruction potential of *P. sulcata* seems to be limited by the fact that it can be found both in benthos and in plankton. Moreover, strong resistance to dissolution and mechanical destruction means that the frustules are vulnerable to lateral transport and redeposition by currents (Lan *et al.* 1995; Kennington *et al.* 1999). Eventually, these intricate conditions can affect the interpretation of the data and the quality of environmental reconstruction.

Based on published data in which changes in abundances of *P. sulcata* have been applied in environmental reconstructions and on our own analysis of *P. sulcata* through the sedimentary facies of cores retrieved from the northern and western part of the South China Sea, this paper aims to summarize the characteristics of environmental parameters during the last glacial cycle of the SE Asia marginal seas. To achieve this aim, we have re-examined the content of *P. sulcata* in sediment cores retrieved from the South China Sea, and the East China Sea shelf and continental slope (Lan *et al.* 1995; Li *et al.* 2002, 2004; Zhi *et al.* 2003; Huang *et al.* 2006, 2009). In this way, we aim to emphasize and highlight the potential of *P. sulcata* in palaeoclimatic, palaeogeographical and palaeoenvironmental reconstructions of the South China and East China seas in the late Pleistocene and Holocene.

Materials and methods

Sampling and diatom analysis and chronology

Piston cores of 83PC (865 cm length) and 196PC (750 cm length) were retrieved during R/V *Haiyang-4* cruises from the northern continental slope of the South China Sea in 2006 and 2008 at a water depth of 1917 and 3200 m, respectively. Clay and clay-silt dominate the lithology of these cores. Samples for stable isotope and diatom analyses were collected at 10 cm interval. The location and basic information on cores is illustrated in Figure 1 and recorded in Table 1.

The chronological framework refers to the oxygen stable isotopes data based on the planktonic foraminifera *Globigerinoides rubber*. Sufficient numbers of clean and intact *G. rubber* shells within the 250–300 μm size (over 20 mg) were selected to test the oxygen isotopes provided using a Finnigan-MAT253 Isotope Mass Spectrometer at the Guangzhou Marine Geological Survey. Calculated $\delta^{18}O$ data were used to graphically interpolate cores with the LR04 Benthic Stack record (Lisiecki & Raymo 2005), as shown in Figure 2. The $\delta^{18}O$ data of the 83PC core have already been published in Yang *et al.* (2009). The limits of Marine Isotope Stages (MIS) 1–4 in core 83PC occur at 100, 220, 490, 560 cm; in core 196PC, however, they occur at 230, 380, 560 and 720 cm. Furthermore, the selected planktonic foraminifer shells from core 83PC were measured using accelerator mass spectrometry (AMS) radiocarbon age dating at the Key Laboratory of Heavy Ion Physics, Beijing University, China, and additional bulk samples were measured at the Poznań Radiocarbon Laboratory, Poland (cf. Table 2). The AMS ^{14}C dating age and the oxygen isotope stratigraphy were used to elucidate the cores age model as shown in Figure 3.

To prepare the diatom slides and to perform the diatomological analysis, 1 g dry wt of each sample was treated with 10% HCl in order to remove calcium carbonate, and with 30% H_2O_2 to mineralize the organic matter. Next, sediments were centrifuged twice with ZnI_2 (heavy liquid) at a density of 2.4 g cm^{-3} to concentrate the diatom abundance and to remove clay particles. An aliquot of about 50 μl of the shaken suspension was transferred onto a cover slip using a pipette. After the suspension had evaporated at room temperature, covers slips were mounted onto labelled slides. Naphrax® was used as a mountant medium. Diatoms were counted in random transects for each sample using a light microscope at magnifications of ×400 and ×1000 (×100 oil immersion lens). Relative abundance, calculated as a percentage of the total diatom assemblage, was measured on the basis of counting more than 200 diatom valves, and was identified at the species or genus level. In this paper, we decided to select *P. sulcata* downcore distribution data in the MIS 1–MIS 4 in both the cores studied. Light microscopy and scanning electron microscopy (SEM) images of *P. sulcata* are illustrated in Figure 4.

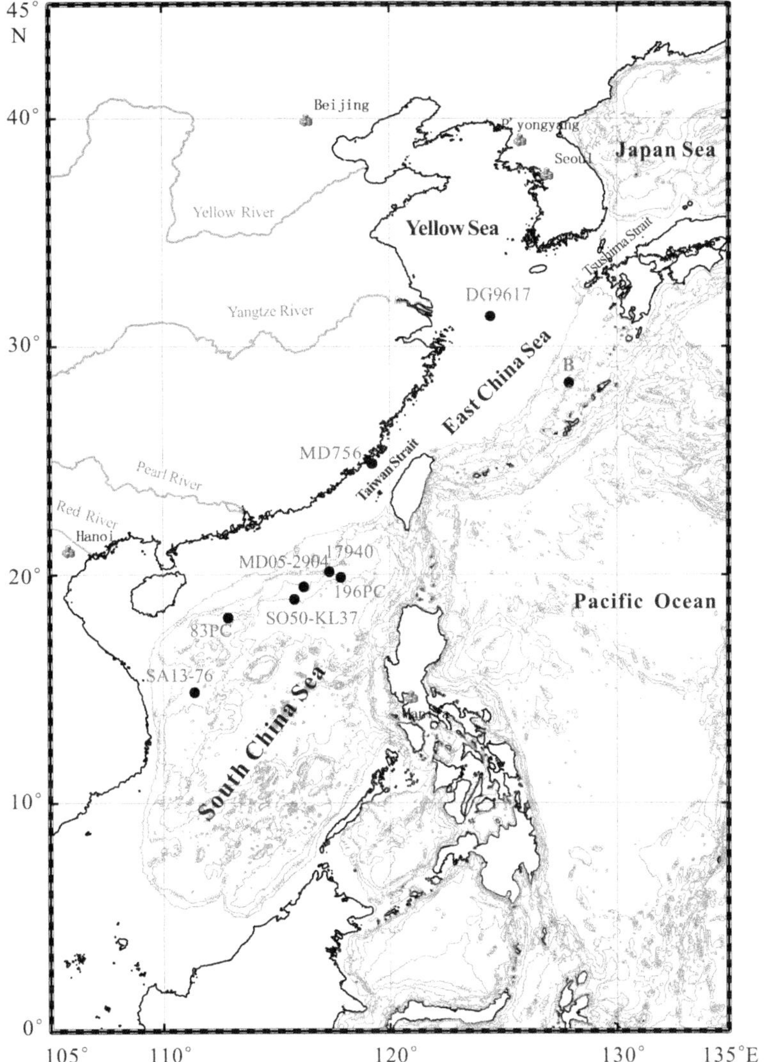

Fig. 1. Location of the cores (ArcGIS software 10.0).

Distribution of Paralia sulcata *in previously published data*

The existing data related to the abundance of *P. sulcata* were extracted from a few publications. Four datasets in our analysis were acquired from the South China Sea, core 17940 (Huang *et al.* 2009), core MD05-2904 (Huang *et al.* 2006), core SO50-KL37 (Lan *et al.* 1995) from the northern South China Sea and core SA13-76 (Sun 2009) of the western South China Sea. In addition, we have included *P. sulcata* data from the continental margin of the East China Sea, core DG9617 and core B (Li *et al.* 2002, 2004), and from the Taiwan Strait, core MD756 (Zhi *et al.* 2003). Detailed information is given in Figures 1 and 4 and Table 1.

Methodological background

In general, the abundance of *P. sulcata* in analysed cores is highest in sediments of shelf areas. Nevertheless, its abundance is relatively low in some sections of the sediment cores collected from the deeper part of these marginal seas. This spatial discrepancy is related to the coastal ecological characteristics of *P. sulcata*. This species favours offshore areas, and waters with decreased temperatures and salinities. Our assumption is that fluctuations in its content

Table 1. *Basic information about the cores*

Location/ core name	Longitude (E)	Latitude (N)	Water depth (m)	Core length (cm)	Age of core	Sea area
83PC	112° 54′	17° 66′	1917	865	*c.* 107 ka	South China Sea
196PC	117° 53′	19° 52′	3200	750	*c.* 74 ka	South China Sea
17940	117° 23′	20° 07′	1728	*c.* 0–820	*c.* 15 ka	South China Sea
MD05-2904	116° 15′	19° 27′	2066	*c.* 0–398	*c.* 15 ka	South China Sea
SA13-76	111° 24′	14° 50′	2801	676	*c.* MIS 5	South China Sea
SO50-KL37	115° 50′	18° 54′	2695	*c.* 0–680	*c.* MIS 4	South China Sea
DG9617	124° 26′	31° 19′	52	848	*c.* 8.5 ka	East China Sea
B	127° 54′	28° 24′	1010	305	*c.* 23 ka	East China Sea
MD756	119° 15′	24° 52′	34	557	*c.* 21 ka	Taiwan Strait

mirror climatic and oceanographic changes on a geological scale. *Paralia sulcata* may, therefore, be applied in reconstructing the distance from the coastline and the water depth, as well as palaeoclimatic and oceanographic changes based on comparisons of the core profiles in these marginal seas.

Results

The abundance of P. sulcata *in sediment cores in the South China Sea*

The relative abundance of *P. sulcata* in core 196PC reached up to 15%, but is very low during the period from *c.* 73 to 16 cal ka BP (the main part of MIS 4–MIS 2), with an average content below 2%, while it increases again to 12% at *c.* 10–8 cal ka BP (the early MIS 1). In the 83PC core, the content of *P. sulcata* reached up to 20%, followed by low values during the period from *c.* 73 to 14 cal ka BP (mid-MIS 4–MIS 2), with a distinct increase in late MIS 4 (59 cal ka BP) to about 20%, and late MIS 3 (*c.* 34 cal ka BP) to around 10%. Throughout the whole of MIS 1, the *P. sulcata* content was relatively high and attained up to 15%.

The relative abundance of *P. sulcata* in the entire core SO50-KL37 was rather low and reached a maximum value of 9%. It revealed increased values during early MIS 4 (*c.* 70 cal ka BP) and reached 5%, and during the period of 11–8 cal ka BP (early MIS 1) rose to 9%. A slight increase in *P. sulcata*

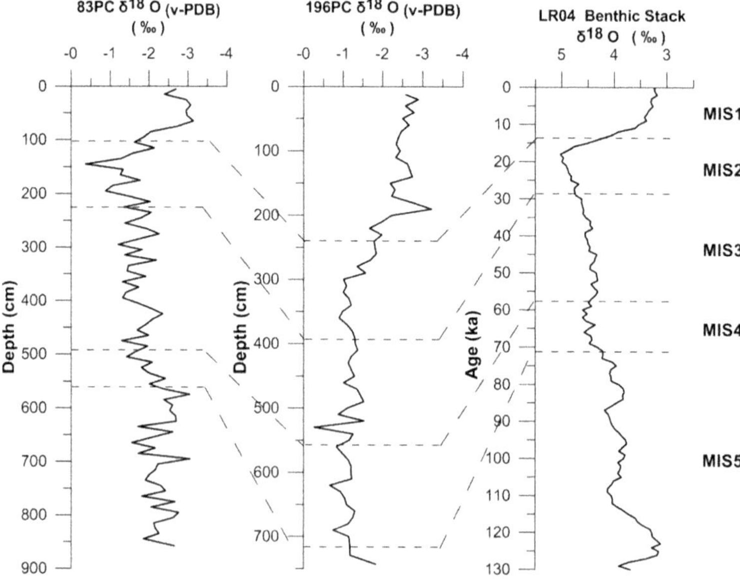

Fig. 2. Oxygen isotope data in the 83PC and 196PC cores correlated to the LR04 benthic stack (Lisiecki & Raymo 2005).

Table 2. *The AMS ^{14}C dating and calibrated age of core 83PC*

83PC depth	Source	Conventional ^{14}C age	Cal. years BP
13 cm	*G. rubber*	4471 ± 37	5160 ± 90*
97 cm	*G. rubber*	11 737 ± 50	13 650 ± 70*
101 cm	bulk	14 100 ± 70	16 524 ± 165†
193 cm	*G. rubber*	19 576 ± 73	23 330 ± 250*
201 cm	bulk	22 230 ± 170	26 183 ± 380†
260 cm	*G. rubber*	27 845 ± 89	32 220 ± 520*
387 cm	*G. rubber*	39 180 ± 170	43 440 ± 470*

*Measured at the Key Laboratory of Heavy Ion Physics, Beijing University, (China) and calibrated with Cal. 5.0 (Yang *et al.* 2009).
†Measured at the Poznań Radiocarbon Laboratory (Poland) and calibrated with CALIB 7 (calibration curve Marine13 and Delta $R = -10 \pm 50$).

content has also been observed in the transition from MIS 4 to MIS 3.

In core SA13-76, the abundance of *P. sulcata* reached up to 40%. Its content was low from 76 to 12 cal ka BP (the main parts of MIS 4–MIS 2), but relatively high in the transition between MIS 5 and MIS 4 at about 72 cal ka BP (*c.* 20%), and in MIS 3 (*c.* 5%) during the period 57–52 cal ka BP. However, the abundances of *P. sulcata* are higher during MIS 1, and attained approximately 40% at about 11–7 cal ka BP.

The abundance of *P. sulcata* in cores 17940 and MD05-2904 was, in general, low and did not exceed 3% between around 15 and 11 cal ka BP of the late MIS 2. However, there is a rapid increase up to approximately 13% in the early MIS 1

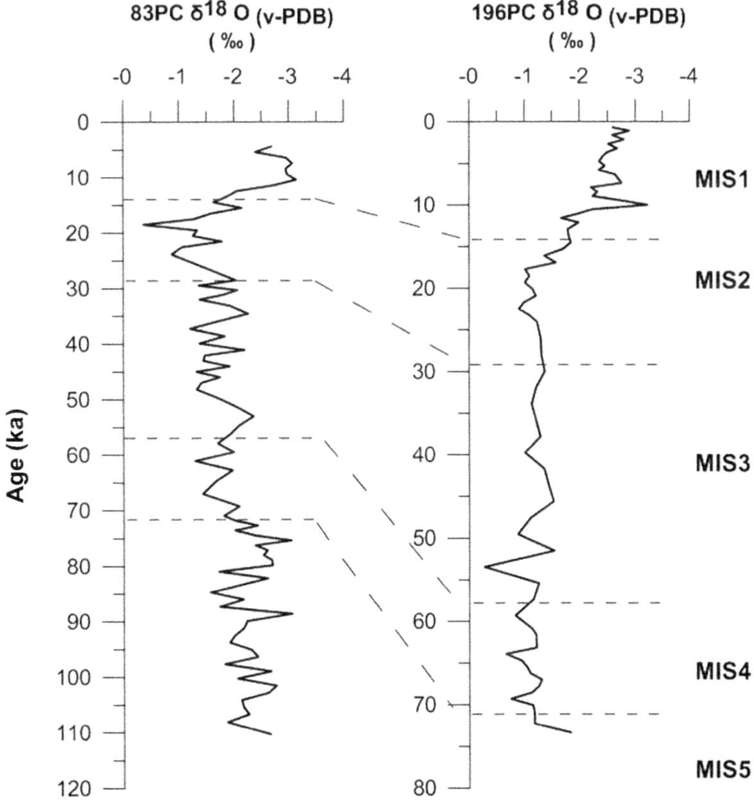

Fig. 3. Chronological framework of the 83PC and 196PC cores.

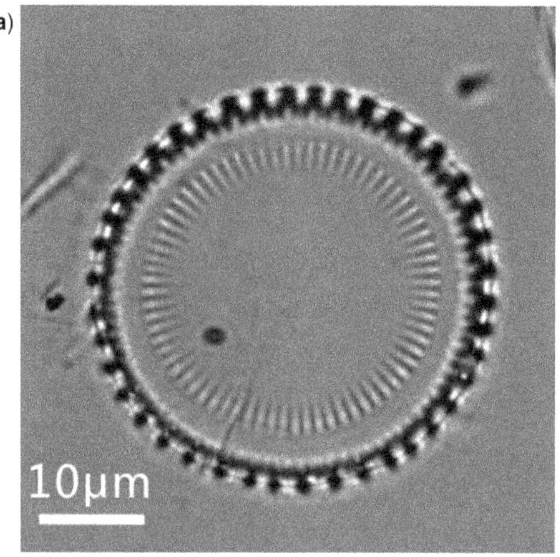

Fig. 4. LM image of (**a**) *Paralia sulcata* from core 83PC and (**b**) the SEM image from littoral of the Yellow Sea.

(*c.* 11–8 cal ka BP), followed by a rapid decrease to less than 2%.

To summarize, in the South China Sea, the abundance of *P. sulcata* shows a general similar tendency, with relatively low values in MIS 4–MIS 2, beginning from *c.* 74 cal ka BP through to 12 cal ka BP, although the detailed values are variable between the cores located on the continental slope. Another common phenomenon shared by *P. sulcata* is that its abundance increased rapidly in the early MIS 1 (11–8 cal ka BP) and was followed by a gradual decrease beginning from about 7 cal ka BP.

The abundance of P. sulcata *in sediment cores from the East China Sea*

The average abundance of *P. sulcata* in core DG9617 from the East China Sea shelf equates to 18%, beginning from about 8 cal ka BP. This ranges from in excess of 35% in the mid-Holocene and down to less than 8% in the late Holocene. In core

B, from the East China Sea continental slope, the highest relative abundances of *P. sulcata* occurred in sediment intervals dated to: *c.* 21–18 cal ka BP (15–42%), *c.* 16–11 cal ka BP (15–45%) and *c.* 9–8 cal ka BP (10–30%). However, in particular time intervals (e.g. *c.* 11–9 and 8–0 cal ka BP), its abundance ranged from 3 to 15%; around 18–16 cal ka BP, the *P. sulcata* content did not exceed 20%.

The abundance of *P. sulcata* in core MD756 in the Taiwan Strait during the period of 21–11 cal ka BP ranged from 30 to 40%. In the following stage, since 10 cal ka BP, it showed a more significant decrease from approximately 30% down to 18%.

Discussion

The palaeoclimatic and palaeoceanographic response of P. sulcata

Fluctuation of P. sulcata *in the last glacial cycle (LGC).* Numerous significant climatic events and environmental fluctuations have been recorded worldwide in the oceanic sediments, especially in marginal seas during the last glacial cycle (LGC) and the Holocene. Abundance of *P. sulcata* in the cores analysed shows stratigraphic variation across the marine oxygen isotopic stages (Fig. 5). On the glacial–interglacial scale, lower abundances of *P. sulcata* occur in the core profiles of the South China Sea during the last glacial stage, with higher values in the post-glacial period. Abundance increases during the early Holocene, with similar trends in the whole South China Sea. However, by contrast, abundances of *P. sulcata* in the cores from the continental slope are higher in the last glacial stage, particularly in the last glacial maximum (LGM) of the East China Sea, but quickly decrease in the early Holocene records. An interesting aspect is that high abundances of *P. sulcata* demonstrate a 'seesaw' effect, with alternate fluctuations in various areas of the China seas (SE Asian marginal seas) since the LGM. These changes are dated to the end of the Pleistocene at *c.* 11.5 cal ka BP.

These discrepancies indicate that the distribution of *P. sulcata* is affected by variations in the preservation in the two locations: namely, the South China Sea continental slope differed from the East China Sea. However, the *P. sulcata* rhythm reflects the dramatic climatic changes of the last glacial and post-glacial period. These discrepancies overlap the resistance to dissolution and further transport characteristics of *P. sulcata* frustules, especially as evidenced by their occurrence in the LGM in a period during a sea-level drop in a very dynamic oceanographic environment (Zhang *et al.* 1990; Ryu *et al.* 2008). Taking into consideration that the East China Sea has a large and broad shelf, it contributed to favourable conditions for *P. sulcata* to form blooms as a coastal species and deposition continued following the shoreward sea-level rise after LGM.

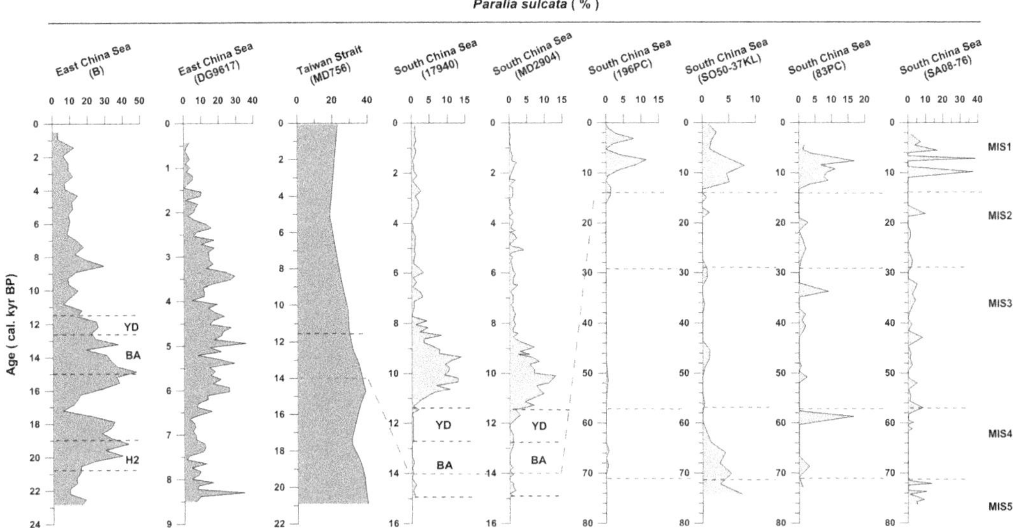

Fig. 5. Percentages of *Paralia sulcata* in cores from marginal seas of SE Asia. Age data and climate event divisions are cited from the original publications (Lan *et al.* 1995; Li *et al.* 2002, 2004; Zhi *et al.* 2003; Huang *et al.* 2006, 2009). YD, Younger Dryas, *c.* 12.6–11.1 cal ka BP; BA, Bølling–Allerød, *c.* 14.8–12.6 cal ka BP; H2, Heinrich-2, *c.* 21–19 cal ka BP; MIS, Marine Isotope Stage.

By contrast, the north South China Sea possesses a relatively narrow shelf zone, which results in poorer environmental conditions for *P. sulcata* during deglaciation and glacial periods. This has resulted in its low abundances in the sediments from the slope, as the currents have redeposited small amounts of *P. sulcata*. This seesaw effect is positively related to discrepancies between ecological conditions and the status of the shelves.

Response of P. sulcata *to palaeoclimatic events.* The *Paralia sulcata* content increased over a short timescale in the early MIS 1, and the early and late MIS 3, in cores MD05-2904, SA13-76, 83PC and SO50-KL37 from the South China Sea in early MIS 1, and early and late MIS 3. This increase can be linked with a global ice-volume change and a variation in the position of the coastline (Wang & Li 2009; Yao *et al.* 2009). An alternative explanation might be that of a short stage with intensified riverine discharge into coastal waters, caused by increased land-based precipitation. Climate changes and sedimentary environment variations in the last glacial period were also recorded in the other research sites from the South China Sea. For instance, calcium carbonate data in the South China Sea core SO49-37, located close to core 83PC, reveal high terrestrial sediment flux events occurring in the MIS 3.1–MIS 3.0 transition periods (Qian 1999). Huang *et al.* (2003) found that the tropical climate was hot and humid in MIS 3, especially in the early and late stage of this warmer, but unstable, period.

The globally recorded Heinrich-2 cold event (H2) took place at the age of 20.9–19 cal ka BP in the East China Sea. This cooling event is reflected in responses of foraminiferal assemblage composition and reduced tropical open-ocean diatoms, corresponding with a decrease in water temperature and weak Kuroshio Current activity (Liu *et al.* 1999; Li *et al.* 2002). The percentage of *P. sulcata* reflected in a sudden increase in core B from the East China Sea indicates an event of warm-water-mass recession (the Kuroshio Current) and an increasing impact of coastal waters. The rapid climate events of the Bølling–Allerød warm period (BA: *c.* 14.8–12.6 cal ka BP) and of the Younger Dryas cold period (YD: *c.* 12.6–11.0 cal ka BP) seem to be characterized by increased abundances of *P. sulcata* in the East China Sea, whereas the transition between the Bølling–Allerød and the Younger Dryas is characterized by relatively lower abundances of this species in the South China Sea.

During deglaciation, a rising sea-level process in the BA and YD period, and the discharge of Yangtze River diluted water, caused the coastal current activity to be reinforced in the East China Sea continental shelf, resulting in *P. sulcata* blooming and an increase in its sedimentation at the continental slope. The content of this species was slightly higher in the Taiwan Strait, but was much lower in the South China Sea during the BA and YD period.

A cold event during 9–8 cal ka BP, representing a rapid climate cooling, was recorded in Greenland and elsewhere in the world (Rohling & Pälike 2005). At this time, an increase in the abundance of *P. sulcata* in the East China Sea was accompanied by similar trend in the Taiwan Strait. By contrast, in the north and west of the South China Sea, *P. sulcata* abundance showed its maximum and stronger response to the climatic events in these regions for the post-glacial period.

Since 8 cal ka BP, the percentage of *P. sulcata* in these marginal seas was generally lower than in the early Holocene. The distance between the cores located on the continental slope and the coast successively increased following the early Holocene climate warming and rise in sea level. This resulted in a reduction of the *P. sulcata* abundance. However, the relative abundance grew in some periods, and this could have been influenced by increased precipitation and stronger land-based runoff that brings more nutrients to a shallow-sea environment.

Palaeogeography and environmental changes in support of significance of P. sulcata

Opening of the Taiwan Strait as indicated by P. sulcata *assemblages. Paralia sulcata* is one of the dominant species of phytoplankton in the Yellow and East China Sea at the present time (Tian *et al.* 2010; Jiang *et al.* 2015), and in benthos of shallow parts of Yellow Sea (Witkowski & Li, unpublished observations). It can, however, proliferate in coastal waters of South China (Jin *et al.* 1982), and is observed abundantly in surface sediments of the continental shelves of the Yellow, East China and South China seas (Jiang 1987). Analysis of *P. sulcata* abundance in the sediments dated to the LGM and last deglaciation period revealed higher values in a core taken from the western part of the Taiwan Strait. Conversely, during the same period, its abundances are lower in the sediments from the northern part of the South China Sea. Such a situation lasted until the early Holocene, when the content of *P. sulcata* increased again. These phenomena are interpreted in terms of palaeoenviromental differences during the last glacial stage in the adjacent sea areas of the Taiwan Strait.

During the LGM, an increase in global ice volume was caused by the expansion of high-latitude ice caps and mountain glaciers. The above phenomena also resulted in a drop in sea level. The East China Sea and the semi-enclosed South China Sea were affected by this event, during which the water level was approximately 120 m lower than present (Wang & Li 2009). Palaeogeographical analysis shows that in the submerged parts of the Taiwan Strait, marine sediments were deposited during the LGM (Lan & Chen 1998). *Paralia sulcata* is recorded in core MD756 during the LGM, with a low-resolution occurrence in sediments that may have been caused by a rapidly increasing water level. However, in the seabed of the western and southern part of the Taiwan Strait, large terrestrial animal fossils dated to the LGM and deglaciation stage were found (Cai 2002). In our opinion, the above fact indicates that shallow areas of the Taiwan Strait were, at least temporarily, exposed to terrestrial conditions during the LGM and last deglaciation period, compared to the sea-level-rise curve created by Zhang *et al.* (1990), as shown in Figure 6. As a result, the palaegeographical conditions of the South China Sea northern shelf and adjacent Taiwan Strait limited the seawater exchange/mixing between the South China Sea and the East China Sea during the last glacial period. Cold and low-salinity water masses heading southwards along the coast of the East China Sea were blocked, causing a decrease in *P. sulcata* abundance in the northern South China Sea. Furthermore, the coastal current in the northern part of the South China Sea was reduced, although the Pearl River discharge into the bay was also limited. Seemingly, such environmental conditions significantly reduced the abundance of *P. sulcata*. In the early Holocene, following the rapid sea-level rise (Zhang *et al.* 1990; Wang 1999), the Taiwan Strait gradually opened, resulting in the renewal of large-scale water exchange between the East China and South China seas. This was triggered by prevailing monsoons, with the eastern China coastal current maintaining a relatively low temperature and low salinity, and the eastern Guangdong coastal current flowing to the south as a result of the increase in precipitation (Wu & Sun 2000). These conditions favoured *P. sulcata* blooms and enhanced deposition in the continental slope sediments. Therefore, the percentage of *P. sulcata* was low in the northern South China Sea during the last glacial period and rapidly

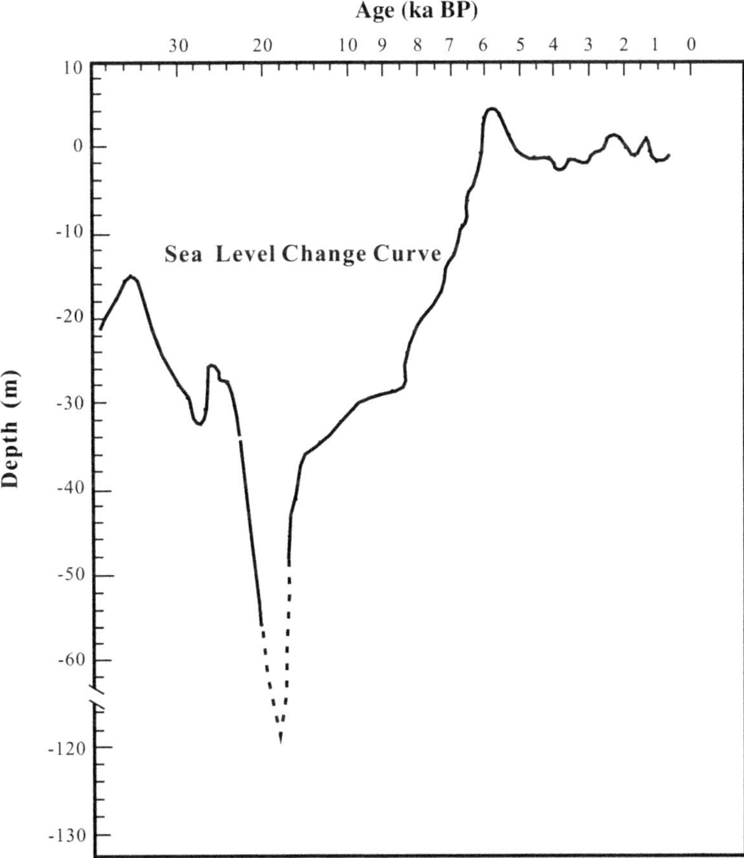

Fig. 6. The sea level change curve in the coastal zone of the South China Sea (modified from Zhang *et al.* 1990).

increased in the early Holocene, and we interpret this as an effect of the closing and opening of the Taiwan Strait.

Conclusions

A comparison of the abundance of *Paralia sulcata* in the sediment cores in the SE Asian marginal seas of the South China Sea and the East China Sea reveals regional and temporal differences with apparent rhythmic fluctuations over the last glacial cycle. *Paralia sulcata* can be considered as a useful proxy reflecting palaeoclimate, palaeogeography and palaeoenvironmental evolutions since the last glaciation in the marginal seas:

- The regional differences in abundance of *P. sulcata* are, in general, related to the distance from the shore and the water depth. The high abundance is associated with sites close to the shore and with shallow-water areas in a dynamic deposition environment. The species abundance shows distinct variations with time in selected coring sites. The content was low in the South China Sea in the MIS 4–MIS 2 samples, whereas high abundances of *P. sulcata* are observed in the East China Sea during the LGM and deglaciation, and in the South China Sea in the early Holocene.
- The abundance of *P. sulcata* reflects climatic variations during the glacial and interglacial periods. This corresponds to palaeoclimate events, including the climate fluctuation in MIS 3, the Heinrich-2 cold event, the Bølling–Allerød warm event, the Younger Drays cold event and, finally, the cold event at 9–8 cal ka BP.
- The abundance of *P. sulcata* is considered to be a good indicator of oceanic circulation in the SE Asian marginal seas. In particular, changes in its abundance are associated with the early Holocene opening of the Taiwan Strait, and the East China Sea southwards expansion of coastal current.
- However, reconstruction potential of this species is limited by the fact that it occurs both in benthos and in plankton. In addition, resistance to dissolution and mechanical destruction of the heavily silicified exoskeleton results in a strong overrepresentation of *P. sulcata* in fossil assemblages and impacts on the interpretation of environmental parameters.

This paper is supported by China Geological Survey Projects (GZH200800502 and 1212010611302). Part of the results presented in this publication were achieved along with a Chinese–Polish research project 'SECEB' funded by the Polish National Science Centre (NCN) allocated on the basis of decision No. DEC-2011/01/N/ST10/07708, and from the Polish Ministry of Science and Higher Education's topical subsidy. The authors wish to express their gratitude to the crew of the R/V *Haiyang-4* cruises who did the sampling. We thank Professor Jan Harff from the University of Szczecin (Poland), who improved the research content. Professor Kevin McCartney from the University of Maine at Presque Isle (USA) and Dr Colin Archibald (Perth, Australia) are acknowledged for their critical reading of the manuscript.

References

Abrantes, F. 1988. Diatom assemblages as upwelling indicators in surface sediments off Portugal. *Marine Geology*, **85**, 15–39.

Abrantes, F., Lopes, C. *et al.* 2007. Diatoms in Southeast Pacific surface sediments reflect environment properties. *Quaternary Science Reviews*, **26**, 155–169.

Blasco, D., Estrada, M. & Johns, B. 1980. Relationship between the phytoplankton distribution and composition and the hydrography in the Northwest African upwelling region near Cabo Carvoeiro. *Deep Sea Research Part A. Oceanographic Research Papers*, **27**, 799–821.

Cai, B. C. 2002. The evidence of dry land of the Taiwan Straits during 360 late Würm glaciation. *Marine Science*, **26**, 51–54 [in Chinese with English abstract].

Guo, X. W., Zhang, Y. S. *et al.* 2010. Characteristics and flux of settling particulate matter in neritic waters: the southern Yellow Sea and the East China Sea. *Deep Sea Research Part II: Topical Studies in Oceanography*, **57**, 1058–1063.

Hasle, G. R. & Syvertsen, E. E. 1997. Marine diatoms. *In*: Tomas, C. R. (ed.) *Identifying Marine Phytoplankton*. Academic Press, San Diego, CA, 5–385.

Huang, Y., Ran, L. H. & Jiang, H. 2006. Diatom from the South China Sea during the latest Pleistocene and their paleoenvironmental significance. *Marine Geology and Quaternary Geology*, **26**, 7–13 [in Chinese with English abstract].

Huang, Y., Jiang, H. *et al.* 2009. Diatom response to changes in palaeoenvironments of the northern South China Sea during the last 15 000 years. *Marine Micropaleontology*, **72**, 99–109.

Huang, Z. G., Zhang, W. Q. & Jiang, L. M. 2003. Climatic characteristics in tropic areas of China during MIS3. *Quaternary Sciences*, **1**, 77–82 [in Chinese with English abstract].

Jiang, H. 1987. Environment analysis of the common fossil diatoms from the sediments of China Sea. *Acata Botanica Sinica*, **1**, 77–82 [in Chinese].

Jiang, Z. B., Chen, J. F. *et al.* 2015. Controlling factors of summer phytoplankton community in the Changjiang (Yangtze River) Estuary and adjacent East China Sea shelf. *Continental Shelf Research*, **101**, 71–84.

Jin, D. X., Cheng, Z. D. *et al.* 1982. *China Marine Benthic Diatoms (The First Volume)*. Ocean Press, Beijing [in Chinese].

Karpuz, N. K. & Schrader, H. 1990. Surface sediment diatom distribution and Holocene paleotemperature variations in the Greenland, Iceland and Norwegian Sea. *Paleoceanography*, **5**, 557–580.

KATSUKI, K., KHIM, B. K. ET AL. 2010. Sea-ice distribution and atmospheric pressure patterns in southwestern Okhotsk Sea since the Last Glacial Maximum. *Global and Planetary Change*, **272**, 99–107.

KENNINGTON, K., HASLETT, S. K. & FUNNELL, B. M. 1999. Offshore transport of neritic diatoms as indicators of surface current and trade wind strength in the Plio-Pleistocene eastern equatorial Pacific. *Paleogeography, Paleoclimatology, Paleoecology*, **149**, 171–181.

LAN, D. Z. & CHEN, C. H. 1998. Sedimentary environment of the Taiwan Strait during Late Wurm glaciation. *Acta Oceanologica Sinica*, **20**, 83–90 [in Chinese with English abstract].

LAN, D. Z., CHEN, Z. D. & LIU, S. C. 1995. *Diatoms in Late Quaternary Sediments from the South China Sea*. Ocean Press, Beijing [in Chinese].

LANGE, C. B., ROMERO, O. E. ET AL. 1998. Offshore influence of coastal upwelling off Mauritania, NW Africa, as recorded by diatoms in sediment traps at 2195 m water depth. *Deep Sea Research Part I: Oceanographic Research Papers*, **45**, 985–1013.

LI, C., LAN, D. Z. & FANG, Q. 2002. Late Quaternary sedimentary diatom from East China Sea continental shelf and its paleoceanographical significance. *Journal of Oceanography in Taiwan Strait*, **21**, 351–359 [in Chinese with English abstract].

LI, C., LAN, D. Z. ET AL. 2004. Late Quaternary diatom biostratigraphic subdivision and correlation in Okinawa Trough. *Journal of Oceanography in Taiwan Strait*, **23**, 507–513 [in Chinese with English abstract].

LISIECKI, L. E. & RAYMO, M. E. 2005. A Pliocene–Pleistocene stack of 57 globally distributed benthic $\delta^{18}O$ records. *Paleoceanography*, **20**, PA1003, 1–17.

LIU, Z. X., SAITO, Y. ET AL. 1999. Late Quaternary paleoceanography research in Okinawa trough on a millennial scale. *Chinese Science Bulletin*, **44**, 883–887 [in Chinese with English abstract].

MCQUOID, M. R. & NORDBERG, K. 2003. The diatom *Paralia sulcata* as an environmental indicator species in coastal sediments. *Estuarine, Coastal and Shelf Science*, **56**, 339–354.

QIAN, J. X. 1999. *Paleoceanography Research of the South China Sea Since the Late Quaternary*. Science Press, Beijing [in Chinese].

ROHLING, E. J. & PÄLIKE, H. 2005. Centennial-scale climate cooling with a sudden cold event around 8200 years ago. *Nature*, **434**, 975–979.

ROUND, F. E., CRAWFORD, R. M. & MANN, D. G. 1990. *The Diatoms*. Cambridge University Press, Cambridge.

RYU, E., LEE, S. J. ET AL. 2008. Palaeoenvironmental studies of the Korean peninsula inferred from diatom assemblages. *Quaternary International*, **177**, 36–45.

SHIGA, K. & KOIZUMI, I. 2000. Latest Quaternary oceanographic changes in the Okhotsk Sea based on diatom records. *Marine Micropaleontology*, **38**, 91–117.

STRAMSKA, M. 2009. Particulate organic carbon in the global ocean derived from Sea WiFS ocean color. *Deep Sea Research Part I: Oceanographic Research Papers*, **56**, 1459–1470.

SUN, M. Q. 2009. *Interpretation of paleoenvironment by diatom in the South China Sea since Late Quaternary*. PhD thesis, Ocean University of China, Qingdao [in Chinese with English abstract].

TANIMURA, Y., SHIMADA, C. & HAGA, M. 2002. Migration of continental mixed-waters preserved in abundance of a diatom species *Paralia sulcata*. Paleoceanography of the Northeastern East China Sea from the Last Glacial through the post glacial. *Quaternary Research*, **41**, 85–93.

TIAN, W., SUN, J. ET AL. 2010. Phytoplankton community in coastal waters of the East China Sea in Spring 2008. *Advances in Marine Science*, **28**, 170–178 [in Chinese with English abstract].

WANG, P. X. 1999. Response of Western Pacific marginal seas to glacial cycles: paleoceanographic and sedimentological features. *Marine Geology*, **156**, 5–39.

WANG, P. X. & LI, Q. Y. 2009. *The South China Sea: Paleoceanography and Sedimentology*. Springer, Berlin.

WANG, K. F., JIANG, H. & ZHANG, Y. L. 1990. *Analysis of Quaternary Palynology and Algae and Environment in the South China Sea and the Coastal Area*. Tongji University Press, Shanghai [in Chinese].

WITON, W., MALMGRENA, B. ET AL. 2006. Holocene marine diatoms from the Faeroe Islands and their paleoceanographic implications. *Palaeogeography, Palaeoclimatology, Palaeoecology*, **239**, 487–509.

WU, G. X. & SUN, X. J. 2000. Late Quaternary organic-wall phytoplankton record in northern slope of South China Sea and its palaeoenvironmental significance. *Marine Geology & Quaternary Geology*, **20**, 57–63 [in Chinese with English abstract].

YANG, X. Q., HELLER, F. ET AL. 2009. Geomagnetic paleointesity dating of South China Sea sediments for the last 130 kyr. *Earth and Planetary Science Letters*, **284**, 258–266.

YAO, Y. T., HARFF, J. ET AL. 2009. Reconstruction of paleocoastlines for the northwestern South China Sea since the Last Glacial Maximum. *Science in China, Series D: Earth Science*, **39**, 753–762.

ZHANG, H. N., CHEN, W. G. ET AL. 1990. *The New Tectonic Activities and Geoenvironment in Coast of the South China*. Seismic Press of China, Beijing, 45–116.

ZHI, C. Y., WANG, K. F. ET AL. 2003. Study on the relationship between diatom assemblage and paleoenvironment of the Late Quaternary in the Taiwan channel and Xiamen Island. *Acta Micropalaeontologica Sinica*, **20**, 244–252 [in Chinese with English abstract].

ZONG, Y. Q. 1997. Implications of *Pararia sulcata* abundance in Scottish isolation basins. *Diatom Research*, **12**, 125–150.

Holocene sedimentary systems on a broad continental shelf with abundant river input: process–product relationships

SHU GAO*, DANDAN WANG, YANG YANG, LIANG ZHOU, YANGYANG ZHAO, WENHUA GAO, ZHUOCHEN HAN, QIAN YU & GAOCONG LI

Ministry of Education Key Laboratory for Coast and Island Development, Nanjing University, Nanjing 210093, China

**Corresponding author (e-mail: shugao@nju.edu.cn)*

Abstract: The region consisting of the Bohai, Yellow and East China seas represents a typical wide continental shelf environment with abundant terrestrial sediment supply. Here, a variety of sedimentary systems have been formed during the Holocene period. These systems have unique characteristics in terms of spatial distribution, material composition, deposition rate, and the timing and duration for their formation, which are related to active sediment-transport processes induced by tides and waves, shelf circulations, and sediment gravity flows. The sedimentary records contained within the deposits have a high temporal resolution, but each with a limited temporal coverage. However, if these records are connected, then they may form a complete archive for environmental change studies. In the field of process–product relationship studies, the mid-Holocene coastal deposits on the Jiangsu coast, the early–middle Holocene sequences of the Hangzhou Bay, the Holocene mud deposits off the Zhejiang–Fujian coasts and the other mud areas over the region are of importance. These systems may be understood by identifying the material supply (from both seabed reworking during the sea-level rise events and river discharges), transport-accumulation processes, the formation of sediment sequences and the future evolution of the sediment systems, for which numerical modelling becomes increasingly important.

Sedimentary archives on continental shelves contain environmental information related to sea-level changes. Ever since the time of Charles Lyell, it has been proposed that the temporal scale of sedimentary sequences can be determined by using the global sea-level curve. For example, Sloss (1963) suggested that there have been six major sea-level cycles since the Cambrian and that these provide a general chronological control for stratigraphic sequences. Subsequently, Vail *et al.* (1977), in an attempt to establish correlations between the sequence and sea-level position on smaller temporal scales, presented a novel global eustatic chart. The most direct application of this new approach is to interpret the stratigraphic characteristics of shallow-marine deposits in terms of time periods covered by the deposits, and to infer possible hiatuses on the basis of known sea-level curves. The sea-level chart has been modified by a number of authors (e.g. Haq *et al.* 1987), to be widely used as a standard tool for sequence stratigraphic analysis.

However, in using the global eustatic chart, only one factor, namely the sea-level position, was emphasized; other important factors such as the initial bathymetry, sediment supply and transport-accumulation processes were largely neglected. As a result, the global eustatic chart is only able to predict the timing of sedimentary sequences, whereas the spatial extent of sedimentary systems cannot be determined. Such an analysis would be sufficient if an investigation were merely connected with macro-scale temporal changes (Kamp & Naish 1998; Paola 2000). However, when smaller time-scales are considered (e.g. the formation and evolution of Holocene shelf–coastal sedimentary systems), then the other factors cannot be neglected. Sloss (1963) actually suggested that, in addition to sea level, the three factors listed above were important, but he did not proceed to analyse the combined influences of all four factors. Over the years, measurements of sediment dynamics on decadal scales were carried out in an attempt to relate the product of the sedimentary environments (i.e. the sedimentary sequences and archives) to the controlling processes (Nittrouer & Wright 1994; Nittrouer *et al.* 2009). Furthermore, scientific research programmes such as Strataform were developed to illuminate process–product relationships (Nittrouer 1999).

In terms of the initial bathymetric conditions at a fixed sea level, the width and bed slope of the shelf is a result of tectonic and sedimentological evolution; while, the terrestrial sediment input influences the sediment budget of the shelf. In combination, these two factors can generate four different shelf types: (1) wide shelves with abundant

From: Clift, P. D., Harff, J., Wu, J. & Qui, Y. (eds) 2016. *River-Dominated Shelf Sediments of East Asian Seas*. Geological Society, London, Special Publications, **429**, 223–259.
First published online September 7, 2015, updated May 26, 2016, http://doi.org/10.1144/SP429.4

sediment supply (e.g. the shelves of the Bohai, Yellow and East China seas); (2) narrow shelves with abundant sediment supply (e.g. the Bengal Bay); (3) wide sediment-starved shelves (e.g. the North Sea in Europe); and (4) narrow shelves with poor sediment supply (e.g. the shelves along the western North American coastline). Furthermore, these two factors also influence the hydrodynamics and, hence, the sediment-transport capacity. Three major processes can be distinguished: tide-induced transport (often in combination with wave action); fine-grained sediment transport due to shelf circulations; and sediment gravity flows (Gao & Collins 2014). Thus, different types of continental shelves will generate different sedimentary systems, resulting in different types of sedimentary archives or records in terms of their resolution, continuity and duration, all of which are important for past global change studies.

The purpose of the present contribution is to synthesize the characteristics of the Holocene sedimentary systems of the Bohai, Yellow and East China seas, including their temporal evolution and spatial distribution patterns, and the resultant sedimentary records, from the point of view of sediment dynamics. The study area, which is characterized by a gentle bed slope and abundant sediment supply, is associated with rich sediment archives for environmental change research. Hence, it would be desirable to elucidate the relationships between the processes and products in order to provide a basis of appropriate stratigraphic interpretation. In addition, some critical scientific problems needing attention in the future are identified.

Regional environmental setting

The region covering the Bohai, Yellow and East China seas is characterized by a wide shelf with abundant terrestrial sediment supply (Fig. 1). Covering a total area of some 1.26×10^6 km^2, the region

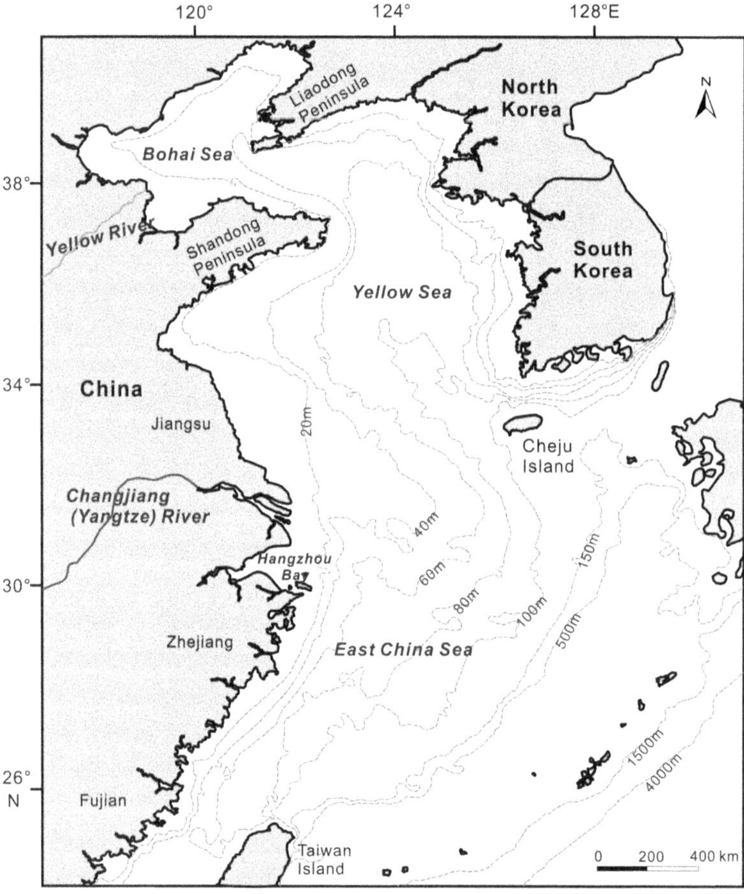

Fig. 1. Geographical setting of the study area (bathymetry in metres).

in consideration receives large sediment inputs from the Yellow and Changjiang rivers, as well as from numerous smaller rivers. As a result, large river deltas and other types of shelf–coastal deposits are formed. The Holocene mud deposits are representative of such an environmental setting (Fig. 2). Here, the river discharges are two magnitudes higher than those of the North Sea in Europe; such a magnitude of sediment supply favours the formation of shelf mud deposits (Dronkers & Miltenburg 1996). The reason for this is that in a closed system the repeated resuspension and settling would not result in overall net deposition, but only redistribution of the sedimentary material within the system; only when there is an external source can net accretion take place. Thus, the high sediment discharge, induced by the regional monsoon climate, meets the condition for the formation of deposits on a large scale.

The Yellow River is well known for its high sediment load over the entire Holocene period. Because this river discharged alternatively into the Bohai and southern Yellow seas during the late Holocene, two large deltas were formed (Chen & Zhu 2012). One is located on the coast of the Bohai Sea, with the most recent part having formed since 1855 and covering an area of around 6×10^3 km^2. The other is the Old Yellow River delta on the Jiangsu coast, which formed in the period from AD 1128 to AD 1855, when the river discharged into the

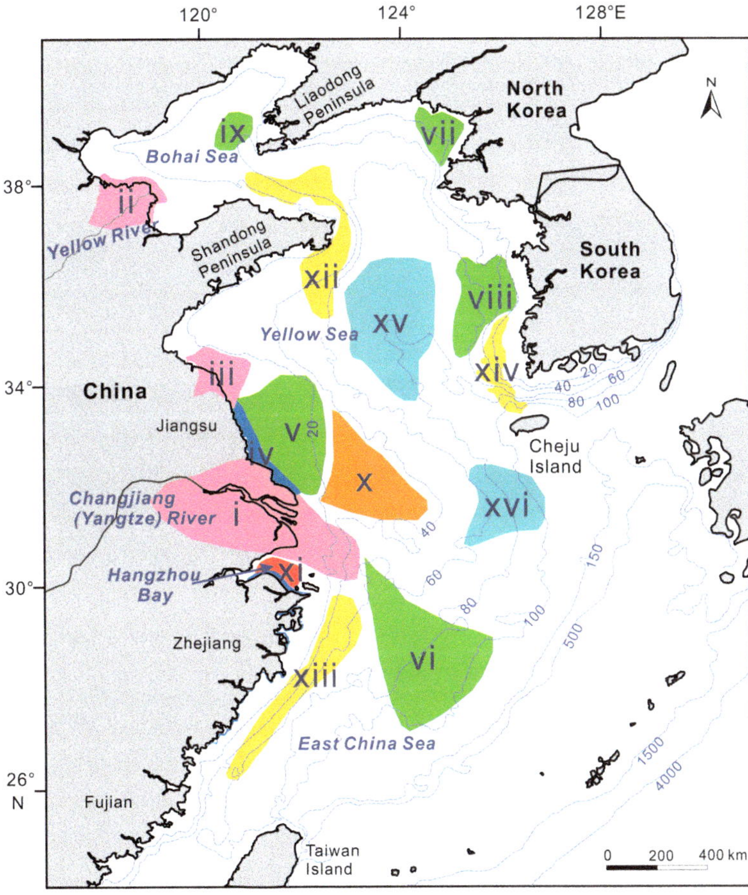

Fig. 2. Distribution of major coastal and shelf Holocene sedimentary systems over the study area: (i) Changjiang River delta; (ii) Yellow River delta; (iii) the abandoned delta associated with the Old Yellow River; (iv) tidal flats on the Jiangsu coast; (v) Jiangsu radial tidal ridges; (vi) palaeo-tidal ridges on the East China Sea shelf; (vii) tidal ridges off the North Korean coast; (viii) tidal ridges off the South Korean coast; (ix) tidal ridges in the Bohai Sea; (x) Yangtze Shoal; (xi) Hangzhou Bay estuarine deposit; (xii) Yellow River distal mud; (xiii) Changjiang distal mud; (xiv) Huskan Mud Belt in the SE Yellow Sea; (xv) isolated mud patch in the central Yellow Sea; and (xvi) isolated mud patch to the south of Cheju Island (modified from Gao & Collins 2014).

southern Yellow Sea; this delta originally covered an area of some 12×10^3 km^2, which has been subject to coastal erosion after the river mouth shifted to the north. The Changjiang River delta has a subaerial area of 25×10^3 km^2 and a subaqueous delta of 10×10^3 km^2 (cf. Gao 2007). In addition, the deposits of adjacent areas (e.g. the Hangzhou Bay) are also influenced by the input of the Changjiang River. It should be noted that the suspended sediment from the large rivers may be transported over long distance to form distal mud deposits such as those around the Shandong Peninsula, northern Yellow Sea, associated with the Yellow River (J. P. Liu *et al.* 2004), and offshore of the Zhejiang–Fujian coast, associated with the Changjiang River (J. P. Liu *et al.* 2007; Xu *et al.* 2009). The fine-grained sediment may be further dispersed towards the middle or outer shelf, where it forms isolated mud patches (Gao & Jia 2003). Along the west coast of Korea, thick Holocene mud deposits were emplaced, the material probably having different sources, although Korean researchers generally favour the small local rivers as the main source (Park *et al.* 2000; Lee & Chu 2001).

As a typical broad, sediment-starved continental shelf environment, the North Sea in Europe is characterized by tidal ridges, tidal sand sheets, tidal flats and barrier island lagoon systems (Flemming & Davis 1994; Collins *et al.* 1995). These Holocene deposits are mainly a product of the reworking of early sedimentary bodies by waves and tidal currents during the post-glacial transgression. Similar deposits are distributed over the eastern China shelf regions, including the tidal-ridge fields in the eastern Bohai Sea and off the Jiangsu coast (the SW Yellow Sea) (Liu *et al.* 1998; Y. Wang *et al.* 2012), as well as off the western coast of Korea (Off 1963), the Yangtze Shoal to the NE of the Changjiang River mouth (Liu 1997) and the palaeo-tidal ridges on the middle–outer shelf of the East China Sea (to the SE of the Changjiang Estuary) (Yang 1989; Z. X. Liu *et al.* 2007). These deposits have resulted from the modification to the underlying strata over the course of the Holocene.

The region is characterized by a monsoon climate: that is, northerly winds prevail in winter but southerly winds are dominant in summer. During the winter, cold outbreaks occur, with winter storms hitting the coasts; in the summer time, the region is often influenced by typhoons, which represent a major natural hazard. The waves are moderate, except during storm events. The tides are mainly semi-diurnal in character, with the largest ranges being observed on the opposite coasts of both sides of the Yellow Sea, the maximum range reaching 9.39 m on the Jiangsu coast, SW Yellow Sea (Ding *et al.* 2014), and in Hangzhou Bay, where large-scale tidal bores occur (Fan *et al.* 2014).

Sediment-transport and accumulation processes

Tidally induced sediment transport

Tides in this region play an important role in sediment transport. Large tidal ranges and strong tidal currents, in combination with shallow water depths, small seabed slopes and wave action, have resulted in active resuspension and transport of fine-grained sediment. Figure 3 shows the spatial and seasonal distribution patterns of suspended sediment over the regions; the high values are mainly due to tidally induced resuspension.

Traditionally, sediment transport in the region was determined from time-series measurements over 26 h at fixed stations, including water depth, current velocity, suspended sediment concentration (SSC), water temperature, salinity and other parameters (Shi 2004). In this way, numerous datasets were collected. Although this method has a large uncertainty in terms of the long-term estimate of sediment fluxes because the measurements may not fully represent the long-term situations, these datasets are invaluable in the analysis of the processes and mechanisms of transport (e.g. resuspension and dispersion/diffusion), and in the modelling of the transport patterns (they provide basic data for parameter determination and model verification). In Table 1, some results of tidal-cycle measurements from inner-shelf areas are listed, which show that the tidal-current velocity is of the order of 1.0 m s^{-1}, the suspended sediment concentration (due to resuspension) has an order of magnitude of 10^{-1}–10^0 kg m^{-3}, and the resultant sediment-transport rate ranges between 10^{-1} and 10^1 kg m^{-1} s^{-1}. In more recent times, longer-term monitoring has been accomplished by means of mooring systems and coastal observatories. For instance, continuous annual measurements of the SSC over the areas adjacent to the Changjiang Estuary have been carried out; the data from several locations indicate that the SSC is controlled by seasonal changes in wave action and river discharges, in addition to tides, which influence the position of the turbidity maxima of the estuarine waters.

The method of numerical modelling has been widely adopted to reveal sediment-transport patterns. For example, suspended-sediment transport has been simulated for the Bohai Sea, the area adjacent to the Yellow River mouth (Jiang *et al.* 2000, 2004; Lu *et al.* 2011), and the Changjiang Estuary and adjoining waters (Chen & Wang 2008; Hu *et al.* 2009). The results obtained indicate that the suspended sediment concentration and its spatial distribution are controlled by tides, but that the long-distance transport of suspended sediment on the shelf is influenced by shelf-circulation patterns.

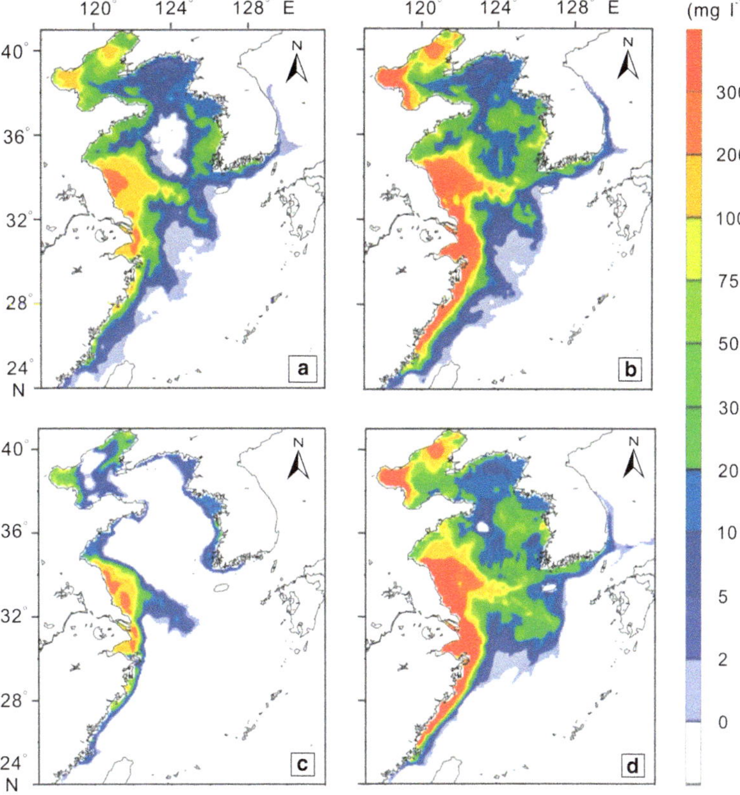

Fig. 3. Model output for suspended sediment concentrations (SSCs) in the study area, where tidal currents are the dominated controlling factor: (**a**) sea-surface SSCs in January; (**b**) bottom SSCs in January; (**c**) sea-surface SSCs in July; and (**d**) bottom SSCs in July (the colour bar denotes the scale of SSCs in mg l^{-1}, modified from Bian 2012).

For the Jiangsu coast, Xing *et al.* (2012) used the model Mike 12 to calculate the suspended sediment concentration in response to the combined action of waves and tidal currents for both the summer and winter seasons, the results being in good agreement with observational data; furthermore, these authors suggest that the enhancement of resuspension in winter is mainly due to the influence of the reduction in seawater temperature, rather than to the wave effect. In contrast, Wang *et al.* (2011) argued that the intensified wave action in winter plays a significant role. The results of their simulation are consistent with the sediment-transport patterns revealed by satellite imagery, which show sediment dispersal towards the Cheju Island. However, further research is required to resolve certain disagreements between simulation and observation.

Important mechanisms associated with tides include the settling–scour lag effects and time–velocity asymmetry (Postma 1961), which result in net landwards transport. The formation of tidal deposits (e.g. tidal flats and tidal ridges) is associated with the tidal current in this region (Gao 2009*a*).

Transport due to currents associated with low-frequency shelf circulations

On the shelf of the Yellow and East China seas, East Asia, which has a width of up to 600 km, circulation includes upwelling induced by the Kuroshio along the shelf edge, monsoon wind-driven currents and water movements related to river discharge (Chen 2009), which results in complicated seasonal patterns (Fig. 4). The general characteristics of the water masses and shelf currents are listed in Table 2. There are a number of mechanisms for the formation of shelf circulations, including: a baroclinic effect (e.g. high-density waters (high salinity, low temperature) intruding onto the shelf in the bottom layer); wind-driven currents (in the case of the monsoon-dominated eastern China seas, with winter and summer winds generating different

Table 1. *Tidal-cycle measurements of suspended sediment concentrations and transport rates at some inner-shelf sites of the region*

Station	Position		Depth (m)	Current speed ($m\ s^{-1}$)	SSC ($kg\ m^{-3}$)	Transport rate ($kg\ m^{-1}\ s^{-1}$)	References
R10	120° 58.2′	33° 39.7′	15.9	0.70	0.45	5.01	Wang (2014)
R11	121° 13.7′	33° 19.5′	11.8	0.60	0.21	1.49	
Y9	120° 51.8′	33° 10.3′	9.2	0.84	0.89	6.88	
Y10	120° 54.63′	33° 11.05′	16.0	0.94	0.55	8.27	
Y11	120° 57.9′	32° 58′	8.9	1.04	0.91	8.42	
Y12	121° 2.95′	32° 54.2′	6.3	1.09	0.98	6.73	
R20	121° 32.6′	32° 57.1′	13.5	0.89	0.10	1.20	
Y + 1	121° 08.8′	32° 35′	15.9	0.81	0.29	3.73	
Y + 2	121° 19.8′	32° 33′	18.1	0.92	0.21	3.50	
R22	121° 44.9′	32° 39.6′	10.8	0.62	0.16	1.07	
R23	121° 37′	32° 39.6′	8.1	0.52	0.13	0.55	
R24	122° 29.69′	32° 39.6′	17.0	0.76	0.10	1.29	
Y13	121° 44.6′	32° 04.6′	5.9	0.58	0.52	1.78	
Y14	121° 48.7′	32° 05.4′	7.6	0.72	0.32	1.75	
Y15	121° 53′	32° 02′	10.7	0.65	0.13	0.90	
Y16	122° 03′	31° 54′	14.7	0.67	0.12	1.18	
E2	123° 06′	34° 30′	77.0	0.10	0.03	0.23	Dong *et al.* (2011)
A08	120° 48.69′	33° 13.24′	0.4	0.15	0.35	0.02	Y. P. Wang *et al.* (2012)
M08	120° 49.29′	33° 13.6′	1.7	0.30	0.38	0.20	
S1	120° 49.9′	33° 13.93′	4.4	1.00	0.12	0.51	
K01	121° 26.58′	32° 53.97′	16.9	0.92	0.16	2.49	Ni *et al.* (2013, 2014)
K02	121° 32.4′	32° 56.06′	9.6	0.61	0.15	0.85	
K03	121° 39′	32° 58.15′	16.4	0.91	0.16	2.39	
K04	121° 48.66′	33° 6.52′	18.6	0.99	0.11	2.08	
K05	121° 50.04′	33° 0.93′	14.2	0.86	0.14	1.68	
K06	121° 54.36′	33° 14.90′	15.8	0.79	0.16	1.99	
K07	121° 57.3′	33° 12.10′	19.2	0.83	0.07	1.14	
K08	122° 1.2′	33° 6.92′	23.1	0.80	0.05	0.97	
K09	122° 3.72′	33° 3.01′	18.1	0.74	0.06	0.77	
P01	121° 29.64′	32° 55.22′	11.4	0.92	0.19	1.96	
P02	121° 34.26′	32° 56.90′	19.2	0.96	0.14	2.49	
YH	122° 34.8′	37° 19.93′	1.5	1.00	0.01	0.02	Jia *et al.* (2003)
DF	120° 48.55′	33° 14.12′	1.8	0.35	0.57	0.36	Li *et al.* (2007)
JB	120° 19.5′	36° 22.98′	7.0	0.30	0.01	0.02	D. L. Yuan *et al.* (2008); Y. Yuan *et al.* (2008)
Tetrapod	121° 51′	31° 12.38′	10.8	1.28	1.30	17.97	Y. Y. Wang *et al.* (2009)
JL	118° 01′	24° 25.49′	8.7	1.10	0.02	0.19	Y. P. Wang *et al.* (2013)
A1	119° 3.42′	38° 0.21′	7.0	0.25	0.06	0.10	Bi *et al.* (2010)
A2	119° 5.76′	38° 1.15′	11.0	0.40	0.02	0.10	
A3	119° 7.5′	38° 1.92′	14.0	0.43	0.22	1.30	
A4	119° 10.14′	38° 3.03′	15.0	0.42	0.03	0.20	
B1	119° 1.26′	37° 25.87′	5.0	0.22	0.05	0.05	
B2	119° 6.3′	37° 28.53′	8.0	0.32	0.04	0.10	
B3	119° 10.98′	37° 30.88′	7.0	0.47	0.12	0.40	
B4	119° 16.32′	37° 33.70′	5.0	0.35	0.29	0.50	
B5	119° 21.18′	37° 36.22′	6.0	0.46	0.40	1.10	
C1	119° 10.98′	37° 46.25′	2.7	0.52	0.47	0.66	Bi *et al.* (2010)
C2	119° 13.92′	37° 46.56′	2.7	0.54	0.96	1.40	
C3	119° 17.22′	37° 48.61′	5.0	0.39	0.31	0.60	
C4	119° 18.78′	37° 49.86′	12.0	0.53	0.14	0.90	
C5	119° 21′	37° 51.58′	15.0	0.41	0.02	0.10	
A1	119° 32.46′	26° 4.85′	14.0	0.86	0.16	1.34	Li *et al.* (2009)
A2	119° 37.2′	26° 3.71′	11.0	0.72	0.73	5.78	
PHW	119° 14.34′	25° 9.92′	4.9	0.65	0.05	0.15	Zhang (2013)

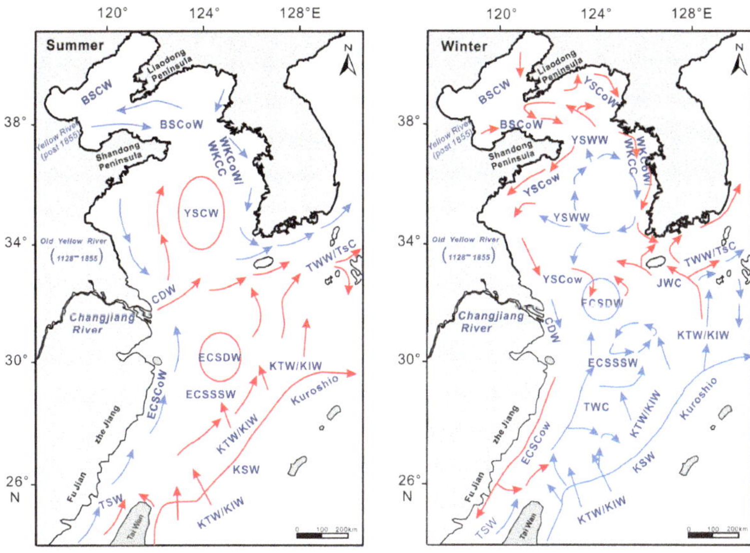

Fig. 4 . Major shelf currents in the summer and winter seasons (compiled on the basis of Chen 2009).

shelf flows); seasonal changes in water temperature that modify the density distributions and thereby trigger water-mass movement; tidal disturbance in shallow waters, generating water masses with different densities from those of the surrounding waters, causing baroclinic movement; and river-derived freshwater inputs, which generate estuarine circulations with river plumes moving seawards in the surface layer.

Fine-grained sediment is transported towards the middle–outer continental shelf, or even across the shelf towards the deep ocean by shelf currents (Liu *et al.* 2006; Yang & Liu 2007). The river-derived sediment can reach the shelf break where it may partially accumulate temporally; when shelf currents permit (e.g. in the East China Sea, where bottom currents flow seawards in winter), the material escapes from the shelf to enter the adjacent

Table 2. *General features of major shelf-current and water-mass systems (based on Liu* et al. *1993; Sun 1995; Hu* et al. *2000; Chen 2009; W. Li* et al. *2012; Liu* et al. *2013)*

List of acronyms	Shelf water mass and current systems	Temperature (°C)		Salinity (psu)	
		Winter	Summer	Winter	Summer
BSCW	Bohai Sea Central Water	<1	25–27	<31	<30
BSCoW	Bohai Sea Coastal Water	<3	24–26	30–31	<31.5
CDW	Changjiang Diluted Water	6–17	27–29	20–31	<32
ECSCow	East China Sea Coastal Water	8–17	26–28	25–31	<33.5
ECSDW	East China Sea Dense Water	9–13	27–28	32–34	31.5–33.5
ECSSSW	East China Sea Shelf Surface Water	13–18	27–29	34–34.7	33–34.2
JWC	Jeju Warm Current	10–15	25–28	33.5–34.5	31–33
KIW	Kuroshio Intermediate Water	16.5–20.5	14.4–19.5	34.4–34.9	34.4–34.8
KSW	Kuroshio Surface Water	20–24	28–29	33.5–34.75	33.5–34.5
TSW	Taiwan Strait Water	18–23	28–29	33.5–34.5	33.5–34
TWC	Taiwan Warm Current	13–18	28–29	33–34.5	32–34
TsC	Tsushima Current	13–15	27–29	34.5–34.75	33.3–34
TWW	Tsushima Warm Water	12–15	26–28	34.5–34.75	32–33
WKCoW	Western Korea Coastal Current	2–5	20–26	31.5–32.5	<30
YSCoW	Yellow Sea Coastal Water	<4	24–26	31–31.5	<31
YSCW	Yellow Sea Cold Water	5–10	8–10	32–33.5	30.5–31.5
YSWW	Yellow Sea Warm Water	5–12	25–27	32–34	30.5–32
YSWC	Yellow Sea Warm Current	8–13	26.7–27	33–33.5	32.5–33

deep waters (Hoshika *et al.* 2003). On the middle shelf of the Yellow Sea, fine-grained sediment originating from the inner shelf and organic material produced in the water column accumulate in an area with weak tidal currents. Numerical modelling shows that the accumulation may be enhanced by upwelling–downwelling patterns (Gao & Jia 2003). It is interesting to observe that, on the inner shelf of the region, sediment suspended in the surface layer is also transported seawards: remote sensing analysis reveals that the turbid water mass expands into the central Yellow Sea in winter (Shi & Wang 2012). This pattern indicates a fundamental difference between the circulation patterns of wide and narrow shelves. On narrow shelves, seawards flow in the bottom layer would imply landwards flow at the surface, according to mass conservation. However, on wide shelves, the movement of the various water masses creates additional possibilities; for instance, the shelf may be divided into sections, within which vertical circulation patterns may differ.

Sediment movement can be caused by shelf circulation resulting from the combined action of wind, sea-surface gradient and baroclinic effects. Once brought into motion, the flow direction will be influenced by the Coriolis force. The horizontal movement of seawater will inevitably lead to sea-surface elevation changes. The variations in water temperature and salinity at different sites cause horizontal and vertical water-mass movement. As a consequence, the shelf-current system in the region is complex; the complexity is further enhanced by the intrusion of the Kuroshio Current and the freshwater discharge from the Changjiang River (G. X. Li *et al.* 2006; Zhang *et al.* 2007; Chen 2009). Although the flow speed associated with the shelf circulation is much lower than that of the tidal currents (i.e. usually of the order of 1–10 cm s^{-1}: Lee & Chao 2003), long-distance transport of fine-grained sediment nevertheless takes place, even across the entire shelf, because of the low settling velocities of the suspended particles (Nittrouer & Wright 1994).

Suspended-sediment transport in response to shelf circulations has been studied by numerical modelling and *in situ* measurements. In physical oceanography, field, observations and numerical simulations have been carried out for the Kuroshio transport (Hsueh *et al.* 1997), as well as for shelf-circulation patterns (Lee *et al.* 2000), in particular those the in the Yellow and East China seas, (Lee *et al.* 2002). On this basis, suspended-sediment transport has been modelled for the entire region (Choi *et al.* 2005).

Along the shelf break of the East China Sea, suspended sediment is transported downslope towards the Okinawa Trough due to localized circulations cells resulting from the interaction between the Kuroshio core flow and Taiwan island (Sheu *et al.* 1999). According to the data collected during 2000–08, the total amount of suspended sediment in the water column of the Yellow and East China seas in summer reaches 0.18×10^9 t (Dong *et al.* 2011). The quantity is further enhanced in winter due to intensified northerly monsoon winds and waves. For instance, the concentration and the transport rate of suspended sediment in the coastal waters adjacent to the Yellow River mouth are much higher in winter than in summer, with the transport rate being 10^0–10^1 kg m^{-1} s^{-1} in winter and 10^{-2}–10^0 kg m^{-1} s^{-1} in summer (Yang *et al.* 2011). Under the various hydrodynamic conditions, sediment supplied by the Yellow River can be transported by the shelf current towards the Yellow Sea through the Bohai Strait (Lu *et al.* 2011). On the outer shelf of the East China Sea, suspended cross-shelf sediment transport takes place towards the Okinawa Trough in the form of bottom turbid layer (Watanabe 2007).

An analysis of the MODIS ocean colour imagery (D. L. Yuan *et al.* 2008; Y. Yuan *et al.* 2008; Zhang *et al.* 2010) indicates that the suspended matter distributed over the Yellow and East China seas in winter is derived from the Jiangsu coast, and the seawards dispersion of the suspended sediment is caused by the shelf currents driven by the winter monsoon (Yuan & Hsueh 2010). Furthermore, analyses of the 2002–08 sea-surface temperature, ocean colour, wind speed and sea-surface elevation anomaly data indicate that turbid water, with a concentration of greater than 50 mg l^{-1}, is expanded seawards during January–April, whilst it is confined to the Jiangsu coastal waters during July–September (Shi & Wang 2010, 2012); remote sensing images display the formation of a suspended sediment plume over the Yangtze Shoal, which reveals a transport pathway from the Jiangsu coast towards the middle shelf in winter.

In terms of the processes and mechanisms responsible for the cross-shelf sediment dispersal towards the deep ocean, indirect evidence has been sought from studies on sediment provenance (Yang *et al.* 2003, 2004; Yang & Youn 2007) and the deposition rate of shelf mud (DeMaster *et al.* 1985; Huh & Su 1999; F. Y. Li *et al.* 2006). The method of tracing the provenance is mainly based on the identification of the source and transport pathways from the material deposited at the accumulation site. Many types of tracers – for example, heavy metals (Cu, Pb, Cd) (Lin *et al.* 2002), rare earth elements (Song & Choi 2009), river-derived magnetic minerals (L. B. Wang *et al.* 2009; Zhang *et al.* 2012), clay minerals (Xu *et al.* 2009; Dou *et al.* 2010), chemical index of alteration (CIA) (Yang *et al.* 2002), ^{210}Pb and ^{137}Cs isotopes

(Chung *et al.* 2003; Du *et al.* 2010), and terrestrial organic matter (Zhu *et al.* 2008) – have been used for this purpose. Thus, the content of Cu, Pb and Cd in the Changjiang sediments has been adopted as tracers to delineate the transport and accumulation on the open shelf and in the Hangzhou Bay, the sedimentary record revealing changes in human activity in the catchment basin (Lin *et al.* 2002). The analysis of clay minerals in the Okinawa Trough, deposited over the last 28 kyr, has revealed the changing patterns of the Changjiang influence: from 28 to 14 ka BP, the sediment is dominated by material supplied by the Changjiang, which is associated with the reworking of shelf deposits during the period of sea-level lowstand from 14 to 8.4 ka BP, followed by material input from Taiwan in the period from 8.4 to 1.5 ka BP and, once again, by material input from the Changjiang over the last 1.5 kyr (Dou *et al.* 2010). Taiwan and the Changjiang clay minerals are both dominated by illite, but the Changjiang sediments have a low crystallinity and a relatively high smectite and kaolinite content; therefore, the sedimentary provenance of the northern Taiwan Strait can be determined based on these differences (Xu *et al.* 2009). After being normalized on the basis of grain-size data, geochemical indices show that the East China Sea material is derived from Chinese rivers; in the Yellow Sea, however, the sediments are supplied from both Chinese and Korean rivers (Lim *et al.* 2006). Many researchers have measured the deposition rate of the shelf mud deposits on the basis of ^{210}Pb and ^{137}Cs dating of cored samples (DeMaster *et al.* 1985; Alexander *et al.* 1991; Huh & Su 1999; F. Y. Li *et al.* 2002; Oguri *et al.* 2003; Lim *et al.* 2007). The results show that near the shoreline the deposition rate is high, reaching levels of 10^0-10^1 cm a^{-1}, owing to abundant material supply; on most of the open shelf region, the deposition rate is of the order of 10^{-1} cm a^{-1}.

The above studies have identified an interesting discrepancy: according to a simplified model of vertical circulation, the winter monsoon would generate landwards movement in the surface layer and seawards movement in the bottom layer in this particular region (Gao *et al.* 2000), but the satellite imagery indicates seawards dispersion in the surface layer, while, on the outer shelf, the bottom layer is also directed towards the shelf edge (Sheu *et al.* 1999). The mechanisms by which the suspended sediment escapes from the coast and enters the cross-shelf transport system require further investigation. In addition to the wind-induced circulation, various forms of barotropic and baroclinic water-mass movements, together with frontal processes, should be given more attention both in nature and in modelling exercises.

Transport associated with sediment gravity flow

With regard to estuarine and continental shelf sediment gravity flows, the earliest research conducted in this area was for the Yellow River subaqueous delta. The Yellow River was well known for its high suspended sediment concentrations, which favour the formation of hyperpycnal flow: that is, the downslope movement of highly concentrated river waters (Wright *et al.* 1988, 1990). However, the suspended sediment load of the Yellow River has been significantly reduced in the recent times; as a result, the conditions for the development of hyperpycnal flows no longer exist. *In situ* observation shows that downslope movement of suspended sediment today dominantly takes place in bottom turbid layers (Wang *et al.* 2007): that is, bottom water masses with enhanced concentrations in response to resuspension induced by tidal currents and waves. This indicates a regime shift with respect to the nature of sediment gravity flows: hyperpycnal flows are today replaced by mobile bottom turbid layers (Fig. 5). Studies carried out elsewhere (Sternberg *et al.* 1996) show that the movement of turbid bottom layers is a frequent and fundamentally different phenomenon from the classical hyperpycnal flow concept proposed by Wright *et al.* (1990). For example, wave-induced resuspension at the bottom will enhance the density of the bottom layer (because of the addition of suspended mater) and, if the density is higher than the surrounding water mass, will cause downslope movement. Although the suspended sediment concentration of Changjiang River is not as high as that of the Yellow River, intense resuspension resulting from the combined action of waves and tidal currents, and extreme events of typhoons and winter storms, make it an ideal area for sediment gravity flow research (Wright & Friedrichs 2006).

Bottom turbid layers in canyon environments have been taken as indicators of sediment gravity flows (Liu & Lin 2004; Oiwane *et al.* 2011), but, on shelves with gentle slopes, the motion of bottom turbid layers may be linked to the shelf circulation rather than to gravity flows. Honda *et al.* (2000) observed that sediment of low ^{14}C content, which is an indicator of old age, is eroded from the bed and transported into the Okinawa Trough, emphasizing the importance of resuspension events and the formation of bottom turbid layers. Hoshika *et al.* (2003) obtained a time series comprising current velocity over spring–neap tidal cycles, suspended sediment concentration, temperature and salinity measured at three mooring stations at a height of 10 m above the bed in the East China Sea; during the observation period, numerous resuspension and settling events occurred, the authors

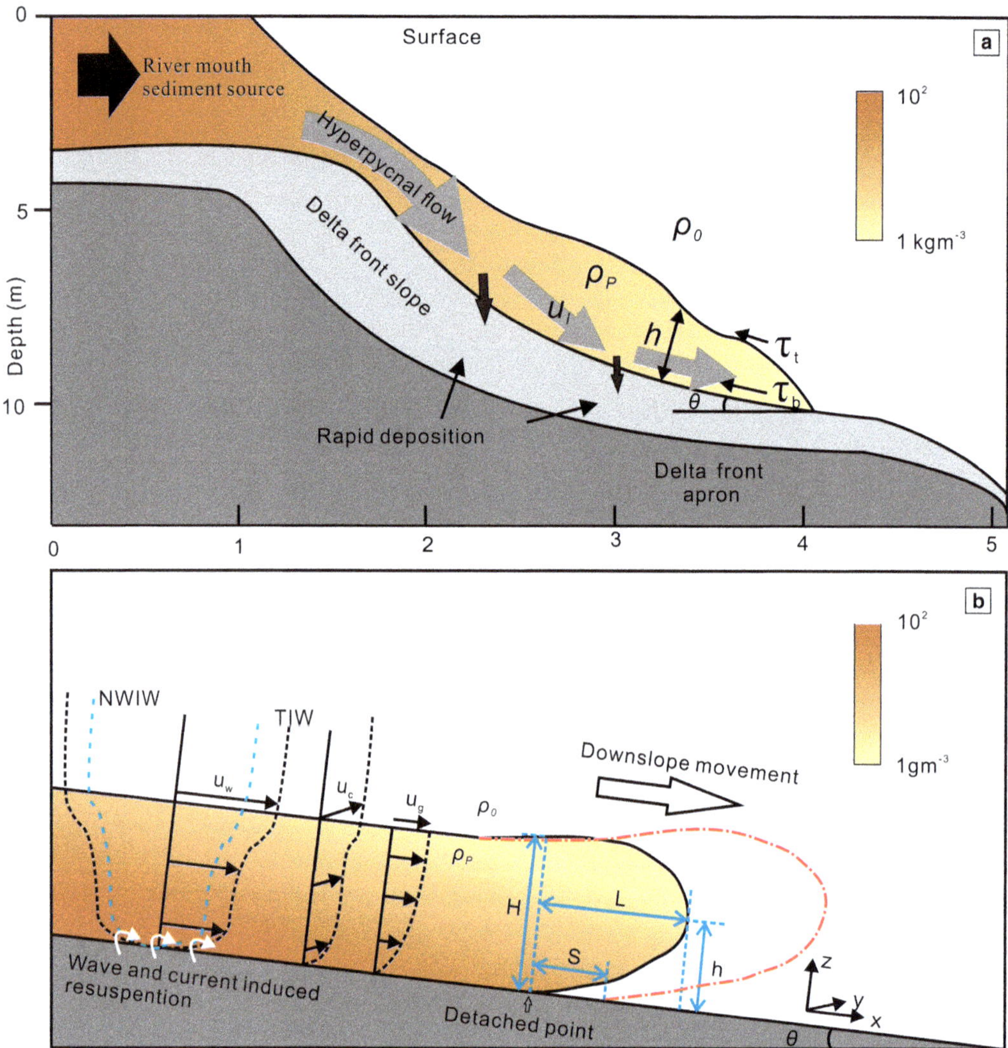

Fig. 5. Types of sediment gravity flows in the study area: (**a**) hyperpycnal flow near the Yellow River mouth (from Wright *et al.* 1990; Wang *et al.* 2007); and (**b**) bottom turbid layers formed in response to resuspension processes on the shelf (from Honda *et al.* 2000; Hoshika *et al.* 2003; Oguri *et al.* 2003; Bian *et al.* 2010; Y. H. Li *et al.* 2012). In (a) the colour concentration only shows the order of magnitude of the suspended sediment concentration from Wang *et al.* (2007). Hyperpycnal flow has the thickness H, the speed u and transport direction (grey arrows) in the bed slope θ. τ_t and τ_b represent upper interface and bed shear stress, respectively. In (b) the velocity of the bottom turbid layer is a combination of wave orbital velocity (U_w), along-shelf current magnitude (U_c) and the speed of gravity current (U_g) (Wright *et al.* 2001). U_w has the different ranges of the normal wind-induced wave velocity (NWIW) and the typhoon-induced wave (TIW). The red line represents the downslope variation trend of the BTL.)

associating downslope motion of the bottom turbid layer with circulation driven by monsoon winds. Oguri *et al.* (2003) have made similar observations; in this case, the bottom turbid layer on the East China Sea shelf had suspended sediment concentrations in excess of 20 mg l^{-1}, the downslope movement once again being associated with wind-driven shelf circulation. However, flow velocities and suspended sediment concentrations measured at mooring stations in the Okinawa Trough and on the continental slope imply that the down-canyon movement of the bottom turbid layer was not associated with tidal currents or other periodic currents, but that the episodic transport was, instead, characteristic of sediment gravity flow (Chung & Hung 2000). In the Okinawa Trough, the sediment flux is enhanced in winter, indicating an overall influence of the monsoon circulation; within the water column, however, the vertical sediment flux at a height of the 600 m above the bed is considerably weaker than in the turbid layer at the bottom (Iseki *et al.* 2003).

On the inner shelf, extreme events such as typhoons can trigger intense resuspension to generate bottom turbid layers. During a typhoon event that occurred in 2009, the suspended sediment concentration was enhanced significantly in the coastal mud area along the Zhejiang–Fujian coastlines (Y. H. Li *et al.* 2012). On another occasion, bottom turbid layer movement was also observed during a typhoon event in the same area (Bian *et al.* 2010). Observations and modelling of suspended-sediment transport for the Yellow River subaqueous delta reproduced similar process, with the model output showing a significant role being played by the bottom turbid layer to transport fine-grained material towards deep areas (A. J. Wang *et al.* 2010). Apparently, these observations and simulation results have demonstrated the importance of sediment gravity flow, but additional analysis is required to distinguish between the transport effects associated with the turbid layer movement and those in relation to shelf horizontal and vertical circulations.

Formation of major coastal-shelf Holocene deposits

The major sedimentary systems

A variety of Holocene sedimentary systems occur in the study area (Fig. 2). They include river deltas, tidal flats, tidal ridges, and coastal and shelf mud deposits. The geometric features and sedimentary characteristics of these systems are listed in Tables 3 and 4, respectively. Some details of these deposits are described below.

River deltas

The intense river discharges of the Changjiang and Yellow rivers have led to the formation of large deltas (Fig. 6). The Changjiang River delta is the largest Holocene deposit in the region. After sea level reached its present position at 6.5 ka BP, sediment infilling of the drowned estuary took 4–5 kyr (Hori *et al.* 2002; Xie *et al.* 2009). Once infilled some 2 kyr ago, the Changjiang sediment began to escape from the estuary on a large scale. One part of this sediment was deposited in the subaqueous delta, which today occupies an area of about 10^4 km^2; the other was transported southwards by shelf currents. This observation implies that a critical level of river discharge exists below which delta growth does not occur. At the present stage, the critical level is between 180 and 300 $\times$ 10^6 t a^{-1} (Gao 2007). It furthermore implies that, under natural conditions, initial delta growth is rapid, but that the growth rate subsequently decreases until the growth limit is reached. Owing to the combined effects of natural processes and human activities, the growth rate of the delta has reduced in recent years, or has even come to an end (Gao *et al.* 2011; Y. H. Wang *et al.* 2013).

The modern Yellow River delta has a short but complex history (Xue 1993; Yu 2002). During the late Quaternary, especially over the last 150 kyr, this river brought a large quantity of sediment towards the sea, forming a vast coastal plain (Ren 2006). However, in northern China, it discharged different courses over the Holocene period (Chen *et al.* 2012). The Yellow River, in northern China, discharged alternatively into the Bohai Sea and the southern Yellow Sea during the Holocene. A large supply of sediments (i.e. 10^9 t a^{-1}, before the 1980s) caused the rapid formation of a delta immediately in the front of the river mouth. In the period from AD 1128 to AD 1855, the river discharged into the southern Yellow Sea, resulting in the formation of a large delta; since 1855, a delta consisting of six lobes and covering an area of 5500 km^2 of land has been formed (Shi *et al.* 2003). Before 1855, it discharged into the southern Yellow Sea on the Jiangsu coast. The Yellow River delta thus merely represents deposition over the last one and half centuries. It occupies an area of some 5600 km^2, with a rate of shoreline advancement of around 10^2 m a^{-1}. Because of the rapid siltation in the channel and near the river mouth, the river frequently changed its route to reach the sea. In this way a number of lobes were formed, the shoreline of abandoned lobes being rapidly eroded (Chu *et al.* 2006).

The abandoned delta of the Yellow River located along the northern Jiangsu coast formed in the period from AD 1128 to AD 1855. In the course of this 727 years, the shoreline advanced by 50–60 km (Gao 2009*a* and the literature cited therein). At present, the abandoned Yellow River delta is eroding and provides the central Jiangsu coast with a certain quantity of fine-grained sediment.

Tidal flats

Tidal flats are distributed widely over the region, especially along the Jiangsu coast (SW Yellow Sea), the Shanghai and Zhejiang coasts, and the western Korean coast. This is consistent with the conditions of abundant fine-grained sediment supply and tidal dominance of the region (Gao

Table 3. *Spatial distribution patterns of the major Holocene deposits*

Sedimentary system	Spatial distribution pattern			Characteristics and architecture	Thickness of deposits	References
	Pattern	Area	Water depth			
Changjiang River delta	Typical funnel-shaped tide-dominated delta with three active elongated isolated river-mouth sandbars	*c.* 52 000 km^2 (subaerial: 23 000 km^2; subaqueous: 29 000 km^2)	*c.* 0–60 m	Thick deltaic deposits composed of tidal sand ridges, prodelta, delta-front, tidal-flats and surface soil deposits. Delta-front deposits: dark-grey silty to fine sand, and thickly interlaminated to thinly interbedded sand and mud. Sand mud couplets and bidirectional cross-lamination. Delta plain: upwards-fining succession and three subfacies: subtidal to lower tidal flat, upper intertidal flat and surface soil	*c.* 30 m (modern river-mouth area); *c.* 60–90 m (incised valley); >60 m with >80 m in incised valley	Li & Li (1983); Chen (1986); Li (1986); Orton & Reading (1993); C. X. Li *et al.* (2000, 2002); Hori *et al.* (2001*a*, 2002); Saito *et al.* (2001)
Yellow River delta	Elongated birdfoot-shaped wave-dominated delta, with a number of lobes	5500 km^2 (since AD 1855)	–	Thin Holocene deltaic deposits with typical overall topset-foreset-bottomset configuration; a steep longitudinal profile in its lower reaches	*c.* 20–30 m	Saito *et al.* (2001); Shi *et al.* (2003)
The abandoned old Yellow River delta	Double clinoform mud wedge with a subaerial –subaqueous delta couplet	Subaerial: 7160 km^2	–	Large-scale clinoform with seawards-dipping lower angled (<0.3°) internal reflectors; a wide, gently inclined topset and a relatively narrow, steeply sloping foreset, stretching seawards about 160 km away from the shoreline	*c.* 4–16 m with a maximum of 20 m	Zhang (1984); Ren (1992); J. P. Liu *et al.* (2007); Liu *et al.* (2013)
Hangzhou Bay estuarine deposit	Funnel-shaped incised-valley fill	8500 km^2	*c.* 8–10 m	270 km long, 100 km wide at the estuary mouth; from bottom to top, the incised valley successions can be grouped into four sedimentary facies: river channel, floodplain estuary, estuary shallow marine and estuary sand bar	*c.* 20–60 m, average 26.5 m, valley up to 120 m	Lin *et al.* (2005); Xie *et al.* (2009)
Tidal flats on the Jiangsu coast	Typically *c.* 5–10 km wide with a significant zonation of salt marsh, mud flat, mixed	>5500 km^2	<5 m	667.5 km long, average bed slope 0.2%, silty sand–mud couplets with horizontal parallel bedding in mixed sand–mud flat and ripples,	*c.* 30 m	Zhu & Xu (1982); Ren *et al.* (1984); Zhang (1984); Ren (1986); Zhu *et al.* (1986, 1998); Yang *et al.*

	sand-mud flat and sand flats			small cross-bedding in sand flats; fining-upwards sequence		(2002); Gao (2009*a*); Y. Wang *et al.* (2012)
Jiangsu Radial tidal ridges	>70 tidal sand ridges, fan-shaped with a central angle of 160°; subaerial and subaqueous parts	*c.* 22 470 km^2	*c.* 0–30 m	Individual sand ridge 10– >100 km long, *c.* 10– 15 km wide, furrows *c.* 10–30 m depth with a maximum of 48 m; flat-topped, comprises estuarine, tidal flat and shoreface deposits	*c.* 30 m	Z. X. Liu *et al.* (1989, 1998, 2007); Zhu & An (1993); Y. Wang *et al.* (1999, 2012); Li *et al.* (2001)
Palaeo-tidal ridges on the East China Sea coast	Moribund sand ridges	*c.* 45–115 m	*c.* 60– 110 m	Asymmetric transverse sections, with steep slopes facing SSW; orientation WNW–ESE, *c.* 10–60 km long with a maximum of 120 km, *c.* 8–14 km wide, *c.* 5–20 m high, average ridge spacing of *c.* 8–14 km; a complicated multiplayer architecture that reflects repeated accumulation–erosion–accumulation cycles	Up to 26 m (crest)	Yang & Sun (1988); Z. X. Liu *et al.* (1998, 2000, 2003, 2007); Saito *et al.* (1998)
Tidal ridges off the South Korean coast	Moribund linear sand ridges on the mid-shelf, modern ridges on the inner shelf	37 000 km^2	*c.* 10–50 m, *c.* 50–90 m	Symmetric transverse profile, *c.* 30–200 km long, *c.* 3–13 wide, *c.* 13–25 m in height, covered by large bedforms (dunes) with wavelengths of *c.* 200–500 m; a thin surface recently reworked sand sheets of <5 m, and a thick transgressive sand ridge deposits of up to 20 m acoustically oblique or prograding clinoforms; composed of sand, sand–mud mixed and mud facies	10 ~ 25 m	Park & Lee (1994); Liu *et al.* (1998); Jin & Chough (2002); Park *et al.* (2003, 2006)
Tidal ridges in the Bohai Sea	Six active, nearly paralleled finger-shaped sand ridges	*c.* 4000 km^2	*c.* 10–36 m	*c.* 8.7–43 km long, *c.* 6–20 m high; ridge spacing *c.* 8.3–16 m; multiple micro-geomorphic features comprise very large, large and medium subaqueous dunes, sand veneers and ribbons, scour furrows and comet-tail marks	*c.* 8–25 m with a maximum of 26 m	Liu *et al.* (1994, 1998); Marsset *et al.* (1996); Liu & Xia (2004)
Yangtze Shoal	An active offshore tidal sand sheet; palaeo-Yangtze submarine delta	*c.* 30 000 km^2	*c.* 25–55 m	270 km-wide, 200 km-long subaqueous dune fields, *c.* 20 000 km^2. Medium- and small-sized bedforms: crestline spacing of *c.* 2.3–13.6 m and	8 ~ 25 m	Liu (1997); Liu *et al.* (1998)

(*Continued*)

Table 3. *Continued*

Sedimentary system	Spatial distribution pattern			Characteristics and architecture	Thickness of deposits	References
	Pattern	Area	Water depth			
				height of *c.* 0.20–0.97 m. Large and very large bedforms: crest-line spacing of *c.* 70–1265 m, heights of *c.* 0.5–2.6 m		
Yellow River distal mud	Complex sigmoidal-oblique clinoform	–	25–40 m	Lowermost (*c.* 2–3 m): retrogradational–aggradational stacking pattern with subhorizontal internal reflectors; Middle parts (*c.* 35 m): a prograding reflection pattern with mostly seawards-stepping reflectors separated by several erosive surfaces; Uppermost (maximum 18 m): an aggradational reflection pattern with sub-parallel reflectors	15–40 m	Liu *et al.* (2004); J. Liu *et al.* (2007)
Changjiang distal mud	Subaqueous mud wedge thins offshore and southwards	–	20 ~ 70 m	Nearly 800 km long, along the inner shelf (*c.* 100 m isobaths) from the Yangtze mouth to Taiwan Strait, across shelf distance <100 km	*c.* 20–40 m	J. P. Liu *et al.* (2006, 2007); Xu *et al.* (2009)
Huskan Mud Belt in the SE Yellow Sea	Southwards-prograding clinoforms	*c.* 8000 km^2	*c.* 10–50 m	2050 km wide, 200 km long; elongated morphology; overlies large-scale dunes	*c.* 10–20 m with maximum 60 m	Park & Yoo (1988); Park *et al.* (1999); Jin & Chough (1998); Chough *et al.* (2002, 2004)
Isolated mud patch in the central Yellow Sea	A relict generally SE-prograding subaqueous mud wedge	–	40–80 m	Rhythmic-stratified, interbedded sand and mud; subunits are parallel to sub-parallel or clinoformal with a sheet or wedge geometry	*c.* 10–40 m	Jin *et al.* (2002); Chough *et al.* (2004)

Note: '–' denotes 'lack of information'.

Table 4. *Sedimentary characteristics of the major Holocene sedimentary systems over the region*

Systems	Sedimentary characteristics			References
	Facies	Sedimentary structure	Sediment type	
Changjiang delta	Prodelta; delta front; subtidal to lower intertidal flat; upper intertidal flat; surface soil	Gradational contact with small-scale cut-and-fill structures; ripple and parallel laminations; thickly interlaminated to thinly interbedded dark-grey sand and mud; bidirectional ripple laminations; abundant plant rootlets and snail shells	Dark-grey silty clay; dark-grey silty to fine sand and sand–mud couplets sand and mud; coarse silt to very fine sand; dull reddish brown to brown clayey silt	Hori *et al.* (2001*b*)
Yellow River delta	Terrestrial/fluvial facies; estuary facies; delta front to prodelta facies; fluvial facies (natural levee and/or floodplain)	Parallel and ripple laminations; parallel lamination with sharp contact or lithological and colour; weathered foraminifer, shell fragments and pollen are found occasionally	Dark reddish brown to greyish brown clayey silt or alternation of clay and silt; dark brown to dark grey clay to very fine sand; greyish to yellowish brown silt to very fine sand	Saito *et al.* (2000)
The abandoned delta	Flood plain; prodelta; delta front	Weak lamination, oblique bedding with muddy intercalation; lamination; ripple and parallel laminations	Mud silt and sand; silt mud and mud silt; silt mud and mud silt	Yuan & Chen (1984)
Hangzhou Bay estuarine deposit	Vertical sequence: palaeo-estuary; shallow marine; estuary	Silt lens, flowing mucky and parallel laminations; silt lens	Greyish brown mud silt; fine silty sand	Wang *et al.* (2006); Lin (1997)
Tidal flat on the Jiangsu coast	Silt flat; mud and sand flat; salt-mash flat; marshland flat	Cross-bedding and graded bedding; current bedding, flaser bedding biological perturbation structure; horizontally laminated bed, root hole, fauna hole; flat bedding, plant root point and worm hole	Silt or fine sand; silt clay or silt; silt; clay silt or silt clay	Zhu *et al.* (1986)
Jiangsu radial tidal ridges	Basement; coast facies; tidal sand facies; littoral tidal-flat facies	Pedogenesis features; shell–sand lenses with low-angle cross-bedding, peat layers; flaser bedding, bidirectional cross-bedding and graded bedding; small cross-bedding, flaser and wavy bedding, abundant foraminiferal tests	Stiff mud; sandy mud; sand, silt–sand and mud of a grey colour; silt and clayey silt	Li *et al.* (2001)
Palaeo-tidal ridges on the East China Sea coast	Floodplain; estuary; shallow marine	Parallel and ripple laminations, discordant contact with underlying stratum; discordant contact with erosion surface; cross-bedding	Silt and mud silt; fine sand	Yang (2002)
Palaeo- and modern tidal ridges off the South Korean coast	Tidal flat and estuarine; shoreface to inner shelf; nearshore	Remnant laminae of silt–clay couplets with intense bioturbation; shell fragments and some gravels; low-angle cross-bedding	Sandy mud; sand; fine sand	Lee & Yoon (1997); Park *et al.* (2006)

(Continued)

Table 4. *Continued*

Systems	Sedimentary characteristics			References
	Facies	Sedimentary structure	Sediment type	
Tidal ridges in the Bohai Sea	Nearshore; shallow marine; tidal sand sheet	Discordant contact with underlying stratum, shell fragments, plant residue; laminated clay layers	Clay silt and silt clay with sand gravel; silt clay; fine sand	S. F. Liu *et al.* (2008)
Yangtze Shoal	Shallow marine	Tidal rhythms, wavy and herringbone cross-beddings	Sand	Ye *et al.* (2004); Liu (1997)
Yellow River distal mud	Salt marsh and coastal plain; subtidal nearshore; deepening shelf; deep shelf	Sandy lenses and clayey flasers, moderate bioturbation with plant fragments; lenticular bedding with shell fragments; clayey silt intercalated with sand layers or lenticular beds, molluscan shell fragments occur sporadically; sandy laminations found occasionally, slight bioturbation with shell fragments	Dark grey to greyish silty clay to silt; silt to sandy silt; dark grey clayey silt; dark greenish grey to dark grey clayey silt	J. P. Liu *et al.* (2007)
Changjiang delta mud	Nearshore; shallow sea shelf	Stacked fining-upwards parasequences of transgression	Light grey sandy silt; light grey and taupe clayey silt and clay	Xiao *et al.* (2006)
Huskan Mud Belt in the SE Yellow Sea	Tidal sand ridges/tidal flat	Parallel-laminated structures, massive bedding and worm hole; homogeneous mud with some shells and shell fragments	Silt and clay; silt sand and sand	Park *et al.* (2000)
Isolated mud patch in the central Yellow Sea	Tidal shallow marine; transition facies; shallow marine	Discontinuous current bedding with shell fragments, rare plant root point and rare worm hole; cross-bedding with rare shell fragments and worm hole: stratification, moderate bioturbation, block structure and shell fragments	Clay silt; clay silt; clay silt	L. B. Wang *et al.* (2009)

Note: '–' denotes 'lack of information'.

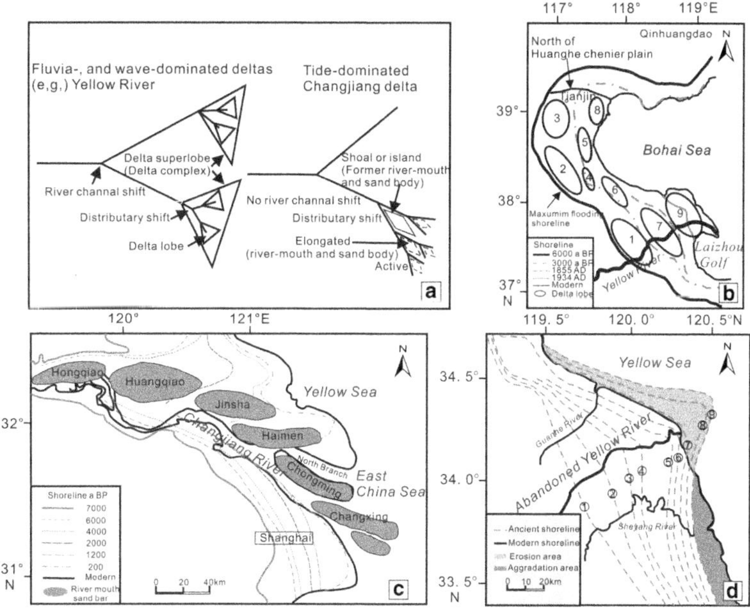

Fig. 6. Characteristics of the river deltas in the study area: (**a**) a summary of the general characteristics of the Yellow River and Changjiang River deltas (after Hori *et al.* 2002); (**b**) the Yellow River delta (age of the various delta lobes: (1) 6–5 ka BP; (2) 5–4.5 ka BP; (3) 4.5–3.4 ka BP; (4) 3.4–3 ka BP; (5) 3 ka BP–602 BC; (6) 602 BC–AD 11; (7) AD 11–AD1048; (8) AD 1048–AD 1128; (9) AD 1855–present (after Xue 1993; Saito *et al.* 2000, 2001); (**c**) the estuarine sand-body ages in the Changjiang delta at: Hongqiao, 7.5–6 ka BP; Huangqiao, 6.5–4 ka BP; Jinsha, 4.5–2 ka BP; Haimen, 2.5–1.2 ka BP; Chongming, 1700–200 years BP; Changxing, 700 years BP (after Li *et al.* 1979; Li & Li 1983; Xu *et al.* 1987); and (**d**) the abandoned (old) Yellow River delta (after Ye 1986; L. Zhang *et al.* 2014; the numbers ①–⑨ indicate shorelines in the years of AD 1194, 1578, 1591, 1700, 1747, 1776, 1803, 1810 and 1855, respectively).

2009*a*). In areas of small river inputs, in contrast, tidal flats are either located in well-sheltered embayments or in estuarine bay head areas along the coastline of Europe (e.g. Evans 1965).

The tidal flats on the Jiangsu coast are on the largest scale in the region, where the intertidal zone occupies an area of more than 5500 km^2 (Ren 1986). Sediment distribution is controlled here by tidal action. Sediment supplied by the Changjiang and Yellow rivers consists of fine-grained sand, silt and clay. The upper parts of the flat system experiences slow changes in water level with weak currents; thus, only muddy material can reach these parts. Over the middle–upper intertidal zone, the tidal current is weak at neap tides, but strong at spring tides. Mud is deposited at neap tide and sandy material at spring tide, thereby forming typical heterolithic tidal bedding. The lower intertidal zone is associated with bedload transport. Therefore, from upper to lower tidal flats, the sediment zonation is characterized by thinly laminated muds, alternating mud and sand layers (i.e. tidal laminae), and cross-bedded sands, respectively. For the region, these zones are referred to as mudflat, mixed mud–sand flat and sand flat, respectively (Zhu & Xu 1982; Zhang 1992). During storms, the combined storm surge and tides enable seawater to reach the supratidal zone; in this case, accretion takes place on the uppermost part, but erosion occurs on the lower intertidal zone (Ren *et al.* 1985).

Because of the sustained sediment supply, the shoreline has continued to advance seawards. Even after the Yellow River switched to its northerly route, erosion of the abandoned delta provides the sediment source that resulted in accretion along the central Jiangsu coast at a rate of 50–150 m a^{-1}. At the same time, the morphology of the tidal flat was maintained by natural processes (Liu *et al.* 2011). In addition to this general trend, the cross-shore tidal-flat profile may be influenced by the presence of tidal creeks (Zhang 1992; Y. P. Wang *et al.* 1999), which modify the time–velocity asymmetry, while also influencing salt marsh growth (Y. P. Wang *et al.* 2012; Gao *et al.* 2014), thereby causing localized changes in the erosion–accretion pattern. The expansion of *Spartina* (an artificially introduced species) had significant effects (Y. P. Wang *et al.* 2012; Gao *et al.* 2014). The original deposition patterns have been changed (Fig. 7): in response to the expansion of *Spartina*, intensified accretion took place on the upper tidal flat, thereby causing rapid shoreline advancement, and, at the same time, the lower tidal flat zone was narrowed, sometimes associated with the formation of salt-marsh cliffs at the marsh–mudflat boundary (Zhao *et al.* 2014).

With the reduction in sediment supply, coastal erosion has been intensified on the northern Jiangsu coast, to the immediate south of the abandoned Yellow River delta. This represents a natural response to the cut-off of sediment supply. When coastal recession occurs, the profile shape of tidal flats cannot be maintained: the lower flat is initially eroded, enhancing the average slope of the entire intertidal zone, followed by erosion of the entire tidal-flat area. This process eventually results in the formation of eroding cliffs along the salt-marsh edge (Zhao *et al.* 2014). In the middle Holocene, a series of cheniers were formed on the Jiangsu coastal plain, now 50–60 km away from the sea dyke (Wang & Ke 1989). These represent the final product of tidal-flat erosion at that time.

The sedimentary record of the tidal flat has a high temporal resolution, although the preservation potential varies between the mudflat, mixed flat and sand flat (Gao 2009*b*). For example, on the central part of the Jiangsu coast, the sediment supply from the Yellow River from AD 1128 to AD 1855 resulted in rapid evolution of the tidal-flat system (see above). The shoreline advanced at a rate of 10^1 m a^{-1}. Even after the northwards shift of the Yellow River mouth to the Bohai Sea, erosion of the old delta continued to supply fine-grained sediment (Yu *et al.* 2014). Thus, a vast coastal plain formed with an up to 20 m-thick Holocene tidal-flat deposit. The high resolution and continuity of the tidal-flat deposit make it an ideal sedimentary

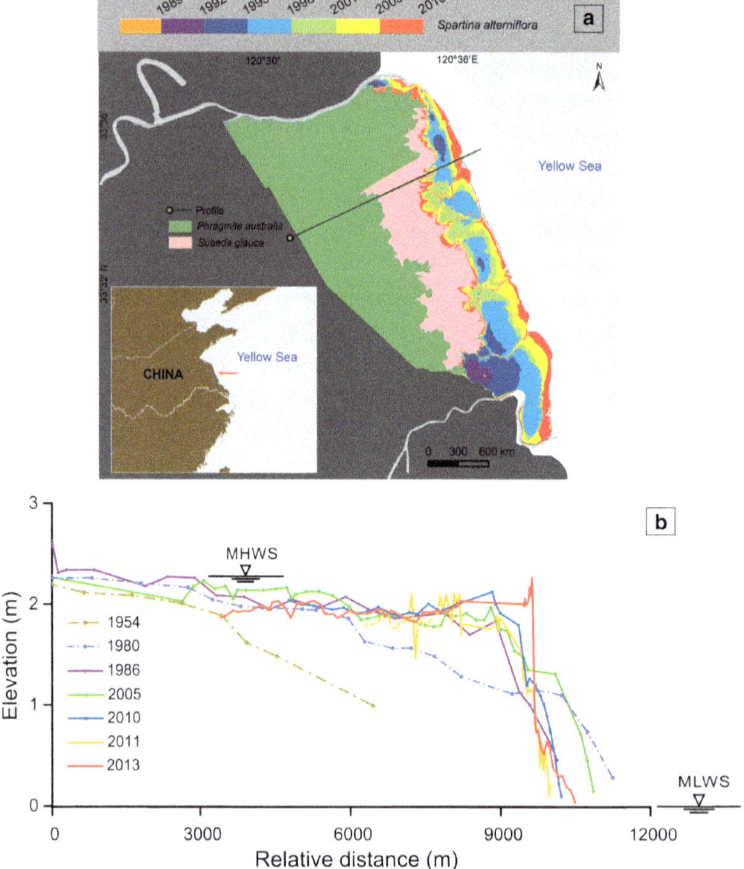

Fig. 7. Tidal-flat morphology on the central Jiangsu coast, before and after colonization by *Spartina alterniflora*: (**a**) location of the coastal profile; and (**b**) coastal profile changes between 1954 and 2013.

record for sea-level, climate, environmental and ecosystem change studies. In particular, the fine-grained materials, including both inorganic materials and organic matter, contain extensive information on past environmental changes.

On the central Jiangsu coast, the shoreline dynamics during the last 6.5 kyr is still poorly understood. From the middle Holocene (*c.* 6.5 ka BP) to the Song Dynasty in AD 1128, the shoreline advanced slowly towards the sea (i.e. 10 km over 5 kyr). However, during the subsequent 727 years, when the Yellow River discharged into the SW Yellow Sea, the shoreline advanced by 50–60 km (Gao 2009*a* and the citations therein). Thus, the rate of shoreline change for these two periods differs by an order of magnitude. At present, the central Jiangsu coast continues to accrete in response to material supply from the eroding former Yellow River delta (Yu *et al.* 2014) and offshore radial sand ridges (Fu & Zhu 1986), but the relative importance of these two sources is unknown. If the offshore areas represent a major sediment source, then it needs to be explained why, during the earlier periods, the shoreline did not prograde at the same rate as today. To answer this question, the shoreline evolution before AD 1127 can be modelled to gain some insight into the changing sediment dynamic conditions associated with the radial tidal ridges, in terms of seabed reworking and landwards transport of fine-grained sediment. Furthermore, the abandoned Yellow River delta and the central Jiangsu coast can be treated as a small-scale source–sink system, for which *in situ* measurements and numerical simulation can be carried out to reveal the evolution pattern of the sedimentary system since the Song Dynasty. Figure 8 shows the shoreline changes since the middle Holocene; in order to explain such a complex changing pattern, the sediment budget in association with the Changjiang and (old) Yellow rivers and the reworking of offshore pre-Holocene strata should be evaluated.

The tidal flats along the coast to the south of the Changjiang Estuary and on the eastern Yellow Sea coast (i.e. the South and North Korean coasts), which are mainly confined to coastal embayment, have similarities to the Jiangsu counterparts. They are also characterized by zonation and fining-upwards sequences. However, because in these places the supply of fine-grained sediment is mainly provided by rivers not directly discharging into the deposition area, the systems are characterized by a lack of sand. As a result, the sand flat in the lower intertidal zone is narrow, or even absent. For instance, the Zhejiang–Fujian coast was originally a rocky coast that was gradually converted into an intertidal flat by the long-distance import of the Changjiang River (Li & Wang 1991). As a result, a narrow coastal plain developed. Because the sediment supply consists only of mud, the zonation of the tidal flats differs from that of the Jiangsu coast. Nevertheless, a general fining-upwards trend is still present. In addition, the width of the intertidal zone is smaller than at Jiangsu.

Tidal ridges and other sandy deposits

Being a tide-dominated region, tidal ridges are well developed. These include the Jiangsu radial tidal ridges, the palaeo-tidal ridges on the East China Sea shelf, the Yellow Sea tidal ridges off the North Korean coast and the tidal ridges in the eastern Bohai Sea (Fig. 9). The coastlines of the ridges are parallel or sub-parallel, aligned to the main tidal flow direction and attain lengths of 100 km, their spacing ranging from 1 to 10 km and inter-ridge channel depths reaching 10–30 m. Typical examples from other regions are the parallel ridge systems of the North Sea in Europe (Berné *et al.* 1994; Trentesaux *et al.* 1999; Deleu *et al.* 2004).

On the Jiangsu coast, China, a large-scale, radially arranged tidal-ridge system is present (Y. Wang *et al.* 2012; Xing *et al.* 2012) (Fig. 9a). Here the tidal current is very strong, reaching up to 3 m s^{-1}. The sediment of the ridges consists mainly of fine to very fine sands derived from the underlying pre-Holocene strata (Li *et al.* 2001). The 70+ ridges cover an area of around $25 \times 10^3 \text{ km}^2$. Compared with the North Sea counterpart (Huthnance 1982*a*, *b*), the local ridges are more mobile (Gao & Collins 2014). For example, an inter-ridge channel was recorded to have formed within 3 years, the water depth increasing from 3 to 11 m (Zhang & Chen 1992). System response is evidently much shorter than in the North Sea. In addition, the crests of the North Sea ridges are not exposed at low tide, whereas the landwards transport of fine-grained sediment along the Jiangsu coast causes the shorewards parts of the ridges to merge with the intertidal flats. These differences result from the fact that the grain size is relatively small and the tidal currents are relatively strong on the Jiangsu coast compared with the North Sea.

The palaeo-tidal ridges on the East China Sea shelf (Fig. 9b) are located in water depths of 40–60 m. Because these were thought to be immobile under the present-day tidal regime, they were referred to as ‘moribund’ ridges. Numerical modelling has indicated that, during the early periods of the Holocene when sea level was lower than present day, the tidal currents were, indeed, much stronger (Uehara *et al.* 2002). The pre-Holocene deposits in this region are composed of sediments rich in mud. It was thus postulated that the ridges were formed in the course of seabed erosion, the sand being transported into the ridges, while the mud was carried away. When comparing the model

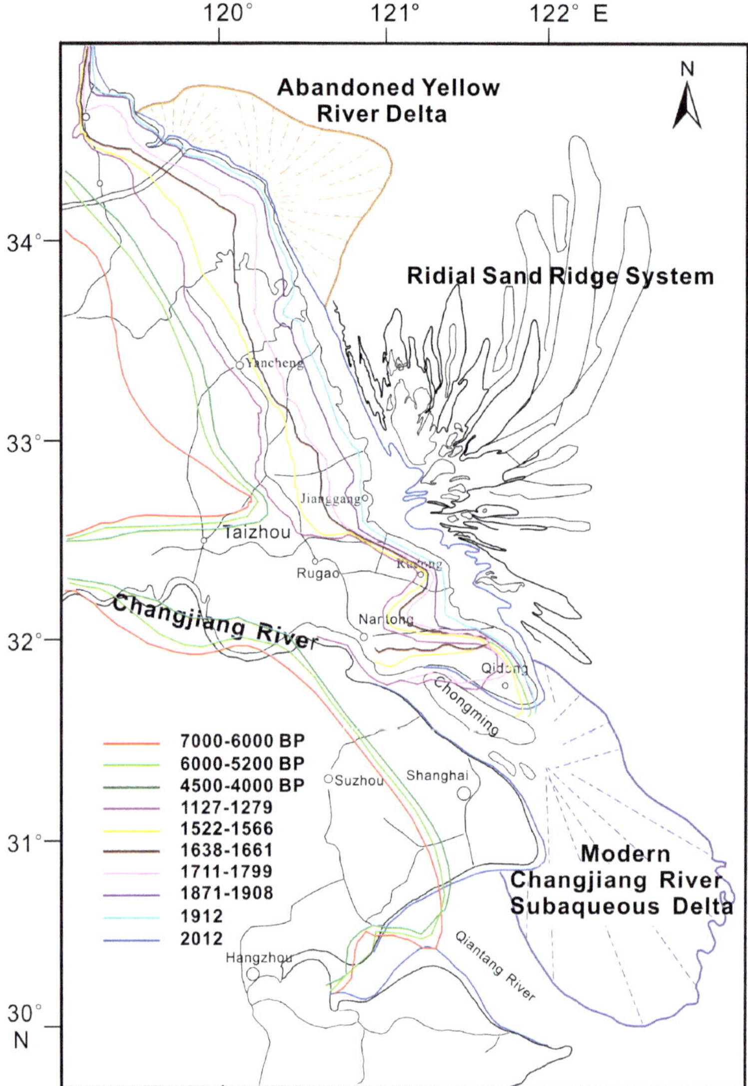

Fig. 8. Shoreline changes on the Jiangsu coast and sediment source–sink for the Holocene deposits (based on Zhu *et al.* 1996; X.X. Zhang *et al.* 2014).

output with the conditions for the tidal-ridge formation, as suggested by Liu *et al.* (1998), the areas suitable for tidal-ridge development at the various stages of sea-level rise during the early Holocene are highly consistent with the occurrence of the old ridges (Fig. 10). However, tidal cycle measurements show that, under the present conditions, the tidal current still exceeds the threshold of bedload motion: that is, the sediment here is mobile (Z. X. Liu *et al.* 2007). Further, seismic surveys and core analyses have indicated that the ridges are migrating. Thus, active ridges may exist in deeper shelf waters, if the tidal current is sufficiently strong.

The tidal ridges in the eastern Bohai Sea (Fig. 9c) are also formed by the reworking of the underlying strata (Z. X. Liu *et al.* 2007). Here, the sedimentary materials were derived from seabed erosion of the NW Bohai Strait, where the tidal current is strong. At first, the seabed was eroded and the material transported into the Bohai Sea, where it later formed the base of the tidal ridges. At later stages, the channel in the Strait was further deepened by tidal

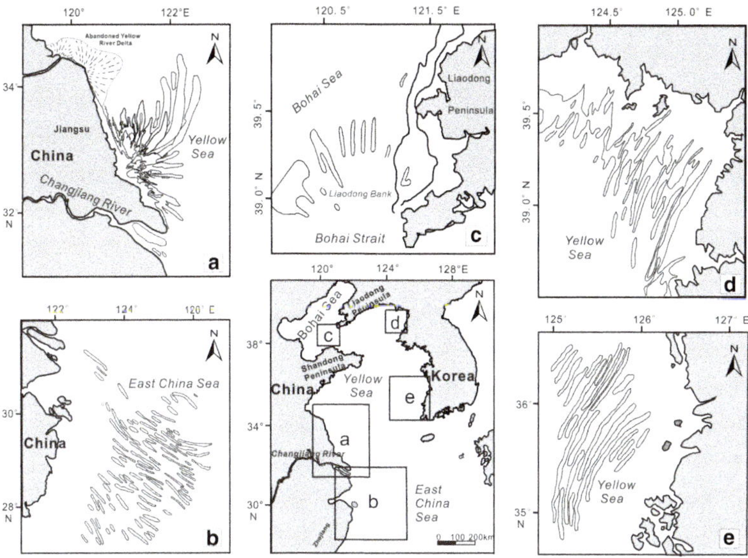

Fig. 9. Tidal-ridge systems of the region: (**a**) Jiangsu radial tidal ridges (from Li *et al.* 2001; Gao 2014); (**b**) palaeo-tidal ridges on the East China Sea shelf (from Z. X. Liu *et al.* 2007); (**c**) tidal ridges in the eastern Bohai Sea (from Liu 1997); (**d**) Yellow Sea tidal ridges off the North Korean coast (from Off 1963); and (**e**) Yellow Sea tidal ridges off the South Korean coast (from Park *et al.* 2006).

scouring, with the sediment being continuously transported into the tidal-ridge area. As verified by ^{14}C dating, the material at the surface is older than that at the base; this is because the dates represent the timing of the biologically generated sediment, rather than the time of deposition (Scholl 1964).

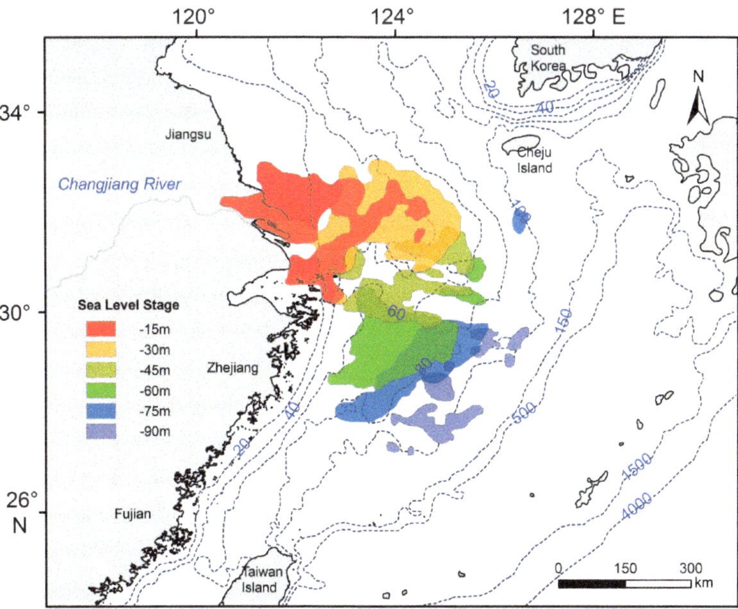

Fig. 10. Model output for the formation of the palaeo-tidal-ridge system on the East China Sea shelf (modified from Uehara *et al.* 2002; Uehara & Saito 2003).

The Yellow Sea tidal ridges off the Korean coast (Fig. 9d, e) follow the classical theory of linear sandbanks, based upon the North Sea studies (Huthnance 1982*a*, *b*). Over the tidal-ridge field off the North Korean coast, the tidal-current speed reaches 1 m s^{-1} and the sediment is derived from the reworking of pre-Holocene strata (Off 1963). The ridges off the South Korean coast consist of two groups: the mid-shelf ridges (water depth of 50–90 m) and the inner-shelf ridges (water depth of <30 m). The former group resulted from the reworking of the pre-Holocene strata during the post-glacial transgression, whilst the shallow-water group is a product of modern tidal processes (Park *et al.* 2006).

If the condition of rectilinear tidal currents is not satisfied, then tidal ridges cannot develop. The critical value of a tidal ellipse ratio (of the lateral component to the longitudinal component of the tidal current) was shown to be 0.4; only at ratios below this value can tidal ridges be formed (Liu 1997). For rotatory currents, in contrast, tidal sand sheets will be formed. For example, the Yangtze Shoal off the Changjiang River mouth is a tidal sand sheet (Liu 1997). Similar to the tidal ridges off the central Jiangsu coast, the sediment composing the sand sheet has also been derived from the underlying pre-Holocene relict deposit. However, it differs from the tidal ridges in that the reworking here is weaker: the sand sheet serves as an armour layer, which now protects the underlying strata from further erosion.

Coastal and shelf mud deposits

The study area is characterized by numerous, widely distributed mud deposits (Fig. 2). In terms of sediment provenance and the transport processes involved, these can be divided into distal mud deposits associated with large rivers, coastal mud belts and isolated mud patches on the shelf.

The distal mud deposits (or remotely distributed clinoform deltas) associated with the Yellow and Changjiang rivers are today well known. The distal mud deposit of the Yellow River is located to the NE of the Shandong Peninsula, up to 600 km away from the river mouth in the western Bohai Sea (Yoo *et al.* 2002; Liu *et al.* 2004, 2006; J. Liu *et al.* 2007); the distal mud deposit of the Changjiang River is located on the inner shelf of the Zhejiang–Fujian coast, up to 800 km away from the river mouth.

Both the distal muds associated with the Changjiang and Yellow rivers are characterized by clinoforms. There are basically two types of clinoforms: convergent and parallel clinoforms (Gao & Collins 2014). The Yellow River distal mud has well-developed convergence and parallel clinoforms, which represent early and later evolutionary stages of distal mud deposition, respectively. At first, the depocentre (i.e. the location with the highest deposition rate) was coincident with the maximum suspended sediment concentration, since the settling flux is proportional to the concentration. As a result, the central part of the mud deposit had the highest elevation, and the sedimentary layers thin from there to the distal edge. Subsequently, as the slope towards the edge increased, the bottom layer moved outwards, under the influence of gravity, to form parallel clinoforms. At the same time, the highest part of the deposit experienced increased current speeds (in response to a reduction in water depth). This interpretation implies that the parallel clinoforms serve as evidence of the effect of sediment gravity flow (Gao & Collins 2014). At the distal mud of the Yellow River, the top of the deposit has a water depth of around 30 m; the two types of clinoforms have been detected on shallow seismic records (Liu *et al.* 2004).

The distal mud of the Changjiang also shows the presence of clinoforms (J. P. Liu *et al.* 2007). However, compared with its Yellow River counterpart, the parallel clinoforms are not yet well developed. As revealed on shallow seismic records, the clinoforms associated with this mud deposit are dominated by the convergence type, only in some places have parallel ones begun to grow. As in the case of the Yellow River counterpart, the elevation of the Changjiang mud deposit is around 30 m. Thus, compared with the Yellow River system, this deposit is less mature in terms of the evolutionary stages. In fact, in southern China, the distal mud of the Pearl River is at an even younger stage: although the distal mud occupies a large area (i.e. >8000 km^2 on the inner shelf of the northern South China Sea, some 600 km away from the river mouth), the mud layer is still rather thin and, hence, lacks clinoform development (Liu *et al.* 2014). The seabed elevation here is located more than 40 m below the sea surface.

The differences in the timing and duration of the three distal mud deposits have been interpreted to be the result of estuarine sediment-infilling processes (Liu *et al.* 2014). The Yellow River did not have a large estuary when sea level reached the high level at 7 ka BP and the sediment discharge was large. Hence, the material discharged by the river was supplied to the distal mud deposit from the start, resulting in continuous accumulation. In the case of the Changjiang, the estuary at 7 ka BP was large and, before infilling was complete, little sediment escaped to the open shelf. It took more than 4 kyr for the estuary to become completely infilled. Therefore, the supply of muddy sediment to the distal mud repository only began about

2 kyr ago. Prior to this, the distal mud was composed of sediment reworked from pre-Holocene shelf deposits. The Pearl River has an even larger estuary and the river discharge is relatively low. Hence, the infilling of the estuary is still ongoing today and only very recently (i.e. some 100 years ago) did the Pearl River sediment begin to escape from the estuary to reach the distal mud area (Liu *et al.* 2014). Hence, the three distal mud systems may be viewed as reflecting different stages of evolution: the Yellow River mud represents a mature stage, the Changjiang River mud a developing stage and the Pearl River mud an early stage (Fig. 11).

It should be noted that the Changjiang distal mud (i.e. the Zhejiang–Fujian shelf mud) may not entirely be composed of modern Changjiang sediments. At present, the Changjiang is, indeed, the source, from where it is transported southwards by shelf currents (J. P. Liu *et al.* 2006, 2007; K. H. Xu *et al.* 2009, 2012). However, borehole analyses (i.e. grain size and ^{14}C age determination) of the Zhejiang–Fujian mud area have revealed a significant 4 kyr hiatus in the sediment sequence (Liu *et al.* 2006; Xu *et al.* 2011): below the upper, younger than 2 kyr layer, the strata are generally older than 6 kyr (Fig. 12). Similar patterns have been identified for Hnagzhou Bay (X. Zhang *et al.* 2014) (see Fig. 12). In terms of the model, this would mean that the uppermost layer was deposited after the Changjiang Estuary had been infilled (see above). The lower part differs from the upper one in that it contains a relatively high sand content (Xiao *et al.* 2006), and the average deposition rate on the basis of ^{14}C dating is lower than the upper layer, indicating discontinuous sediment accumulation (Xu *et al.* 2012). These characteristics indicate that the sediment source changed since early Holocene times. The tidal-ridge systems on the East China Sea shelf may have served as a major sediment source for the formation of the lower layers. As described above, to form these ridges a large amount of fine-grained sediment would have to have been reworked (Z. X. Liu *et al.* 2007), stronger than present tidal currents providing the hydrodynamic forcing (Uehara *et al.* 2002), which resulted in landwards transport of the fine-grained

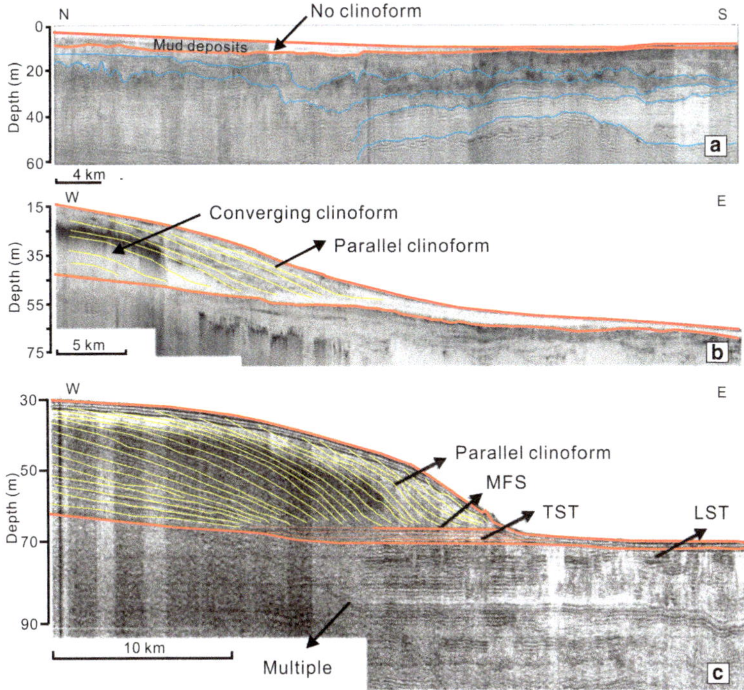

Fig. 11. Evolution stages of the distal mud deposits associated with large rivers: (**a**) early stage, represented by the situation on the northern shelf of the South China Sea, associated with the Pearl River (Liu *et al.* 2014); (**b**) developing stage, represented by the frontal part of the Yellow River distal mud and the Zhejiang–Fujian coastal mud associated with the Changjiang (Liu *et al.* 2006); and (**c**) mature stage, represented by the Yellow River distal mud in the eastern Bohai Strait (Liu *et al.* 2004). LST, lowstand systems tract; MFS, maximum flooding surface; TST, transgressive systems tract.

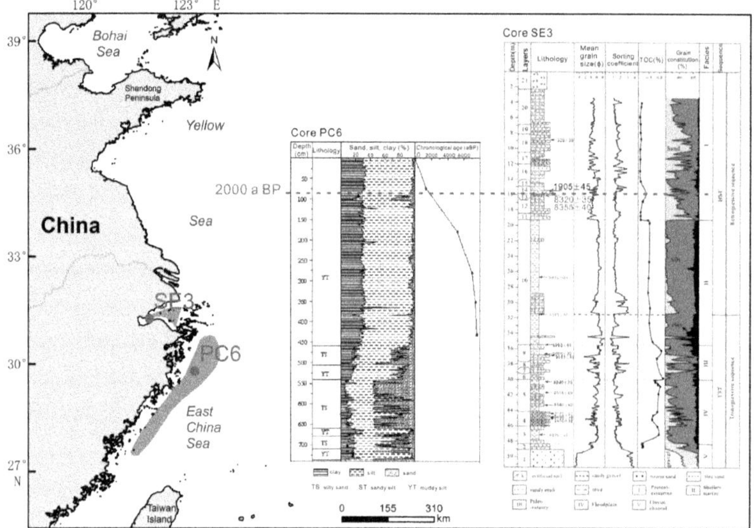

Fig. 12. Hypothesized source–sink characteristics of the Changjiang distal mud and the Holocene sequences of Hangzhou Bay (after Liu *et al.* 2006; Xiao *et al.* 2006; Xu *et al.* 2011; X. Zhang *et al.* 2014). 2000 a BP = 2 ka BP.

material. This material was mixed with the inner-shelf sediment (and partly the river-derived sediment); as a result, the ^{14}C-derived sediment age would be older than the materials in the source area. This hypothesis can explain the sediment sequences shown in Figure 12, but further studies on sediment provenance and sediment-transport modelling are required to test this hypothesis.

The Holocene Hangzhou Bay estuarine incised valley fill has a thickness of up to 100 m (Lin *et al.* 2005). Of this sequence, the modern Changjiang sediments (supplied over the last 2 kyr) only occupy the uppermost 20 m (Yu *et al.* 2012; X. Zhang *et al.* 2014). Because before 2 ka BP not much Changjiang sediment reached Hangzhou Bay, the older sediments may also have had an offshore source, similar to the Zhejiang–Fujian shelf mud (see above). Hence, we propose the same hypothesis for the Hangzhou Bay mud deposits: the reworking of pre-Holocene strata and subsequent tidally induced transport were responsible for the formation of the thick mud deposits. Once again, this hypothesis should be tested by further investigations.

While large amounts of sediment are trapped on the inner shelf of the East China and Yellow seas, some material has been dispersed by shelf currents. Generally, the mud deposits formed in this way are associated with a low deposition rate and low temporal resolution. The mud deposit over the central Yellow Sea and to the south of Cheju Island in the southern Yellow Sea belongs to this category (Lim *et al.* 2007). Here, the tidal currents are weak (i.e. below the threshold for the initiation of sediment movement). Fine-grained sediment from various sources (including reworked and modern-day river input) has accumulated in this region, together with biogenic particles (Gao & Jia 2003). The sediment was transported by shelf currents (Park *et al.* 2001) before finally settling to the bed. Vertical circulation may, in addition, contribute organic material and thereby enhance deposition. In this way, a 1–10 m-thick sediment layer formed in the course of the late Holocene. Because the water depth lies within the range of wave action (Graber *et al.* 1989), resuspension may occur under extreme conditions. A mechanism related to an upwelling or coupled upwelling–downwelling system has been proposed to explain the deposition patterns and the enhancement of the organic matter content (Fig. 13): particles generated by biological activity will be concentrated in the core area of the system; as a result, the accumulation rate may be enhanced, and the dissolved nutrient content may become higher than that of the surrounding waters owing to degeneration of the organic particles (Gao & Jia 2003).

Discussion

Process–product relationships for the Holocene deposits

As outlined above, the broad shelf and an abundant terrestrial sediment supply determined the spatial

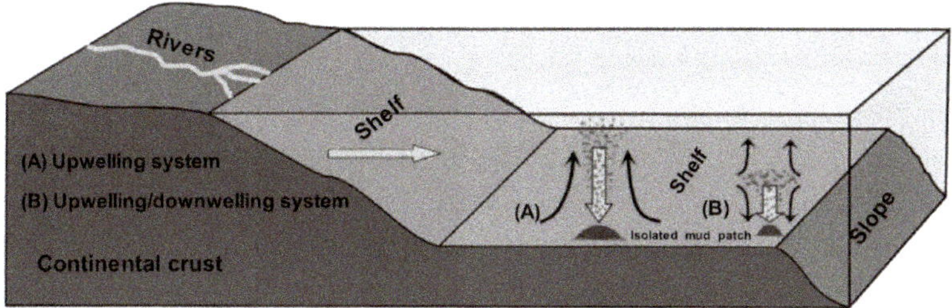

Fig. 13. Water-column processes for the formation of the isolated mud patches in the central Yellow Sea and to the south of Cheju Island (after Gao & Jia 2003; base diagram from Frank 2011).

distribution and temporal evolution of the various Holocene sedimentary systems in the region considered here. From the point of sediment dynamics, an understanding of the transport and accumulation processes will lead to a sound prediction of the product (i.e. the Holocene deposits), while including the tidally dominated or combined tide–wave-dominated deposits, shelf-circulation-generated deposits and the deposits influenced by sediment gravity flows.

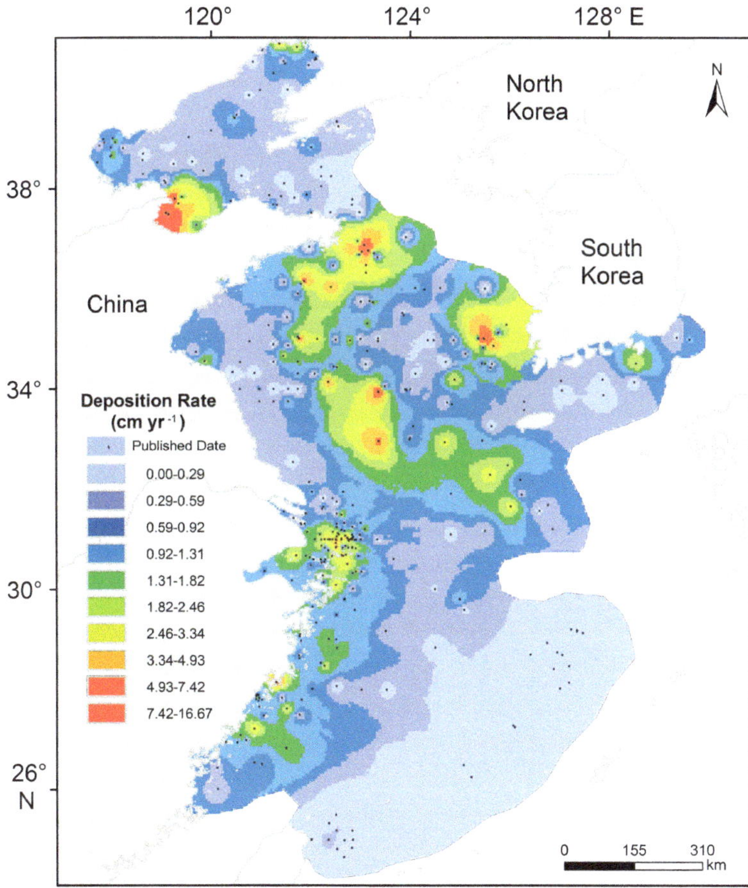

Fig. 14. Contemporary deposition rates of the region, as derived from ^{210}Pb and ^{137}Cs analysis (compiled on the basis of DeMaster *et al.* 1985; Alexander *et al.* 1991; Huh & Su 1999; F.Y. Li *et al.* 2002, 2006; Oguri *et al.* 2003; Lim *et al.* 2007).

Table 5. *Temporal resolution, continuity and time coverage of the sedimentary records associated with the major Holocene deposits over the region*

Sedimentary system	Starting time	Temporal resolution	Duration	Mechanisms/factors influencing continuity	References
Changjiang River delta	*c.* 8.5–6 ka BP	100 years	*c.* 6 ka–present	Progressive and lateral shift towards the SE	Li (1986); Chen & Stanley (1995); C. X. Li *et al.* (2000, 2002); Saito *et al.* (2001); Hori & Saito 2007; Hori *et al.* (2001*b*, 2002); Gao (2007)
Yellow River delta	*c.* 7–6 ka BP	100 years	LGM– *c.* AD 1128 *c.* AD 1855–present	Lateral channel avulsion and migration	Zhang (1984); Saito *et al.* (2001); Liu *et al.* (2004); Xue *et al.* (2004)
Abandoned Old Yellow River delta	AD 1128	10–100 years	*c.* AD 1128–AD 1855	Coastal erosion	Zhang (1984); Liu *et al.* (2013)
Hangzhou Bay estuarine deposit	LGM 2 ka	10 years	LGM– *c.* 6 ka *c.* 2 ka–now	Hydrodynamic disturbance and weak bioturbation	Feng *et al.* (1990); Lin *et al.* (2005)
Tidal flats on the Jiangsu coast	*c.* AD 1500	<1 year–1 day	*c.* AD 1500– present	Vertical mixing by bioturbation, storms and frequent migration of tidal creeks	Ren *et al.* (1984); Zhang (1984); X. Y. Liu *et al.* (2008); J. Wang *et al.* (2010); Li *et al.* (2011); Y. Wang *et al.* (2012)
Jiangsu radial tidal ridges	*c.* 7 ka BP	–	*c.* 2 ka to present	Shift of tidal channels; bed erosion during storm events	Zhang *et al.* (1998); Li *et al.* (2001); Z. X. Liu *et al.* (2007); Y. Wang *et al.* (2012)
Palaeo-tidal ridges on the East China Sea coast	*c.* 11–9 ka BP	–	*c.* 11–6 ka BP	Lateral sand ridge migration towards the SW, and new ridges continually replacing old ones	Uehara *et al.* (2002); Z. X. Liu *et al.* (2003, 2007)

Palaeo-tidal ridges off the South Korean coast	*c.* 14–9.5 ka BP	–	*c.* 14– < 8 ka BP	Lateral migration	Park & Lee (1994); Liu *et al.* (1998); Jin & Chough (2002); Park *et al.* (2003, 2006)
Modern tidal ridges off the South Korean coast	*c.* 7 ka BP	–	–	Lateral migration	Park *et al.* (2006)
Tidal ridges in the Bohai Sea	8 ka BP *c.* 9–8 ka BP	–	*c.* 8 ka BP–present	Material supply from the adjacent Bohai Strait	Ren & Shi (1986); Xia *et al.* (1995); Liu *et al.* (1998)
Yangtze Shoal	*c.* 14–10 ka BP	–	–	Seabed erosion, bioturbation	Chen (1986); Liu (1987); Qin *et al.* (1987); Jin (1992); Liu *et al.* (2004)
Yellow River distal mud	*c.* 11.6–9 ka BP	10–100 years	–	Water depth and sediment transport due to currents and gravity flows	Liu *et al.* (2002, 2004); J. Liu *et al.* (2007)
Changjiang distal mud	*c.* 7 ka	10 years	–	Water depth and sediment transport due to currents and gravity flows	J. P. Liu *et al.* (2006, 2007); Liu *et al.* (2010)
Huskan Mud Belt in the SE Yellow Sea	Since 8 ka BP	100 years	–	Intense bioturbation, seabed erosion	Chough *et al.* (2004); Lim *et al.* (2007)
Isolated mud patch in the central Yellow Sea	*c.* 9.5–7.5 ka BP	100 years	–	Intense bioturbation, seabed disturbance during storms	Milliman *et al.* (1987, 1989); Jin *et al.* (2002); Lim *et al.* (2007)

Note: '–' denotes 'lack of information'.

Although much effort has been expended to understand the sediment-transport patterns, the question of how these relate to the depositional characteristics of the Holocene sedimentary systems has received much less attention. In the first place, the magnitude of the deposition rate can be explained by sediment dynamic observations. Over the years, a large dataset has accumulated on the basis of ^{210}Pb and ^{137}Cs analyses, and a relatively good spatial coverage (Fig. 14). In the mud area of the central Yellow Sea, the deposition rate is of the order of 1 mm a^{-1}. This deposition rate must be seen in relation to a shelf-circulation velocity of 0.1 m s^{-1}, a suspended sediment concentration of 0.01 kg m^{-3}, a water depth of 50 m and net sediment-transport rates of the order of 0.1 kg m^{-1} s^{-1}. Keeping the spatial scale of 10^2 km in mind, the continuity equation for sediment transport results in an accumulation rate of the order of 1 mm a^{-1}, which is consistent with the value from isotope determinations. This demonstrates that the sediment dynamic method can be applied to determine the temporal scale of mud deposits.

Furthermore, the numerical models for hydrodynamics and sediment dynamics provide insights into the evolution of the sedimentary systems. An example is the evolution of the tidal ridges on the East China Sea continental shelf. As outlined above, these ridges are distributed in different depths (Yang 1989; Berné *et al.* 2002; Z. X. Liu *et al.* 2007). As shown in Figure 10, numerical simulations successfully reconstructed the dynamic environment for the tidal-ridge field during the Holocene (Uehara *et al.* 2002; Uehara & Saito 2003). The hydrodynamic condition in terms of palaeo-tidal currents has also been revealed by other researchers (Zhu & Chang 2000; Chen & Zhu 2012). If the sediment-transport patterns are integrated with the tidal current and seabed sediment data, then the history of the tidal-ridge system evolution can be reproduced.

From the study of Holocene deposition in the East China Sea, the diversity of depositional systems in space and time becomes apparent, indicating that the evolution of shelf sedimentary systems is not controlled by sea-level changes alone. A combination of *in situ* observations and numerical modelling will be able to establish the process–product relationship. Recently, such a forward-modelling approach for the formation of sedimentary sequences has been adopted in studies of geomorphological evolution, in addition to process studies. For the present study areas, the early Holocene deposits can be used as a test to the model output: for example, those in the Changjiang Estuary, Hangzhou Bay and the inner-shelf region off the Zhejiang–Fujian coast. The model output may be compared with the sequences, so that the downcore ^{14}C profile (Fig. 12) may be explained in terms of the mixing between the reworked sediment and the coastal or land-derived sediment.

The nature of the Holocene sedimentary records

Continental shelf sedimentary records or archives are important in the investigations of past environmental changes (e.g. climate, sea level, seabed morphology, ecosystems). For an appropriate interpretation of the records, three criteria are important: temporal resolution, vertical continuity and time coverage. The general characteristics of the Holocene sedimentary records of the region are summarized in Table 5.

Compared with their counterparts in deep oceans, these deposits have a relatively high resolution. The temporal resolution is a function of the deposition rate and the intensity of vertical mixing (Gao & Collins 2014). In the deep ocean, the deposition rate is often of the order of 10^{-1}–10^{-2} mm a^{-1}, which may be disturbed easily by vertical material exchange induced by physical or biological processes. Thus, 1 kyr may be considered as a high resolution. However, when a temporal scale of 10 kyr (i.e. the entire Holocene) is concerned, a resolution of better than within 100 years is required. In this respect, the shelf–coastal deposits represent ideal sedimentary archives. In the areas considered here, the river deltas, tidal flats and shelf mud deposits are associated with a deposition rate of 10^1 mm a^{-1} (e.g. Gao 2009*b*; Fan *et al.* 2011). Thus, the resolution can be as high as 1 year (e.g. the deposits of the upper tidal flat).

The second criterion for a good sedimentary record is continuity. In many shelf–coastal systems, both accretion and erosion may occur at the same site. Extreme events, such as a typhoon storm surges, may destroy large parts of the sedimentary record. Nevertheless, the shelf mud deposits of the region tend to have relatively continuous records. In the future, numerical modelling may be used to recover missing records. For instance, based on process-orientated modelling, using the existing sedimentary records as calibration material, the environmental conditions can be reconstructed. Then, data simulation (once again, using the existing records) can be carried out in order to identify the missing parts of the record.

The third criterion for high-quality sedimentary records is the temporal coverage, or duration, of the record. Any single sedimentary record from a specific sedimentary system will cover a limited time period of the Holocene period. For example, the deposits of the abandoned Yellow River delta contain a record of 727 years, starting from AD

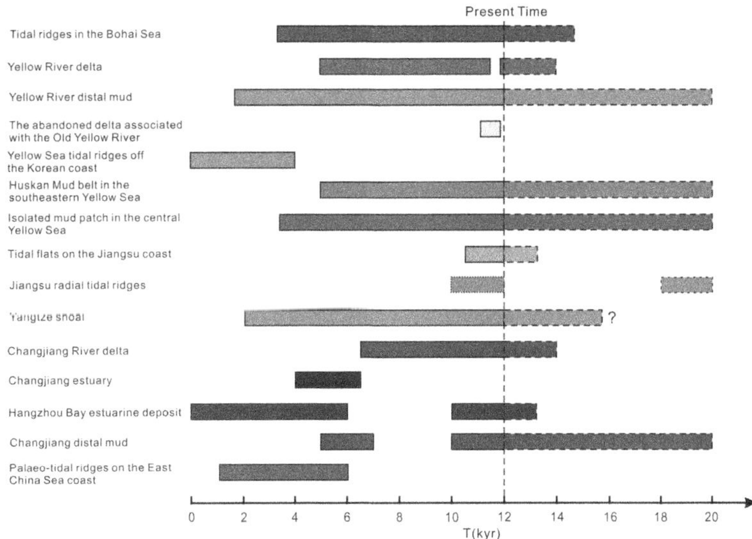

Fig. 15. Sedimentary records: present status and future evolution (based on the datasets in Table 5).

1128 and ending in AD 1855. Likewise, the distal mud deposits have different timings for their initiation, as indicated by the distal muds associated with the Yellow, Changjiang and Pearl rivers. However, these individual sedimentary records can be connected to form an integrated record covering more extensive periods.

Shelves with large widths and abundant sediment supply have the best chance of preserving complete records of the Holocene (Gao & Collins 2014). Using the available information, an attempt was made to reconstruct the temporal coverage of the various types of sedimentary records. As shown in Figure 15, it was assumed that the Holocene commenced 12 kyr ago, the beginning being marked as 0 and some future time limit as 20 kyr. Within this temporal framework, the records of the sedimentary systems described in Figure 2 cover different time periods. For the river deltas, parts of both the beginning and the final stages of the Holocene records are missing. Coastal systems such as tidal flats would be short lived because they developed over short time periods when sufficient sediment was supplied by seabed erosion or river input. The tidal-ridges systems are subjected to constant reworking; as a result, the original record formulated may be destroyed by subsequent lateral migration. In the case of the radial tidal ridges off the Jiangsu coast, less than 2 kyr have been preserved because of such lateral migration (Yin Yong pers. comm.). It is thus possible that future records will also not record more than 2 kyr. The distal mud deposits associated with large rivers, and the shelf mud patches, have the highest potential for extended records. During the early and intermediate stages, the settling of suspended sediment provides material for continuous accumulation; during later stages, the transport mechanisms are dominated by the movement of bottom turbid layers, the presence of 'parallel clinoforms' implying continuous deposition. Therefore, these records may extend well into the future, as long as the river input continues. The prediction of the future sedimentary record is a research direction that requires further efforts. Once again, the sediment dynamics will be a useful tool for dealing with the critical issues of the timing for the onset of deposition, the spatial distribution of sedimentary systems with high-resolution records and the overall duration of sediment accumulation.

Conclusions

- In the presence of a broad shelf and abundant terrestrial sediment supply, sediment dynamic processes determine the spatial distribution and temporal evolution of the various Holocene sedimentary systems in the Bohai, Yellow and East China seas. According to the transport and accumulation processes, the Holocene deposits can be classified into different categories: deposits formed by tidal currents or combined current–wave interactions (e.g. tidal river deltas, tidal flats and tidal ridges); shelf-circulation-dominated systems, (e.g. distal muds and isolated shelf mud patches); and systems influenced by sediment gravity flow (i.e. movement of bottom turbid layers), as indicated by the presence of parallel clinoforms in distal mud deposits at a mature stage.

- The inherent interrelationship between the transport processes and the deposits implies that forward modelling of process–product relationships is an important research direction. Within such a framework, more and better information on aspects such as the Holocene sedimentation along the Jiangsu coast, as well as sediment sources and transport pathways for the Hangzhou Bay and Zhejiang–Fujian inner-shelf sedimentary systems during the early-middle Holocene, can be obtained. The modelling of the formation of sedimentary systems and records is an additional important scientific issue.
- In such an environmental setting, the Holocene sedimentary systems provide 'high-resolution time slices' on temporal scales better than 100 years. In addition, in response to continuous sediment supply, new sedimentary systems will be formed in the future, especially shelf mud deposits. If the spatial distribution patterns of the records, which change with time, can be identified correctly, then it becomes possible to connect high-resolution time slices to generate complete Holocene sedimentary records useful in climate, environment and ecosystem change studies.

A part of the research reported here was presented at the 34th International Geological Congress (Brisbane, Australia, 5–10 August, 2012). Subsequently, the study has been financially supported by the Ministry of Science and Technology, China, through the project 'Land–Ocean Boundary Processes and Their Impacts on the Formation of the Yangtze Deposition System' (grant number 2013CB956500). Mr Niu Zhan-sheng is thanked for his help with the preparation of some of the figures. Professor B. W. Flemming is thanked for his comments and revision suggestions.

References

Alexander, C. R., DeMaster, D. J. & Nittrouer, C. A. 1991. Sediment accumulation in a modern epicontinental-shelf setting: the Yellow Sea. *Marine Geology*, **98**, 51–72.

Berné, S., Trentesaux, A., Stolk, A., Missiaen, T. & Debatist, M. 1994. Architecture and long-term evolution of a tidal sandbank – the Middelkerke Bank (Southern North-Sea). *Marine Geology*, **121**, 57–72.

Berné, S., Vagner, P. *et al.* 2002. Pleistocene forced regressions and tidal sand ridges in the East China Sea. *Marine Geology*, **188**, 293–315.

Bi, N. S., Yang, Z. S., Wang, H. J., Hu, B. Q. & Ji, Y. J. 2010. Sediment dispersion pattern off the present Huanghe (Yellow River) subdelta and its dynamic mechanism during normal river discharge period. *Estuarine, Coastal and Shelf Science*, **86**, 352–362.

Bian, C. W. 2012. *Chinese Coastal Sediment Transport in the Bohai Sea, Yellow Sea and East China Sea*. Doctorial thesis, Ocean University of China, 100 [in Chinese].

Bian, C. W., Jiang, W. S. & Song, D. H. 2010. Terrigenous transportation to the Okinawa Trough and the influence of typhoons on suspended sediment concentration. *Continental Shelf Research*, **30**, 1189–1199.

Chen, B. & Wang, K. 2008. Suspended sediment transport in the offshore near Yangtze Estuary. *Journal of Hydrodynamics*, **20**, 373–381.

Chen, C. T. A. 2009. Chemical and physical fronts in the Bohai, Yellow and East China seas. *Journal of Marine Systems*, **78**, 394–410.

Chen, Q. Q. & Zhu, Y. R. 2012. Holocene evolution of bottom sediment distribution on the continental shelves of the Bohai Sea, Yellow Sea and East China Sea. *Sedimentary Geology*, **273–274**, 58–72.

Chen, Y. Z., Syvitski, J. P. M., Gao, S., Overeem, I. & Kettner, A. J. 2012. Socio-economic impacts on flooding: a 4000 year history of the Yellow River, China. *AMBIO*, **41**, 682–698.

Chen, Z. Y. 1986. Subaqueous topography and sediments off modern Changjiang Estuary. *Donghai Marine Science*, **4**, 28–37 [in Chinese with English abstract].

Chen, Z. Y. & Stanley, D. J. 1995. Quaternary subsidence and river channel migration in the Yangtze Delta Plain, Eastern China. *Journal of Coastal Research*, **11**, 927–945.

Choi, B. H., Mun, J. Y., Ko, J. S. & Yuk, J. H. 2005. Simulation of Suspended Sediment in the Yellow and East China Seas. *China Ocean Engineering*, **19**, 235–250.

Chough, S. K., Kim, J. W., Lee, S. H., Shinn, Y. J., Jin, J. H., Suh, M. C. & Lee, J. S. 2002. High-resolution acoustic characteristics of epicontinental sea deposits, central-eastern Yellow Sea. *Marine Geology*, **188**, 317–331.

Chough, S. K., Lee, H. J., Chun, S. S. & Shinn, Y. J. 2004. Depositional processes of late Quaternary sediments in the Yellow Sea: a review. *Geosciences Journal*, **8**, 211–264.

Chu, Z. X., Sun, X. G., Zhai, S. K. & Xu, K. H. 2006. Changing pattern of accretion/erosion of the modern Yellow River (Huanghe) subaerial delta, China: based on remote sensing images. *Marine Geology*, **227**, 13–30.

Chung, Y. C. & Hung, G. W. 2000. Particulate fluxes and transports on the slope between the southern East China Sea and the South Okinawa Trough. *Continental Shelf Research*, **20**, 571–597.

Chung, Y. C., Chung, K., Chang, H. C., Wang, L. W., Yu, C. M. & Hung, G. W. 2003. Variabilities of particulate flux and ^{210}Pb in the southern East China Sea and western South Okinawa Trough. *Deep-Sea Research II*, **50**, 1163–1178.

Collins, M. B., Shimwell, S. J., Gao, S., Powell, H., Hewitson, C. & Taylor, J. A. 1995. Water and sediment movement in the vicinity of linear sandbanks: the Norfolk Banks, southern North Sea. *Marine Geology*, **123**, 125–142.

Deleu, S., Van lancker, V., van Den eynde, D. & Moerkerke, G. 2004. Morphodynamic evolution of the kink of an offshore tidal sandbank: the Westhinder

Bank (Southern North Sea). *Continental Shelf Research*, **24**, 1587–1610.

DeMaster, D. J., McKee, B. A., Nittrouer, C. A., Qian, J. C. & Cheng, G. D. 1985. Rates of sediment accumulation and particle reworking based on radiochemical measurements from continental shelf deposits in the East China Sea. *Continental Shelf Research*, **4**, 143–158.

Ding, X. R., Kang, Y. Y., Mao, Z. B., Sun, Y. L., Li, S., Gao, X. & Zhao, X. X. 2014. Analysis of largest tidal range in radial sand ridges southern Yellow Sea. *Acta Oceanological Sinica*, **36**, 12–20 [in Chinese with English abstract].

Dong, L. X., Guan, W. B., Chen, Q., Li, X. H., Liu, X. H. & Zeng, X. M. 2011. Sediment transport in the Yellow Sea and East China Sea. *Estuarine, Coastal and Shelf Science*, **93**, 248–258.

Dou, Y. G., Yang, S. Y. et al. 2010. Clay mineral evolution in the central Okinawa Trough since 28 ka: implications for sediment provenance and paleoenvironmental change. *Palaeogeography, Palaeoclimatology, Palaeoecology*, **288**, 108–117.

Dronkers, J. & Miltenburg, A. G. 1996. Fine sediment deposits in shelf seas. *Journal of Marine Systems*, **7**, 119–131.

Du, J. Z., Wu, Y. F., Huang, D. K. & Zhang, J. 2010. Use of ^{7}Be, ^{210}Pb and ^{137}Cs tracers to the transport of surface sediments of the Changjiang Estuary, China. *Journal of Marine Systems*, **82**, 286–294.

Evans, G. 1965. Intertidal flat sediments and their environments of deposition in the Wash. *Quarterly Journal of the Geological Society of London*, **121**, 209–240, http://doi.org/10.1144/gsjgs.121.1.0209

Fan, D. D., Tu, J. B., Shang, S. & Cai, G. F. 2014. Characteristics of tidal-bore deposits and facies associations in the Qiantang Estuary, China. *Marine Geology*, **348**, 1–14.

Fan, D. J., Qi, H. Y., Sun, X. X., Liu, Y. & Yang, Z. S. 2011. Annual lamination and its sedimentary implications in the Yangtze River delta inferred from High-resolution biogenic silica and sensitive grain-size records. *Continental Shelf Research*, **31**, 129–137.

Feng, Y. J., Li, Y., Xie, Q. C. & Zhang, L. R. 1990. Morphology and activity of sedimentary interfaces of the Hangzhou Bay. *Acta Oceanologica Sinica*, **12**, 213–223.

Flemming, B. W. & Davis, R. A., Jr. 1994. Holocene evolution, morphodynamics and sedimentology of the Spiekeroog Barrier Island system (southern North Sea). *Senkenbergiana Maritima*, **24**, 117–155.

Frank, M. 2011. Oceanography Chemical twins, separated. *Nature Geoscience*, **4**, 220–221.

Fu, M. Z. & Zhu, D. K. 1986. The sediment sources of the offshore submarine sand ridge field of the coast of Jiangsu province. *Journal of Nanjing University (Natural Sciences Edition)*, **22**, 536–544 [in Chinese with English abstract].

Gao, S. 2007. Modeling the growth limit of the Changjiang Delta. *Geomorphology*, **85**, 225–236.

Gao, S. 2009*a*. Geomorphology and sedimentology of tidal flats. *In*: Perillo, G. M. E., Wolanski, E., Cahoon, D. & Brinson, M. (eds) *Coastal Wetlands: an Ecosystem Integrated Approach*. Elsevier, Amsterdam, 295–316.

Gao, S. 2009*b*. Modeling the preservation potential of tidal flat sedimentary records, Jiangsu coast, eastern China. *Continental Shelf Research*, **29**, 1927–1936.

Gao, S. 2014. Sediment dynamic processes, mechanisms and evolution trends of the radial linear sandbanks, Jiangsu coast. *In*: Wang, Y. (ed.) *Environmental and Resource Characteristics of the Southern Yellow Sea Radial Sand Ridges*. China Ocean Press, Beijing, 275–293.

Gao, S. & Collins, M. 2014. Holocene sedimentary systems on continental shelves. *Marine Geology*, **352**, 268–294.

Gao, S. & Jia, J. J. 2003. Modeling suspended sediment distribution in continental shelf upwellinig/downwelling settings. *Geo-Marine Letters*, **22**, 218–226.

Gao, S., Cheng, P., Wang, Y. P. & Cao, Q. Y. 2000. Characteristics of suspended sediment concentrations over the areas adjacent to the Changjiang River estuary, the summer of 1998. *Marine Science Bulletin*, **2**, 14–24.

Gao, S., Wang, Y. P. & Gao, J. H. 2011. Sediment retention at the Changjiang sub-aqueous delta over a 57 year period, in response to catchment changes. *Estuarine, Coastal and Shelf Science*, **95**, 29–38.

Gao, S., Du, Y. F., Xie, W. J., Gao, W. H., Wang, D. D. & Wu, X. D. 2014. Environment-ecosystem dynamic processes of *Spartina alterniflora* salt-marshes along the eastern China coastlines. *Science in China (Earth Sciences)*, **57**, 2567–2586.

Graber, H. C., Beardsley, R. C. & Grant, W. D. 1989. Storm-generated surface-waves and sediment resuspension in the East China and Yellow Seas. *Journal of Physical Oceanography*, **19**, 1039–1059.

Haq, B. U., Hardenbol, J. & Vail, P. R. 1987. Chronology of fluctuating sea levels since the Triassic. *Science*, **235**, 1156–1167.

Honda, M. C., Kusakabe, M., Nakabayashi, S. & Katagiri, M. 2000. Radiocarbon of sediment trap samples from the Okinawa trough: lateral transport of ^{14}C-poor sediment from the continental slope. *Marine Chemistry*, **68**, 231–247.

Hori, K. & Saito, Y. 2007. An early Holocene sea-level jump and delta initiation. *Geophysical Research Letters*, **34**, L18401, http://doi.org/10.1029/2007GL031029

Hori, K., Saito, Y., Zhao, Q. H., Cheng, X. R., Wang, P. X., Sato, Y. & Li, C. X. 2001*a*. Sedimentary facies and Holocene progradation rates of the Changjiang (Yangtze) delta, China. *Geomorphology*, **41**, 233–248.

Hori, K., Saito, Y., Zhao, Q. H., Cheng, X. R., Wang, P. X., Sato, Y. & Li, C. X. 2001*b*. Sedimentary facies of the tide-dominated paleo-Changjiang (Yangtze) estuary during the last transgression. *Marine Geology*, **177**, 331–351.

Hori, K., Saito, Y., Zhao, Q. H. & Wang, P. X. 2002. Architecture and evolution of the tide-dominated Changjiang (Yangtze) River delta, China. *Sedimentary Geology*, **146**, 249–264.

Hoshika, A., Tanimoto, T., Mishima, Y., Iseki, K. & Okamura, K. 2003. Variation of turbidity and particle transport in the bottom layer of the East China Sea. *Deep-Sea Research Part II: Topical Studies in Oceanography*, **50**, 443–455.

Hsueh, Y., Schultz, J. R. & Holland, W. R. 1997. The Kuroshio flow-through in the East China Sea: a numerical model. *Progress in Oceanography*, **39**, 79–108.

Hu, J. Y., Kawamura, H., Hong, H. S. & Qi, Y. Q. 2000. A review on the currents in the South China Sea: seasonal circulation, South China Sea warm current and Kuroshio intrusion. *Journal of Oceanography*, **56**, 607–624.

Hu, K. L., Ding, P. X., Wang, Z. B. & Yang, S. L. 2009. A 2D/3D hydrodynamic and sediment transport model for the Yangtze Estuary, China. *Journal of Marine Systems*, **77**, 114–136.

Huh, C. A. & Su, C. C. 1999. Sedimentation dynamics in the East China Sea elucidated from ^{210}Pb, ^{137}Cs and ^{239,240}Pu. *Marine Geology*, **160**, 183–196.

Huthnance, J. M. 1982*a*. On the formation of sand banks of finite extent. *Estuarine Coastal and Shelf Science*, **15**, 277–299.

Huthnance, J. M. 1982*b*. One mechanism forming linear sand banks. *Estuarine Coastal and Shelf Science*, **14**, 79–99.

Iseki, K., Okamura, K. & Kiyomoto, Y. 2003. Seasonality and composition of downward particulate fluxes at the continental shelf and Okinawa Trough in the East China Sea. *Deep-Sea Research II: Topical Studies in Oceanography*, **50**, 457–473.

Jia, J. J., Gao, S. & Xue, Y. C. 2003. Sediment dynamic processes of the Yuehu inlet system, Shandong Peninsula, China. *Estuarine, Coastal and Shelf Science*, **57**, 783–801.

Jiang, W. S., Pohlmann, T., Sündermann, J. & Feng, S. Z. 2000. A modelling study of SPM transport in the Bohai Sea. *Journal of Marine Systems*, **24**, 175–200.

Jiang, W. S., Pohlmann, T., Sun, J. & Starke, A. 2004. SPM transport in the Bohai Sea: field experiments and numerical modeling. *Journal of Marine Systems*, **44**, 175–188.

Jin, J. H. & Chough, S. K. 1998. Partitioning of transgressive deposits in the southeastern Yellow Sea: a sequence stratigraphic interpretation. *Marine Geology*, **149**, 79–92.

Jin, J. H. & Chough, S. K. 2002. Erosional shelf sand ridges in the mid-eastern Yellow Sea. *Geo-Marine Letters*, **21**, 219–225.

Jin, J. H., Chough, S. K. & Rvang, W. H. 2002. Sequence aggradation and systems tracts partitioning in the mid-eastern Yellow Sea: roles of glacio-eustasy, subsidence and tidal dynamics. *Marine Geology*, **184**, 249–271.

Jin, X. L. (ed.). 1992. *Marine Geology of the East China Sea*. China Ocean Press, Beijing [in Chinese].

Kamp, P. J. J. & Naish, T. 1998. Forward modelling of the sequence stratigraphic architecture of shelf cyclothems: application to Late Pliocene sequences, Wanganui Basin (New Zealand). *Sedimentary Geology*, **116**, 57–80.

Lee, H. J. & Chao, S. Y. 2003. A climatological description of circulation in and around the East China Sea. *Deep-Sea Research II: Topical Studies in Oceanography*, **50**, 1065–1084.

Lee, H. J. & Chu, Y. S. 2001. Origin of inner-shelf mud deposit in the southeastern Yellow Sea: Huskan Mud Belt. *Journal of Sedimentary Research*, **71**, 144–154.

Lee, H. J. & Yoon, S. H. 1997. Development of stratigraphy and sediment distribution in the northeastern Yellow Sea during Holocene sea-level rise. *Journal of Sedimentary Research*, **67**, 341–349.

Lee, H. J., Jung, K. T., Foreman, M. G. G. & Chung, J. Y. 2000. A three-dimensional mixed finite-difference Galerkin function model for the oceanic circulation in the Yellow Sea and the East China Sea. *Continental Shelf Research*, **20**, 863–895.

Lee, H. J., Jung, K. T., So, J. K. & Chung, J. Y. 2002. A three-dimensional mixed finite-difference Galerkin function model for the oceanic circulation in the Yellow Sea and the East China Sea in the presence of M2 tide. *Continental Shelf Research*, **22**, 67–91.

Li, C. X. 1986. Deltaic sedimentation. *In*: Ren, M. E. (ed.) *Modern Sedimentation in Coastal and Nearshore Zone of China*. China Ocean Press, Beijing/Springer, Berlin, 253–378.

Li, C. X. & Li, P. 1983. The characteristics and distribution of Holocene sand bodies in the Changjiang delta area. *Acta Oceanologica Sinica*, **2**, 54–66.

Li, C. X. & Wang, P. 1991. Stratigraphy of the Late Quaternary barrier lagoon depositional systems along the coast of China. *Sedimentary Geology*, **72**, 189–200.

Li, C. X., Guo, X. M., Xu, S. Y., Wang, J. T. & Li, P. 1979. The characteristics and distribution of Holocene sand bodies in Changjiang, River Delta area. *Acta Oceanologica Sinica*, **1**, 252–268 [in Chinese with English abstract].

Li, C. X., Chen, Q. Q., Zhang, J. Q., Yang, S. L. & Fan, D. D. 2000. Stratigraphy and paleoenvironmental changes in the Yangtze Delta during the Late Quaternary. *Journal of Asian Earth Sciences*, **18**, 453–469.

Li, C. X., Zhang, J. Q., Fan, D. D. & Deng, B. 2001. Holocene regression and the tidal radial sand ridge system formation in the Jiangsu coastal zone, east China. *Marine Geology*, **173**, 97–120.

Li, C. X., Wang, P., Sun, H. P., Zhang, J. Q., Fan, D. D. & Deng, B. 2002. Late Quaternary incised-valley fill of the Yangtze delta (China): its stratigraphic framework and evolution. *Sedimentary Geology*, **152**, 133–158.

Li, D. Y., Chen, J., Wang, A. J. & Yu, X. G. 2009. Suspended sediment characteristics and transport in the Minjiang estuary during flood seasons. *Ocean Engineering*, **27**, 70–80 [in Chinese with English abstract].

Li, F. Y., Gao, S., Jia, J. J. & Zhao, Y. Y. 2002. Contemporary deposition rates of fine-grained sediment in the Bohai and Yellow Seas. *Oceanologia et Limnologia Sinica*, **33**, 364–369 [in Chinese with English Abstract].

Li, F. Y., Li, X. G., Song, J. M., Wang, G. Z., Cheng, P. & Gao, S. 2006. Sediment flux and source in northern Yellow Sea by ^{210}Pb technique. *Chinese Journal of Oceanology and Limnology*, **24**, 255–263.

Li, G. X., Han, X. B., Yue, S. H., Wen, G. Y., Yang, R. M. & Kusky, T. M. 2006. Monthly variations of water masses in the East China Seas. *Continental Shelf Research*, **26**, 1954–1970.

Li, H. Q., Yin, Y., Shi, Y., He, H. C. & Liu, X. Y. 2011. Micro-morphology and contemporary sedimentation rate of tidal flat in Rudong, Jiangsu Province. *Journal of Palaeogeography*, **13**, 150–160 [in Chinese with English abstract].

Li, W., Wang, Y. H., Wang, J. N. & Wei, H. 2012. Distributions of water masses and hydrographic structures in the Yellow Sea and East China Sea in the spring and summer 2011. *Oceanologia et Limnologia Sinica*, **3**, 615–623 [in Chinese with English abstract].

Li, Y. H., Wang, A. J., Qiao, L., Fang, J. Y. & Chen, J. 2012. The impact of typhoon Morakot on the modern sedimentary environment of the mud deposition center off the Zhejiang-Fujian coast, China. *Continental Shelf Research*, **37**, 92–100.

Li, Z. H., Gao, S. & Chen, S. L. 2007. Characteristics of tide-induced bottom boundary layers over the Dafeng intertidal flats, Jiangsu Province, China. *Ocean Engineering*, **25**, 53–60 [in Chinese with English abstract].

Lim, D. I., Jung, H. S., Choi, J. Y., Yang, S. & Ahn, K. S. 2006. Geochemical compositions of river and shelf sediments in the Yellow Sea: grain-size normalization and sediment provenance. *Continental Shelf Research*, **26**, 15–24.

Lim, D. I., Choi, J. Y., Jung, H. S., Rho, K. C. & Ahn, K. S. 2007. Recent sediment accumulation and origin of shelf mud deposits in the Yellow and East China Seas. *Progress in Oceanography*, **73**, 145–159.

Lin, C. M. 1997. Sequence stratigraphic study on the Hangzhou Bay since 15 000 a BP. *Geological Review*, **43**, 273–280 [in Chinese with English abstract].

Lin, C. M., Zhuo, H. C. & Gao, S. 2005. Sedimentary facies and evolution in the Qiantang River incised valley, eastern China. *Marine Geology*, **219**, 235–259.

Lin, S., Hsieh, I. J., Huang, K. M. & Wang, C. H. 2002. Influence of the Yangtze River and grain size on the spatial variations of heavy metals and organic carbon in the East China Sea continental shelf sediments. *Chemical Geology*, **182**, 377–394.

Liu, J., Saito, Y., Wang, H., Yang, Z. G. & Nakashima, R. 2007. Sedimentary evolution of the Holocene subaqueous clinoform off the Shandong Peninsula in the Yellow Sea. *Marine Geology*, **236**, 165–187.

Liu, J., Saito, Y., Kong, X. H., Wang, H., Xiang, L. H., Wen, C. & Nakashima, R. 2010. Sedimentary record of environmental evolution off the Yangtze River estuary, East China Sea, during the last ~13 000 years, with special reference to the influence of the Yellow River on the Yangtze River delta during the last 600 years. *Quaternary Science Reviews*, **29**, 2424–2438.

Liu, J., Kong, X. H., Saito, Y., Liu, J. P., Yang, Z. S. & Wen, C. 2013. Subaqueous deltaic formation of the Old Yellow River (AD 1128–1855) on the western South Yellow Sea. *Marine Geology*, **344**, 19–33.

Liu, J. P., Milliman, J. D. & Gao, S. 2002. The Shandong mud wedge and postglacial sediment accumulation in the Yellow Sea. *Geo-Marine Letters*, **21**, 212–218.

Liu, J. P., Milliman, J. D., Gao, S. & Cheng, P. 2004. Holocene development of the Yellow River's subaqueous delta, North Yellow Sea. *Marine Geology*, **209**, 45–67.

Liu, J. P., Li, A. *et al.* 2006. Sedimentary features of the Yangtze River-derived along-shelf clinoform deposit in the East China Sea. *Continental Shelf Research*, **26**, 2141–2156.

Liu, J. P., Xu, K., Li, A. C., Milliman, J. D., Velozzi, D. M., Xiao, S. B. & Yang, Z. S. 2007. Flux and fate of Yangtze River sediment delivered to the East China Sea. *Geomorphology*, **85**, 208–224.

Liu, J. T. & Lin, H. L. 2004. Sediment dynamics in a submarine canyon: a case of river–sea interaction. *Marine Geology*, **207**, 55–81.

Liu, S. F., Zhuang, Z. Y. & Long, H. Y. 2008. Environmental evolution and sheet sedimentation in late Quaternary in the east Bohai Sea. *Marine Geology & Quaternary Geology*, **28**, 25–31 [in Chinese with English abstract].

Liu, S. X., Shen, X. Q., Wang, Y. Q. & Han, S. X. 1993. The averaged distributions and variability of water masses in the Bohai, Yellow and East China seas. *Acta Oceanologica Sinica*, **15**, 3–7 [in Chinese with English abstract].

Liu, X. J., Gao, S. & Wang, Y. P. 2011. Modeling profile shape evolution for accreting tidal flats composed of mud and sand: a case study of the central Jiangsu coast, China. *Continental Shelf Research*, **31**, 1750–1760.

Liu, X. Q. 1987. Relict sediments in China continental shelf. *Marine Geology & Quaternary Geology*, **7**, 1–14 [in Chinese].

Liu, X. Y., Gao, J. H., Bai, F. L., Liu, Z. Y. & Pan, S. M. 2008. Grain size information in different evolution periods of Xinyanggang tidal flat in Jiangsu Province. *Marine Geology & Quaternary Geology*, **28**, 27–35 [in Chinese with English abstract].

Liu, Y. L., Gao, S., Wang, Y. P., Yang, Y., Long, J. P., Zhang, Y. Z. & Wu, X. D. 2014. Distal mud deposits associated with the Pearl River over the northwestern continental shelf of the South China Sea. *Marine Geology*, **347**, 43–57.

Liu, Z. X. 1997. Yangtze Shoal – a modern tidal sand sheet in the northwestern part of the East China Sea. *Marine Geology*, **137**, 321–330.

Liu, Z. X. & Xia, D. X. 2004. *Tidal Sands in the China Seas*. China Ocean Press, Beijing.

Liu, Z. X., Huang, Y. C. & Zhang, Q. A. 1989. Tidal current ridges in the southwestern Yellow Sea. *Journal of Sedimentary Petrology*, **59**, 432–437.

Liu, Z. X., Xia, D. X., Tang, Y. X., Wang, K. Y., Berne, S., Marsset, T. & Bourillet, J. F. 1994. Tidal deposit system in the Eastern Bohai Sea in the Holocene. *Science in China (Series B)*, **24**, 1331–1338 [in Chinese with English abstract].

Liu, Z. X., Xia, D. X., Berne, S., Wang, K. Y., Marsset, T., Tang, Y. X. & Bourillet, J. F. 1998. Tidal depositional systems of China's continental shelf, with special reference to the eastern Bohai Sea. *Marine Geology*, **145**, 225–253.

Liu, Z. X., Berné, S., Saito, Y., Lericolais, G. & Marsset, T. 2000. Quaternary seismic stratigraphy and paleoenvironments on the continental shelf of the East China Sea. *Journal of Asian Earth Sciences*, **18**, 441–452.

Liu, Z. X., Yin, P., Xiong, Y. Q., Berné, S., Trentesaux, A. & Li, C. X. 2003. Quaternary transgressive and regressive depositional sequences in the East China Sea. *Chinese Science Bulletin*, **48**, 81–87.

Liu, Z. X., Berné, S. *et al.* 2007. Internal architecture and mobility of tidal sand ridges in the East China Sea. *Continental Shelf Research*, **27**, 1820–1834.

Lu, J., Qiao, F. L., Wang, X. H., Wang, Y. G., Teng, Y. & Xia, C. S. 2011. A numerical study of transport dynamics and seasonal variability of the Yellow River sediment in the Bohai and Yellow seas. *Estuarine, Coastal and Shelf Science*, **95**, 39–51.

Marsset, T., Xia, D., Berné, S., Liu, Z., Bourillet, J. F. & Wang, K. 1996. Stratigraphy and sedimentary environment since the Late Quaternary in the Eastern Bohai Sea. *Marine Geology*, **135**, 97–114.

Milliman, J. D., Qin, Y. S., Ren, M. E. & Saito, Y. 1987. Man's influence on the erosion and transport by Asian rivers: the Yellow River (Huanghe) example. *Journal of Geology*, **95**, 751–762.

Milliman, J. D., Qin, Y. S. & Park, Y. A. 1989. Sediments and sedimentary processes in the Yellow and East China Seas. *In*: Taira, A. & Masuda, F. (eds) *Sedimentary Facies in the Active Plate Margin*. Terra Scientific, Tokyo, 233–249.

Ni, W. F., Wang, Y. P., Zou, X. Q. & Gao, J. H. 2013. Study on the sediment dynamics and stability of the Kushuiyang tidal channel at radial sand ridges in the southern Yellow Sea. *Marine Science Bulletin*, **32**, 668–677 [in Chinese with English abstract].

Ni, W. F., Wang, Y. P., Zou, X. Q., Zhang, J. C. & Gao, J. H. 2014. Sediment dynamics in an offshore tidal channel in the southern Yellow Sea. *International Journal of Sediment Research*, **29**, 246–259.

Nittrouer, C. A. 1999. STRATAFORM: overview of its design and synthesis of its results. *Marine Geology*, **154**, 3–12.

Nittrouer, C. A. & Wright, L. D. 1994. Transport of particles across continental shelves. *Reviews of Geophysics*, **32**, 85–113.

Nittrouer, C. A., Austin, J. A., Field, M. E., Kravitz, J. H., Syvitski, J. P. M. & Wiberg, P. L. (eds). 2009. *Continental Margin Sedimentation: from Sediment Transport to Sequence Stratigraphy*. International Association of Sedimentologists, Special Publications, **37**.

Off, T. 1963. Rhythmic linear sand bodies caused by tidal currents. *Bulletin of the American Association of Petroleum Geologists*, **47**, 324–341.

Oguri, K., Matsumoto, E., Yamada, M., Saito, Y. & Iiseki, K. 2003. Sediment accumulation rates and budgets of depositing particles of the East China Sea. *Deep Sea Research Part II: Topical Studies in Oceanography*, **50**, 513–528.

Oiwane, H., Tonai, S., Kiyokawa, S., Nakamura, Y., Suganuma, Y. & Tokuyama, H. 2011. Geomorphological development of the Goto Submarine Canyon, northeastern East China Sea. *Marine Geology*, **288**, 49–60.

Orton, G. J. & Reading, H. G. 1993. Variability of deltaic processes in terms of sediment supply, with particular emphasis on grain size. *Sedimentology*, **40**, 475–512.

Paola, C. 2000. Quantitative models of sedimentary basin filling. *Sedimentology*, **47** (Suppl. 1), 121–178.

Park, S. C. & Lee, B. H. 1994. Depositional patterns of sand ridges in tide dominated shallow water environments: Yellow Sea coast and South Sea of Korea. *Marine Geology*, **120**, 65–75.

Park, S. C. & Yoo, D. G. 1988. Depositional history of Quaternary sediments on the continental shelf off southeastern coast of Korea (Korea Strait). *Marine Geology*, **79**, 65–75.

Park, S. C., Yoo, D. G., Lee, K. W. & Lee, H. H. 1999. Accumulation of recent muds associated with coastal circulations, southeastern Korea Sea (Korea Strait). *Continental Shelf Research*, **19**, 589–608.

Park, S. C., Lee, H. H., Han, H. S., Lee, G. H., Kim, D. C. & Yoo, D. G. 2000. Evolution of late Quaternary mud deposits and recent sediment budget in the southeastern Yellow Sea. *Marine Geology*, **170**, 271–288.

Park, S. C., Han, H. H. & Yoo, D. G. 2003. Transgressive sand ridges on the mid-shelf of the southern sea of Korea (Korea Strait): formation and development in high-energy environments. *Marine Geology*, **193**, 1–18.

Park, S. C., Lee, B. H., Han, H. S., Yoo, D. G. & Lee, C. W. 2006. Late Quaternary stratigraphy and development of tidal sand ridges in the eastern Yellow Sea. *Journal of Sedimentary Research*, **76**, 1093–1105.

Park, Y. A., Choi, J. Y. & Gao, S. 2001. Spatial variation of suspended particulate matter in the Yellow Sea. *Geo-Marine Letters*, **20**, 196–200.

Postma, H. 1961. Transport and accumulation of suspended matter in the Dutch Wadden Sea. *Netherlands Journal of Sea Research*, **1**, 148–190.

Qin, Y. S., Zhao, Y. Y., Chen, L. R. & Zhao, S. L. (eds). 1987. *Geology of the East China Sea*. Science Press, Beijing, 290 [in Chinese].

Ren, M. E. (ed.). 1986. *Comprehensive Investigation of Coastal Zone and Tidal Flat Resources, Jiangsu Province*. China Ocean Press, Beijing [in Chinese].

Ren, M. E. 1992. Human impact on coastal landform and sedimentation – The Yellow River example. *GeoJournal*, **28**, 443–448.

Ren, M. E. 2006. Sediment discharge of the Yellow River, China: past, present and future – a synthesis. *Advance in Earth Sciences*, **21**, 551–563.

Ren, M. E. & Shi, Y. L. 1986. Sediment discharge of the Yellow River (China) and its effect on the sedimentation of the Bohai and the Yellow Sea. *Continental Shelf Research*, **6**, 785–810.

Ren, M. E., Zhang, R. S. & Yang, J. H. 1984. Sedimentation on tidal mud flat in Wanggang area, Jiangsu Province, China. *Marine Science Bulletin*, **3**, 40–54 [in Chinese with English abstract].

Ren, M. E., Zhang, R. S. & Yang, J. H. 1985. Effect of typhoon No.8114 on coastal morphology and sedimentation of Jiangsu Province, Peoples' Republic of China. *Journal of Coastal Research*, **1**, 21–28.

Saito, Y., Katayama, K. *et al.* 1998. Transgressive and highstand systems tracts and post-glacial transgression, the East China Sea. *Sedimentary Geology*, **122**, 217–232.

Saito, Y., Wei, H. L., Zhou, Y. Q., Nishimura, A., Sato, Y. & Yokota, S. 2000. Delta progradation and chenier formation in the Huanghe (Yellow River) Delta, China. *Journal of Asian Earth Sciences*, **18**, 489–497.

Saito, Y., Yang, Z. S. & Hori, K. 2001. The Huanghe (Yellow River) and Changjiang (Yangtze River) deltas: a review on their characteristics, evolution and sediment discharge during the Holocene. *Geomorphology*, **41**, 219–231.

Scholl, D. W. 1964. Recent sedimentary record in mangrove swamps and rise in sea level over the southwestern coast of Florida. *Marine Geology*, **1**, 344–366.

SHEU, D. D., HOU, W. C., CHUNG, Y. C., TANG, T. Y. & HUNG, J. J. 1999. Geochemical and carbon isotopic characterization of particles collected in sediment traps from the East China Sea continental slope and the Okinawa Trough northeast of Taiwan. *Continental Shelf Research*, **19**, 183–203.

SHI, C., ZHANG, D. D. & YOU, L. 2003. Sediment budget of the Yellow River delta, China: the importance of dry bulk density and implications to understanding of sediment dispersal. *Marine Geology*, **199**, 13–25.

SHI, W. & WANG, M. H. 2010. Satellite observations of the seasonal sediment plume in central East China Sea. *Journal of Marine Systems*, **82**, 280–285.

SHI, W. & WANG, M. H. 2012. Satellite views of the Bohai Sea, Yellow Sea, and East China Sea. *Progress in Oceanography*, **104**, 30–45.

SHI, Z. 2004. Behaviour of fine suspended sediment at the North passage of the Changjiang Estuary, China. *Journal of Hydrology*, **293**, 180–190.

SLOSS, L. L. 1963. Sequences in the cratonic interior of North America. *Geological Society of America Bulletin*, **74**, 93–113.

SONG, Y. H. & CHOI, M. S. 2009. REE geochemistry of fine-grained sediments from major rivers around the Yellow Sea. *Chemical Geology*, **266**, 328–342.

STERNBERG, R. W., CACCHIONE, D. A., PAULSON, B., KINEKE, G. C. & DRAKE, D. E. 1996. Observations of sediment transport on the Amazon subaqueous delta. *Continental Shelf Research*, **16**, 697–715.

SUN, X. P. 1995. Features of temperature-salinity of the Tsushima warm current in the East China Sea in summer and differences of its salinity in the Summers of 1987 and 1989. *Journal of Oceanography of Huanghai & Bohai Seas*, **13**, 1–10 [in Chinese with English abstract].

TRENTESAUX, A., STOLK, A. & BERNE, S. 1999. Sedimentology and stratigraphy of a tidal sand bank in the southern North Sea. *Marine Geology*, **159**, 253–272.

UEHARA, K. & SAITO, Y. 2003. Late Quaternary evolution of the Yellow/East China Sea tidal regime and its impacts on sediments dispersal and seafloor morphology. *Sedimentary Geology*, **162**, 25–38.

UEHARA, K., SAITO, Y. & HORI, K. 2002. Paleotidal regime in the Changjiang (Yangtze) Estuary, the East China Sea, and the Yellow Sea at 6 ka and 10 ka estimated from a numerical model. *Marine Geology*, **183**, 179–192.

VAIL, P. R., MITCHUM, R. M. ET AL. 1977. Seismic stratigraphy and global changes of sea level. *In*: PAYTON, C. E. (ed.) *Seismic Stratigraphy – Applications to Hydrocarbon Exploration*. American Association of Petroleum Geologists, Memoirs, **26**, 49–212.

WANG, A. J., PAN, S. M., ZHANG, Y. Z. & LIU, Z. Y. 2010. Modern sedimentation rate of the submarine delta of the Changjiang River. *Marine Geology & Quaternary Geology*, **30**, 1–6 [in Chinese with English abstract].

WANG, H. J., YANG, Z. S., LI, Y. H., GUO, Z. G., SUN, X. X. & WANG, Y. 2007. Dispersal pattern of suspended sediment in the shear frontal zone off the Huanghe (Yellow River) mouth. *Continental Shelf Research*, **27**, 854–871.

WANG, J., BAI, C. G., XU, Y. H. & BAI, S. B. 2010. Tidal couplet formation and preservation, and criteria for discriminating storm-surge sedimentation on the tidal flats of central Jiangsu Province, China. *Journal of Coastal Research*, **26**, 976–981.

WANG, J. H., ZHOU, Y. ET AL. 2006. Late Quaternary sediments and paleoenvironmental evolution in Hangzhou Bay. *Journal of Palaeogeography*, **8**, 551–558 [in Chinese with English abstract].

WANG, L. B., YANG, Z. S., ZHAO, X. H., XING, L., ZHAO, M. X., SAITO, Y. & FAN, D. J. 2009. Sedimentary characteristics of core YE-2 from the central mud area in the south Yellow Sea during last 8400 years and its interspace coarse layers. *Marine Geology & Quaternary Geology*, **29**, 1–11 [in Chinese with English abstract].

WANG, X. H., QIAO, F. L., LU, J. & GONG, F. 2011. The turbidity maxima of the northern Jiangsu shoal-water in the Yellow Sea, China. *Estuarine, Coastal and Shelf Science*, **93**, 201–211.

WANG, Y. 2014. *The Environment and Resources of Radiate Sand Ridges of South Yellow Sea*. Ocean Press, Beijing [in Chinese].

WANG, Y. & KE, X. K. 1989. Cheniers on the east coastal-plain of China. *Marine Geology*, **90**, 321–335.

WANG, Y., ZHU, D. K., YOU, K. Y., PAN, S. M., ZHU, X. D., ZOU, X. Q. & ZHANG, Y. Z. 1999. Evolution of radiate sand ridge field of the South Yellow Sea and its sedimentary characteristics. *Science in China (Series D)*, **42**, 97–112.

WANG, Y., ZHANG, Y. Z., ZOU, X. Q., ZHU, D. K. & PIPER, D. 2012. The sand ridge field of the South Yellow Sea: origin by river-sea interaction. *Marine Geology*, **291–294**, 132–146.

WANG, Y. H., DONG, P., OGUCHI, T., CHEN, S. L. & SHEN, H. T. 2013. Long-term (1842–2006) morphological change and equilibrium state of the Changjiang (Yangtze) Estuary, China. *Continental Shelf Research*, **56**, 71–81.

WANG, Y. P., ZHANG, R. S. & GAO, S. 1999. Geomorphic and hydrodynamic responses in salt marsh-tidal creek systems, Jiangsu, China. *Chinese Science Bulletin*, **44**, 544–549.

WANG, Y. P., GAO, S., JIA, J. J., THOMPSON, C., GAO, J. H. & YANG, Y. 2012. Sediment transport over an accretional intertidal flat with influences of reclamation, Jiangsu coast, China. *Marine Geology*, **291**, 147–161.

WANG, Y. P., VOULGARIS, G., LI, Y., YANG, Y., GAO, J. H., CHEN, J. & GAO, S. 2013. Sediment resuspension, flocculation, and settling in a macrotidal estuary. *Journal of Geophysical Research: Oceans*, **118**, 5591–5608.

WANG, Y. Y., HE, Q. & LIU, H. 2009. Variation of near-bed suspended sediment concentration in South Passage of the Changjiang Estuary. *Journal of Sediment Research*, **6**, 6–13 [in Chinese with English abstract].

WATANABE, M. 2007. Simulation of temperature, salinity and suspended matter distributions induced by the discharge into the East China Sea during the 1998 flood of the Yangtze River. *Estuarine, Coastal and Shelf Science*, **71**, 81–97.

WRIGHT, L. D. & FRIEDRICHS, C. T. 2006. Gravity-driven sediment transport on continental shelves: a status report. *Continental Shelf Research*, **26**, 2092–2107.

WRIGHT, L. D., WISEMAN, W. J. *ET AL.* 1988. Marine dispersal and deposition of Yellow River silts by gravity-driven underflows. *Nature*, **332**, 629–632.

WRIGHT, L. D., WISEMAN, J. R., YANG, Z. S., BORNHOLD, B. D., KELLER, G. H., PRIOR, D. B. & SUHAYDA, J. N. 1990. Processes of marine dispersal and deposition of suspended silts off the modern mouth of the Huanghe (Yellow River). *Continental Shelf Research*, **10**, 1–40.

WRIGHT, L. D., FRIEDRICHS, C., KIM, S. & SCULLY, M. 2001. Effects of ambient currents and waves on gravity-driven sediment transport on continental shelves. *Marine Geology*, **175**, 25–45.

XIA, D. X., LIU, Z. X. & WANG, K. Y. 1995. Stratigraphy and sedimentary environments in the eastern Bohai Sea since the Late Pleistocene. *Acta Oceanologica Sinica*, **17**, 86–92 [in Chinese with English abstract].

XIAO, S. B., LI, A. C. *ET AL.* 2006. Coherence between solar activity and the East Asian winter monsoon variability in the past 8000 years from Yangtze River-derived mud in the East China Sea. *Palaeogeography, Palaeoclimatology, Palaeoecology*, **237**, 293–304.

XIE, D. F., WANG, Z. B., GAO, S. & DE VRIEND, H. J. 2009. Modeling the tidal channel morphodynamics in a macro-tidal embayment, Hangzhou Bay, China. *Continental Shelf Research*, **29**, 1757–1767.

XING, F., WANG, Y. P. & WANG, H. V. 2012. Tidal hydrodynamics and fine-grained sediment transport on the radial sand ridge system in the southern Yellow Sea. *Marine Geology*, **291–294**, 192–210.

XU, F. J., LI, A. C., LI, T. G., XU, K. H., CHEN, S. Y., QIU, L. W. & CAO, Y. C. 2011. Rare earth element geochemistry in the inner shelf of the East China Sea and its implication to sediment provenances. *Journal of Rare Earths*, **29**, 702–709.

XU, K. H., MILLIMAN, J. D., LI, A. C., LIU, J. P., KAO, S. J. & WANG, S. M. 2009. Yangtze- and Taiwan-derived sediments on the inner shelf of East China Sea. *Continental Shelf Research*, **29**, 2240–2256.

XU, K. H., LI, A. C. *ET AL.* 2012. Provenance, structure, and formation of the mud wedge along inner continental shelf of the East China Sea: a synthesis of the Yangtze dispersal system. *Marine Geology*, **291–294**, 176–191.

XU, S., WANG, J. & LI, P. 1987. Developing processes and sand body features of modern Changjiang River Delta. *In*: YANG, Q. & XU, S. (eds) *Recent Yangtze Delta Deposits*. East China Normal University Press, Shanghai, 264–277 [in Chinese with English abstract].

XUE, C. T. 1993. Historical changes in the Yellow River delta, China. *Marine Geology*, **113**, 321–329.

XUE, C. T., ZHOU, Y. Q. & ZHU, X. H. 2004. The Huanghe River course and delta from end of Late Pleistocene to the 7th century BC. *Acta Oceanologica Sinica*, **26**, 48–61 [in Chinese with English abstract].

YANG, C. & SUN, J. 1988. Tidal sand ridges on the East China Sea shelf. *In*: DE BOER, P. L., VAN GELDER, A. & NIO, S. D. (eds) *Tide-influenced Sedimentary Environments and Facies*. Reidel, Dordrecht, 23–38.

YANG, C. S. 1989. Active, moribund and buried tidal sand ridges in the East China Sea and the Southern Yellow Sea. *Marine Geology*, **88**, 97–116.

YANG, G. S., SHI, Y. F. & JI, Z. X. 2002. The morphological response of typical mud flat to sea level change in Jiangsu coastal plain. *Acta Geographica Sinica*, **57**, 76–84 [in Chinese with English abstract].

YANG, S. Y. & YOUN, J. S. 2007. Geochemical compositions and provenance discrimination of the central south Yellow Sea sediments. *Marine Geology*, **243**, 229–241.

YANG, S. Y., JUNG, H. S., LIMB, D. I. & LI, C. X. 2003. A review on the provenance discrimination of sediments in the Yellow Sea. *Earth-Science Reviews*, **63**, 93–120.

YANG, S. Y., LIMB, D. I., JUNG, H. S. & OH, B. C. 2004. Geochemical composition and provenance discrimination of coastal sediments around Cheju Island in the southeastern Yellow Sea. *Marine Geology*, **206**, 41–53.

YANG, W. D. 2002. Structure and sedimentary environment for submarine dune ridges in the East China Sea. *Marine Geology & Quaternary Geology*, **22**, 9–16.

YANG, Z. S. & LIU, J. P. 2007. A unique Yellow River-derived distal subaqueous delta in the Yellow Sea. *Marine Geology*, **240**, 169–176.

YANG, Z. S., JI, Y. J., BI, N. S., LEI, K. & WANG, H. J. 2011. Sediment transport off the Huanghe (Yellow River) delta and in the adjacent Bohai Sea in winter and seasonal comparison. *Estuarine, Coastal and Shelf Science*, **93**, 173–181.

YE, Q. C. 1986. On the development of the abandoned Yellow River delta in Northern Jiangsu province. *Acta Geographica Sinica*, **41**, 112–122 [in Chinese with English abstract].

YE, Y. C., ZHUANG, Z. Y., LAI, X. H., LIU, K. & CHEN, X. L. 2004. A study of sandy bedforms on the Yangtze Shoal in the East China Sea. *Periodical of Ocean University of China*, **34**, 1057–1062 [in Chinese with English abstract].

YOO, D. G., LEE, C. W., KIM, S. P., JIN, H. M., KIM, J. K. & HAN, H. C. 2002. Late Quaternary transgressive and highstand systems tracts in the northern East China Sea mid-shelf. *Marine Geology*, **187**, 313–328.

YU, L. S. 2002. The Huanghe (Yellow) River: a review of its development, characteristics, and future management issues. *Continental Shelf Research*, **22**, 389–403.

YU, Q., WANG, Y. P., GAO, S. & FLEMMING, B. 2012. Modeling the formation of a sand bar within a large funnel-shaped, tide-dominated estuary: Qiantangjiang Estuary, China. *Marine Geology*, **299–302**, 63–76.

YU, Q., WANG, Y. W. & GAO, S. 2014. Tide and continental shelf circulation induced suspended sediment transport on the Jiangsu coast: winter observations out of Xinyanggang. *Journal of Nanjing University (Natural Sciences Edition)*, **50**, 626–535 [in Chinese with English abstract].

YUAN, D. L. & HSUEH, Y. 2010. Dynamics of the cross-shelf circulation in the Yellow and East China Seas in winter. *Deep-Sea Research II: Topical Studies in Oceanography*, **57**, 1745–1761.

YUAN, D. L., ZHU, J. R., LI, C. Y. & HU, D. X. 2008. Cross-shelf circulation in the Yellow and East China Seas indicated by MODIS satellite observations. *Journal of Marine Systems*, **70**, 134–149.

YUAN, Y., WEL, H., ZHAO, L. & JIANG, W. S. 2008. Observations of sediment resuspension and settling off the mouth of Jiaozhou Bay, Yellow Sea. *Continental Shelf Research*, **28**, 2630–2643.

YUAN, Y. R. & CHEN, Q. 1984. Sediments and sedimentary facies of the abandoned Yellow River underwater delta in the south Yellow Sea. *Marine Geology & Quaternary Geology*, **4**, 35–43.

ZHANG, D. S., ZHANG, J. L., ZHANG, C. K. & WANG, Z. 1998. Tidal currents develop-storm surges destroy-tidal currents restore – A preliminary explanation for the dynamic mechanism of formation and evolution of radiate sand ridges in Yellow Sea. *Science in China (Series D)*, **28**, 394–402 [in Chinese].

ZHANG, H. F. 2013. Deposition kinetic characteristics and seabed stability analysis of Pinghai Bay, Fujian. *Journal of Applied Oceanography*, **32**, 36–45 [in Chinese with English abstract].

ZHANG, L., LIU, Z., ZHANG, J., HONG, G. H., PARK, Y. & ZHANG, H. F. 2007. Reevaluation of mixing among multiple water masses in the shelf: an example from the East China Sea. *Continental Shelf Research*, **27**, 1969–1979.

ZHANG, L., CHEN, S. L. & LIU, X. X. 2014. Evolution of the abandoned Huanghe (Yellow) River delta in north Jiangsu province in 800 years. *Acta Geographica Sinica*, **45**, 626–636 [in Chinese with English abstract].

ZHANG, M. W., TANG, J. W., DONG, Q., SONG, Q. T. & DING, J. 2010. Retrieval of total suspended matter concentration in the Yellow and East China Seas from MODIS imagery. *Remote Sensing of Environment*, **114**, 392–403.

ZHANG, R. S. 1984. Land-forming history of the Huanghe River delta and coastal plain of north Jiangsu. *Acta Geographica Sinica*, **39**, 173–184 [in Chinese with English abstract].

ZHANG, R. S. 1992. Suspended sediment transport processes on tidal mud flat in Jiangsu Province, China. *Estuarine Coastal and Shelf Science*, **35**, 225–233.

ZHANG, R. S. & CHEN, C. J. 1992. *A Study on the Evolution of the Jiangsu Offshore Linear Sand Ridges and the Trends of Sandbank Merging into the Intertidal Land.* China Ocean Press, Beijing [in Chinese].

ZHANG, W. G., MA, H. L., YE, L. P., DONG, C. Y., YU, L. Z. & FENG, H. 2012. Magnetic and geochemical evidence of Yellow and Yangtze River influence on tidal flat deposits in northern Jiangsu Plain, China. *Marine Geology*, **319–322**, 47–56.

ZHANG, X., LIN, C. M., DALRYMPLE, R. W., GAO, S. & LI, Y. L. 2014. Facies architecture and depositional model of a macrotidal incised-valley succession (Qiantang River estuary, eastern China), and differences from other macrotidal systems. *Geological Society of America Bulletin*, **126**, 499–522.

ZHANG, X. X., WANG, W. W., YAN, C. C., YAN, W. B., DAI, Y. X., XU, P. & ZHU, C. X. 2014. Historical coastline spatio-temporal evolution analysis in Jiangsu coastal area during the past 1000 years. *Scientia Geographica Sinica*, **34**, 344–351 [in Chinese with English abstract].

ZHAO, Y. Y., GAO, S., WANG, D. D., XU, Z., ZHU, D. & WANG, X. F. 2014. Characteristics and formation mechanisms of the rhythmic morphology of salt-marsh edge cliffs. *Acta Geographica Sinica*, **69**, 378–390 [in Chinese with English abstract].

ZHU, C., CHENG, P., LU, C. C. & WANG, W. 1996. The evolution analysis on Yangtze River delta and the coastal areas of northern Jiangsu Province, since 7000 years ago. *Scientia Geographica Sinica*, **16**, 207–213 [in Chinese].

ZHU, C., XUE, B., PAN, J. M., ZHANG, H. S., WAGNER, T. & PANCOST, R. D. 2008. The dispersal of sedimentary terrestrial organic matter in the East China Sea (ECS) as revealed by biomarkers and hydro-chemical characteristics. *Organic Geochemistry*, **39**, 952–957.

ZHU, D. K. & XU, T. G. 1982. Wetland evolution and management issues of the central Jiangsu coast. *Journal of Nanjing University (Natural Sciences Edition)*, **18**, 799–818 [in Chinese with English abstract].

ZHU, D. K. & AN, Z. S. 1993. Formation and Evolution of radial sand ridges in Jiangsu offshore zones. *In*: *Paper Collection of Geography on Celebrating 80th Birthday of Prof. Ren Mei-e*. Nanjing University Press, Nanjing, 142–147 [in Chinese].

ZHU, D. K., KE, X. K. & GAO, S. 1986. Tidal flat sedimentation of Jiangsu coast. *Oceanography of Huanghai and Bohai Seas*, **4**, 19–27 [in Chinese with English abstract].

ZHU, D. K., MARTINI, L. P. & BROOKFIELD, M. E. 1998. Morphology and land-use of coastal zone of the north Jiangsu Plain, Jiangsu Province, eastern China. *Journal of Coastal Research*, **14**, 591–599.

ZHU, Y. & CHANG, R. 2000. Preliminary study of the dynamic origin of the distribution pattern of bottom sediments on the continental shelves of the Bohai Sea, Yellow Sea and East China Sea. *Estuarine, Coastal and Shelf Science*, **51**, 663–680.

Index

Page numbers in *italics* refer to Figures. Page numbers in **bold** refer to Tables.

acoustic Doppler current profiler (ADCP) 156, 182
agriculture, impact on weathering 48
anthropogenic impact 5–6, 15
  Diaokou lobe 191–192, **192**
  Lingdingyang Bay 182
  Pearl River delta sediments 48
apatite fission track (AFT), provenance evidence 18–20, 23
$^{40}Ar/^{39}Ar$ dating, role in provenance studies 20–21

Baiyun Sag 32, 33
Bashi Straits 9, *10*
^{10}Be date 199
Beibu Gulf *100*, 101
  geological setting 88–90
  location 87
  map *88*
  sea-level change post glacial 101–102
  sediment movement 2
  southern mud depocentre *89*
    sediment study
      methods 90
      results
        ^{14}C dating *92*, **92**, 95, 96
        geochemistry 93–95, *94*
        lithology 93
        seismic profiles 90, *91*, 92–93
      results discussed 96–97
Beijiang River 32, 33
Beipanjiang River 32, 33
Bohai Basin 186
Bohai Bay 187
Bohai Sea *224*, *225*
  environmental setting 224–225, 226
  sedimentary process-product relationship 246, **249**, 250–251
  sedimentary processes
    current circulations 227, 229, *229*, **229**, 230–231
    gravity flows 231–233
    tidal 226–227, **228**
  sedimentary record *251*
  sedimentary systems *225*
    deltas 233
    muds 244–246, *245*
    tidal flats 233, 239–241
    tidal ridges **235**, **238**, 241–244
Bohai Sea Mud *88*
Bohai Strait *245*
Bolivina spp. **202**, **205**, *206*
Borneo 11–12, 14
bottom stress and flow, Yangtze River plume 163–164, *165*, *166*, *167*
Bulimina spp. 202, **202**, **205**, *206*
butterfly delta 2, 90

^{13}C isotope analysis 198
^{14}C age measurement
  Beibu Gulf *92*, **92**, 95, 96
    southern mud depocentre 90
  Diaokou lobe sediments 188
  Gulf of Tonkin 75
  offshore cores 212, **215**
  offshore Hainan Island
    method 102
    results 104, *104*
Cangzhou rise 186
carbonate
  offshore Hainan Island sediment study
    methods 104
    results *110*
  South China Sea sediment *200*
Cassidulina carinata **202**, **205**, *206*
Cathaysia Block 123, *124*
Cathaysia (South China) terrane 11, *12*, 19
Central Yellow Sea Mud *88*
Changjiang (Yangtze) River *88*, *224*, 225, *225*, 226
  catchment climate 140
  catchment geology 139–140
  delta sedimentary system 233, **234**, **236**, **237**, **248**, **249**
  discharge pattern *156*, 230, 231
  gravity flow 231
  history of inputs to East China Sea 140
    deglaciation 144–145
    Last Glacial Maximum evidence 140–141, *142*, 143–144, *143*
    mid-to late Holocene 145–146
    modern mud patches 146–148
  sediment plume to East China Sea 154–155
    data collection 155–157
    methods of analysis 157
    results
      bottom boundary layer flow 163–164
      complex water masses 159–160
      echogram *162*
      interface microstructures 160, 162–163
      mean flows and residuals 160, *161*
      salinity *159*, 160, *160*, *161*
      sediment gravity currents 158–159
      seismic data 157–158, *157*
      temperature *159*, 160, *160*, *161*
      turbidity maxima 158
    results discussed 164–167
  sedimentary record *251*
  setting 137, *138*

Cheju Island (SW) Mud *88*
chemical index of alteration (CIA) 16–17, 46
Pearl River delta study 56, **57**, **58**, **59**, **60**, 62–63, *65*, *69*
chemical weathering, impact of 3, 4–5, 45
clay minerals 14–15
proxies 46–47
Pearl River delta sediment study *48*
methods
analyses 50–51
sampling 49–50
results
chemical index of alteration 56, **57**, **58**, **59**, **60**, 62–63, **65**
clay mineralogy *49*, **53**, 54, *55*, **57**, **58**, **59**, **60**, 61, 62
geochemistry **52**, **53**, 54–55, *55*, *64*
grain size 63, 65, *66*
Sr isotope ratio 56, **57**, **58**, **59**, **60**, *63*
weathering proxies 48–49
results discussed 65, 67–69
Chilostomella **202**, 203, **205**, *206*
chlorite 14, *15*, 47, 67–68, 69
Chuanbi Channel 178
Cibicides wuellerstorfi 202, **202**, **205**, *206*
Clavulina spp. 202, **202**
clay minerals
offshore Hainan Island sediment study
methods 103
results *110*
Pearl River delta study
methods 51
results *49*, **53**, 54, *55*, **57**, **58**, **59**, **60**, 61, 62
results discussed 67–68, 69
provenance indicators 13–15
South China Sea study 33
weathering proxies 47
climate, forcing factors 1
conductivity-temperature-depth (CTD) sensor 156
Coriolis force 230
current systems *229*, **229**
impact on sediment transport 5
longshore 2, *3*

Dangerous Grounds *10*, 12, *13*
dating
^{40}Ar/^{39}Ar, role in provenance studies 20–21
^{10}Be 199
^{14}C
Beibu Gulf
southern mud depocentre sediment study
method 90
results *92*, **92**, 95, 96
Diaokou lobe sediments 188
Gulf of Tonkin 75
offshore cores 212, **215**
offshore Hainan Island
method 102
results 104, *104*
optically stimulated luminescence (OSL) 102, 104, *104*
zircon U–Pb dating
role in provenance studies 21–23, *22*
Yangtze River sediments 143
delta lobes *see* Diaokou Lobe
Diaokou Channel *186*, 187
Diaokou Lobe, Yellow (Huanghe) River
location *186*
sediment study
methods
^{14}C dating 188
coring 187
magnetic properties 187–188
particle size analysis 187
^{210}Pb 188
results
facies analysis 188–190
stratigraphy 190–191
results discussed
correlations *193*
lobe profile *193*
sediment ages 191, **192**
stratigraphic evolution 191, 192, 194
summary 194
diatoms, *Paralia sulcata* study
application in palaeoenvironmental reconstruction
methods 212
results
East China Sea 216–217
South China Sea 214–216
results discussed
palaeoclimatic significance 217–218
palaeogeographic significance 218–220
characteristics 211–212
distribution 213–214
environmental indicator properties 212
SEM image *216*
Dongge Cave speleothem study 46, 48–49
location map *48*
weathering record 48, *49*
Dongjiang River 32, 33

East Asia marginal seas *213*
see East China Sea; South China Sea; Yellow Sea
East Asia monsoon
characteristics 197, 225, 226, 227
circulation patterns 74, 154, 211, 231
history of 1, 9
impacts on
clay minerals 14, 47
fluvial sediment *3*, 4, 6
foraminifera productivity 203, 205, 207
weathering 46, 47

reconstruction of palaeoclimatic impact with diatoms
methods 212
results
East China Sea 216–217
South China Sea 214–216
results discussed
palaeoclimatic significance 217–218
palaeogeographic significance 218–220
characteristics 211–212
distribution 213–214
environmental indicator properties 212
SEM image *216*
East China Sea *224*, *225*
diatom distribution 216–217
environmental setting 224–225, 226
ocean and shelf current circulation 138, 154
sedimentary process-product relationship 246–247, **248**, 250–251
sedimentary processes
current circulation 227, 229, *229*, **229**, 230–231
gravity flow 231–233
tidal 226–227, **228**
sedimentary systems *139*, *225*
deltas 233
muds 244–246
tidal flats 233, 239–241
tidal ridges **235**, **237**, 241–244
setting 137, *138*
Yangtze (Changjiang) River inputs
history of inputs to East China Sea 140
deglaciation 144–145
Last Glacial Maximum evidence 140–141, *142*, 143–144, *143*
mid-to late Holocene 145–146
modern mud patches 146–148
plume to East China Sea 154–155
data collection 155–157
methods of analysis 157
results
bottom boundary layer flow 163–164
complex water masses 159–160
echogram *162*
interface microstructures 160, 162–163
mean flows and residuals 160, *161*
salinity *159*, 160, *160*, *161*
sediment gravity currents 158–159
seismic data 157–158, *157*
temperature *159*, 160, *160*, *161*
turbidity maxima 158
results discussed 164–167
environment 6–7
see palaeoenvironmental analysis
europium (Eu) anomaly 17, 34
eustasy, effect on sedimentation 3–4

Fanshi Channel 172, *172*, 178
Fanshi Shoal 177
fission track methods *see* apatite fission track
fluvial sediment, factors affecting *3*
foraminifera
offshore Hainan Island
methods 103
results 105
South China Sea (northern) 198
methods 198
results
carbonate *200*
ecological significance 200–203, **202**, **205**
$\delta^{18}O$ *200*
species diversity 199–200, *201*, *206*
stratigraphy 199
results discussed 203, *204*, 205, 207

Gd/Yb 17
geochemistry and provenance 15–17
Beibu Gulf southern mud depocentre
methods 90
results 93–95, *94*
Pearl River delta study
methods 51
results **52**, **53**, 54, *55*
Pearl River sediment study
methods 35
results 35
Tonkin, Gulf of 75–76, **77**, *78*, **79**
Yangtze (Changjiang) River offshore mud patches 146, *147*
Globigerinoides ruber 212, **215**
Globobulimina spp. **202**, 203, **205**, *206*
Globocassidulina subglobulosa 202, **202**, **205**, *206*
grain size
offshore Hainan Island sediment study
methods 104
results **108**, *110*
Pearl River delta sediment study 63, 65, *66*
gravity flows and currents, sediment 153, 165, 167, 231–233, *232*
Guijiang River 32, 33
Gunagdong Coastal Mud *88*

Hainan Island *10*, 11, *100*, 101
geology 74
Last Glacial Cycle offshore sediment study
methods of analysis
^{14}C dating 102
carbonate 104
clay minerals 103
foraminifera 103
grain size 104
$\delta^{18}O$ 103
OSL dating 102
seismic profiling 102
results
ages 104, *104*
facies analysis 107–111
foraminifera 105

Hainan Island (*Continued*)
$\delta^{18}O$ 105–106, *105*, *106*
sedimentology 104–105, 106–107
seismic architecture 111–113
stratigraphy 104–105
synthesis 113–117
results discussed 117–119
Han River deltaic basin
age of sequence 130
geological setting 127, *128*
sediment dates **127**
tectonic subsidence 129
Hang Hau Formation 50, *50*
Hangzhou Bay *224*, *225*, 231, **234**, **237**, *246*, *251*
Hanjiang River 31
Hanoi Trough 124
hard isothermal remanent magnetization (HIRM) 146
head-water capture 1, *3*, 9
heavy mineral analysis
Pearl River sediment study
methods 34
results 35
results discussed 37–39
Hengmen Outlet 171, *172*, 180, 181
Hoeglundina spp. 201, **202**, **205**, *206*
Hongqili Outlet 171, *172*, 179, 181
Hongshuihe River 33
Huanghe (Yellow) River *88*, 137, 140, 146, *224*, *225*
delta *see* Diaokou lobe
delta sedimentary system 233, **234**, **236**, **237**, **248**, **249**
discharge 230
discharge rate 187
estuary tides and currents 187
features 185
geographical setting 186–187
gravity flows 231, *232*
history 225–226
sedimentary record *251*
human impact *see* anthropogenic impact
Humen Outlet 171, *172*, 179
humidity proxy 47
Hupijiao Rise *138*, 140
Huskan Mud Belt **236**, **238**, *251*
hyperpycnal waters 153, 231

illite 14, *15*, 47, 67–68, 69, 231
India-Asia tectonic collision 9
Indosinian Orogeny 11, 20
isotopes
fractionation 47
role in provenance studies 17–18

Jaomen Outlet 171, *172*, 179, 181
Jeju Warm Current *229*, **229**
Jiangsu coast *224*, *225*, 227, 230
sediment systems **234**, **235**, **237**, 239, *240*, 241, *243*
sedimentary record *251*

kaolinite 14, *15*, 47, 67–68, 69, 231
Kaoping River 11
Khorat Plateau 19, *22*
Kontum Massif 19, *19*
Kuroshio Current 2, 138, *155*, 187, 227, *229*, **229**, 230

La/Yb 17
laser ablation ICP-MS 21
Last Glacial Cycle
diatom response 217–218
sediments offshore Hainan Island
methods of analysis
^{14}C dating 102
carbonate 104
clay minerals 103
foraminifera 103
grain size 104
$\delta^{18}O$ 103
OSL dating 102
seismic profiling 102
results
ages 104, *104*
facies analysis 107–111
foraminifera 105
$\delta^{18}O$ 105–106, *105*, *106*
sedimentology 104–105, 106–107
seismic architecture 111–113
stratigraphy 104–105
synthesis 113–117
results discussed 117–119
Last Glacial Maximum (LGM) 88, 138, 186
history of Yangtze River sedimentation 140–141, *142*, *143*
impact in Gulf of Tonkin 74
sea-level change 218–219
weathering characteristics 14
latitude, effect on weathering 3
leaching 46–47
Leizhou Peninsula 74, *100*
Lenticulina spp. 201, **202**, *206*
Lhasa Block *12*
Lingding Channel 172, *172*
Lingdingyang Bay
geographical setting 171–173, *172*
geomorphological study
methods
GIS 175–176
remote sensing 174–175, *175*
surveying 173–174, *174*
results 176–177
bathymetry 177–178, *178*, *179*, *180*
hydrodynamics 182
shoreline changes *177*, 178–182
tidal currents *181*
location 173
Liujang River 32
longshore currents 2
Luofushan-Dongjiang Fault 125, 126

Luzon Arc 11, *13*
sediment studies
AFT evidence 18
clay mineralogy 14, 15, *15*
isotope evidence 17, *18*
REE 17

Mekong River 9, *10*, 11, 12, 16
sediment studies
AFT evidence 19
$^{40}Ar/^{39}Ar$ dating 20, *20*
CIA evidence 16, *16*
clay mineralogy 14
isotope character 17
zircon U–Pb dating 21, *22*
Melonis barleeanus 202, **202**, **205**, *206*
mineralogy, role in provenance study 13–15
Molengraaff River 23
monazite 17
monsoon *see* East Asia monsoon
mud belts, development 153–154
muscovite $^{40}Ar/^{39}Ar$ dating, role in provenance studies 20–21

Nanliu River 2
Nanpanjiang River 32, 33
Nd evidence 17, *18*, 23, 33
Nodosaria spp. 201, **202**, *206*
North Yellow Sea Mud *88*

$\delta^{18}O$
in offshore cores 212, *214*, *215*
Dongge Cave speleothems 46, 48–49, *48*, *49*
sediments offshore Hainan Island
methods 103
results 105–106, *106*
South China Sea *200*
Okinawa Trough 137, *138*, 141, 143, 144, 148, 231
optical backscatter sensor (OBS) 156
optically stimulated luminescence (OSL) dating
offshore Hainan Island
method 102
results 104, *104*
optimal interpolation sea surface temperature (OI-SST) 156–157
oxygen minimum zone (OMZ) 198, 203, 207

palaeoenvironmental analysis
see diatoms; foraminifera
see also sediment record under named seas
Paralia sulcata
application in palaeoenvironmental reconstruction
methods 212
results
East China Sea 216–217
South China Sea 214–216
results discussed
palaeoclimatic significance 217–218
palaeogeographic significance 218–220
characteristics 211–212
distribution 213–214
environmental indicator properties 212
SEM image *216*
particle size distribution
offshore Hainan Island sediment study
methods 104
results **108**, *110*
Pearl River delta sediment study 63, 65, *66*
^{210}Pb, Diaokou lobe sediments 188
Pearl River 9, *10*, 11, 12, 16, *88*
characteristics
sediment transport 2
shoreline history *176*
delta sediment chemical weathering study *48*
methods
analyses 50–51
sampling 49–50
results
chemical index of alteration 56, **57**, **58**, **59**, **60**, 62–63, **65**
clay mineralogy *49*, **53**, 54, *55*, **57**, **58**, **59**, **60**, 61, 62
geochemistry **52**, **53**, 54–55, *55*, *64*
grain size 63, 65, *66*
Sr isotope ratio 56, **57**, **58**, **59**, **60**, *63*
weathering proxies 48–49
results discussed 65, 67–69
deltaic plain
age of sequence 130
geological setting 125–127, *126*
lithostratigraphy 131–133
sediment dates **127**
tectonic subsidence 129
discharge to Gulf of Tonkin 74
estuary system 171, *172*
interaction with Lingdingyang Bay
geographical setting 171–173, *172*
geomorphological study 173
methods
GIS 175–176
remote sensing 174–175, *175*
surveying 173–174, *174*
results 176–177
bathymetry 177–178, *178*, *179*, *180*
hydrodynamics 182
shoreline changes *177*, 178–182
tidal currents *181*
sediment characterization
drainage pattern 31, 32, *32*, 33
methods 34
results
heavy minerals 35, *37*
major element geochemistry 35, *38*
rare earth elements 34–35

Pearl River (*Continued*)
results discussed
geochemistry 39, 41
heavy minerals 37–39
rare earth elements 35–37, *36*, *39*
sampling strategy 33–34
sediment provenance study
AFT evidence 20
CIA evidence 16, *16*
clay mineral assemblage 14, 15, *15*, 47–48
isotope evidence 17, *18*
Pearl River Distal Mud *88*, *245*
Pearl River Proximal Mud *88*
Philippine Islands 11
physical weathering and erosion 47
provenance
apatite fission track evidence 18–20
bulk geochemistry evidence 15–17
CIA evidence 16, *16*
clay mineral evidence 13–15, 47–48
impact on shelf sediments 4
importance of recognition 9
isotope evidence 17–18
muscovite $^{40}Ar/^{39}Ar$ evidence 20–21
role of reworking 12–13
terrane factors 11, *12*
zircon U–Pb dating evidence 21–23
Pyrgo spp. 201, **202**, **205**, *206*

Qaidam Block *12*
Qiangtang Block *12*, *20*, 21
Qingshui Channel 186, 187
Qiongdongnan Basin 31, *100*, 101
Qiongzhou Strait 2, 74, *74*, *88*, *100*

rare earth elements (REE)
Oligocene sediment 33
Pearl River sediment study
methods 34
results 34–35, *36*
results discussed 35–37
role in provenance studies 17
Tonkin, Gulf of 76, **77**, 78, **80**, *81*, *82*
Red River Fault Zone *20*, 101, 123, *124*
Red (Song Hong) River 9, *10*, 11, 12, 16, *88*, 89
AFT evidence 19
$^{40}Ar/^{39}Ar$ dating evidence 20, *20*
CIA evidence 16, *16*
clay mineral assemblage 14
deltaic basin
age 130–131
geological setting 124, *125*
sediment dates **127**
tectonic subsidence 127–129
discharge to Gulf of Tonkin 73–74
drainage basin 31
input to Gulf of Tonkin 73–74
isotope evidence 17
REE evidence 17
sediment 2, 101
zircon U–Pb dating evidence 21, *22*
Richardson number 162, *163*, *164*
river dispersal system 153

salinity, Yangtze River plume *159*, *160*, *161*, *165*, *166*, *167*
sea-level change
Bohai Basin 186
impact on weathering 3, 3–4, 4–5
Post Glacial 101–102
South China Sea *219*
sea-level cycles 223
sediment channel 165, 167
sediment dispersal, rivers 153
seismic stratigraphic profiles
Beibu Gulf southern mud depocentre 90, *91*
offshore Hainan Island
methods 102
results *112*, *113*, *114*, **114**, *115*, **117**
Yangtze River plume study
methods, CHIRP 157
results 157–158, *158*
sequence stratigraphy 3–4
Sham Wat Formation 50, *50*, 51, **52**
Shandong Coastal Mud *88*
Shandong Peninsula *224*, *225*, 226
Shawan–Ningdingyang Fault 125
shelf sea classification 223–224
Shenxian Channel 185–186, 187
Sibumasu terrane *12*, 21
smectite 14, *15*, 47, 67–68, 69, 231
Oligocene sediment 33
soil mobilization 48
Song Chay Fault 124, *125*
Song Hong (Red) River 9, *10*, 11, 12, 16, *88*, 89
AFT evidence 19
$^{40}Ar/^{39}Ar$ dating evidence 20, *20*
CIA evidence 16, *16*
clay mineral assemblage 14
deltaic basin
age 130–131
geological setting 124, *125*
sediment dates **127**
tectonic subsidence 127–129
discharge to Gulf of Tonkin 73–74
drainage basin 31
input to Gulf of Tonkin 73–74
isotope evidence 17
REE evidence 17
sediment 2, 101
zircon U–Pb dating evidence 21, *22*
Song Hong (Yinggehai) Basin 31, *100*, 101
Song Lo Fault 124, *125*
Songpan Garze Terrane 11, *12*, *20*

South China Sea 9
diatom distribution 214–216
distal muds *245*
geological setting 10–11
location 31
map *10*
palaeogeography *13*
sea-level change *219*
sediment input characterization 11–13
apatite fission track 18–20
bulk geochemistry 15–17
bulk isotopes 17 18
clay minerals 13–15
muscovite $^{40}Ar/^{39}Ar$ 20–21
zircon U–Pb 21–23
sediment record 46
sediment weathering proxies for Pearl River delta *48*
methods
analyses 50–51
sampling 49–50
results
chemical index of alteration 56, **57**, **58**, **59**, **60**, 62–63, **65**
clay mineralogy *49*, **53**, 54, *55*, **57**, **58**, **59**, **60**, 61, 62
geochemistry **52**, **53**, 54–55, *55*, *64*
grain size 63, 65, *66*
Sr isotope ratio 56, **57**, **58**, **59**, **60**, *63*
weathering proxies 48–49
results discussed 65, 67–69
tectonic setting 10–11
see also sediment input studies
Beibu Gulf
Hainan Island
Tonkin, Gulf of
Taiwan
South China Sea (northern)
basement 31, *32*
deltaic inputs
age 130–131
Han River 127, *128*
lithostratigraphy 131–133
Pearl River 125–127, *126*
Red River 124, *125*
foraminifera community study 198
methods 198
results
carbonate *200*
ecological significance 200–203, **202**, **205**
$\delta^{18}O$ *200*
species diversity 199–200, *201*, *206*
stratigraphy 199
results discussed 203, *204*, 205, 207
oxygen minimum zone 198
primary production 197
Pearl River sediment input characterization
drainage pattern 31, 32, *32*, 33
methods 34
results
heavy minerals 35, *37*
major element geochemistry 35, *38*
rare earth elements 34–35
results discussed
geochemistry 39, 41
heavy minerals 37–39
rare earth elements 35–37, *36*, *39*
sampling strategy 33–34
tectonic subsidence
Han River deltaic basin 129
Pearl River deltaic basin 129
Red River basin 127–129
South-East Yellow Sea Mud *88*
South-West Yellow Sea Mud *88*
Spartina 239–240
speleothems and climate records
Dongge Cave study 46, 48–49
location map *48*
weathering record 48, *49*
Sr isotope ratios 17, *18*, 23, 47
Pearl River delta sediment study
methods 51
results **51**, 56, **57**, **58**, **59**, **60**, *63*
results discussed 68
stratification processes 154
Yangtze River plume
methods of study 156
results *163*
Sunda Shelf 23
Sundaland *13*
Sundaland Block 123, *124*
suspended sediment *see* turbidity

Taiwan *10*, 11
river systems 137, *139*
sediment characterization
clay mineralogy 14, *15*, 231
isotope evidence 17, *18*
sediment erosion 31
sediment inputs to South China Sea 46, 50, 54, **58**, **59**, **60**, *61*, *62*, *63*
Taiwan Strait, opening 218–220
Taiwan Warm Current 138, 154, *155*, *229*, **229**
Tancheng–Luujiang Fault 186
tectonic forcing factors 1
temperature, Yangtze River plume *159*, *160*, *161*, *165*, *166*, *167*
thermochronology *see* apatite fission track (AFT)
Three Gorges Dam 145
Tonkin, Gulf of 2, *10*
geological setting 73–74
sediment geochemistry study
methods 74–75
results
lithology 75

Tonkin, Gulf of (*Continued*)
major/trace element geochemistry 75–76, **77**, *78*, **79**
rare earth elements 76, **77**, 78, **80**, *81*, *82*
results discussed 78, 81–83
significance of environmental change 83
Tsengwen River 11
Tsushima Current *229*, **229**
turbidity measurements, Yangtze River plume 156, 158, *164*, *165*, *166*, *167*
turbulence, methods of study 157
Typhoon Herb 11

U–Pb zircon dating
role in provenance studies 21–23, *22*
Yangtze River sediments 143
uplift of Tibetan Plateau, impact of 1, *3*, 9
upwelling 165, 227
Uvigerina spp. 202, **202**, **205**, *206*

volcanic provinces 11

weathering *see* chemical weathering
Western Korea Coastal Current *229*, **229**

Xianxain rise 186
Xihu depression *138*, 140
Xijiang Fault 125, 126
Xijiang River 32
XRD 14

Yangtze (Changjiang) River *88*, *224*, 225, *225*, 226
catchment climate 140
catchment geology 139–140
delta sedimentary system 233, **234**, **236**, **237**, **248**, **249**
discharge pattern *156*, 230, 231
gravity flows 231
history of inputs to East China Sea 140
deglaciation 144–145
Last Glacial Maximum evidence 140–141, *142*, 143–144, *143*
mid-to late Holocene 145–146
modern mud patches 146–148
plume to East China Sea 154–155
data collection 155–157
methods of analysis 157
results
bottom boundary layer flow 163–164
complex water masses 159–160
echograms *162*
interface microstructures 160, 162–163
mean flows and residuals 160, *161*
salinity *159*, 160, *160*, *161*
sediment gravity currents 158–159
seismic data 157–158, *157*
temperature *159*, 160, *160*, *161*
turbidity maxima 158
results discussed 164–167
sedimentary record *251*
setting 137, *138*
Yangtze Craton 11, *12*, *22*
Yangtze River Proximal Mud *88*
Yangtze Shoal 141, *156*, **235**, **238**, **249**, *251*
Yellow (Huanghe) River *88*, 137, 140, 146, *224*, *225*
delta *see* Diaokou lobe
delta sedimentary system 233, **234**, **236**, **237**, **248**, **249**
discharge 230
discharge rate 187
estuary tides and currents 187
features 185
geographical setting 186–187
gravity flows 231, *232*
history 225–226
sedimentary record *251*
Yellow Sea *224*, *225*
currents systems *229*, **229**
environmental setting 224–225, 226
sedimentary process-product relationship 246–247, **249**, 250–251
sedimentary processes
current circulations 227, 229, *229*, **229**, 230–231
gravity flows 231–233
tidal 226–227, **228**
sedimentary systems *225*
deltas 233
muds **236**, **238**, 244–246
tidal flats 233, 239–241
tidal ridges 241–244
Yinggehai (Song Hong) Basin 31, *100*, 101
Youjiang River 33
Yunnan–Guizhou Plateau 31

Zhe-Min Coastal Current 138
Zhedong depression *138*, 140
Zhejiang–Fujian Coastal Current (ZFCC) 154, *155*
Zhejiang–Fujian Coastal Mud Belt *88*, 154, *155*, *245*
Zhoushan *156*
zircon U–Pb dating
role in provenance studies 21–23, *22*
Yangtze River sediments 143
Zuojiang River 33